KT-598-827

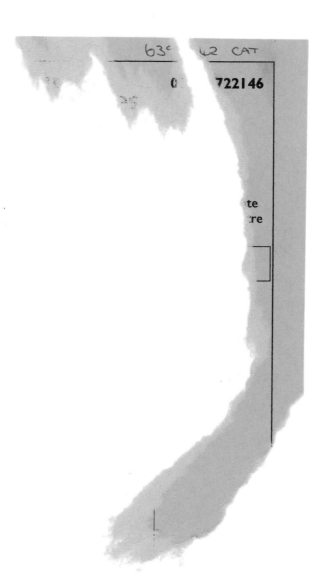

63° 42 CAT

722146
te
re

DR003459

CORNWALL COLLEGE

MARINE ORNAMENTAL SPECIES

COLLECTION, CULTURE & CONSERVATION

MARINE ORNAMENTAL SPECIES

COLLECTION, CULTURE & CONSERVATION

Edited by

JAMES C. CATO

CHRISTOPHER L. BROWN

Iowa State Press
A Blackwell Publishing Company

James C. Cato, Ph.D., is Director of the Florida Sea Grant College Program and Professor at the University of Florida. He earned B.S. and M.S. degrees at Texas Tech University and his Ph.D. in Food and Resource Economics at the University of Florida. Dr. Cato has organized or participated in about 100 conferences and workshops, has given 96 papers and presentations, and has authored or co-authored over 150 publications. His academic achievements cover many aspects of coastal and marine economics including boating, marinas, fisheries, aquaculture, and the value of safer seafood. Recent priority areas of the Florida Sea Grant College Program include marine biotechnology, fisheries, aquaculture (including marine ornamental fish), seafood safety, sustainable water-dependent businesses, water quality, coastal ecosystems and habitats and coastal storms.

Christopher L. Brown, Ph.D., is Director of the Marine Biology Program and Associate Chairman of the Department of Biological Science at Florida International University. He graduated with honors from Union College, New York, earned his Ph.D. at the University of Delaware, and had postdoctoral training at the University of California, Berkeley. Dr. Brown's first faculty appointment was at the University of Hawaii, where he developed a research focus on ornamental fish culture and proposed the first international meeting on marine ornamentals. His bibliography includes 50 publications on fish developmental biology and aquaculture.

© 2003 Iowa State Press
A Blackwell Publishing Company
All rights reserved

Iowa State Press
2121 State Avenue, Ames, Iowa 50014

Orders:	1-800-862-6657
Office:	1-515-292-0140
Fax:	1-515-292-3348
Web site:	www.iowastatepress.com

Authorization to photocopy items for internal or personal use, or the internal or personal use of specific clients, is granted by Iowa State Press, provided that the base fee of $.10 per copy is paid directly to the Copyright Clearance Center, 222 Rosewood Drive, Danvers, MA 01923. For those organizations that have been granted a photocopy license by CCC, a separate system of payments has been arranged. The fee code for users of the Transactional Reporting Service is 0-8138-2987-9/2003 $.10.

♾ Printed on acid-free paper in the United States of America

First edition, 2003

Library of Congress Cataloging-in-Publication Data

Marine ornamental species: collection, culture, and conservation /
edited by James C. Cato and Christopher L. Brown
    p. cm.
Includes bibliographical references and index.
  ISBN 0-8138-2987-9 (alk. paper)
  1. Ornamental fishes. 2. Marine aquarium fishes. 3. Ornamental fish trade.
  I. Cato, James C. II. Brown, Christopher L. 1952-
  SF457.1 M36 2003
  639.34′2—dc21   2002154431

The last digit is the print number: 9 8 7 6 5 4 3 2 1

Contents

V. Stakeholder Perspectives

A. Museums and Public Aquariums

B. Collectors

C. Hobbyists

D. Government

E. NGOs/Environmental Management

Preface

By all accounts, Marine Ornamentals 2001: Collection, Culture and Conservation, was a highly successful meeting. The November 2001 gathering in Orlando, Florida, was attended by more than 300 individuals from 23 countries, representing a broad cross-section of interests in the marine ornamental fish and invertebrate industry, including scientists, hobbyists, students, commercial collectors, and representatives of government agencies. This, the second assembly of its kind, yielded evidence of technical progress, new approaches to old problems, and the attainment of a collective industry voice on matters of universal concern.

The group that assembled for Marine Ornamentals 2001 represents an international body of people that, until 1999, never really had a dedicated forum for discussion. For many years prior to that, academic, regulatory, and business practitioners certainly had ample reasons to get together, and there is nothing new about the potential benefits of developing an industrywide consensus about the best way to approach thorny conservation and economic issues. Nevertheless, and despite the magnitude of the economic scale of the marine ornamentals trade, the marine segment of the international aquarium industry had only the options of assembling as a subgroup among broader groups that included the overwhelmingly larger freshwater component of the industry, or with smaller groups that focused on marine ornamental issues but not necessarily on an all-inclusive, multidisciplinary global scale. In part this may have been the result of the recreational nature of the consumer market that ornamental fish serve, which has had the tendency to obscure the seriousness of reef conservation and economic matters. Marine Ornamentals 1999 filled this vacancy, on something of a trial basis, beginning with a four-day conference in Kona, Hawaii. That conference attracted a multiprofessional group of people that made an initial attempt to come to grips with the interests, concerns, and scientific progress of the marine ornamentals trade. The published products of that conference include a special 240-page volume of *Aquarium Sciences and Conservation*, with 19 chapters representing a diverse range of interests, and a 127-page Hawaii Sea Grant volume of contributed papers. The conference was organized around the headings collection, culture, and conservation, and among the key themes emerging at that conference was the desire for the participants in this trade to regulate themselves in order to set and maintain a high standard of environmental accountability in their approach to the use and enjoyment of the world's reef resources. With that meeting, baseline economic and biological data began to emerge, and many of the important conservation and economic issues began to come into focus. Although there were sharply differing points of view on a variety of details, a certain commonality of interest became apparent among the participating representatives of governing agencies, public aquariums, aquaculture researchers, collectors, businesspeople, hobbyists, and journalists. The common denominator is that we all share in the love of reef organisms and have an abiding interest in their perpetuation. Among the most important products of Marine Ornamentals 1999 was a resounding vote in favor of a second conference, which was to be held in 2001 in Florida.

The 2001 meeting held in Orlando, Florida, drew worldwide registration and followed a program that explored and expanded upon themes identified in the inaugural meeting in 1999. Keynote and plenary speakers represented six different countries, and papers were contributed by authors from 13 countries. The concept of sustainability had come into prominence, and sessions stretching over two days dealt specifically with the need for this industry to eliminate exploitative and short-sighted elements and cultivate sustainability. Clearly, there are several ways for the industry to attain a higher degree of sustainability, including the establishment of an industrywide standard for collection and shipping practices, leading to reduced mortality and waste of reef resources. In addition, the continued development of methods to culture marine ornamental species is also viewed as an eventual solution, or at least a partial solution, to the currently heavy dependence on the harvest of wild stocks.

Among the more exciting events at the Marine Ornamentals 2001 conference was the announcement by the Marine Aquarium Council of the launch of a new standard for certification ("MAC Certification") for environmentally conscious practitioners of reef fish harvest, transport, and marketing. This is a historic development, and one that addresses one of the marine ornamentals industry's most volatile flash-points—the perceived choice in developing nations between continued reef destruction and starvation in coastal communities that have come to depend on reef products. This announcement infused the conference with an optimistic tone and a sense that one of the difficult socioeconomic problems of this industry is being tackled. Another major buzz went through the conference with the news of breakthroughs in the larval rearing of the pygmy angelfishes by two aquaculture pioneers in Hawaii—Frank Baensch of Reef Culture Technologies and Charles Laidley of the Oceanic Institute. As has so often been the case in the biological sciences, a healthy competition has yielded what appears to be a photo finish. It appears that tank-raised pygmy angelfishes will be available to consumers in the foreseeable future and that the rearing technology used for these important species will un-doubtedly be adapted to other fishes as the aquaculture side of the industry progresses.

Yet another new direction came to light at the Marine Ornamentals 2001 conference, with a significantly enriched program in the area of fish health management; we are coming to recognize that marine ornamental fish and invertebrates can benefit from an organized and professional approach to their health needs, through research, veterinary care, and education. A plenary presentation by Michael Stoskopf and three full sessions of contributed papers on nutrition and disease raised awareness and expectations of a rising standard of health care for marine ornamentals.

With the emerging sense of community that an interdisciplinary meeting of this sort fosters, there is a distinct perception that it becomes possible that the sum of the impacts of these individual accomplishments can become larger than its component parts. Real advancement is best made by driven individuals focusing and expending their creative energies on problem solving, but synergy comes into play through coordination with others that have shared or partially overlapping interests. This industry is already showing signs of becoming stronger, more mature, and more responsible as a result of the networking and open exchange of ideas that began with this pair of conferences, Marine Ornamentals 1999 and Marine Ornamentals 2001. The industry is taking the initial steps to redefine itself through a coordinated effort to eliminate the destructive and short-sighted practices associated with profiteering, to establish safe and effective new methods of collecting and marketing fish, to develop sensible and reliable alternatives to the near-complete dependence of this industry on wild animals, to establish a professional standard of health care for our marine ornamental animals, and to educate the public, consumers, and regulatory bodies.

Acknowledgments

Each of the authors and coauthors of the chapters in this book are gratefully acknowledged for taking the time to write the chapters and to respond quickly to our comments and suggestions. Marine Ornamentals '01 was held in late November 2001, and by August 2002, the book was completed and submitted to the publisher. This occurred only because of the dedication of each contributor. Ms. Jacquelyn Whitehouse is greatly acknowledged for her contribution and assistance in processing and organizing the draft chapter manuscripts.

We also thank the 336 participants from 23 countries who attended Marine Ornamentals '01. Their attendance and the questions they asked during presentations not only made the conference possible but also added to the final quality of the chapters in this book. Major sponsors for Marine Ornamentals '01 were Florida Sea Grant; Florida Department of Agriculture and Consumer Affairs Division of Aquaculture; and the National Sea Grant College Program. Cosponsors were: Hawaii, Mississippi/Alabama, Oregon, and Virginia Sea Grant; Living Seas at Walt Disney World®; Tropical Fish Hobbyist Magazine; and the University of Florida's Institute of Food and Agricultural Sciences. Contributors were Louisiana, Maryland, North Carolina, Texas, and New York Sea Grant; Florida International University Marine Biology Program; Ornamental Fish International; and the Fisheries Industry Division of the United Nations Food and Agriculture Organization. Trade show participants were Aquarium Systems/Marineland; Aquatic Eco-Systems, Inc.; Dolphin Fiberglass Products, Inc.; Frigid Units, Inc.; Keeton Industries; Marine Aquarium Council; Marine Biotech, Inc.; Marine Specialties International; New Life International, Inc.; Ocean Dreams, Inc.; Pacific Aqua Farms; Segrest Farms; and Signature Coral Co. Whereas the sponsors and trade show participants did not contribute directly to the completion of this book, they, like the attendees, made Marine Ornamentals '01 possible and thus indirectly caused the book to be written. The academic, agency, NGO, and industry sponsors also allowed the conference to maintain a nice balance of scientific and applied focus as is also reflected throughout the chapters of the book.

Finally, the editors are grateful to Florida International University and the University of Florida for allowing us to use our time to serve as editors.

Contributors

Chapter numbers appear in parentheses.

Harry Ako (19)
College of Tropical and Subtropical Agriculture
Molecular Bioscience and Biosystems
  Engineering
St. John 511
University of Hawaii at Manoa
Honolulu, Hawaii 96822 USA
hako@hawaii.edu

Joy Alban (24)
International Marinelife Alliance
83 West Capitol Drive, Bo. Kapitolyo
Pasig City (Metro Manila)
1601 Philippines
joyalban@hotmail.com

Ricardo Araújo (15)
Estação de Biologia Marinha do Funchal
Cais do Carvão
Promenade de Orla Marítimado Funchal,
  Gorgulho
9000-107 Funchal, Madeira
Portugal
ricardo.araujo@mail.cm-funchal.pt

Cynthia Armstrong (14)
Florida Marine Research Institute
14495 Harllee Road
Port Manatee, Florida 34221-9620 USA
cynthia.armstrong@fwc.state.fl.us

Michael Arvedlund (16)
Research Fellow, Danaq Consulting, Ltd.
Department of Coral Ecology and Aquaculture
Rådmand Steins Alle 16C, 619
2000 Fredericksberg, Denmark
ma@danaq.dk

Christina M. Balboa (5,13)
Coastal and Marine Research Associates
Biological Resources Program
World Resources Institute
10 G Street, NE Suite 800
Washington, DC 20002 USA
cristina.balboa@yale.edu

Austin Bowden-Kerby (11)
Project Scientist, Coral Gardens Initiative
Counterport International / Foundation of
  the Peoples of the South Pacific
P.O. Box 14447
Suva, Fiji Islands
bowdenkerby@connect.com.fj

Christopher L. Brown (Ed.,1,18)
Director, Marine Biology Program
Fellow, Honors College
Florida International University
Academic Building 1, Room 318
3000 NE 151st Street
North Miami, Florida 33181 USA
brownch@fiu.edu

Andrew W. Bruckner (12)
NOAA Fisheries
Office of Protected Resources
1315 East West Highway
Silver Spring, Maryland 20910 USA
andy.bruckner@noaa.gov

Bruce W. Bunting (9)
Vice President
Center for Conservation Finance
World Wildlife Fund U.S.
1250 24th Street NW
Washington, DC 20037 USA
bruce.bunting@wwfus.org

ROSEWARNE LEARNING CENTRE

Ricardo Calado (15)
Laboratório Marítimo da Guia
Faculdade Ciências Universidade Lisboa
Estrada do Guincho
2750-642 Cascais
Portugal
rjcalado@hotmail.com

James C. Cato (Ed.,1)
Professor, Food and Resource Economics
Director, Florida Sea Grant College Program
University of Florida
P.O. Box 110400
Gainesville, Florida 32611-0400 USA
jcato@ifas.ufl.edu

Billy D. Causey (26)
Superintendent
Florida Keys National Marine Sanctuary
P.O. Box 500368
Marathon, Florida 33050 USA
billy.causey@noaa.gov

John Claussen (28)
Conservation and Community Investment
  Forum
423 Washington Street, 4th Floor
San Francisco, California 94111 USA
john@cea.sfex.com

Scott E. Clement (25)
Research Assistant
Fisheries and Illinois Aquaculture Center
Southern Illinois University
Carbondale, Illinois 62901 USA
sclement@siu.edu

Brian Cole (18)
2510 Riverbend Drive
Ruskin, Florida 33570 USA
BAcole4@juno.com

John S. Corbin (1)
Manager
Aquaculture Development Program
Hawaii Department of Agriculture
1177 Alacea Street, Room 400
Honolulu, Hawaii 96813 USA
aquacult@aloha.com

Jamie Craggs (16)
London Aquarium
London, United Kingdom
aquatic@londonaquarium.co.uk

John Dawes (27)
Secretary General
Ornamental Fish International
Apartado de Correos 129
29692 Sabinillas, Manilva
Málaga, Spain
secretariat@ornamental-fish-int.org

Chris de Bodisco (10)
Research Economist
Food and Resource Economics Department
University of Florida
P.O. Box 110240
Gainesville, Florida 32611-0240 USA
cdebodisco@mail.ifas.ufl.edu

Robert L. Degner (10)
Professor
Food and Resource Economics Department
University of Florida
P.O. Box 110240
Gainesville, Florida 32611-0240 USA
RLDegner@mail.ifas.ufl.edu

J. Nicholas Ehringer (14)
Professor
Hillsborough Community College
10414 E. Columbus Drive
Tampa, Florida 33619-7865 USA
nehringer@hcc.cc.fl.us

Theo Espero (24)
International Marinelife Alliance
83 West Capitol Drive, Bo. Kapitolyo
Pasig City (Metro Manila)
1601 Philippines
t_espero@hotmail.com

William W. Falls (14)
Aquaculture Program Manager and Associate
  Professor
Hillsborough Community College
10414 E. Columbus Drive
Tampa, Florida 33619-7856 USA
wfalls@hcc.cc.fl.us

Claudia Farfan (18)
Department of Zoology
2538 The Mall
Edmondson Hall
Honolulu, Hawaii 96822 USA
cfarfan@cicese.mx

Ruth Francis-Floyd (7,8)
Professor
Department of Large Animal Clinical Science
and Department of Fisheries and Aquatic
Science
University of Florida
P.O. Box 100136
Gainesville, Florida 32610-0136
rff@ifas.ufl.edu

Todd Gardner (22)
Biology Department
Hofstra University
Hempstead, New York 11549 USA
fishtail22@aol.com

Edmund Green (3)
Marine and Coastal Program
UNEP World Conservation Monitoring Centre
219 Huntingdon Road
Cambridge, CB3 ODL
United Kingdom
ed.green@unep-wcmc.org

Heather Hall (23)
Zoological Society of London
Regent's Park
London NWI 4RY
United Kingdom
heather.hall@zsl.org

Darlene Haverkamp (14)
Florida Marine Research Institute
14495 Harllee Road
Port Manatee, Florida 34221-9629 USA
darlene.haverkamp@fwc.state.fl.us

Roy Herndon (14)
Ocean Dreams, Inc.
13005 Sea Critters Lane
Dover, Florida 33527-3648 USA
oceandreamsmarti@hotmail.com

Teresa Herndon (14)
Ocean Dreams, Inc.
13005 Sea Critters Lane
Dover, Florida 33527-3648 USA
seacritters@compuserve.com

G. Joan Holt (17)
Marine Science Institute
750 Channel View Drive
Port Aransas, Texas 78373 USA
joan@utmsi.utexas.edu

Paul Holthus (9)
Executive Director
Marine Aquarium Council
923 Nuuana Avenue
Honolulu, Hawaii 96817 USA
paul.holthus@aquariumcouncil.org

RuthEllen Klinger (7,8)
Biological Scientist
Department of Large Animal Clinical Sciences
University of Florida
P.O. Box 100136
Gainesville, Florida 32610-0136 USA
klingerr@ufl.edu

Sherry L. Larkin (6,10)
Assistant Professor
Food and Resource Economics Department
University of Florida
P.O. Box 110240
Gainesville, Florida 32611-0240 USA
slarkin@ufl.edu

Junda Lin (15)
Professor
Department of Biological Sciences
Florida Institute of Technology
150 University Boulevard
Melbourne, Florida 32901 USA
jlin@fit.edu

Benita Manipula (24)
International Marinelife Alliance
83 West Capitol Drive, Bo. Kapitolyo
Pasig City (Metro Manila)
1601 Philippines
bench_manipula@hotmail.com

Bryan McCullough (24)
International Marinelife Alliance
119 Merchant Street, Suite 610
Honolulu, Hawaii 96813 USA
brian@marine.org

Andreas Merkl (28)
Conservation and Community Investment Forum
423 Washington Street, 4th Floor
San Francisco, California 94111 USA
andreas@cea.sfex.com

Martin A. Moe, Jr. (2)
Green Turtle Publications
222 Gulfview Drive
Islamorada, Florida 33036 USA
martin_moe@yahoo.com

Luís Narciso (15)
Laboratório Marítimo da Guia
Faculdade Ciências Universidade Lisboa
Estrada do Guincho
2750-642 Cascais
Portugal
lnarciso@fc.ul.pt

Sandy Nettles, P.G. (14)
N.S. Nettles and Associates, Inc.
201 E. Roosevelt Blvd.
Tarpon Springs, Florida 32689 USA
nettlesinc@earthlink.net

Michael Nichols (14)
Triton Marine
P.O. Box 0012
Ozona, Florida 34660 USA
ron@liverocks.com

Katia Olivier (4)
Consultant
FAO/Fishery Industries Division
Fish Utilization and Marketing Service
Viale delle Terme di Caracalla
00100 Rome, Italy
katia.Olivier@free.fr

John E. Parks (13)
Coastal and Marine Research Associate
Biological Resources Program
World Resources Institute
10 G Street N.E.
Washington, D.C. 20002 USA
jparks@wri.org

Michael F. Payne (21)
Kalbarri Seahorse Centre
P.O. Box 512
Kalbarri, WA 6536
Australia
kalseahorse@westnet.com.au

Joe Pecorelli (16)
London Aquarium
London, United Kingdom
aquatic@londonaquarium.co.uk

Robert S. Pomeroy (13)
Coastal and Marine Research Associate
Biological Resources Program
World Resources Institute
10 G Street N.E.
Washington, D.C. 20002 USA
rpomeroy@wri.org

Vaughan R. Pratt (24)
International Marinelife Alliance
110 Merchant Street, Suite 610
Honolulu, Hawaii 96813 USA
vpratt@marine.org

Peter J. Rubec (24)
Senior Scientist
International Marinelife Alliance
2800 4th Street North
Suite 123
St. Petersburg, Florida 33704 USA
peterrubec@cs.com

Vernon T. Sato (19)
Division of Aquatic Resources
Avenue Fisheries Research Center
1039 Sand Island Parkway
Honolulu, Hawaii 96819 USA
vtsato@compuserve.com

Sylvia Spalding (9)
Communications Director
Marine Aquarium Council
923 Nuuana Avenue
Honolulu, Hawaii 96817 USA
sylvia.spalding@aquariumcouncil.org

Emma R. Suplido (24)
International Marinelife Alliance
83 West Capitol Drive, Bo. Kapitolyo
Pasig City (Metro Manila)
1601 Philippines
maree_cs@hotmail.com

Clyde S. Tamaru (18,19)
Hawaii Sea Grant College Program
School of Ocean and Earth Science and
   Technology
University of Hawaii at Manoa
Honolulu, Hawaii 96822 USA
ctamaru@hawaii.edu

Heather F. Thompson (28)
Conservation and Community Investment
 Forum
423 Washington Street, 4th Floor
San Francisco, California 94111 USA
heather@cea.sfex.com

G. Christopher Tilghman (8)
Graduate Research Assistant
Fisheries and Aquatic Sciences Department
7922 NW 71st Avenue
Gainesville, Florida 32611-0600 USA
fishkill@ufl.edu

Douglas Warmolts (23)
Assistant Director of Living Collections
Columbus Zoo and Aquarium
9990 Riverside Drive
Powell, Ohio 43065 USA
dwarmolt@colszoo.org

Ronald P. Weidenbach (19)
Co-owner/Manager
Hawaii Fish Company
P.O. Box 1039
Waialua, Hawaii 96791-1039 USA
hawaiifish@msn.com

Darcy L. Wheeles (28)
Conservation and Community Investment
 Forum
423 Washington Street, 4th Floor
San Francisco, California 94111 USA
darcy@cea.sfex.com

Chris M.C. Woods (20)
National Institute of Water and Atmospheric
 Research
P.O. Box 14-901
Wellington
New Zealand
c.woods@niwa.cri.nz

Introduction

This volume consists of a collection of 28 chapters resulting from papers presented at the Marine Ornamentals 2001: Collection, Culture and Conservation conference, in keeping with the goal of documenting the status of the marine ornamental fish and invertebrate trade. The chapters are written by some of the invited keynote speakers and by some of the speakers who submitted contributed papers. The perspectives are mixed and include representative contributions from researchers, economists, and various industry stakeholders. It is the intent of the editors that this will serve as a snapshot of the status of issues in the marine ornamentals world, which we consider to be something of an action shot. It is a rapidly changing industry with fast-moving elements.

The marine aquarium hobby has grown steadily, in large measure as a result of improvements in reef tank technology. It is now possible for a moderately experienced aquarist to enjoy a thriving marine aquarium with reasonable expectations of survival among its occupants. Technology has driven growth in demand, but only a small part of this demand is being met by farmed fishes and invertebrates. Estimates of the percentage of wild fish marketed vary but still hover well above 90%.

Part I, Introduction, of this book includes two benchmark chapters that are both packed with original survey data. Issues of sustainability are at the core of the first chapter, presented by a group representing the organizers of the Marine Ornamentals 2001 conference, led by John Corbin. An electronic survey was conducted during the course of that conference, instantaneously tabulating the responses of participants to a series of simple questions about complex issues. More than an exercise in measuring opinions, this survey established and prioritized recommendations for industry positions on a range of issues, including some that are contentious. It also provided some insight into the demographics of conference attendees, among which researchers from the United States were best represented, but many other occupations and nationalities were represented, including people who had a wide range of experience in the marine ornamentals industry. A total of eighteen priority recommendations under six headings were set, establishing a well-defined catalog of industry concerns, preferences, and perceptions. These range from endorsement of the new Marine Aquarium Council (MAC) certification procedures to the development and broad dissemination of appropriate technology to end users. Clear preferences also emerged with regard to the goals for future conferences.

The second chapter is by one of the people most closely identified with and in large measure responsible for progress in the culture of marine ornamentals—Martin Moe. Moe's chapter presents results of a survey in which more than 400 respondents characterized their experience with the culture of marine fish and invertebrates, as well as perceptions about the status, constraints, and future of aquaculture of these species. Although this is a casual survey, as opposed to a statistical survey, numerous trends are apparent in the results. In general, hobbyists prefer cultured organisms to wild organisms, support the goal of a conservation-based industry, and believe in the continued development of aquaculture technology. The commercial sector is struggling with the culture of marine ornamentals, both technically and financially, but 83% of respondents believe cultured marine ornamentals will eventually be a major factor in the market. Replies from researchers indicated that research funding is either difficult to find or unavailable and confirm the axiomatic view that larval feeding and survival are believed to be the rate-limiting constraints. The remainder of the volume is organized into four parts, each with chapters falling under several thematic subheadings.

Within Part II, Progress and Current Trends in Marine Ornamentals, we include the topics economics, health management, certification, and management. This section represents the state of the art of the marine ornamentals industry, as seen in the advances being made and patterns that emerge in the data.

Four chapters under Part II, A., Trade, Marketing, and Economics establish a current economic profile of the industry. Contributions in this section establish the most complete and current statistics and trends on trade in marine ornamental species in the United States (Balboa, Larkin) and on a global scale (Green, Olivier). Collectively, chapters 3 to 6 establish an economic framework sufficient to understand the forces that drive the industry. The analysis of global perspectives on the trade by Green takes a circumspect view of available databases and recent analyses, raising important questions about common assumptions and conclusions that have been reached. An analysis of the strengths, weaknesses, and trends to be found within the Global Marine Aquarium Database (GMAD) is a particular focus of this chapter. Supply-side trends are the primary focus of Olivier's analysis of United Nations Food and Agriculture Organization (FAO) statistics. These international trade data reveal patterns of production, consumption, and shifting market shares, and also bring to light relevant labor and environmental issues. Larkin's chapter uses a combination of data from landings and surveys to establish an organized overview of marine ornamental product sources and destinations within the United States. Balboa's analysis compiles U.S. Fish and Wildlife and U.S. Customs declaration data, resulting in a summary of economic trends over time that includes, among other things, the relative value per piece of marine fish and the growth of marine as opposed to freshwater imports.

The next section of this volume, Part II, B., presents selected contributions from the Health Management symposia, representing two of the most promising areas of current interest in the care of marine fish: disease and nutrition. Thankfully, the unfortunate image of the aquarium hobby as a rather short one-way trip for fish from the aquarium shop to a porcelain burial-at-sea is becoming an outdated concept,

thanks in part to quantum advances in fish care. Increasingly, consumers are able to provide long-term management and even set up home fish-breeding operations, drawing upon the contributions of both commercial innovators and scientific investigators. Support for disease diagnosis among marine ornamental species is a relatively new development; it has not been uncommon for marine ornamental fish reference texts to provide a cursory description of a few well-known pathogens and some precautionary comments about aquarium management and nutrition as the best means of avoiding the lot of them, and heaven help you if they strike.

Dramatic changes have occurred in the precision of diagnostic techniques within the past several years, as methodically explained by Francis-Floyd and Klinger in chapter 7. Much of the guesswork is evaporating as a consequence of a rapidly growing knowledge base. This chapter, based on the 129 marine ornamental fish disease case histories compiled at the University of Florida's fish disease laboratory, allows the reader to gain insights into the frequency and characteristics of particular diseases, their symptoms, and etiologies as they vary among a range of affected species. Nutritional status has long been recognized as a determinant of disease resistance in aquarium fish, but again older generalities are giving way to specific knowledge. In chapter 8, Tilghman and co-workers have presented a controlled study of disease development in a model organism, the herbivorous surgeonfish, in which nutritional variables had a distinct impact on the maintenance of body weight and resistance during an outbreak of cryptokaryosis.

Among the most troubling of issues in the marine ornamental trade is the fact that historically, short-term economic gains have been made at the expense of reef health on an ecosystem scale, sometimes with disastrous results. The resolution of the problem of environmental indifference by those who would exploit the reefs for financial gain rose among the priority recommendations of the Marine Ornamentals 1999 conference, and was identifiable as one critical issue upon which the future of the industry hinges. In order for environmental sensitivity and sustainability to be an industry standard rather than a pair of overused buzzwords, comprehensive and industrywide methods of accountability must be put in place. The long-anticipated establishment of a certification protocol for reef fish collection and shipment is the subject of chapters 9 and 10, Part II, C. First, Bunting and coauthors make a convincing case that reef use and conservation are not mutually exclusive. Reef use is characterized from the points of view of various and sometimes competing users, and the impacts of human use are illustrated. The Marine Aquarium Council's certification process is introduced, and the organization behind this process is spelled out. MAC certification is a potentially powerful device, which could begin to address the complex and urgent sustainability issues, in part by setting up a verifiable method of directing market preferences to organisms that have been collected and transported in an environmentally responsible way. Ultimately, the success or failure of MAC certification will have a great deal to do with the validity of the underlying economic assumptions. The competitiveness of MAC-certified fish with potentially cheaper alternatives may have the final say as to whether the conservation goals of the Marine Aquarium Council will be met. The environmental friendliness of the MAC-certification approach may have some costs to consumers, although it is also arguable that the certification process will improve economic returns, through reduced losses and as a result of public perceptions that certified fish have increased value relative to uncertified competitors. Larkin and coauthors model the economics of MAC-certification, using the Queen Angelfish, *Holocanthus ciliaris.* This model predicts anticipated costs and benefits of MAC certification and establishes a set of break-even prices that are dependant in part on the size of specimens being marketed.

Reef policy—particularly with regard to the harvest of ornamental fish and invertebrates—is set differently in different parts of the world, and the inclination to guide this policy is variably influenced by economic and conservation motives. Two chapters in Part II, D., Management, explore the mechanisms of reef management policy.

Chapter 11 discusses a unique, community-based method of reef management that is in place in Fiji (Bowden-Kerby). Resource management in this model is tightly intertwined with social conventions, as opposed to a European law-based set of standards left in place throughout much of the Pacific Islands by colonialists. Village-level users of the reefs in Fiji have a recognized legal entitlement to reef use, along with some responsibilities; these individuals are recognized as reef custodians. Comanagement of the reef, by the government and by reef custodians, is a delicate balance in which a fair amount of authority resides with village chiefs. This is a balance that is also influenced by the interests of nongovernmental organizations (NGOs) as well as recognized economic engines sharing the coastal community (e.g., resorts). This is a complex management model that offers some advantages over the hard and fast rule of law, in the application of community interests and sound conservation principles to common-sense decision making. This model is laid out in such a way that the community-based principles of government may, in part or as a whole, be considered for application elsewhere.

A slightly more Westernized view of reef management is the subject of chapter 12, by Bruckner. The long-term viability of the approach to management of coral fisheries (or their sustainability) is presented as an approach requiring some degree of control over elements that include the extent and methods of harvesting allowed and the impact that extractive methods can be allowed to have on the reef. In contrast with the social or community-based approach discussed above, the ecosystem-based approach uses the application of biological principles to industry regulation. The establishment of best management practices for coral harvests may involve the establishment of species quotas, or localized restrictions that rotate over time, in order to minimize the collection pressure on any one particular reef habitat. A scientifically based decision-making process of this sort requires baseline data on the distribution and biology of corals, often calling for ongoing technical support in the management process.

Aquaculture technology for corals is developing at a fast pace, and captive rearing is expected to relieve increasing amounts of pressure on wild reef corals and other invertebrates. Part III, The Invertebrates, A., Live Rock Cultivation, addresses progress in the area of cultivation of live rock—the colorful assemblages of rock or calcium carbonate and settled organisms that are in high demand among hobbyists. In chapter 13, Parks and colleagues present the results of an economic analysis of the costs and returns of coral and live rock culture in Florida, Fiji, the Solomon Islands, and the Philippines. This chapter includes the sort of business analyses that are required by entrepreneurs or investors, specifically the sort that predict the extent of investment and the number of years required for a venture to become profitable. The scale of the venture and the costs of labor are important determinants of profitability. As assumed in the modeling of economics for certified fish, discussed above, it appears that environmentally friendly products could eventually carry a premium in the marine invertebrate markets, although no such certification is in place. Overall, the financial risks of coral and live rock cultivation remain high, and the likelihood of profitability, at least in the near term and under present market conditions, remains low. The potential viability of such operations may improve as changes in the regulations or enforcement of regulations of the harvest or transportation of live, wild invertebrates occur. The culture of live rock may thrive on a smaller, local scale in areas where the harvest of wild material has been prohibited, as described in Florida by Falls and coauthors in chapter 14. This chapter examines the history and trends of live rock cultivation, concluding that economics is not the only factor inhibiting entrepreneurship, but that bureaucracy can also inhibit the process. Both the slow returns on investments and the restrictive nature of regulations have limited the growth of live rock cultivation businesses in Florida, although slow growth is in fact evident.

The update on invertebrate aquaculture continues in Part III, B., chapter 15, with an overview of the aquaculture of marine ornamental shrimp by Calado et al. Invertebrate shrimp cultivation has been the beneficiary of technology developed in support of the culture of Penaeid shrimps as an edible crop and has also been the subject of a dedicated research effort focused on the life cycles and biological requirements of the ornamental species. Ornamental shrimps are not especially difficult to keep in captivity, although the rearing of larvae is by no means an intuitive process. As in the case of so many other marine species, it is the successful cultivation of larvae, or in other words, the reliable production of juveniles that limits the development of commercial applica-

tions. Most people feel that this is a matter of larval nutrition. Calado and coauthors summarize the rearing systems available, the species under investigation, and the problems and issues that remain the subjects of active study.

The cultivation of corals has been through a fairly exciting upswing recently, as the use of simple fragmentation techniques has become a recognized method of propagating corals in sufficient quantities to support commercial operations, at least on a small scale. In Part III, C., chapter 16, Arvedlund and colleagues assess the present and predict the future of coral cultivation in a comprehensive overview of the asexual approach to coral propagation, considering both the captive propagation of corals in closed systems, some of which are not even near the ocean, and in situ in the marine environment. Selected case studies are used to illustrate taxonomically based trends in the propagation methods and potentials of many of the most desired species or groups of corals. The scientific content of reports of coral cultivation is lagging somewhat behind the descriptive work, according to Arvedlund et al., and this chapter underscores the need for more peer-reviewed basic science in the biology of corals.

Ornamental fish are the mainstays of the ornamental industry, both in the marine and freshwater sectors. Part IV, Reef Fishes, covers the aquaculture of marine ornamental fish, with subsections focused on A. Hatchery Methods, B. Feeding and Nutrition, and C. Seahorses and their cultivation and issues of sustainability.

A research group at the University of Texas led by Joan Holt has maintained a long history of productivity in the culture of marine ornamental fish and invertebrates, including the development of methods of inducing spawning and, especially the investigation of suitable larval rearing methods. In chapter 17, Holt provides a brief review of the status of hatchery technology for ornamental marine fish species, including discussions of culture systems and the all-important matter of larval nutrition. The use of diets sufficiently enriched in fatty acids or other diets in which some food is wasted introduces water quality concerns, so much of our current larval feeding technology can be viewed as a series of choices among trade-offs. This is clearly a report on an ongoing process, one that will be of use to anyone wishing to try his or her hand at rearing marine ornamental larvae or wishing to make a contribution to the basic biology of these organisms.

Chapter 18, by Brown and coauthors, addresses a unique aspect of culture, using a suitable freshwater model to take a predictive look at some of the technology that is likely to be transferred before long to marine ornamental fish. Usually, once the life cycle is closed and an adequate means of rearing larvae has been established, attention turns to the production of larvae year-round, on demand. Using techniques developed in the culture of marine fish for human consumption, photoperiod, temperature, and dietary manipulations were used to trigger reproductive maturation several months ahead of schedule in the freshwater ornamental fish, the rainbow shark (*Epalzeorhynchus frenatus*). These fish were then induced to spawn with hormones, demonstrating the applicability of environmental and endocrine manipulations to a species that has generally been constrained to spawning at just one time of year. These authors predict that this technology will be transferred yet again, from freshwater aquaculture back to marine aquaculture, at some time in the near future when production is limited not by larval feeding, but by seasonal patterns of gamete production.

The feeding of larval fishes with the most advanced hatchery technology relies heavily on rotifers, especially as an initial diet. Tamaru et al. examine the biology, nutritional value, and culture of rotifers in chapter 19 that combines a review of the status of this field and a set of original data documenting performance parameters of rotifers used in marine fish culture. These include practical data on the nutritional content of rotifers enriched with six different nutritional supplements, and other data that will be useful to the larviculturist. Experimental results include depictions of rotifer consumption rates and larval weights as functions of feed (rotifer) densities, impacts of enrichment media on rotifer mortality, and stress tolerance of fish that have consumed enriched versus unenriched rotifers.

The seahorses have been the focus of a sustained aquaculture effort, and one in which some considerable ground is being gained. Several species of seahorses have been cultured, with varying degrees of success, and much of the current attention is focused on the refinement of protocols, reduction of culture costs, and improvement of culture efficiency and consistency. Some urgency exists about seahorse cultivation, because of the combined effect of a continued demand not only for the ornamental trade but also for use in traditional Chinese medicine. These and perhaps other factors have led to an apparent crisis in the availability of wild seahorses for harvest. Three chapters, 20 through 22, provide insight into the continuing progress being made in seahorse aquaculture.

Hippocampus abdominalis is a seahorse species that has presented some significant problems with larval mortality in its preliminary rearing efforts. Woods reports in his chapter that steady advancement is being made in the rearing of the larvae of this species. A set of controlled experiments is reported in which optimal water temperature is determined, variation in stocking density is examined as it affects growth, and the use of inert foods as a cost containment measure was explored. Viewed collectively, these results cast an image of this species of seahorse as a prime candidate for cultivation, in which high survival rates and, consequently, high yields can be expected, and mass cultivation appears to be just around the corner.

Payne has also focused on the subject of balancing live and inert diets in his culture subject, *H. barbouri*. This is a species of which tank-reared specimens are already commonly available, so much of the current effort is aimed at making production costs more manageable by replacing live feeds with frozen dietary components, ideally without any compromise in product quality. In this chapter, Payne presents original data comparing the growth and survival of juvenile *H. barbouri* on diets consisting of live fatty-acid enriched *Artemia*, and enriched or unenriched frozen copepods. These studies clearly demonstrate that this species of seahorse can thrive on a frozen diet, raising hopes that commercial enterprises can thrive as well, with a production cost burden that can be reduced by eliminating the need for continuous culture of live feed organisms.

The need to develop more cost-effective diets was also the motivation of an experimental study by Gardner, who is seeking to refine culture methods for *Hippocampus erectus*, especially by reducing the cost of feeds for young juveniles. These fish prosper on copepod-intensive diets, but the cost of providing these is prohibitively high for commercialization. Again, the issue is squarely oriented on the minimization of expensive live feeds culture, although in this particular study the comparison was one of the relative efficacy of live *Artemia* nauplii and the more labor-intensive and hence expensive wild-collected copepod-dominated wild plankton. The goal of this study was to determine the minimum number of days of copepod feeding needed to obtain optimal growth and survival. Results provide novel insight into the beneficial effects of inclusion of copepod-intensive wild plankton even for a short time, although the downsides of this sort of diet have been documented here and elsewhere.

The final segment of this book, Part V, Stakeholder Perspectives, includes contributions from a range of stakeholders—users of reef resources, consumers, and others who have an interest in keeping the world's coral reefs viable. These chapters are written by participants in the marine ornamentals dialogue who represent a specific group or constituency, and in some cases the presentation differs from the customary scientific style of presentation. Nevertheless, these contributions are part of the ongoing discussion, and these groups are entitled to a say in the debate about how to approach and manage our coral reefs and our marine ornamentals industry.

Public aquariums have traditionally played a prominent role in elevating consciousness about conservation issues, and increasingly they have become fully involved in the process of developing appropriate technologies both for reef management and for the aquaculture of reef species. In chapter 23, Hall and Warmolts outline the role and responsibility of public aquariums in reef conservation, citing and detailing numerous ongoing initiatives among the public aquariums that have made public demonstrations of their research programs, including a number of captive culture efforts. Collectively, these programs represent a significant share of the culture and conservation work now underway, including studies on the captive propagation of elasmobranches and jellyfishes not mentioned elsewhere in this volume. In addition to the scientific role that public aquariums increasingly fulfill, the important contribution of these institutions as educational entities is emphasized in this chapter.

One of the most actively debated topics in the evaluation of reef-related concerns is the use of destructive collection practices, such as dynamite fishing and cyanide. The extent of this set of problems is discussed widely—most often in the most highly speculative of terms. Among the only agencies that are approaching this issue quantitatively—cyanide use in particular—is the International Marinelife Alliance group consisting of chapter author Peter Rubec and coworkers. Cyanide has unfortunately been used as a convenient shortcut to obtaining marketable reef fishes, with disregard for its toxicity. Consequently, the use of cyanide to collect fish typically results in mortality among many of the collected fishes along with a share of the nearby fish and invertebrates unintentionally exposed to this toxic compound. Clearly, the economic incentive is one of short-term gain at the expense of reef health. Rubec and associates have set up a sampling and testing procedure in the Philippines, where this problem has been especially acute. This program, in place since 1996, has produced data on the presence of cyanide residues in marine food and aquarium fish exported from the Philippines, including specific data on a family-by-family basis. As this program has grown, the number of fishes sampled annually has increased, and the distribution of species and families tested has been undergoing changes. Some trends that have emerged in the patterns of cyanide use are discussed in chapter 24.

The driving force in the aquarium trade is consumer demand; in this industry, the stocking of home aquariums is the process that has people scrambling through reefs throughout the tropics to collect fish and invertebrates for export. If it were not for the interests of consumers, coral reefs might be viewed as just an environmental curiosity and would almost certainly not be the subject of such intense collection pressure. For this reason, we have included chapter 25, by Clement, representing the interests and perspectives of Joe Six-Pack, the guy who takes enormous pride in his fish tank, the fellow who desperately wants to use his Visa card to buy some more reef fish. Clement provides a rare insight into the mind of the consumer, including the desires that drive supply and demand, as well as the perceptions that will either make or break MAC certification.

Sorting out the multiple and sometimes directly contradictory views of reef use and conservation is the job of the reef management official. Providing a lifetime of personal insight into the management of Florida Keys coral reef reserves is reef manager Billy Causey, who provides a long-term view of the trends and interests that have affected the continuing balance of reef use and conservation interests. Intelligible management plans must be simple enough for the lowest common denominator of citizen to comprehend but must take into account the biological and health concerns of some 6,000 species of reef inhabitants. It is a daunting task, to say the least, and a compounded series of compromises, without a doubt. Chapter 26 outlines degrees of conservation protection applied to reef systems, including the establishment of sanctuaries, reserves, Special Use Areas, and Wildlife Management Areas, among other conservation tools. It is a sensible guide to management tools from someone for whom management is an occupation.

The large-scale capture of wild marine species has been and is likely to remain for at least the immediate future, the primary source of animals for the marine aquarium trade. In chapter 27, John Dawes, of the Nongovernmental Organization (NGO) Ornamental Fish International, presents his analysis of the underlying beliefs and assumptions behind the continuing practice of reef harvesting. The desire to breed large numbers of ornamental species is called into question, and our techni-

cal capability of doing so is also subjected to some scrutiny. The accuracy of figures propagated by media sources is challenged and categorized as a "culture of misinformation" by the author. These and other illustrations of problems in industry perceptions are raised with the intention of supporting the trend favoring constructive changes in the industry.

The perspectives of a second NGO, the Conservation and Community Investment Forum, are represented in the concluding chapter (28) by Merkl and coauthors, addressing a business-oriented view of the changes taking place in the marine ornamentals industry. This group seeks to establish a private-sector initiated reform of the reef fish trade in the Indo-Pacific region, with an aim of setting up sustainable practices in the place of present practices that include exploitative ones. This chapter summarizes observations made during a six-week field trip to Indonesia, leading to the conclusion that Indo-Pacific reefs need a coordinated system of protective reform before reefs have been irreversibly damaged. This chapter follows the money, from consumer to collector, illustrating the trends that perpetuate destructive practices and that lead to a disregard for losses due to mortality. This group views reef conservation problems as largely financial in their origin and proposes solutions and reform, which also are financially based.

PART I
Introduction

1

Marine Ornamentals Industry 2001: Priority Recommendations for a Sustainable Future

John S. Corbin, James C. Cato, and Christopher L. Brown

Introduction

Today's Industry

The marine ornamentals industry around the world encompasses a broad array of disciplines, interests, and activities. Simply describing the chain of product distribution (chain of custody) includes mention of collectors and culturists, wholesalers and transshippers, distributors and retailers small and large, and of course hobbyists. When we add to this the equipment and supplies manufacturers, government resource managers and regulators, researchers and extension agents, educators and public aquariums, various media and international conservation organizations, the list of stakeholders in the future of the industry becomes daunting.

Indeed, as the twenty-first century dawns, interest in marine ornamentals is growing rapidly but perhaps not for the best of reasons. Reportedly, between 15 and 30 million marine fish from among approximately 1,000 species enter the trade every year, though accurate statistics are lacking (Wood 2001). Hundreds of species of marine invertebrates are also sold. Currently, as much as 98% of the marine ornamental species marketed are wild animals collected from coral reefs, mostly in tropical developing countries, for example, the Philippines and Indonesia (Moe 2001; Dawes 2001a). Major problems with collecting in the wild have been described by numerous recent reports, including the widespread use of chemicals and other destructive collection methods that damage coral reefs, negative social and economic impacts on rural coastal communities, and inadequate handling and shipping procedures that cause unnecessary stock mortalities (Baquero 1999; Wood 2001; Cruz 2001; Dawes 2001a).

Aquacultured sources in the marine ornamentals trade account for less than 2% of the supply, and sources of commercial quantities of product have been slow to develop (Moe 2001). The life histories of many economically important marine reef fish and invertebrates are extremely complex and difficult to control (Ziemann 2001; Brown 2001). Moreover, the marketplace has yet to appreciate fully the advantages of cultured species over wild-caught species and therefore accept the higher prices charged (Stime 2001).

The Marine Ornamentals Conference Series

In 1999, a group of state government, university, and private interests in Hawaii conceived and organized an international conference on marine ornamentals to bring together for the first time the diverse components of the industry for a comprehensive assessment of its status and discussion of the future. Delegates to "Marine Ornamentals '99: Collection, Culture and Conservation" came from 21 countries and strived to develop holistically a definitive understanding of the industry, its current challenges, and the actions needed to realize a sustainable future in the twenty-first century (Corbin 2001). The conference, held in Waikoloa, Hawaii, November 16–19, 1999, was underwritten by the University of Hawaii Sea Grant College Program, with additional sponsorships from sea grant programs in other states and many organizations in the public and private sectors.

The success of the first conference led the Florida Sea Grant College Program at the University of Florida to build on the concept and organize the second international conference, "Marine Ornamentals 2001: Collection, Culture and Conservation" in Orlando, Florida, November 26 to December 1, 2001. An outstanding collection of plenary speakers and paper sessions were presented to an international audience with representatives from 23 countries. Again, the intent was to bring together all facets of the marine aquarium industry to identify and discuss the critical issues that will affect the broad goal of creating an economically and environmentally sustainable future for all stakeholders (see Corbin 2001).

Both conferences were highlighted by systematic surveying of attendees to identify important industry issues, gauge concerns, and develop a consensus by voting to prioritize recommendations for further expansion. The results of the Marine Ornamentals 2001 process are reported here.

Conference Survey Process

Organizers of both meetings desired to take full advantage of the diversity and depth of the expertise of hundreds of delegates and create a forum for dialogue on all manner of critical industry concerns and problems. As a general approach, attendees were asked to consider the information they gained during the conference and their own unique experiences to identify the issues that are most critical to sustainable industry expansion in the coming years. Then attendees were asked to participate in a process to recommend specific priority actions and vote for the most important recommendations.

Marine Ornamentals '99 Process

The Marine Ornamentals '99 (MO '99) process consisted of an ending plenary session in which delegates, through a facilitated panel and audience "brainstorming" session, generated 59 recommendations for further consideration. These wide-ranging recommendations were voted on by attendees in a post-conference balloting to reduce the number to 20 top priority items.

These MO '99 action items encompassed the following categories: (1) government; (2) research and education; (3) Marine Aquarium Council; (4) incentives and certification; (5) resource management; (6) communication and marketing, and (7) general industry development guidelines (Corbin 2001).

Marine Ornamentals 2001 Process

The Marine Ornamentals 2001 (MO 2001) process was similar in concept to MO '99 but differed significantly in details of implementation. As with the MO '99 survey, suggested issues could address any dimension of the industry, for example, environmental impacts, industry cooperation and partnerships, social and cultural concerns, public education and outreach, international trade, regulation, economics and marketing, and aquaculture research needs. Issue items were again framed as industry recommendations for voting purposes.

A three-part survey vehicle was developed to collect information for use with a novel instantaneous electronic polling process. Part One consisted of nine questions that requested background information on the voter and responses that would help plan Marine Ornamentals 2003. Part Two allowed attendees to reprioritize the 20 priority recommendations from MO '99 in Hawaii because it was judged that only limited progress had been made and most were still important today. Part Three allowed attendees to suggest new recommendations and prioritize them. At an evening facilitated discussion during the conference, attendees generated the 40 new recommendations for consideration, which were distributed to all attendees before voting took place at the end of the conference.

Attendee Voting

Collection of background information, Part One, and voting on the priority recommendations for Part Two and Part Three, were carried out during the afternoon of the final day of the Marine Ornamentals '01 conference, using an individualized, proprietary electronic voting system that allows instantaneous audience polling and immediate projection of results (Padgett Communications, Inc.). Each audience member was given an electronic response pad (touch key pad) at his or her seat. Questions or lists of recommendations were projected on a large screen, and audience members were directed through the voting by a facilitator. The number of participants varied from 85 to 92 voters throughout the electronic survey.

Electronic voting to determine priority recommendations was carried out sequentially on six groups of ten recommendations. Groups One and Two consisted of priority recommendations from MO '99, while Groups Three through Six consisted of the forty new recommendations generated at MO 2001, which were randomly assigned to groups of ten. Participants were allowed to choose their top priority recommendation in each group, in each of three successive votes, so each audience member had the opportunity to choose three priorities.

An on-site computer then integrated the three votes to display a single numerical value for each item and ranked the recommendations from those receiving the highest value (most votes) to those receiving the lowest value (least votes). The top three vote getters from within each group of ten were considered top priority for the group and were added to make a total of 18 priority recommendations for MO 2001. This approach to establishing priorities in a group of participants is known as the $N/3$ Technique, in which N is the number of items to be prioritized in each group and the result is the number of votes given to each person in the group and the number of priorities to be set (Ching, University of Hawaii, personal communication 2001).

Voting Audience Information

Of the nine queries by Part One of the survey, three were designed to create a statistical characterization of the MO 2001 attendees. Question 1 asked about the attendee's role in the industry, that is, (1) aquaculture producer; (2) research scientist; (3) retailer; (4) wholesaler/importer; (5) collector; (6) hobbyist; (7) nongovernmental organization; (8) government agency; (9) trade association; and (10) other.

Results indicated that all role categories listed were represented in the voting except wholesaler/importer, suggesting that the conference goal of bringing together a diverse group of interests was achieved to a significant degree. The group most represented was research scientists at 35% of voters. Next was aquaculture produc-

ers and "other" at 15% each. Next were hobbyists and nongovernmental organizations at 11% each, and then government agencies at 9%. Relatively minor representation was logged for retailers, collectors, and trade associations at 1% each.

Question 2 asked about the location of the attendee's business or organization. Areas listed for response were: (1) United States; (2) Canada; (3) Europe; (4) Middle East; (5) Africa; (6) Asia; (7) Pacific Islands; (8) Caribbean Islands; (9) Central America; and (10) South America. As might be expected, the United States accounted for 74% of voters. The next highest attendance came from Europe at 11%, with Canada, Pacific Islands, and the Caribbean Islands at 3% each. Asia had 2%, and the Middle East and South America each had 1% of the voting audience. Africa and Central America were not represented. Results suggest that conference organizers for MO 2003 need to make a greater effort to attract marine ornamental interests from outside the United States, though it can be stated that important contemporary source and market locations for marine species were well represented at MO 2001.

Question 3 asked how long the attendee had been involved in the marine ornamentals industry, that is, 5 years or less, 5 to 10 years, 10 to 15 years, 15 to 20 years, and more than 20 years. Results indicate that all levels of industry experience were represented in the voting. Attendees who had the least involvement made up to 45% of the audience, while 10, 15, and 20 years or less experience accounted for 18%, 16%, and 7%, respectively. Those people who had the longest involvement in the industry accounted for 12% of the voting audience. Data indicate that a broad cross section of experience was present at the conference and for the voting.

Priority Recommendations from MO 2001

The 18 top priority recommendations selected by the voting audience were grouped for discussion purposes under six headings: (1) resource management and product certification; (2) research and public/private partnerships; (3) formal and hobbyist education; (4) marketing; (5) enhanced communication; and (6) future conferences. Priorities are presented and discussed below. Because the voting database can identify patterns of individual voters, comments will also be made on patterns of voting based on the participant's role in the industry, geographic location, and time spent in the industry.

Resource Management and Product Certification Priority Recommendations

- Develop reliable trade and biological data for marine ornamentals.
- The Marine Aquarium Council (MAC) should promote methods that allow a distinction between cultured and wild-caught marine ornamentals in the marketplace.
- The marine ornamentals industry should adopt the goal of sustainably producing a quality product at affordable prices.

Discussion: Data availability and sustainability of producing quality products are continuing concerns of industry members. The adequacy and reliability of available trade and scientific data to manage and conserve coral reef resources is perceived as unsatisfactory, though the situation is improving with the recent development of a Global Marine Aquarium Data Base by the United Nations Environmental Program, World Conservation Monitoring Centre, and the Marine Aquarium Council (Lem 2001; Green 2001).

Notably MAC, formed in 1998 to address various problems with sourcing and distribution in the industry, unveiled its logo at the 2001 conference and announced that the logo would be used in the launch of its program of product certification. Plans call for encouraging widespread certification of the components of the chain of custody so that consumers can have a means of supporting businesses that operate in an environmentally sensitive and sustainable way (Bunting 2001).

Eventually, sustainable cultured species sources will also be certified (Holthus 2001). Affordability by the hobbyist of higher priced, sustainably produced, and certified products remains a major impediment, however, as indicated by many discussions during the conference. It can be said, however, that the long-term goal of utilization of sustainable collection, culture, and handling practices throughout the marine

ornamentals industry clearly had strong support from attendees (for example, see Smith 2001).

Research and Public/Private Partnership Priority Recommendations

- Investigate mechanisms for the scientific community and the marine ornamentals industry to partner on research in order to accelerate scientific progress.
- International and federal research funding sources should give the highest priority to projects involving the advancement of marine ornamentals aquaculture and reef preservation, and this recommendation should be forwarded to all appropriate organizations.
- Develop and publicize standardized approaches to spawning and rearing more marine ornamental fish species.
- Develop a priority list of research topics on marine ornamentals for public funding organizations.
- MAC should provide a clearinghouse for sources of funding (i.e., grants) available to hobbyists, students, and so on.

Discussion: Attendees at MO 2001 strongly endorsed the need for accelerated research on the biology and culture of marine aquarium species and the ecology and management of coral reef environments around the world. The majority of marine life supplies to the marketplace will come from wild-caught sources for the foreseeable future, so proper management is necessary to maintain the reef resources (Dawes 2001b). Moreover, bottlenecks to commercial-scale culture of economically important species abound, as evidenced by the limited numbers and variety of cultured species currently coming to market (Moe 2003).

Several speakers addressed common bottlenecks to culture (e.g., Moe 2003; Holt 2001). Controlled spawning and first feeding, when marine larvae convert from absorption of their yolk sacs to exogenous feeds, was highlighted as a crucial area for increased effort (Holt 2001; Laidley 2001). The design of hatchery rearing systems suitable for mass rearing of small and delicate marine larvae also needs more work (Holt 2001). Disease diagnosis and management were highlighted as another major area for targeted research dollars (Stoskopf 2001; Moe 2003).

It was suggested that much greater progress could be made if scientists and industry members cooperated on identifying critical areas of research and carrying out specific projects. Moreover, the marine ornamentals industry needs to determine and publicize its research priorities to national and international funding sources in government and the nonprofit, conservation sector, but there is currently no formally organized entity charged with this task. To some extent, the series of Marine Ornamentals meetings has defined a de facto body of concerned individuals that has assumed this role. This group has identified as a priority the establishment of a centralized clearinghouse of information to allow hobbyists, students, and others interested in marine ornamentals research to find funding so these groups could pursue their own ideas or join with others to expand the overall effort.

Formal and Hobbyist Education Priority Recommendations

- Promote mechanisms to transfer technical information from the scientist to the hobbyist.
- Develop formal education programs at the undergraduate and graduate levels in marine ornamentals aquaculture and aquarium science.

Discussion: A clear message from the previous marine ornamentals conference carried over to MO 2001, namely that hobbyists and others in the industry need open access to more and better information to increase sustainable sources and the number of marketable species (Brown 2001). Whereas a number of university scientists, a few trade magazines, and an assortment of public aquariums are doing excellent jobs in information transfer and public education, it is widely perceived that a more focused effort to share information between marine ornamentals researchers and hobbyists is needed to advance the industry. Further, as the aquaculture component expands and greater numbers of marine hobbyists are fostered, a significant effort is needed to create formal educational opportunities at colleges and universities to improve the skill of commercial culturists and hobbyists.

Marketing Priority Recommendations

- The marine ornamentals industry must develop a greater consumer demand for fish aquacultured and/or collected in a sustainable manner.
- The marine ornamentals industry should encourage the notion that aquacultured animals and plants are bred to be better adapted to the aquarium environment and therefore have higher value.
- The marine ornamentals industry should accept and endorse sustainable collection and sustainable cultured sources and adopt a policy of expanding the market for both sources together.

Discussion Attendees at MO 2001 recognized that the implementation of sustainable collection and culture practices for various species and certification of sustainably produced products to harness market forces (ecolabeling), will mean these products cost more in the marketplace than stock from unmanaged, wild-caught, and sometimes exploitative sources. Increasing consumer acceptance for higher priced, sustainably produced products will be key to product acceptance and reducing pressures on wild stocks (Spalding 2001; Hoff 2001). In particular, discussions indicated market advantages of cultured products—for example, predictable supply, increased quality and survival, and predictable size, finage, and coloration—need to be emphasized in the marketplace. Attendees suggested that much industrywide discussion needs to occur to adopt a policy of comarketing sustainably collected and cultured products, but efforts by MAC are moving rapidly in the right direction and were strongly endorsed by attendees (Bunting 2001).

Enhanced Communication Priority Recommendations

- Develop mechanisms to disseminate culture information for marine ornamentals species (new and old), such as a centralized information base.
- Establish a website for global interests in marine ornamentals to exchange scientific and nonscientific information on breeding, species biology, economics, and so on.

• Develop mechanisms and linkages for communication and information exchange between the marine ornamentals community and the zoological and public aquarium communities.

Discussion: Enhancing communication among and between marine ornamentals interests was a recurring theme of discussions at the conference. Centralizing marine ornamentals information into readily accessible electronic databases and development of one or more interactive websites for hobbyists to exchange information were emphasized as important needs. Apparently, existing web resources are more focused on the freshwater portion of the hobby, and opportunities exist to better serve marine enthusiasts.

The plenary presentation on the role of public aquariums in the conservation and sustainability of the marine ornamentals trade was particularly well received (Hall and Warmolts 2001). As stated, public aquariums are a highly visible component of the trade and have a large responsibility to encourage and support sustainable resource use. They do so today through a variety of efforts, from individual institutional projects to multi-institutional global programs, such as Project Seahorse. Brief exposure at MO 2001 to the diversity of activities of the world's public aquariums with marine ornamental and reef conservation programs clearly left attendees wanting to know how to engage this network of scientists and educators for greater communication and information exchange.

Priority Recommendations for Future Conferences

• More governmental agencies, retailers, and international representatives should be involved in the discussions at MO 2003.
• The MO 2003 conference should have more sessions on culture.

Discussion: The last priority recommendations address some desired improvements to future conferences, though evaluations at both MO '99 and MO 2001 were quite favorable. Attendees suggested that a greater number of representatives from government agencies, the retail sector, and international organizations would enhance the discussions at the next conference. Survey results generally support this notion because government, retail, and international groups were clearly under-represented at MO 2001. It was also emphasized that MO 2003 conference organizers should consider ways to increase the number and quality of sessions on marine ornamentals culture, which should be easily accommodated because research efforts around the world are increasing. Strong support for the expansion of commercial culture was heard during the entire conference.

Conclusions

The marine ornamentals industry is faced with a host of complex environmental, economic, social, cultural, and political issues on an international scale. Ultimately, survival of the hobby and sustainability of the livelihoods of millions of people are at stake.

MO 2001 was highly successful in providing a forum for discussing the variety of complex issues facing the industry today and deciding what to do about them. Plenary and paper sessions included 70 presentations by attendees from 21 different countries. An excellent cross-section of industry components and experiences was present, as evidenced by the range of results of the survey on attendee role in the industry, location of the organization, and years of experience with marine ornamentals issues. It is clear, however, that organizers of MO 2003 and future conferences should strive for a more balanced and diverse mix of involved organizations.

The 18 industry priorities rising to the top of attendee concerns touched on six general areas: (1) resource management and product certification; (2) research and public/private partnerships; (3) formal and hobbyist education; (4) marketing; (5) enhanced communication; and (6) future conferences. The recommendations provide both near-term and long-term direction for the efforts of industry proponents to expand. The priority solicitation and voting process in MO 2001 was a marked improvement over MO '99, with twice the participation, but process improvement and consensus building should be ongoing goals for future organizers. Further, conference organizers hope and encourage all components of the marine ornamentals industry to review these results and incorporate the priority action items into their short-term and long-term strategic plans. Lastly, readers are encouraged to look for notification of future conferences in the continuing series and are also urged to attend in order to share ideas and participate in successfully guiding the industry forward.

References

Baquero, Jaime 1999. Marine Ornamentals Trade, Quality and Sustainability for the Pacific Region. South Pacific Forum, May 1999, 52 pp.

Brown, Stanley 2001. Information exchange and captive propagation. In Proceedings of Marine Ornamentals 99, University of Hawaii Sea Grant College Program, pp. 9–12.

Bunting, Bruce 2001. Buy a fish, buy a coral, save a reef: The importance of economic incentives to sustain conservation. In Marine Ornamentals 2001, Program and Abstracts, University of Florida Sea Grant College Program, pp. 3–4.

Corbin, John S. 2001. Marine Ornamentals 99, conference highlights and priority recommendations. *Aquarium Sciences and Conservation*, Vol. 3, No. 1–3, pp. 3–11.

Cruz, Ferdinand P. 2001. Supplying the demand for sustainability: Stories from the field. In Marine Ornamentals 2001, Program and Abstracts, University of Florida Sea Grant College Program, pp. 13–14.

Dawes, John 2001a. Resource management and regulation: Current status and future trends. In Proceedings of Marine Ornamentals 99, University of Hawaii Sea Grant College Program, pp. 21–32.

Dawes, John 2001b. Wild-caught marines and the ornamental aquatic industry. In Marine Ornamentals 2001, Program and Abstracts, University of Florida Sea Grant College Program, p. 15.

Green, Edmund 2001. Separating fish facts from fishy fiction (not forgetting invertebrates). In Marine Ornamentals 2001, Program and Abstracts, University of Florida Sea Grant College Program, pp. 16–17.

Hall, Heather, and Douglas Warmolts. 2001. The role of public aquaria in the conservation and sustainability of the marine ornamental trade. In Marine Ornamentals 2001, Program and Abstracts, University of Florida Sea Grant College Program, p. 18.

Hoff, Frank 2001. Future of marine ornamental fish culture. In Marine Ornamentals 2001, Program and Abstracts, University of Florida Sea Grant College Program, p. 55.

Holt, G. Joan 2001. Research on culturing the early life stages of marine ornamental species. In Marine Ornamentals 2001, Program and Abstracts, University of Florida Sea Grant College Program, pp. 19–20.

Holthus, Paul. 2001. From reef to retail: Marine ornamental certification for sustainability is here. In Marine Ornamentals 2001, Program and Abstracts, University of Florida Sea Grant College Program, pp. 21–23.

Laidley, Charles W. 2001. Captive reproduction of yellow tang and pygmy angel fishes at the Oceanic Institute. In Marine Ornamentals 2001, Program and Abstracts, University of Florida Sea Grant College Program, pp. 60–62.

Lem, Audun 2001. International trade in ornamental fish. In Marine Ornamentals 2001, Program and Abstracts, University of Florida Sea Grant College Program, p. 26.

Moe, Martin A. 2001. Marine ornamentals: The industry and the hobby. In Proceedings of Marine Ornamentals 99, University of Hawaii Sea Grant College Program, pp. 53–63.

Moe, Martin A. 2003. Culture of marine ornamentals: For love, for money, and for science. In Marine Ornamental Species: Collection, Culture, and Conservation. Ed. J.C. Cato and C.L. Brown. Ames: Iowa State Press.

Smith, Walt 2001. Responsibilities for collection and opportunities in aquaculture for developing countries through the marine aquarium trade. In Marine Ornamentals 2001, Program and Abstracts, University of Florida Sea Grant College Program, pp. 167–168.

Spalding, Sylvia 2001. Creating consumer demand for MAC certified marine ornamentals. In Marine Ornamentals 2001, Program and Abstracts, University of Florida Sea Grant College Program, p. 134.

Stime, Jim Jr. 2001. Hobbyist perspectives, uninformed or blissfully naïve? In Proceedings of Marine Ornamentals 99, University of Hawaii Sea Grant College Program, pp. 73–78.

Stoskopf, Michael K. 2001. Current issues in disease control in marine ornamentals. In Marine Ornamentals 2001, Program and Abstracts, University of Florida Sea Grant College Program, p. 31.

Wood, Elizabeth 2001. Global advances in conservation and management of marine ornamentals resources. *Aquarium Sciences and Conservation*, Vol. 3, No. 1–3, pp. 65–77.

Ziemann, David A. 2001. The potential for the restoration of marine ornamentals fish populations through hatchery releases. *Aquarium Sciences and Conservation*, Vol. 3, No. 1–3, pp. 107–117.

2

Culture of Marine Ornamentals: For Love, for Money, and for Science

Martin A. Moe, Jr.

Introduction

The hobby of keeping marine aquariums and the associated industries of marine ornamental organism collection, captive culture, aquarium product manufacture, distribution, and wholesale and retail trade have changed and grown rapidly in the past 15 years. The engine of this growth and change is the developing ability of hobbyists and small, captive propagation businesses to maintain and culture living corals in small marine aquarium systems. The growth of this hobby/industry in coral keeping has also stimulated the collection and propagation of other aquarium organisms such as fish, live rock, live sand, and a wide variety of invertebrate animals. Strong environmental concerns about the sustainability of wild stocks of collected organisms have also developed as marine live animal collection and trade have expanded and as environmental and anthropogenic stresses on tropical marine ecosystems have greatly increased.

Hobbyists are growing corals almost as a gardener grows flowers, and small commercial coral culture businesses are developing in all areas of the United States and in other countries as well. Increasing market value, the concerns about increasing restrictions on collection of wild stocks, and new successes in the culture of marine ornamental fish are stimulating the interest and efforts of commercial breeders, public aquariums, and scientists, as well as hobbyists, in the culture of marine ornamental organisms. The accelerating growth of biological understanding, and the technology that this knowledge has spawned, is rapidly changing the social and economic structure of the hobby. These changes and developments, although intuitively understood by hobbyists and professionals, are graphically expressed in this chapter using results from a survey conducted during the last half of 2001.

The motivations, attitudes, concerns, and practices of hobbyists, commercial breeders, and scientists who culture marine ornamentals may differ greatly, but the roots of these endeavors are parallel. It is important to know the diverse interests, efforts, successes, failures, and attitudes of those engaged in these pursuits to understand the present and plan for the future of the ornamental marine life hobby and industry. The survey on the culture of marine ornamentals prepared for the Marine Ornamentals 2001 conference attempts to collect and analyze this basic information through a questionnaire distributed to hobbyists, scientists, and commercial breeders all over the world.

By mid-November 2001, 408 hobbyists, scientists, and commercial breeders had responded to the questionnaire. Although the survey was not designed to reach a statistical sample from a known population of hobbyists, scientists, and commercial breeders, the results provide insight into the state of culture in the marine hobby/industry at this time. The questions were designed to provide information on the organisms that have been cultured successfully and those whose culture has been attempted and failed. It focused on the organisms that provide the best return for commercial culture and those whose successful culture is most desired but not possible at this time. A total of 325 hobbyists reported on the organisms that they most frequently propagate and what they do with home-bred marine ornamentals. The attitudes of hobbyists toward culture and the hobby itself are also explored. Thirty-four commercial breeders reported on the status of their business, on their greatest difficulties, and the organisms with which they are most financially successful. Forty-nine scientists reported on their sources of funding, the organisms they are working with now, those they have worked with in the past, and those that they have successfully propagated. Their failures are also reported, as is the basic purpose of their research.

Development, Distribution, and Analysis of the Survey Questionnaire

The questionnaire was composed of 31 questions divided into three sections. The first question identified the respondent as a hobbyist, a commercial breeder, or as a scientist or professional aquarist. The hobbyist was directed to questions 2 through 10, the commercial breeder to questions 11 through 20, and the scientist/professional aquarist to questions 21 through 31. Thus, each individual respondent had only about 10 questions to answer, and the questionnaire could be completed fairly quickly. The questions in each category were designed to provide some insight into the scope of activities, motivations, attitudes, successes, failures, and financial importance of the culture efforts of respondents in each category. Although most of the questions required single or multiple responses, there was also ample opportunity for expanded expression of ideas, attitudes, and opinions.

The questionnaire was distributed almost exclusively by computer through the Internet from April through November of 2001. Respondents were solicited from hobbyist bulletin boards, marine aquarium clubs and societies, professional aquarist organizations, commercial web sites, and through personal email requests. A total of 408 individuals responded to the questionnaire, including hobbyists (325), commercial breeders (34), and scientists/professional aquarists (49).

Most of the questions required the respondent to choose one or more answers, but several questions in each section asked for a list of organisms or a comment. Thus, compilation of the results required that data be collected and tallied from each questionnaire, as well as compilation of the responses from questions requiring comment.

Many of the questions elicited a multiple response; the respondents often chose two or more of the options presented for a single question. It was important that the percentage of response to each answer represented the number of individuals choosing that answer rather than the percentage of distribution of all the answers to that particular question. Therefore, the percentage of response reported for each option is the percentage of respondents who chose that answer and not the percentage of the total response that that answer elicited. For example, question 10 in the hobbyist section had six possible answers. Respondents were encouraged to select all the choices with which they were in agreement. There were 1,157 responses from 324 respondents. Of the 324 respondents, 260 selected choice A as one of their options. The 260 responses to choice A represented only 22% of the total 1,157 responses. However, of the 324 hobbyists who responded to this question, 260, representing 80% of the respondents, selected choice A as one of their answers to this question. So the correct interpretation of the data is that 80% of the hobbyists responding to the questionnaire supported the work of the Marine Aquarium Council (choice A). Thus, the percentage analyses for those questions with multiple responses were reported as a percentage of the respondents answering that question.

The Hobbyist Response: Those Who Culture for Love

Table 2.1 presents data recovered from the 324 respondents to the hobbyist section of the questionnaire. The survey questions and the answer selections are abbreviated in this table.

Questions 2 through 10 pertained to hobbyists. There were 325 responses from hobbyists from all over the world, mostly from the United States. Not all respondents answered all questions. In the hobbyist questionnaire, all questions except number 2 elicited multiple answers from most respondents. The responses covered the range of intensity of effort in this hobby, all the way from the casual aquarist, to those to whom the word *obsession* just might be applicable.

Question 2 functioned to separate hobbyists into three groups—those who do nothing extra to propagate marine organisms (20%), those who actively propagate marine organisms but

Table 2.1   Hobbyist questionnaire results

Question and response data	Responses to each option	Percent of respondents choosing each option
2. Culture effort		
324 respondents (no multiple responses)		
A. Do not culture	64	20
B. Minimal culture effort	177	55
C. Strong culture effort	83	26
3. Disposal of cultured organisms		
318 respondents (no multiple responses)		
A. Give to other hobbyists	151	47
B. Sell to other hobbyists	107	34
C. Do not produce excess organisms	60	19
6. Major problems in culture of marine organisms		
324 respondents (451 responses)		
A. Disease	17	5
B. Lack of time	145	45
C. Lack of facilities	159	49
D. Lack of knowledge	102	31
E. Other	28	9
7. Future intentions toward culture		
324 respondents (372 responses)		
A. Continuous effort	261	80
B. Abandon culture efforts	7	2
C. A current source of income	26	8
D. An anticipated source of income	78	24
9. Attitude toward purchase of cultured organisms		
324 respondents (383 responses)		
A. No concern for cultured vs. wild	51	16
B. Buy cultured organisms when possible	210	65
C. Request cultured organisms	101	31
D. Buy only cultured organisms	21	6
10. Issues and opinions		
324 respondents (1157 responses)		
A. Support the Marine Aquarium Council	260	80
B. Support improvement, doubt success	161	50
C. Favor legislation	136	42
D. Favor voluntary efforts	230	71
E. Feel that culture is necessary	236	73
F. Feel that wild collection is necessary	134	41

on a casual basis (55%), and those who make a strong effort at propagation (26%). The results indicated that a large number of hobbyists, 81%, now propagate marine organisms, and about 26% make a serious effort at propagation. This indicates that the biomass of captive marine organisms, mostly corals, that resides in hobbyists' tanks is growing rapidly. The growing availability of cultured marine organisms, mostly corals, now present and growing rapidly

within the hobby without dependence on wild collection or typical avenues of commercial distribution also signals widespread and fundamental changes occurring within the hobby.

Question 3 was designed to learn a little about the economics of the hobby at the level of the hobbyist. Of the 318 hobbyists who answered that question, 60 indicated that they both gave away and sold excess organisms to other hobbyists. The purpose of the question was to develop an indication of the proportion of hobbyists who cultured and distributed marine organisms without economic gain and those who obtained some recompense from their activity. Therefore all 60 responses that indicated both sales and gifts were assigned to B (sales to other hobbyists). Fifteen respondents also indicated that they traded organisms with other hobbyists, and fifteen also indicated that they sold or traded organisms with local fish stores in addition to sales to other hobbyists. These 30 responses were not listed separately, and these respondents were included only under the sales category B, where they reflect some economic return to the hobbyist.

Widespread propagation of ornamental marine organisms at the hobbyist level was essentially nonexistent until the mid-1990s. The response to this question shows what has happened in just a few years: 47% of hobbyists only give organisms to other hobbyists, 34% sell to other hobbyists or otherwise gain materially from their efforts, and only 19% do not produce excess organisms. About 80% of hobbyists give, exchange, or sell excess organisms, almost all of which are corals. This may be one of the reasons that 65% of the commercial respondents indicated that their business was either "staying alive" (50%) or was in a questionable status (15%). One of the things that fueled the growth and development of the freshwater aquarium hobby is propagation of fish by hobbyists. The propagation of corals by marine hobbyists seems to be on track for fulfilling the same niche on the marine side of the hobby. Culture and propagation of corals and some fish are expanding rapidly, and the hobby may be entering an increasing rate of growth. Soon the infrastructure of the marine ornamental hobby/industry may look much like that of its big brother, the freshwater aquarium hobby/industry.

Interestingly, in question 6 (324 respondents and 451 responses), most hobbyists reported lack of facilities (49%), time (45%), and knowledge (31%) as their major problems, while only 5% listed disease as a major problem. This indicates that disease, although certainly a problem at one time or another to most hobbyists, is not a major concern.

On question 7, there were 372 responses from the 324 respondents. Apparently, culture and propagation are a major part of the hobby for most hobbyists because 80% of the respondents indicated that culture would be a continuing effort in their practice of the hobby. Only 2% indicated that they would soon cease culture efforts. The 48 multiple responses were divided between choice A, a continuing effort, and choices C and D, a current source of income (8% of the respondents) and an anticipated source of income (26% of the respondents). The unequal division of responses between C and D indicates that, as in most areas of human endeavor, planning and anticipation greatly exceed actual performance.

Question 9 explored hobbyists' attitudes toward cultured organisms in the marketplace, and question 10 looked at issues and opinions about the future of the hobby. Many insights may be gleaned from these two questions, more than can be explored in this limited space. From question 9, however, it is obvious that a strong bias exists toward the purchase of cultured organisms. Only 16% of the respondents indicated no discrimination between cultured or collected organisms. The great majority expressed some preference for cultured organisms. Question 10 showed that support for the efforts of the Marine Aquarium Council is quite strong (80%), that hobbyists favor voluntary efforts (71%) over legislation (42%) to improve the quality of the trade, and that most (73%) feel that expansion and development of culture is immediately necessary in the hobby. A significant number (41%), however, feel that wild collection will be the foundation of the hobby for many years into the future.

Question 4 requested a listing of the species of organisms that hobbyists have been successful at propagating, and question 5 requested a listing of those species that have failed in propagation. Altogether, hobbyists reported successful propagation of about 230 species of ornamental marine organisms. This number is highly approximate since many hobbyists reported in only broad categories, by genus or common name, and the list reflects this ambiguity. A complete listing of organisms reported to be propagated by scientists and professional aquarists is presented in the discussion of the scientists' section of the questionnaire. Hobbyists also reported most of these species. Table 2.2 presents a summary of the most common organisms, usually only by genus, that were reported to have been successfully propagated by

Table 2.2 Organisms successfully propagated by hobbyists

Organism category		Organisms commonly propagated	Number of reports
Live rock/sand		Live sand	11
Algae		All algae species reports	39
		Macro algae	9
		Caulerpa spp.	16
		Halimedia spp.	6
Sponges/tunicates		Sponges	5
Anemones		All anemone species reports	15
		Entacmea quadricolor, bubble tip	12
Corals	All coral species reports		1023
	All soft coral species reports		701
	Mushroom corals, all species		124
		Discosoma spp.	16
		Ricordea spp.	8
	Gorgonians, all species reports		27
	Leather corals, all species reports		169
		Sarcophyton spp.	74
		Cladiella spp.	33
		Lobophytum spp.	25
		Sinularia spp.	30
	Xenia spp.	All *Xenia* spp. reports	134
Other soft corals		*Anthelia* spp.	20
		Capnella spp.	17
		Nephthea spp.	12
		Clavularia spp.	168
		Briaerium spp.	33
		Pacylavularia violacea	15
		Zooanthus spp.	35
	All stony coral species reports		313
		Acropora spp.	87
		Euphyllia spp.	32
		Montipora spp.	71
		Pocillopora spp.	19
		Seriatopora spp.	12
Annelids		Various polychaetes	16
Mollusks		Mostly gastropods	34
Crustaceans		Mostly shrimp	25
		Lysmata wurdemanni	5
Echinoderms		Mostly brittle stars	12
Fish		All fish species	111
		Amphiprion/Premnas spp.	55
		Gobiosoma spp.	4
		Hippocamphus spp.	10
		Pseudochromis spp.	7
		Pterapogon kauderni	24

hobbyists. Hobbyists reported successful propagation of corals with 1,023 responses, fish with 111 responses, and crustaceans with only 25. This reflects the ease of culture and the great interest in propagation of corals. Propagation of fish and crustaceans requires more facilities and greater daily care than corals, and although hobbyists' interest in fish and crustacean propagation is high, few can expend the required effort.

Interestingly, the response to question 5, a request for a list of species where propagation was attempted and failed, was similar to the listing of species successfully propagated by hobbyists. Apparently for almost every species that is successfully propagated by one hobbyist, another fails at the same attempt. The same organisms that typically thwart commercial and scientist breeders—anemones, particularly carpet anemones, *Stichodactyla* spp.; soft corals, *Dendronephthya* spp.; hard corals, *Goniopora* spp. and *Catalaphyllia* sp.; cleaner shrimps, *Lysmata* spp.; pigmy angelfish, *Centropyge* spp.; and dottyback basslets, *Pseudochromis* spp.—were also commonly reported as unsuccessful by hobbyists as well.

Some hobbyists were quite perfunctory with the questionnaire, indicating an answer with just a mark and providing no comments whatsoever. Others answered every question with an essay. But most did provide a comment or two at various points and in response to questions 8 and 10, and these comments were collected and listed under the appropriate categories. Question 8 was a broad request for comments on what would improve the hobby for the hobbyist. Question 10, and the other questions as well, solicited opinions on the state of hobby and what is wrong, and right, and needs improvement. Some of these comments are insightful, some entertaining, and some revealing.

Under question 8, most hobbyists wanted more time, money, facilities, and equipment to pursue their hobby. High-quality livestock for propagation was a common need, as were sources of plankton or planktonlike foods for larval rearing and coral feeding. There was also a strong call for pertinent information, up-to-date books and articles, and knowledge exchange with other hobbyists, scientists, and professional breeders.

Many questions, especially question 10, elicited expansive comments on the place of cultured organisms in the hobby, trade practices, environmental concerns, propagation problems, and many other issues, such as certification of dealers and the need for education. Many hobbyists expressed great concern about the collection and sale of organisms known to do poorly in captive situations, and many would favor legislation to prevent such practices. On the other hand, many comments were strongly against any legislation directed toward management of the trade in marine organisms. The common practices of retailers, and to a lesser extent wholesalers, and the quality of the instruction and information provided by retailers garnered much criticism from many hobbyists. There was a persistent call for licensing of retailers to assure that they knew and practiced the best techniques for handling and maintenance of marine organisms. The culture of marine organisms was almost universally heralded as essential to the hobby and very important to environmental preservation. The call was also strong for sustainable practices in the wild collection of ornamental fish and coral fisheries.

The Commercial Response: Those Who Culture for Money

Questions 11 through 20 were directed toward commercial breeders, those who conduct a business, either full time or part time, in propagation of marine ornamental organisms. This definition includes live rock and live sand, which are essential living elements in most modern marine aquarium systems.

There were only 34 commercial responses. Many commercial breeders declined to respond to the survey despite a personal invitation to participate. It is understandable that commercial breeders were the group most reluctant to participate and were the least open in their responses because commercial operations have the most to gain by keeping successful technology proprietary and the most to lose by revealing information that may contribute to their commercial success. The commercial breeders who responded to the questionnaire specialized in either corals or fish. Of the 34 respondents, 4 (12%) produced neither corals nor fish but worked with live rock or *Artemia* (brine shrimp), 14 (41%) breed fish and/or crustaceans (cleaner shrimps), and 16 (47%) breed corals.

Table 2.3 presents data recovered from the 34 respondents to the commercial breeder sec-

Table 2.3   Commercial questionnaire results

Question and response data	Responses to each option	Percent of respondents choosing each option
11. Business status		
34 respondents (no multiple responses)		
A. Doing well	12	35
B. Staying alive	17	50
C. Survival questionable	5	15
12. Site and water use		
34 respondents (no multiple responses)		
A. Open sea or open systems	4	12
B. Inland with closed systems	22	65
C. Both open and closed systems	8	24
15. Income level from culture of marine organisms		
33 respondents (no multiple responses)		
A. Negative factor	3	9
B. Break even	4	12
C. 5 to 25%	7	21
D. 26 to 50%	3	9
E. 51 to 75%	4	12
F. 76 to 100%	12	36
16. Major problem areas		
34 respondents (62 responses)		
A. Biological	14	41
B. Financial	20	59
C. Personnel	13	38
D. Facilities	15	44
17. Status of culture in the marine life industry		
34 respondents (38 responses)		
A. Small and struggling	20	59
B. Important and growing	15	44
C. Already significant	3	9
18 Future of culture in the marine life industry		
34 respondents (no multiple responses)		
A. Small and restricted	6	18
B. Slow development into a major factor	22	65
C. Rapid development into a major factor	6	18

tion of the questionnaire. The survey questions and the answer selections are abbreviated in this table.

Question 11 inquired about the status of the respondent's business. Without being too intrusive, the goal was to determine if the business was financially strong, marginal, or in danger of failing. The results, 35% doing well, 50% marginal (the survey question was worded as "staying alive"), and 15% questionable ("checking the want ads"), indicated that there is a strong nucleus doing well with commercial culture and propagation, but the majority of ornamental culture concerns may be living on the edge.

Commercial breeders face commercial competition from wild collection and consumer competition from hobbyist propagation. Many hobbyists are part-time commercial propagators of specialized and often rare and high-quality corals. Their continued existence as a hobbyist/commercial breeder is usually part time or hobby based and is not dependent on a particular sales volume. It is difficult for most full-time small business operations to compete with both ends of this commercial spectrum. It takes a great deal of hard work, technical expertise, and business acumen to be successful. As the hobby continues to grow and the trade becomes more dependent on captive propagation, however, the expanding market will ease financial pressures on small businesses.

Question 12 inquired about location with respect to natural waters and, interestingly, the great majority of commercial concerns (65%) were located inland, relying completely on artificial conditions. Facility size and concomitant production capacity are constrained by operational expenses, that is, artificial seawater, available space, lighting, heating, and electrical costs, at inland locations, but proximity to markets and ease of product transport are advantages. Locations on tropical shores are usually remote, thus the advantages of natural seawater, abundant sunlight, and wild collection of desirable brood stock organisms are typically offset by transportation and labor problems.

Question 15 explored the contribution that cultured organisms contributed to the business income. The culture of marine organisms is undertaken by a variety of businesses. For some, it is only a portion of the business and for others it represents the total effort of their operation. There were two peaks in the responses, 21% reported 5% to 25% and 36% reported 76% to 100% of the business came from cultured organisms. Thus, most of the responding concerns were solely culture based, but of those who had other sources of income, most reported that culture makes up about 20%, or less, of their trade.

Question 16 tried to identify the areas that the respondents felt were their major problem areas. Probably predictably, 59% of the respondents selected financial as a problem area. Facilities were selected as a problem area by 44% of the respondents, biological problems were selected by 41%, and personnel difficulties were selected by 38%. Although facilities, biological problems, and personnel were major problem areas, a little over one third of the respondents indicated that each of these three areas harbored difficulties. The overriding problem area as indicated by almost two thirds of the respondents is in the financial area. This corresponds well with question 11, where only 35% of the commercial breeders indicated that finances were not a problem.

Question 17, with 34 respondents and 38 responses, elicited an opinion on the status of culture and propagation in the current state of the hobby/industry. Few (9%) thought that culture currently occupies a significant place in the industry. Most (59%) thought that culture was still small and struggling and, interestingly, 44% considered that culture was an important and growing sector of the industry.

Commercial propagation of both fish and corals requires an intimate knowledge of the biology and requirements for captive culture of these animals, but fish and crustacean culture is more labor intensive and biologically complex than coral culture. It was interesting, therefore, even though the number of respondents was quite small, to compare the responses of fish and coral breeders to questions 17 and 18. In question 17, 64% of the fish breeders and 50% of the coral culturists considered propagation of marine organisms as a small and struggling sector of a large and varied industry. Thirty-six percent of fish breeders and 38% of coral culturists considered propagation an important and growing sector of the industry, and none of the fish breeders and 13% of the coral culturists considered propagation already to be a significant sector of the industry. Coral culturists have a more positive and confident opinion on the current status of culture in the marine aquarium industry than do the fish breeders.

Question 18 required the 34 commercial breeders to look into the future and predict the eventual status of culture and propagation in the marine aquarium industry. A majority, 65%, thought that over the long term culture will become a major part of the industry, 18% thought that culture is quickly going to attain a major status, and the same number (18%) thought that culture would remain as only a small and restricted part of the industry picture.

When the responses of the fish breeders and coral culturists were separated, 21% of the fish breeders and 13% of the coral culturists thought that in the future culture would be a small, restricted part of the total industry; 71% of the fish breeders and 56% of the coral culturists thought that culture would slowly develop into a major sector of the industry; and 7% of the fish breeders and 31% of the coral culturists thought that culture would become a major part of the industry in the near future. Again, the coral culturists have a more optimistic view of the future of marine organism propagation than do the fish breeders.

Question 13 explored the species that commercial breeders successfully propagate, and question 14 those species that commercial breeders do not now but would like to propagate.

The 34 commercial breeders provided 201 responses, representing about 99 species, and reported successful propagation of about 110 various species of corals, 5 crustaceans, and 74 species of fish. Table 2.4 presents a summary of the most common organisms, usually only by genus, that were reported successfully propagated by commercial breeders. The number of species (99) reported successfully propagated is highly approximate since many commercial breeders reported in only broad categories, by genus or common name, and the list reflects this ambiguity. Scientists and public aquarists also propagated all the species reported as successfully propagated by commercial breeders. Thus, the complete listing of organisms reported propagated by scientific researchers presented in the science section of the questionnaire includes those species successfully commercially propagated.

Commercial breeders responding to question 14, those species most desired for new culture, listed about 4 species of soft corals, 14 species of hard corals, 3 groups of mollusks, 6 crustaceans, 1 echinoderm, and 33 species of fish. Other commercial breeders, hobbyists, and scientific researchers have successfully cultured many of the species listed, so appearance on this list does not mean that the species has not been, or cannot be, bred. Soft corals desired for culture included *Dendronepthea* spp. and *Cespitularia* sp.; stony corals included *Blastomussa wellsii*, *Plerogyra sinuosa*, *Catalaphyllia* sp., and *Trachyphyllia* spp.; *Lysmata debelius* and *Enoplometopus* spp. were among the crustaceans; *Diadema antillarum* was the echinoderm; and the fish included various *Acanthurus* and *Zebrasomas* species, *Centropyge* spp., and other angelfish and butterfly fish.

Question 19 asked the respondent to state the breakthrough development in marine culture that they would most like to see. New and better food organisms for larval culture was the most frequently mentioned development. Second was more affordable lighting for coral culture, and third was the successful culture of marine fish with small pelagic eggs (which has recently been accomplished). Sustainable sexual reproduction of corals and more effective disease control were next on the list.

Along the same line as question 19, question 20 inquired about the major impediments to financial success. One of the main themes in question 19 was strongly repeated in the comments under question 20. The greatest impediment to success was seen as the difficulty of competing in the market with the lower price of wild-collected organisms. A second but related theme was the need to increase awareness of the importance and value of cultured organisms to hobbyists and retailers. It was also suggested that restriction and regulation of collected organisms would improve the market potential for cultured organisms. The low price and larger size of collected specimens makes competition in the marketplace difficult. Technical difficulties such as food culture, location, propagation problems, and production costs were mentioned but were greatly overshadowed by marketing problems and price competition with wild-collected stock. Limited facilities and personnel and the need for funding were also mentioned as major difficulties.

The Scientist's Response: Those Who Culture for Knowledge

Questions 21 through 31 were directed toward scientific researchers who conduct research into captive propagation of marine ornamental organisms. The line between scientific research and commercial propagation is sometimes blurred, because the organisms produced by scientific research may occasionally be sold, and most commercial propagators engage in some form of scientific experimentation to improve their products. Each individual, however, knows if his or her work is primarily a scientific or commercial effort, and so each respondent determined the categorization of his or her position. Individuals in the scientific category included university professors and students, researchers at government and private facilities, independent scientists, and professional aquarists at public and private aquariums.

Table 2.5 presents data recovered from the 49 respondents to the science section of the questionnaire. The survey questions and the answer selections are abbreviated in this table.

Question 21 explored affiliation among the 49 respondents to the science section of the questionnaire. Student and government affilia-

Table 2.4 Organisms successfully propagated by commercial breeders (34 respondents)

Organism category		Organisms commonly propagated	Number of reports
Live rock/sand		Live rock	4
Algae		All algae species reports Macro algae *Caulerpa* spp.	2
Sponges/tunicates		Sponges	0
Anemones		All anemone species reports *Entacmea quadricolor*, bubble tip	1
Corals	All coral species reports		110
	All soft coral species reports		59
	Mushroom corals, all species		10
		Actinodiscus spp./*Discomoa* spp.	4
		Ricordea spp.	1
	Gorgonians, all species		4
	Leather corals, all species		16
		Sarcophyton spp.	7
		Cladiella spp.	2
		Lobophytum spp.	1
		Sinularia spp.	4
	Xenia spp.	Various *Xenia* spp. reports	13
	Other soft corals	*Capnella* spp.	1
		Nephthea spp.	4
		Clavularia spp.	1
		Pacylavularia violacea	1
		Zooanthus spp.	5
	All stony coral species reports		51
		Acropora spp.	21
		Euphyllia spp.	2
		Montipora spp.	8
		Pocillopora spp.	2
		Hydnophora spp.	2
Annelids			0
Mollusks		*Strombus gigas*	1
		Trochus spp.	1
Crustaceans		All species reports	10
		Hymenocera picta	3
		Lysmata amboinensis	1
		Lysmata wurdemanni	5
		Stenopus hispidus	1
Echinoderms			0
Fish		All fish species	72
		Amphiprion/Premnas spp.	47
		Gobiosoma spp.	4
		Hippocampus spp.	2
		Pseudochromis spp.	9
		Pterapogon kauderni	3

Table 2.5 Science questionnaire results

Question and response data	Responses to each option	Percent of respondents choosing each option
21. Affiliation		
49 respondents (52 responses)		
A. Student	2	4
B. Government	2	4
C. University	15	31
D. Commercial	9	18
E. Public aquarium	16	33
F. Independent	8	16
22. Funding sources		
49 respondents (61 responses)		
A. Self-funded	12	24
B. Nonprofit grants	16	33
C. Commercial support	10	20
D. Part of aquarist duties	23	47
23. Funding availability		
49 respondents (61 responses)		
A. Readily available	8	16
B. Available but difficult to find	32	65
C. Not available	10	20
27. Major problem areas		
49 respondents (148 responses)		
A. Financial	4	8
B. Facilities	20	41
C. Breeding (spawning)	10	20
D. Obtaining viable larvae	9	18
E. Feeding early larvae	29	59
F. Survival of late larvae	17	35
G. Loss at metamorphosis	11	22
H. Loss of early juveniles	11	22
I. Disease	14	29
J. Other	10	20
28. Work objective		
49 respondents (68 responses)		
A. Applied research	25	51
B. Basic research	19	39
C. Professional aquarist	24	49
29. Publication		
49 respondents (69 responses)		
A. Thesis or school report	5	10
B. Hobbyist literature	24	49
C. Scientific paper	30	61
D. Confidential report	4	8
E. Do not publish	6	12
30. Value of the hobbyist literature		
49 respondents (53 responses)		
A. Useful	31	63
B. Read but seldom cite	18	37
C. Do not read	4	8

tions were few, and actually student responses could fit into the university category, and most government responses might well be in the public aquarium category. Those respondents that split their work between two or more categories accounted for the three multiple responses. One would expect that most science respondents would be in public aquariums (33%) and universities (31%) and that those in the commercial arena (18%) would be fewer. Perhaps the most interesting result was that 16% were independent scientific researchers, which indicated that many in the scientific sector shared motivations with hobbyists.

Question 22 explored the source of funding for research. There were 61 total responses from the 49 respondents, indicating that up to 12 respondents received funding from one or more categories. Professional aquarists (aquarists at public or private aquariums) receive the strongest support (47%), but most of their activities were within the range of their jobs as aquarists. Nonprofit grants (33%) exceeded commercial support (20%), and self-funded research (24%) indicated that many scientists are actually in the "for love" category.

Question 23 inquired as to whether funding was readily available, difficult to find, or not at all available. The majority response, 65%, indicated that funding could be found but not easily, and 20% reported that funding simply was not available.

Question 27 explored the major problems reported by scientific researchers and the 49 respondents reported 148 responses to this question. Feeding early larvae elicited the largest number of responses at 29 individuals representing 59% of the respondents. Lack of facilities, 20 individuals (41%) also received a large percentage of the response. Survival of late larvae received 17 (35%) responses, and that combined with the large response to feeding early larvae indicates that larval culture is a major problem for scientific researchers. Other stages of propagation included breeding, obtaining viable larvae, loss at metamorphosis, and loss of early juveniles, with each accounting for about 20% of the response. Note, however, that most of the reported scientific research is with fish, not corals; thus, this response reflects the particular difficulties of fish culture. Only 8% of the respondents, four individuals, reported the financial area as a major problem. This does not

correspond with the results of question 23, availability of funding. Perhaps, unlike the commercial sector, scientists maintain a greater separation between technical and financial problems.

Question 28 indicates that the objective of most captive culture research is fairly evenly spread among applied research, 51% of the respondents, basic research, 39% of the respondents, and the daily work of professional aquarists, 49% of the respondents. The 49 respondents provided 68 responses to this question. Thus, a large number of the respondents, around 40%, considered their work as falling into more than one category. The majority of the responding researchers, 51%, put their work in the applied category, thus indicating a connection with some commercial application.

Questions 29, 30, and 31 explored the product of most scientific research, the publication of results. Of particular interest was assessment of the connections between scientific research and hobbyist literature.

Question 29 examined the publication practices of scientific researchers, and although most, 61% of the respondents, publish or plan to publish in the scientific literature, many, 49%, report publication in the hobbyist literature. Most of the multiple responses, 20 of the 69 total responses, were assigned to both the hobbyist and the scientific literature. The large percentage, 49%, of the respondents who publish in the hobbyist literature reveals the closeness to the hobby that many scientists have established. A significant portion of the scientific research that is done in the field of propagation of marine ornamentals is secret. Four of the respondents indicated that their work resulted in confidential reports, and six additionally reported that they do not publish their work. These 10 reports suggest that the work of about 20% of the science respondents is directed toward private applications and may never be published.

Question 31 explores the attitude of scientists toward the hobbyist literature. Most, 63% of the respondents, found the hobby literature useful, and 37% indicated that they read the hobbyist literature but seldom cite this literature since it is not peer reviewed. Only four, 8% of the respondents, reported that they do not read the hobbyist literature.

Question 30 requested a list of the publications where that respondent publishes. A total of 58 publications were listed, with 16 hobbyist publications and 42 scientific publications mentioned. The most commonly listed hobbyist publications were *Aquarium Fish Magazine, Freshwater and Marine Aquarium Magazine, Marine and Reef USA Annual,* and *Tropical Fish Hobbyist Magazine.* The most commonly listed scientific publications were *Aquaculture, Journal of MaquaCulture, Bulletin of Marine Science, Copeia, Drum and Croaker, Journal of Applied Aquaculture, Journal of the World Aquaculture Society, Malacologia, Marine Biology, Today's Aquarist,* and *Veliger.*

Questions 24, 25, and 26 inquired which species are currently under culture, which have been successfully cultured, and which species have been attempted and failed in culture. The listing for question 25, those species reported as successfully cultured by the 49 scientific respondents, was the most extensive of any of the species lists. It included most of the species

from all species lists from all categories of aquarists. There were 432 responses to this question that included about 275 species. This list is included in its entirety in Table 2.6 to provide a complete list of marine organisms reported successfully cultured by scientific researchers in this survey.

The number after each group or species indicates the number of responses that reported that species or group. No number indicates one report. There were some responses that included just common names, some with just scientific names (some spelled correctly and some not), and some listed as just broad categories. Note that this list does not include every species that has been captive bred and some of the species on the list may not meet a stringent criteria for captive bred. Although certainly not complete, this list provides a summary of the species that have been successfully bred by scientists and aquarists.

Table 2.6 Organisms successfully propagated by scientific research (49 respondents)

Live Rock / Sand
Live rock, 5

Algae/rotifers
Red macro algae
Micro algae, 2
Rotifers 3

Sponges/Tunicates

Cnidarians
Jellyfish (total responses, 19)
Actinia equina
Aurellia aurita, 7 moon jelly
Cassiopea sp.
Catostylus tagi
Chrysaora achlyos, 2 black jelly
Chrysaora fuscescens, sea nettle
Chrysaora quinquechirra
Cyanea capillata, 2 lion's mane jelly
Pelagia colorata, purple-striped jelly
Phyllorhiza punctata
Proboscidactyla flavicirrata

Anemones (total responses, 8)
Aiptasia diaphana, small rock anemone
Condylactis gigantean
Corynactis californica, strawberry
Entacmaea quadricolor, bulb anemone
Metridium gigantea
Metridium senile white plumose anemone,
Stoichactis helianthus
Urticina crassicornis

Corals (total responses, 38)

Soft corals (total responses, 18)
Actinodiscus spp., mushroom anemone
Clavularia spp., star polyp,
Corallimorph spp., mushroom anemone
Lithophyton spp.
Lobophytum spp.
Nephthea spp.
Parazoanthus gracilis, yellow polyps
Plexaura spp.
Ricordia spp.
Sarcophyton spp.
Sarcophyton lobulatum, 2 leather coral
Sinularia, spp.
Zooanthus pulchellus, green button polyps
Zooanthus sociatus, brown button polyps

Stony corals (total responses, 20)
Acropora spp., (Red Sea)
Acropora spp., 3 staghorn coral
Agaricia spp., lettuce coral
Fungia spp., 2 disc coral
Hydnophora spp., horn coral
Lithophyton spp., finger coral
Manicina areolata, rose coral
Millepora alcicornis, fire coral
Montipora spp., staghorn coral
Pocillopora damicornis, 2 cauliflower
Pocillorpora spp., (Red Sea)
Seriatopora spp., (Red Sea)
Sinularia spp., finger coral
Stylophora spp., (Red Sea)

Continued

Table 2.6 *(Continued)*

Annelids (total responses, 5)
Hydroides dianthus
Phragmatopoma lapidosa
Platynereis bicanaliculata
Sabellariidae, sandcastle worms
Serpula vermicularis

Mollusks (total responses, 22)
Aplysia californica
Calliostoma ligatum
Chlamys hastate
Euprymna sp.
Haliotis kamtschatkana
Haminoea callideginata, paper bubble sea slug
Haliotis rufescens, red abalone
Katharina tunicata
Mytilus trosilus
Neptunea lyrata
Neptunea pribiloffensis
Neptunea tabulata
Octopus bimaculatus, two-spotted octopus
Octopus spp. 2
Oenopota levidensis
Sepia officianalis 3, European cuttlefish
Sepia spp. (16 species)
Sepioteuthis sepioides, reef squid

Crustaceans (total responses, 32)
Ampelisca abdita, amphipod
Cancer magister, dungeness crab
copepods 2
Emerita analoga, mole crab
Hawaiian red shrimp
Hyalella azteca, amphipod (freshwater)
Hymenocera picta, 3 harlequin shrimp
Lysmata debelius, 3 fire shrimp
Lysmata rathbunae, 2
Lysmata wurdemanni, 6 peppermint shrimp
Lysmata, californica
Lysmata amboinensis, 3 scarlet cleaner shrimp
Mastigias papua
Palaemon serratus
Palaeomonetes pugio, grass shrimp
Petrolisthes cinctipes, porcelain shore crab
Rhynchocynetes durbanensis, camel shrimp
Tozeuma carolinensis, arrow shrimp

Echinoderms (total responses, 13)
Arbacia punctulata, sea urchin
Cucumaria miniata, orange sea cucumber
Cucumaria pseudocurata, brooding sea cucumber
Dendraster excentricus
Diadema antillarum, longspine sea urchin
Florametra serratissima
Linychinus spp., urchins
Parastichopus californicus, 2 spiny sea cucumber
Pisaster ochraceous
Strongylocentrotus droebachiensis
Strongylocentrotus pallidus

Fish (total responses, 283)
Abedefduf spp., damsels
Acanthochromis polyacanthus
Amblyeleotris randalli, 2 Randall's prawn goby
Amblygobious phalaena, brownbarred goby
Amblygobious rainfordi, Rainford's goby
Amphprion spp., 16 clownfish
Amphiprion allardi, Allard's
Amphiprion clarkii, 6 Clark's
Amphiprion frenatus, 4 tomato
Amphiprion latezonatus, wide-banded
Amphiprion leucokranos, white bonnet
Amphiprion melanopus, 5 red and black
Amphiprion melanopus, (Fiji)
Amphiprion ocellaris, 9 common
Amphiprion percula, 4 percula
Amphiprion perideraion, pink skunk
Amphiprion polymnus, saddleback
Anarrhichthys ocellatus, Wolf eel
Anisotremus virginicus, 2 porkfish
Antennarius hispidus, frogfish
Atherinops affinis, topsmelt
Aulorhynchus flavidus
Bluefin trevally
Bovichthys argentinus,
Calloplesiops altivelis, 3 comet
Canary Blenny (*Melacanthus oualanensis* ?)
Careproctus trachysoma
Centropyge fisheri, Fisher's angel
Centropyge loriculus, flame angel
Cephalocyllium ventriosum, swell shark
Chaetodipterus faber, Atlantic spadefish
Chanos chanos, milkfish
Chasmodes suburrae, Florida blenny
Chiloscyllium plagiosum, 2 white-spotted bamboo
   shark
Chiloscyllium punctatum, 2 bamboo shark
Chrysiptera prasema, 2
Coryphaena hippurus, 2 mahimahi
Cyclopterus lumpus, 2 lumpfish
Cymatogaster aggregata, shiner perch
Cynoscion nebulosus, spotted sea trout
Cyprinodon variegatus, 2 sheephead minnow
Dascyllus albisella, Hawaiian threespot damsel
Dascyllus trimaculatus, threespot damsel
Diplodus argenteus
Doryrhamphus excisus, blue striped pipefish
Doryrhamphus multiannulatus, banded pipefish
E. lineatus, lined perch
Embiotoca jacksoni, black perch
Enoplosus armatus
Equetus acuminatus, 2 hi hat
Equetus lanceolatus, 6 jacknife fish
Equetus punctatus, spotted drum
Equetus umbrosus, 2 cubbyu
flying fish
Gnathanodon speciosus, golden trevally
Gobiodon citrinus, citron goby
Gobiodon okinawae, yellow coral goby
Gobiosoma bosc

Table 2.6 *(Continued)*

Gobiosoma chancei
Gobiosoma evelynae, 3 sharknosed goby
Gobiosoma figaro
Gobiosoma genie
Gobiosoma horsti, yellowline goby
Gobiosoma multifasciatum, 3 greenband goby
Gobiosoma oceanops, 4 neon goby
Gobiosoma prochilus
Gobiosoma puncticulatus, redhead goby
Gopiosoma spp. (Elactinus per VanTassell, 2000)
Gramma loreto, 4 royal gramma
Gramma melacara, black cap basslet
Haemulon flavolineatum
Haemulon plumieri, white grunt
Haploblepharus edwardsii
Haploblepharus pictus
Heteroconger canabus, Cortez garden eels
Heterodintus francisci, horn shark
Hippocampus abdominalis, 3 pot-bellied sea
  horse
Hippocampus barbouri, 3 Barbour's seahorse
Hippocampus breviceps
Hippocampus capensis, South African seahorse
Hippocampus coronatus, 2
Hippocampus erectus, 9 lined seahorse
Hippocampus fisheri
Hippocampus fuscus
Hippocampus ingens, 2 Pacific seahorse
Hippocampus kuda complex, seahorses
Hippocampus kuda, 5 spotted seahorse
Hippocampus reidi, 6 longsnout seahorse
Hippocampus spp. 2 seahorses
Hippocampus spinosissimus, hedgehog seahorse
Hippocampus subelongatus, West Australia seahorse
Hippocampus zostrae, 5 dwarf seahorse
Hippocampus whitei, White's seahorse
Hypleurochilus geminatus
Hypoplectus unicolor, hamlet
Hypsyrus caryi, rainbow perch
Katsuwonis pelamis, tuna
Lachnolaimus maximus, hogfish
Leuresthes tenius, California grunion
Lutjanus analis, mutton snapper
Lutjanus decussates, checkered seaperch
Lutjanus grissus, gray snapper
Lythrypnus dalli, Catalina goby
Lythrypnus pulchellus, gorgeous goby
Micrometrus minimus, dwarf perch
Microspathodon chrysurus, jewelfish, yellow-tailed
  reef fish
Menidia beryllina, silverside
Monodactylus argenteus, silver batfish
Mugil sp., striped mullet
Myliobatis aquila
Myoxocephalus aenus, grubby sculpin
Myoxocephalus scorpius, short horned sculpin
Ocyurus chrysurus, yellowtail snapper

Opistognathus aurifrons, 2 yellowhead jawfish
Oreochromis mossambicus, tilapia
Pagrus major
Paralichthys lethostigma, southern flounder
Pheusochromis aldbrensis
Pholis laeta, crescent gunnel
Phyllopteryx taeniolatus
Plectorhinchus gibbosus, brown sweetlips
Plectorhinchus picus, dotted sweetlips
Pleuronectes platessa
Polydactylus sp., Pacific threadfin
Pomacanthus arcuatus, grey angelfish
Pomacanthus maculosus, half moon
Pomacanthus paru, French angelfish
Poroderma pantherium
Premnas biaculeatus, 6 maroon clown
Premnas sp., all forms
Psammoperca waigiensis, sand bass
Pseudochromis aldabrensis
Pseudochromis flavivertex
Pseudochromis fridmani, 4 orchid dottyback
Pseudochromis novaehollandae
Pseudochromis olivaceus
Pseudochromis paccagnellae
Pseudochromis sankeyi
Pseudochromis spp., dottybacks
Pseudochromis springeri
Pterapogon kauderni, 8 Banggai cardinalfish
Puffers
Rafa clavata, 2 skate
Raja erinacea, little skate
Raja ocellatus, winter skate
Rivulus marmoratus
Sarda orientalis
Sciaenops ocellatus, 2 red drum
Scyliorhinus canicula
Seriola dorsalis, jack
Seriola dumdumerili, greater amberjack
soles
Strongylura marina, Atlantic needlefish
Synchiropus splendidus, green mandarinfish
Syngnathus californiensis, kelp pipefish
Syngnathus floridae
Syngnathus fuscus
Syngnathus leptorhynchus, bay pipefish
Syngnathus lousianide
Syngnathus scovelli, Gulf pipefish
Takifugu niphobles
Thunnus thynnus
Tigrigobius macrodon
Tigrigobius multifasciatum
Tigrigobius puncticulatum
Trachinotus carolinus, pompano
Trachinotus falcatus, permit
Triakis semifasciata, leopard shark
Urolophus halleri, round stingray
Urophycis chuss, red hake

Question 24 requested science respondents to list the species that were currently under culture. There were 249 responses to this question that included about 173 species. The list included 8 species of jellyfish, 2 to 4 anemone species, about 30 species of corals, 9 species of mollusks, 13 species of crustaceans, 2 echinoderms, and about 102 species of fish.

Question 26 requested the respondents to list the species where culture was attempted but failed. There were 145 reports that included about 109 species. There was only one report of failed coral culture and that was sexual reproduction of *Acropora* sp. The list included 14 crustaceans, mostly *Lysmata* sp., and about 96 species of fish. The fish most commonly reported as failed in culture included angelfish, butterfly fish, triggerfish, wrasses, tangs, and damselfish.

The numbers of species reported successfully cultured in this survey are only rough approximations. Many responses consisted of just a genus of grouped species; *Acropora* spp., for example, was often reported, and this represents a genus that contains a large number of individual species, many of which are difficult to identify to species level. Also, although the 49 respondents to this survey in the scientific section indicated successful propagation of only 28 species of corals, this is not an accurate estimate of the total success that researchers and professional aquarists have had in this arena. Charles Delbeek (2001) reports that the Waikiki Aquarium holds more than 170 species of stony and soft corals, and the Pittsburgh Aqua Zoo maintains more than 150 species of stony corals. It is probably true, however, that most public aquariums do not culture more than a few species of corals; thus, the broad trends indicated by the data may be reasonably accurate.

Contrasts and Parallels in Culture Efforts of Hobbyists, Breeders, and Scientists

The strongest parallel among hobbyists, commercial breeders, and scientists is that they work with most of the same species of marine ornamental organisms. Their motivations and the intensity of their culture efforts differ greatly, however. Scientists work at culture and propagation for many reasons—to investigate the basic biology and ecology of marine organisms, to advance the capabilities of commercial companies, to provide alternatives to wild collection, and to better understand the marine environment in order to enhance the displays and the mission of public aquariums. Commercial breeders share these motivations, but above all is the necessity to be profitable and to be commercially viable. If this goal is not met, then the days of commercial operation are certainly numbered. Hobbyists, on the other hand, have the luxury to be part scientist and part entrepreneur and to propagate just for the fun and satisfaction of creating new worlds in a rare and beautiful environment. Also, it is now possible for successful hobbyists to sometimes become commercial breeders, or at least develop into part-time entrepreneurs, and use their propagation skills to fund much if not all of the structure of their hobby. When this occurs, hobbyists then take on many of the motivations of commercial breeders. In the marine aquarium hobby, as in many other science-oriented hobbies, one finds scientist/hobbyists, individuals who have education and training in the marine sciences that blend their vocational interests into a hobby of keeping and propagating marine ornamental organisms. Of course, both commercial breeders and scientists frequently invade each other's territory as circumstances and opportunities dictate. Thus, although their basic motivations differ, the lines among these three groups are indistinct and often overlap.

Scientists work first to understand reproductive and larval biology of marine organisms and the biology and dynamics of coral reefs, and second to develop and enhance commercial scale production of marine ornamentals, a motivation that they share with commercial breeders. The major interest of commercial breeders is to produce a commercial quantity of marketable, high-quality animals. The basic considerations for the species they propagate are ease of culture, high market demand, and a market price sufficient to make the rearing effort profitable. Unlike both hobbyists and scientists, commercial breeders must concentrate on mass culture of certain desirable organisms. The strongest contrast between the endeavors of hobbyists and scientists and those of commercial breeders is the magnitude of scale of culture. Whereas scientists or hobbyists may satisfy their motivations through culture of a relatively small number of individuals, commercial breeders must, by definition, rear large numbers of individuals. Thus, they must develop culture techniques that differ in many ways from those of the science laboratory and the small-scale aquarium systems of the hobbyist. Hobbyists may have an agenda that includes, for any number of reasons, culture efforts expended on a particular species. Most culture efforts by hobbyists are driven by chance, however. The species that do well in their systems and propagate themselves with little effort from the hobbyist are those that the hobbyist will propagate.

Hobbyists culture corals (131 species) to a much greater extent than they do fish (38 species), while science generally cultures fish (170 species) to a much greater extent than corals (28 species). Commercial breeders also culture corals (55 species) to a greater extent than they do fish (35 species), but the number of species of corals and fish is relatively even. This pattern is certainly a result of the great differences between coral and fish culture.

Captive propagation, whether for love, money, or science, is separated into two quite different endeavors: the vegetative coral culture and the culture of fish and other organisms through sexual reproduction. The former is similar to gardening, and the latter is basically controlled animal propagation from the spawning of brood stock, through larval culture, and into the juvenile and adult stages. Both take place in a marine aquatic environment, but the similarity of these endeavors does not extend beyond physical and chemical control of the captive marine environment. In the world of the hobbyist and the scientist, however, this distinction is imprecise. Fish and corals are usually kept together in the same systems, and propagation is often a function of vegetative ecosystem growth rather than culture of a specific species. In contrast, the commercial breeder must separate the species under culture and engage in growing individual coral colonies or in the spawning and controlled culture of marine animals. The efforts of scientists and hobbyists differ not so much in scale but in purpose, intensity, manipulation, and documentation.

Summary

The survey, although not designed for scientific analysis, provides some insight into current efforts, trends, and attitudes of those who culture marine ornamental organisms.

A total of 324 hobbyists responded to the survey. The results indicate that a large proportion of hobbyists, perhaps as high as 80%, now culture marine organisms, primarily corals, and many, about 80%, give or sell excess organisms, mostly corals, back into the hobby. Many hobbyists, 45% and 49%, respectively, consider lack of time and lack of facilities to be the major problems they encounter in the hobby. Eighty percent plan to continue culturing marine organisms, and 24% anticipate the culture of marine organisms to be a future source of additional income. Most marine hobbyists, 65%, buy cultured organisms when possible; only 16% did not care or did not distinguish between wild-collected and cultured organisms. Also, most hobbyists, 73%, felt that cultured organisms were necessary for the future of the hobby.

A total of 34 commercial breeders responded to the survey. Only 35% of the commercial breeders who responded reported that they were doing well in a business sense. Fifty percent reported their business success to be marginal, and 15% reported that business failure was close. The majority of commercial breeders, 65%, were operating at inland locations and were dependent on closed artificial seawater systems and artificial lighting. Income from the culture of marine organisms for commercial culturists varied from a negative factor to the entirety of their income. Thirty-six percent reported their income at 76% to 100% dependent on cultured organisms, with 21% reporting 5% to 25% of their income derived from culture. Fifty-nine percent of the respondents reported finances as a major problem area, with facilities, 44% of those responding, next in line. The majority of commercial breeders, 59%, considered the status of culture in the industry as small and struggling, and most, 65%, thought that captive propagation of marine organisms would slowly develop into a major factor in the marine aquarium industry.

Most of the 49 scientific researchers who responded to the survey were affiliated with public aquariums, 33%, or universities, 31%. The greatest source of funding, 47%, came from employment as professional aquarists; other sources of funding—self-funding, 24%, nonprofit grants, 33%, and commercial support, 20%—were rather evenly distributed. The major problem areas that were reported were technical and biological: feeding early larvae, 59%, survival of late larvae, 35%, loss at metamorphosis, 22%, and other culture problems. Lack of facilities reported by 41% was important, however. The majority of the respondents, 51%, reported that applied research was the purpose of their work, with 39% reporting that they were engaged in basic research, and 49% of the respondents performed their culture work as part of their duties as a professional aquarist. The majority of the respondents, 61%, indicated that they publish their work in the scientific literature, but a large percentage, 49%, also or exclusively published in the hobbyist literature. Most scientists, 63%, found the hobbyist literature useful, and 37% read the hobbyist literature but do not cite it in their publications.

This survey illustrated that there is understanding and strong support from hobbyists, and acknowledgment and directed effort from scientists and commercial breeders, toward captive propagation of marine ornamental organisms. Most of those engaged in the hobby and the industry feel that culture will be a very important part of the future of the hobby and its associated industries.

There is no question now that the world's coral reefs and their associated fauna are in decline. The new *World Atlas of Coral Reefs*, the first comprehensive map of Earth's coral reefs, indicates that they collectively cover only about 110,000 square miles, an area about the size of the state of Nevada. That is only about half the size of previous estimates. Given the current rate of decline, marine scientists have predicted that coral reefs as we know them could be gone in as few as 50 years.

This is not the fault of the marine aquarium industry, although some would bend public perception toward this view. At this time of environmental crisis, the culture of marine organisms should be a positive force in the hobby, the marine aquarium industry, and in the public mind. The message should be that the marine aquarium hobby and the industry that serves it strive to understand, protect, and preserve the coral reef environment. The marine hobby should not be perceived as a major contributor to the decline of coral reefs.

The culture efforts of hobbyists, commercial breeders, and scientists will provide an intimate knowledge of the biology and ecology of coral fish and coral reefs that foster a better understanding of the factors contributing to their decline. It is possible that just as current captive propagation of endangered terrestrial animals provides a reserve stock for reintroduction into the wild, so might captive stocks of corals and fish also serve this purpose at some future time. Coral reef areas close to large population centers suffer the greatest species loss and environmental stress. Commercial culture in these areas can provide jobs, maintain the marine aquarium hobby and industry, and serve as a reservoir of information and livestock to support the natural environment. It is imperative that the hobby and the industry continue to improve the support for captive propagation of marine ornamental organisms.

References

Delbeek, Charles J. 2001. Coral farming; past, present and future trends. *Aquarium Sciences and Conservation* 3(1-3):171–181.

PART II

Progress and Current Trends in Marine Ornamentals

A. *Trade, Marketing, and Economics*

3

International Trade in Marine Aquarium Species: Using the Global Marine Aquarium Database

Edmund Green

Global Perspectives on the Aquarium Trade, 2002

The State of Coral Reefs

The opening paragraph of the last global report on the status of the world's coral reefs from the Global Coral Reef Monitoring Network (GCRMN 2000) makes depressing reading, especially for anyone trying to sustain a livelihood from a reef:

> Coral reefs of the world have continued to decline since the previous GCRMN report in 1998. Assessments to late 2000 are that 27% of the world's reefs have been effectively lost, with the largest single cause being the massive climate-related coral bleaching event of 1998. This destroyed about 16% of the coral reefs of the world in 9 months during the largest El Niño and La Niña climate changes ever recorded. While there is a good chance that many of the 16% of damaged reefs will recover slowly, probably half of these reefs will never adequately recover. These will add to the 11% of the world's reefs already lost due to human impacts such as sediment and nutrient pollution, over-exploitation and mining of sand and rock and development on, and "reclamation" of, coral reefs.

Debate over the Role of the Aquarium Trade in the Current Decline of Coral Reefs

An estimated 1.5 to 2.0 million people worldwide keep marine aquariums, approximately half in the United States and a quarter in Europe. For the most part, these are hobbyists who maintain tropical fish stocks in home aquariums. Dedicated enthusiasts are able to propagate many species of coral and fish, but most aquariums are stocked from wild-caught species.

In recent years the aquarium industry has attracted much controversy. Opponents to the trade draw attention to the damaging techniques sometimes used to collect fish and invertebrates and to the high levels of mortality associated with insensitive shipping and poor husbandry along the supply chain. Aquarium species are typically gathered by local fishers using live capture techniques (such as slurp guns or barrier and hand nets) or chemical stupefactants such as sodium cyanide. The latter is nonselective and adversely affects the overall health of the specimens, as well as killing nontargeted organisms. Consequently, the marine aquarium trade is frequently referred to as a major contributing cause to the global decline of coral reefs.

Supporters of the aquarium industry maintain that it is potentially highly sustainable, that proper collection techniques have minimal impact on the coral reef, and that the industry is relatively low volume but very high value. There is little disagreement about the latter. A kilogram of aquarium fish from one island country was valued at almost US$500 in 2000, whereas reef fish harvested for food were worth only US$6. Aquarium species are a high-value source of income in many coastal communities that have limited resources, with the actual value to the fishers determined largely by market access. In Fiji many collectors pay an access fee to the villages to collect on their reefs. By selling directly to exporters, local people can have incomes many times the national average. By contrast, in the Philippines many middlemen are part of the trade, and collectors themselves typically earn only around US$50 per month. The controversy over the benefits and costs of the trade, in terms of environmental impact, persists largely because of a lack of quantitative data on the global trade.

Key Questions Requiring Answers

Everyone who has an interest in marine conservation would like to know if the trade in marine aquarium species is sustainable at the source. Naturally, the answer is not easily obtainable and requires two key questions to be addressed initially.

First, which species are involved? The trade is constructed around individual species. End consumers make most of their purchasing decisions on particular species, dictated by the constraints of their aquarium system and the other organisms in it, their experience and skill level, or by fashion and impulse. Species that are naturally rare or have highly restricted ranges can reasonably be expected to be vulnerable to overcollection. For example, some species in trade such as the Moorish idol, *Zanclus cornutus*, are widely distributed across most of the Pacific Ocean. It is therefore unlikely that trade is a threat to this species. Other species of fish in trade, such as the Indonesian cardinalfish, *Pterapogon kauderni*, have extremely limited biogeographical ranges, and so there may be justifiable cause for concern if this species is traded in large quantities. With more than 1,200 species of fish and 300 species of invertebrates in trade, a dearth of knowledge on the ecology of individual species of either group (but especially invertebrates) exists. It is just as important to know the life history characteristics such as reproduction, larval dispersal, and age at collection, as well as collection methods for groups of species. Those fish that are hermaphroditic and maintain harems, whose populations are dependent on infrequent larval recruitment from

occasional strong year-classes, and that are harvested as adults have life history characteristics that may render them particularly vulnerable to overexploitation, if traded in large numbers. Conversely, common species may naturally occur in numbers sufficiently large to sustain a relatively small collection.

Second, how many of each species are traded? The history of the commercial fishing industry provides too many depressing examples of previously numerous populations of fish being decimated through overexploitation. Even abundant species of aquarium organisms cannot therefore be assumed to be safe if the quantities in which they are collected are unknown. In 2001 a pair of angelfish, collected from depths greater than 100 meters with the use of specialist deep diving techniques, sold for $5,000 each, an event that attracted a good deal of attention. These fish are believed to be rare, and justifiable concern existed that such high prices would fuel collection targeted at a vulnerable species. There is no indication, however, that this was anything other than an unusual sale to a highly specialized customer. If only a handful of these fish are collected and sold each year, then, even with little idea of the biology or population dynamics of the species, it is unlikely to be threatened by trade. Clearly an understanding of the quantities in which the species in trade are collected and sold is essential.

If certain aquarium species are identified as being threatened through trade, through a combination of their life history characteristics and the quantities in which they are collected, then appropriate action will need to be taken in the interests of both trade and conservation. Options for action would be restricted to blanket global bans if the basis for targeted measures at the national level—where the organisms were coming from and where they were being shipped—remained unknown. Ideally, trade data will also indicate any species for which cause for concern may arise in the future. In other words, data collected over time will be valuable to identify trends in species not yet at risk from trade but potentially vulnerable.

Questions of lesser immediate importance concern the number of cultured and certified organisms in trade, so that the contribution of these practices towards sustainability might be assessed.

Existing Sources of Data on the Aquarium Trade

Statistics generated by national governments, either through customs or other officials who have responsibility for monitoring imports and exports, should be a good source of information on species in the aquarium trade, but unfortunately, with a few exceptions this is not the case. Commonly no distinction is made in the records between marine and freshwater organisms, and such general terms as "fish" may be used that in practice also includes invertebrates. Occasionally other organisms such as reptiles may be reported in the same category.

Further complications arise through the use of different units. Some trade is recorded in monetary value (e.g. U.S. Customs and imports to the nations of the European Union) and other trade by weight. In the former the value of exports is usually recorded as free-on-board (the value of the organisms without packaging, freight, tax, or transport), while import value is expressed as the carriage, insurance, and freight cost. In cases where the weight of shipments is monitored, the data recorded nearly always include packaging and water, substantially overestimating the volume of live material, especially for fish and noncoral invertebrates. To add further to the confusion, sources within the industry unofficially advise that these types of statistics are unreliable because many operators will, at times, overstate quantities on their invoices for the purposes of insurance or, on other occasions, understate quantities to reduce the duty payable. Of course, the only way of checking would be for officials to open and inspect every shipment, something that is clearly impossible.

Few governments record exports by number. Singapore recorded exports of 1,294,200 fish in 1998 (Wood 2001) but without species level data. The Maldives government records the total number of individual species exported. The governments of Vanuatu, Tonga, and the Solomon Islands require collectors to submit records of their exports, in terms of the number of each species shipped, as a condition of their licenses. State governments in Australia appear to be unique in requiring their collectors to register catch data as opposed to export data. Clearly any post-catch mortality would not be recorded in export data. State governments collect the data.

So, there is no standardized system of reporting used by governments, nor is there ever likely to be one. In seeking answers to the questions concerning the international trade in aquarium species, however, we must use what data are available while acknowledging the constraints under which the data are useful. The data gathered by national government officials are nonetheless a useful source of information, more so for assessing the value of the trade than the quantity, and therefore environmental impact. Table 3.1 shows the value of imports of marine ornamental fish into European Community countries from 1992 to 1998.

Work under way at the World Resources Institute (WRI) in Washington, D.C., has compiled data taken from invoices collected by the U.S. Fish and Wildlife Service (USFWS). The USFWS has responsibility for monitoring shipments of wildlife into the United States and, among other duties, is responsible for enforcing national (e.g. the Lacey Act) and international (e.g. Convention on International Trade in Endangered Species of Flora and Fauna, or CITES; see http://www.cites.org/) legislation pertaining to wildlife trade. A copy of the invoice, which accompanies every shipment, is logged and stored with the USFWS, which therefore has a species-specific record of every single import of aquarium organisms into the United States. Staff at WRI have obtained copies of these invoices for the month of October 2000, a monumental task necessitating the filing of large numbers of freedom of information requests to obtain the data, the cataloguing of thousands of pieces of paper, the separation of freshwater and marine species, and data interpretation, standardization, and storage. The data will be compared with imports for October 1992, when information was similarly obtained and studied to examine possible changes in imports across those eight years.

CITES is another major source of data on components of the aquarium trade. CITES attempts to assess the trade in species, listed in Appendix II of the Convention, which are believed to be vulnerable to exploitation but not yet at risk of extinction. All species of hard coral and giant clams are listed under Appendix II of CITES, and parties to the convention are then obliged to produce annual reports specifying the quantity of trade that has taken place in

Table 3.1 Value of imports (carriage, insurance, and freight) of marine ornamental fish into European Community countries from 1992 to 1998

Supplying country	Maximum and minimum annual import value US$000	Import value 1992 US$000	Import value 1998 US$000
Indonesia	2,108–5,116	2,108	4,990
Philippines	1,300–1,746	1,300	1,746
USA: mainly Florida and Hawaii	920–1,561	1,089	1,561
Sri Lanka	1,145–1,328	1,193	1,025
Singapore[a]	510–1829	1,739	510
Kenya	338–497	451	497
Maldives	232–412	363	357
Brazil	26–304	69	302
Yemen	0–217	0	217
Egypt	116–221	117	177
Australia	58–154	84	117
Saudi Arabia	19–119	19	105
Fiji	26–78	26	78
Curacao	10–102	102	78
Eritrea	0–73	0	73
Malaysia	4–56	4	56
Costa Rica	0–44	44	37
Cuba	2–37	0	37
Mauritius	0–31	0	30
Belize	2–27	18	27
Iran	0–46	0	20
Thailand	5–29	10	16
Dominican Republic	0–37	0	13
Barbados	11–36	34	11
Colombia	2–12	3	11
India	0–8	0	8
Haiti	0–17	0	7
Hong Kong[a]	1–6	1	7
Solomon Islands	0–9	0	7
United Arab Emirates	0–5	3	5
Tanzania	0–4	0	4
Belau (Palau)	0–2	0	2
Ethiopia	0–11	0	2
South Africa	0–2	4	2
Martinique	58–85	60	0
Bahrain	0–62	18	0
Djibouti	0–32	5	0
Guadeloupe	0–25	25	0
Israel	1–4	1	0
Japan	0–11	0	0
Madagascar	0–6	6	0
Marshall Islands	0–1	1	0
Taiwan	0–4	1	0
Total		8,903	12,135

Source: Wood, 2001, with permission.

[a] re-export (not a producer).

each listed species. The magnitude and taxonomic composition of the international trade can then be calculated (Green and Shirley 1999).

The reporting process for corals is not without considerable problems. Trade is either recorded numerically or in units of weight. There is no universally accepted factor to convert numbers of coral to weight, and vice versa, and it is probable that a single conversion factor would not be appropriate as coral is traded internationally for a variety of purposes other than live for aquariums. Conversion factors range from 0.2 to 1.0 kg for live and all corals in trade, respectively. Quantities traded are confounded because shipments of coral rock, sand, and gravel also require CITES permits, but perhaps the most intractable problem is that of species identification. Parties to CITES are obliged under Article VIII, paragraph 6(b), to identify specimens to the species level. With species being identified in a very small proportion (2% in 1997) of records, there is a widespread failure to fulfill this obligation for corals. The failure to record species is most likely symptomatic of the real difficulties in identifying coral species in taxonomically complex genera such as *Porites* and *Acropora* (Green and Hendry 1999). These problems are universal, however, and CITES does generate data that are uniform, easily available, and amenable to analysis, despite difficulties, for global or regional trends. Before 2001 nothing comparable existed for any other marine organisms traded within the aquarium industry.

Existing sources of data do not provide good quality, quantitative information on the numbers, origin, and destination of the full range of individual species in trade. As a result, the most frequently encountered information is qualitative. In some notable cases (Wood 2001), the quality of the data has been measured and the data represents an extremely useful contribution to the debate on the aquarium trade. In others it has led to proposed measures for regulation of the aquarium trade being compared in importance to those proposed to control global climate change (McManus 2001), as well as confusion in the media between the aquarium trade and the live food fish industry.

Recent Policy Decisions Affecting the Trade and the Need for Good Data

The marine aquarium trade continues to receive the attention of politicians and conservation organizations alike, attracted by accounts of destructive collection practices, the introduction of alien species, overexploitation, and the threat of extinction of target species. Some regulation has already been established; more is being called for (McManus 2001, Moore and Best 2001) and may follow. The U.S. government is considering "taking appropriate action to ensure that international trade in coral reef species for use in USA aquariums does not threaten the sustainability of coral reef species" (United States Coral Reef Task Force 2000). The European Union has already suspended imports of some corals from Indonesia (Table 3.2).

Both the United States and the European Union are carrying out consultation exercises,

Table 3.2 European Union suspension actions affecting imports of some corals from Indonesia on the basis that the trade was not demonstrably sustainable

Genus/species	Regulation	Suspension active from
Catalaphyllia jardinei	1	Data unavailable
Blastomussa merleti	2	September 16, 1999
Cynaria lacrymalis	2	September 16, 1999
Euphyllia glabrascens	2	September 16, 1999
Euphyllia divisa	2	September 16, 1999
Plerogyra simplex	2	September 16, 1999
Trachyphyllia geoffroyi	2	September 16, 1999
Euphyllia spp.	2	July 11, 2000
Plerogyra spp.	2	July 11, 2000
Hydnophora excesa	2	July 11, 2000
Hydnophora microconos	2	July 11, 2000

Note: 1. Suspended under Article 4.6(b) of Commission Regulation (EC) No. 191/2001 (Published in the Official Journal, L 29/12 on January 1, 2001).

2. Following negative opinions of the Scientific Review Group, suspended under Article 4.2(a) of Council Regulation (EC) No. 338/97.

yet the arguments for and against these measures have taken place in a near vacuum of information on the extent, and therefore the impact, of the aquarium trade. Anyone disagreeing with the conclusion published recently in a scientific journal that "its [the Banggai cardinalfish] status in the wild is precarious, with heavy collecting [for aquariums] continuing" (Hawkins and Roberts 2000) would find it difficult to produce a counterargument because no one has any idea how many Banggai cardinalfish are traded. Some sympathy is surely owed to those people whose task it is to set policies that balance the need for conservation with the regulation of a potentially sustainable industry that employs thousands of people and provides high incentives for reef stewardship. These decision makers need good quality, quantitative, representative information on the trade in aquarium species. It is no argument that a host of other activities may be having a greater impact than the aquarium trade, which is therefore not worthy of special attention. In the Philippines many reefs have been destroyed by dynamite fishing and smothered under sediments washing out from clear-cut forests, yet collectors may be threatening the survival of rare fish by removing them from these already stressed ecosystems. This claim is made in the same article (Hawkins and Roberts 2000) for the bluespotted angelfish (*Chaetodontoplus caeruleopunctatus*). The gravity of the situation for everyone involved in the aquarium industry should not be underestimated, for there is nothing the public understands more readily than the extinction of a beautiful creature. The case for caution and banning the trade in bluespotted angelfish would be appealing to any policy maker seeking an easy option.

Progress in a debate, which at times has been vociferous and contentious, remains constrained by the lack of quantitative and unbiased information. At stake is the employment of thousands of people, especially in source nations, and the high incentives for coral reef stewardship, which the marine aquarium trade is capable of providing. An ill-informed decision could therefore either encourage the continuation of an environmentally destructive practice or deprive economically disadvantaged communities of a much needed source of income, compelling them to turn to perhaps even more destructive, short-term, activities in order to feed their families.

Structure of the Global Aquarium Trade and the Role of Wholesalers

Local fishers operating at the village or community level collect the vast majority of marine ornamentals. Their catch may pass through a network of middlemen (as in the Philippines) or be passed directly to a wholesale company (as in Fiji), but it is ultimately exported from the country of origin by only a few wholesale exporting companies.

The organisms are air freighted to wholesale import companies in the market nations, which in turn supply networks of retail outlets for the home aquarist (Figure 3.1). Nearly all marine ornamentals therefore pass through a limited number of wholesale export companies in the source nations and a similar number of wholesale import companies in the market nations.

The best source of quantitative data on the aquarium trade is the wholesale import and export companies that link the supply and retail ends of the business. As a matter of routine business practice, companies keep records of their sales, either as paper copies of their invoices or on company computer databases. The exact nature of these records varies, but all record the quantity of any individual species bought or sold, the date of each transaction, and the source and destination of the shipment. Company sales records are therefore an excellent source of data on marine aquarium species in trade and the only source for species not recorded under any other process (e.g., CITES).

Development of the Global Marine Aquarium Database (GMAD)

Over the past two years, good working relationships have been established with wholesale import and export companies from all around the world, with the help of the Marine Aquarium Council and such trade organizations as the Asosiasi Koral Kerang dan Ikan Indonesia (AKKII [Indonesia Coral, Shell and Ornamental Fish Association]), the Philippine Tropical Fish Exporter's Association (PTFEA), the Singapore Aquarium Fish Exporter's Association (SAFEA), the Ornamental Fish International (OFI), and the Ornamental Aquatic Trade Association (OATA). The companies have provided access to their sales records, and these data

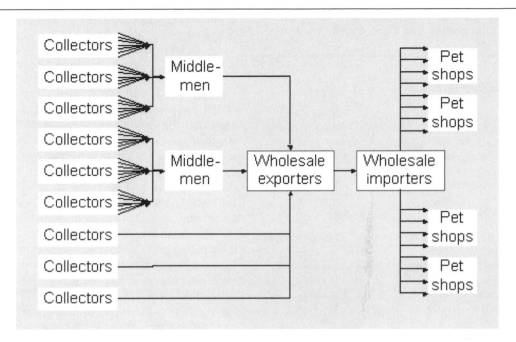

Fig. 3.1   A simplified schematic of the structure of the global marine aquarium trade.

have been through a careful and methodical period of data conversion (e.g., paper-based records have been computerized) and formatting (e.g., data from different electronic systems have been placed into a single standardized format).

Contributors of Data to GMAD

Data from 45 representative wholesale exporters and importers of marine aquarium species (Table 3.3) have been harmonized by this process into a single publicly available Global Marine Aquarium Database (GMAD).

Structure of GMAD and Use of the Data

GMAD contains in excess of 50,000 trade records. Each trade record is a total number traded for a unique combination of species name (fish, corals, invertebrates other than corals), country of export, country of import, and date (year). These trade data are linked to two external databases: (1) FishBase (Froese and Pauly 2002) for photographs of fish species and fish taxonomy, and (2) the Species Conservation Database (UNEP-WCMC 2002) for invertebrate taxonomy and legislation.

The trade data cannot simply be pooled because some of the contributing importers trade with some of the contributing exporters, and so pooling data would create duplications. Therefore, GMAD has been designed so that import or export data can be queried only separately. For example, the trade in any (or all) species between any pair of countries must be calculated from either import data or export data.

For example, if you are interested in the number of powder brown tang (*Acanthurus japonicus*) exported from the Philippines to the Netherlands, then you calculate two numbers: the first based on export data, the second on import data. In September 2001 GMAD contained data from five Philippine exporters and import data from two Dutch importers, and this query would have given you 2,151 fish exported and 425 imported. Given the respective sources of the data, it is probable that the figure based on exports is most representative of the trade in *Acanthurus japonicus* between these two countries. There are, of course, companies in the Philippines and the Netherlands trading in *Acanthurus japonicus* other than those that have contributed their data to GMAD, and therefore this figure of 2,151 fish is just a quan-

Table 3.3 Forty companies and four management authorities that have contributed data to GMAD

Company Name	Country
Queensland Fisheries Service[a]	Australia
Bahrain Aqualife Centre	Bahrain
Belau Aquaculture	Belau
Cook Islands (Chip Boyle)	Cook Islands
Ocean 2000	Fiji
Walt Smith	Fiji
Amblard Overseas Trading	France
von Wussow Importe	Germany
Anexa Tirta Sukya	Indonesia
Aqua Marindo	Indonesia
Asia Pacific Aquatics	Indonesia
Bali Blue International	Indonesia
Banyu Biru	Indonesia
Bekael Eska Gemilang	Indonesia
Cahaya Baru	Indonesia
Dharma Intipermal	Indonesia
Dinar	Indonesia
Gloria	Indonesia
Golden Marindo	Indonesia
Inti Samudra Lestari	Indonesia
Pacific Aneka Mina	Indonesia
Sangputra Wimasjaya	Indonesia
Segatama	Indonesia
Vivaria Indonesia	Indonesia
Ministry of Fisheries and Agriculture[a]	Maldives
Marshall Islands Marine Resource Authority[a]	Marshall Islands
Azoo-Mexico	Mexico
De Jong Marinelife	Netherlands
Waterweelde	Netherlands
Aquarium Habitat	Philippines
Aquascapes	Philippines
Brem Marine	Philippines
Land Mark Trading Corporation	Philippines
Tai-Lin Marine Product	Philippines
Ministry of Agriculture and Water[a]	Saudi Arabia
Aquarium Arts	Solomon Islands
Keells Aquariums	Sri Lanka
Tropical Fish Lanka	Sri Lanka
Independent Aquatics Imports	U.K.
Tropical Marine Centre	U.K.
Swallow Aquatics	U.K.
Quality Marine	U.S.A.
Sea Dwelling Creatures	U.S.A.
Segrest	U.S.A.

[a] Government management authority.

titative total on which to base estimates of the whole trade in this species.

A consequence of this is that GMAD cannot be used to calculate the absolute volume of trade in any one species or between any pair of countries. Such calculations, however, are based on quantitative data, and the degree to which these data are indicative of the trade depends in part on the proportion of operational wholesale export and import companies contributing data to GMAD.

If the assumption can be made that companies supplying the same markets do not differ substantially in the species or quantities that they trade, then GMAD data may be used to calculate the relative contribution of any species (and indeed country or region) to the trade with more certainty. To continue the previous example, seven species of *Acanthurus* are exported from the Philippines and *A. japonicus* constitutes 65% of trade in this genus and 1% of the trade in all fish from the Philippines. If it can be assumed that the Philippine export companies that have not contributed data to GMAD do not trade in substantially different quantities of this species, then these percentages may be taken as representative.

Strengths and Weaknesses of the Database

GMAD is the only source of quantitative data on the trade in marine aquarium fish and invertebrates other than corals. Although these data will never be complete in the sense of providing a complete global coverage, something that would be obtained only from the sales data of every wholesale trader in marine aquarium species, they do provide a quantitative basis for estimates of national, regional, and global trade, which can therefore be revised as the data improve. GMAD also provides quantitative data on the trade in corals, which can be compared to data collected under the CITES reporting process. Another significant strength of the trade data in GMAD is that it is species specific, though this is largely dependant on the quality of standard reference taxonomies and as such is most reliable for fish, less so for corals (especially given the difficulty of identifying corals), and perhaps is least reliable for the less well-known invertebrate groups.

Another major advantage of GMAD is that the data can be used to examine the trade in marine ornamental species from both ends of the chain of supply. Data have been collected from exporters in nine different countries and from importers in five. Because these companies export to and import from many others, GMAD contains data on the trade in marine aquarium species involving 30 source and 44 destination countries. For some countries of origin, such as Yemen, Brazil, Vietnam, and Papua New Guinea, information is based only on import data, and likewise for some countries of destination, such as Israel, Malaysia, Switzerland, and even the Vatican, information is based only on export data. For the most important source and market countries GMAD contains both export and import data that can be cross-referenced.

The principal weaknesses of GMAD are that in 2001, most data had been provided by companies and management authorities in source countries, so there is a bias toward export data. More data are needed from importers, and efforts are underway to obtain these. Currently (in 2001) it is not possible to make year-to-year comparisons of trends in the trade, as data have been provided from different years, for incomplete periods, and only for the most recent months. Efforts are also under way, therefore, to collect a complete year's worth of data from all providers for 2000 and 2001.

Internet access to GMAD is updated every six months and available at the following address: http://www.unep-wcmc.org/marine/GMAD/

Description of the Global Marine Aquarium Trade

The data within GMAD allow a description, based on quantitative data for the first time, of the trade links among different regions of the world, the trade profile of any individual nation, and the taxonomy of the trade in fish, corals, and other invertebrates. It is also possible to refine previous global estimates of the total number of fish in trade and corals and to produce a new estimate of the number of invertebrates in trade.

Regional Trade Links

The vast majority (85%) of all organisms in the aquarium trade originate from the Western Pacific. North America imports two thirds of all organisms in trade, and the volume of trade between these two regions accounts for just over half of the global total. Table 3.4 indicates that the proportion of world trade in aquarium organisms entering the European Union and Southeast Asia is approximately equal, at around 13%. Most other trade accounts for small proportions. The trade in Caribbean

Table 3.4 Regional trade links, expressed as percentage of the total number of all species in the global trade

	North America	Southeast Asia	European Union	Unknown	Wider Europe	Middle East	Australasia	South America	Africa	Total
Western Pacific	52.0	14.0	12.8	4.9	0.6	0.2	0.1	0.1	<0.1	84.8
Caribbean	5.8	—	<0.1	—	—	—	—	—	—	5.8
North America	4.3	—	0.2	—	—	—	—	—	—	4.6
European Union	1.4	—	0.1	—	—	—	—	—	—	1.5
Southeast Asia	0.7	—	0.2	—	—	—	—	—	—	0.9
Red Sea	0.7	—	<0.1	—	—	—	—	—	—	0.7
Australasia	0.4	—	<0.1	0.5	—	—	—	—	—	1.0
Africa	0.2	—	—	—	—	—	—	—	—	0.2
Eastern Pacific	0.1	—	—	—	—	—	—	—	—	0.1
South America	<0.1	—	0.2	—	—	—	—	—	—	0.2
Indian Ocean	<0.1	—	<0.1	—	—	—	—	—	—	<0.1
Middle East	—	—	<0.1	0.3	—	—	—	—	—	0.3
Total	65.7	14.0	13.5	5.7	0.6	0.2	0.1	0.1	<0.1	100

Source: Calculated from the Global Marine Aquarium Database.

Table 3.5   The ten species of fish, coral, and other invertebrates that constitute the greatest portion of exports from the Western Pacific

Fish species	Percent of regional total	Coral species	Percent of regional total	Invertebrate species	Percent of regional total
Zebrasoma flavescens	20.4	*Trachyphillia geoffroyi*	10.2	*Stenopus hispidus*	10.8
Amphiprion ocellaris	4.9	*Cataraphyllia jardinei*	8.5	*Radianthus malu*	9.3
Chrysiptera taupou	7.1	*Heliofungia actiniformis*	6.8	*Stoichactis kenti*	5.5
Dascyllus aruanus	6.5	*Euphyllia glabrescens*	5.9	*Hetrodactyla hemprichii*	5.1
Chromis viridis	3.0	*Goniopora minor*	5.7	*Diadema savignyi*	4.5
Amphiprion percula	3.0	*Goniopora lobata*	5.4	*Rhynchocinetes durbanensis*	4.5
Dascyllus albisella	2.9	*Euphyllia divisa*	5.2	*Mespilia globulus*	3.9
Dascyllus trimaculatus	2.7	*Goniopora stokesi*	4.9	*Linkia laevigata*	3.8
Centropyge bicolor	4.5	*Euphyllia cristata*	2.8	*Calcinus elegans*	3.5
Centropyge bispinosus	4.4	*Dendrophyllia fistula*	2.5	*Astropecten polyacanthus*	3.2

Source: Calculated from the Global Marine Aquarium Database.

species is almost 6% of the global aquarium exports, an amount slightly greater than the exports originating from North America (almost entirely the U.S. states of Florida and Hawaii). Trade links between other regions are negligible or nonexistent.

Fish   The overall pattern of regional trade links for fish is similar to the global picture. The Western Pacific exports 85% and the United States imports 68% of fish in global trade. The Caribbean exports 8%, and the European Union and Southeast Asia import similar quantities (11% of the fish in global trade).

Corals   A very different picture emerges when the export data for corals are examined. Almost all (99.5%) corals in trade originate from the Western Pacific, with species exports from other regions being negligible. Again, the majority, two thirds, is imported into North America. The European Union appears to take more corals than Southeast Asia, 18% of the corals in global trade compared with 13%.

Invertebrates   Fewer of the invertebrates other than corals are exported from the Western Pacific, although it is still, by far, the major source

of these organisms, accounting for 75% of global exports. North America produces 17% of the invertebrates in global trade, with smaller amounts coming from Southeast Asia (4%) and the Caribbean (3%). Not surprisingly, North America imports the majority of invertebrates, but at 59% they are a smaller portion of the global trade than either fish or corals. Southeast Asia and the European Union account for less, but similarly sized, amounts of trade, with 20% and 18% of the invertebrates in global trade being imported into these regions, respectively.

Clearly the Western Pacific is the major source of organisms in each of these three broad taxonomic groups. Table 3.5 lists those species of fish, coral, and invertebrates that account for the greatest portion of exports from the region: in each case 10 species account for more than 50% of the total exports.

National Trade Profiles

GMAD contains trade records from 30 exporting and 44 importing nations. For any of these countries it is possible to generate a profile of species in trade and to rank these species in terms of the proportion of exports or imports. To illustrate the type of information that can be

Table 3.6 The top 10 invertebrate imports to the United States

Species	Percent
Tectus niloticus	4.6
Stenopus hispidus	2.6
Rhynchocinetes uritai	1.8
Lima scabra	1.5
Condylactis passiflora	1.2
Lysmata amboinensis	1.1
Linkia laevigata	0.9
Lysmata californica	0.9
Tridacna derasa	0.8
Radianthus malu	0.7

Source: Calculated from the Global Marine Aquarium Database.

Table 3.7 The top 10 fish imports to Japan

Species	Percent
Chromis viridis	8.6
Chrysiptera parasema	5.4
Amphiprion ocellaris	5.4
Chromis caerulea	3.3
Nemateleotris magnifica	3.1
Thalassoma amblycephalum	1.8
Chrysiptera springeri	1.8
Paracanthurus hepatus	2.2
Amphiprion clarkii	2.2
Dascyllus albisella	1.8

Source: Calculated from the Global Marine Aquarium Database.

generated, Table 3.6 lists the chief imports of invertebrates, a group in which the trade is poorly recorded, into the United States, a market about which much is already known. The taxonomy of these principal imports is quite mixed. Four of these species are crustaceans, all various types of shrimp. Three are molluscs of different types, with one gastropod species (*Tectus niloticus*) and two bivalves (*Lima scabra* and *Tridacna derasa*). *Condylactis passiflora* and *Radianthus malu* are both cnidarians (anemones), and *Linkia laevigata* is an echinoderm (starfish). Overall, the United States imports organisms from 209 invertebrate genera from 18 countries around the world, with 310 known species being recorded in trade.

By contrast, Table 3.7 lists the chief imports of a relatively well-known group of organisms, coral reef fish, into Japan, a market about which little is known. Fish from 200 genera and 425 known species are imported to Japan from five countries. These are the Cook Islands, Indonesia, the Maldives, the Philippines, and Sri Lanka, although this list certainly reflects more the spread of data within GMAD rather than a complete list of nations supplying the Japanese market for marine aquarium fish.

Taxonomy of the Marine Aquarium Trade

As the previous example of fish imports to Japan illustrates, trade profiles can be generated taxonomically (Table 3.7): seven of the top imports to Japan are Pomacentridae (damselfish). *Nemateleotris magnifica* is a goby, *Paracanthu-*

rus hepatus a species of surgeonfish, and *Thalassoma amblycephalum* a wrasse (family: Labridae). In total 1,196 species of fish are traded globally, substantially more than previous estimates of up to 1,000 (Wood 2001). These 1,196 species come from 406 genera, and trade in small numbers in unidentified species of fish come from another 26 genera and 12 higher groups. Most species—806—are associated with coral reefs during the majority of their life history and presumably are collected from reef areas, but the relative number of traded species that are associated with habitats other than reefs (e.g., mud flats, seagrass beds, and mangroves) is surprisingly high. Approximately one third of all species of fish in trade fall into this category, although those traded in most numbers are coral reef species. In total 26% of all known coral reef fish species are traded live for aquariums.

Fish The taxonomic profile of global trade is shown in Figure 3.2 and is similar to that for Japan. Species of Pomacentridae dominate, accounting for 42% of all fish sold internationally. This is most probably due to a combination of their small size, attractive colors, and general ease of care. This family also includes the anemone fish, which are highly popular and which are cultivated in commercial quantities. Acanthuridae (surgeonfish), Pomacanthidae (angelfish), and Labridae (wrasses) are also very popular. Species in the 10 families in Table 3.8 account for 83% of the international trade in aquarium marine fish.

Indonesia and the Philippines are the center of fish species diversity, and as expected more

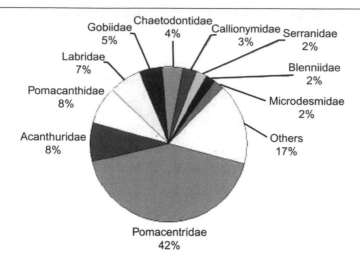

Fig. 3.2   The taxonomy (family) of the global trade in fish.

species are exported from these two countries than any others (Table 3.9). In total, 27 endemic species are exported from source nations, the majority (17) from Australia and a significant number (7) coming out of Hawaii.

Corals   There are 102 species of coral in trade, though some doubt must be placed on the full species list generated from GMAD data, given the complexities of coral taxonomy, which in some cases requires positive identification from dead skeletons and not live specimens. These 102 species come from 52 genera, and additional trade in unidentified species comes from 20 genera and one higher group.

Invertebrates   Conclusions are even more difficult to draw for other invertebrates, given the lack of a standard reference system of taxonomy. Although every effort has been made to check and remove them, it is likely that the list of invertebrate species names within GMAD contains some misspellings and synonyms. With this caveat probably indicating that true totals are slightly lower, we can state that a total of 293 invertebrate species other than corals are known to be traded for marine aquariums, with these species taken from 113 genera. There is trade in unidentified species from another 93 genera (because many invertebrates are simply identified to genus, for practical reasons), and 19 groups higher than family.

Table 3.8   The relative number of species in trade of the ten major families in Figure 3.2

Family	English name	No. of species	No. of species traded
Pomacentridae	Damselfish	329	112
Acanthuridae	Surgeonfish	73	32
Pomacanthidae	Angelfishes	84	67
Labridae	Wrasse	392	82
Gobiidae	Gobies	1006	26
Chaetodontidae	Butterfly fish	124	53
Callionymidae	Dragonets	102	15
Serranidae	Groupers and fairy basslets	449	80
Blenniidae	Combtooth blennies	345	33
Microdesmidae	Wormfish	60	10

Source: Calculated from the Global Marine Aquarium Database.

Table 3.9 National trade profiles: number of species of fish exported from individual countries, including the number of endemics

Country	Number of fish species	Endemics
Indonesia	630	
Philippines	584	1
Fiji	291	
Sri Lanka	260	
Palau	225	
Solomon Islands	186	
Maldives	148	
Brazil	137	1
Saudi Arabia	104	
Australia	101	17
United States	65	8[b]
Netherlands	53	
Vanuatu	39	
Marshall Islands	34	
Mexico	31	
Cook Islands	24	
Singapore	23	
Kenya	22	
Yemen	21	
New Caledonia	18	
Papua New Guinea	14	
Bahrain	13	
Egypt	11	
United Kingdom[a]	6	
Costa Rica	4	
Haiti	4	
Christmas Island	3	
Tonga	2	
Japan	1	

Source: Calculated from the Global Marine Aquarium Database.

[a] Most likely re-exports.

[b] Seven from Hawaii.

GMAD data are particularly useful in describing trade in groups defined by common use but without a uniform taxonomic basis. One example is the trade in ornamental shrimps. These "shrimps" are from three distinct families—the Hippolytidae (cleaner shrimps), the Stenopodidae (boxing shrimps), and Rhynchocinetidae (dancing shrimps). In 1999, a total of 48,852 shrimp from 15 species in these families was traded internationally, with *Stenopus hispidus* being the most popular (24%), followed by three species of *Lysmata* (*L. grabhami*, 18%; *L. multicissa*, 15%; and *L. amboinensis*, 15%). Smaller numbers of *Rhynchocinetes, Saron,* and *Thor* spp. were also traded.

Table 3.10 Exports of *Chaetodon falcula* from the Maldives in 2000

Destination	Percent	No. in 2000
United States	77.8	2823
Sri Lanka	16.0	579
Germany	1.7	60
United Kingdom	1.6	59
Italy	1.4	50
France	0.5	19
Hong Kong	0.4	15
Japan	0.3	10
Netherlands	0.2	6
Belgium	0.1	2

Source: Calculated from the Global Marine Aquarium Database.

Estimates of the Global Trade in Marine Ornamental Organisms

Table 3.10 provides the answer to the question, what happens to *Chaetodon falcula* exported from the Maldives? Most are exported to the United States, and a sizeable majority is shipped to Sri Lanka, most probably for re-export to Europe. Much smaller quantities are sent directly elsewhere, mostly to Europe. In 2000 a total of 3,623 *Chaetodon falcula* were exported.

This is a good example because, in the case of exports from the Maldives, a complete year's worth of data has been received from the Ministry of Fisheries & Agriculture, which collects it from exporters as a condition of their licenses. Collection of data from all exporters in a country for a complete year is, however, exceptional. In the majority of cases data have been collected from some, but not all, exporters and importers, and the data usually represent less than a full year's worth of trading. At issue then is the degree to which the data within GMAD are representative of global trade and whether global trade can be estimated from them.

Looking at the trade in coral into the European Union illustrates the dilemma further. According to the available import data for 1999, the European Union brought in 16,500 pieces of coral from around the world, but the available export data indicate a trade nearly three times that size, 47,500 pieces. It is not clear what these numbers mean, and they do not answer the simple question, how many pieces of coral did the European Union take in 1999? In a theoretical situation, if we know that there are two exporters in a single country and that in 1999 one sent 1,000 pieces of coral to the European Union, then one option is to assume that both companies are the same size and deal in similar quantities of coral. Thus, trade would be estimated at 2,000 pieces per year. Similarly, if the total number of wholesale companies operating within the industry worldwide is known, it would be possible to estimate the global trade by using simple multipliers on the known quantities in trade (i.e., 2 × 1,000).

There is no basis for this assumption, however. Indeed, common sense would suggest that all companies are not equal in size and that there will most likely be a few companies shipping large quantities of fish, corals, and other invertebrates. This is borne out by the company-specific data that have been collected. A few very large companies handle between one and ten million fish per year, business that is six orders of magnitude greater than the smallest, which deal in less than 100 fish per year (Figure 3.3). However, both very small and very large companies are not the norm: most companies

appear to handle an annual total of between 10,000 and 100,000 fish per year.

Figure 3.3 is based on sales data supplied by the 40 commercial companies listed in Table 3.3. For each group of organisms (fish, coral, and invertebrates), GMAD trade data can be used to calculate the typical annual sales for each size category of company as shown in Table 3.11.

A survey of wholesale exporters and importers based on the membership of trade organizations such as AKKII, PTFEA, SAFEA , OFI, and OATA, the MAC network, and a thorough Internet search found 276 companies in operation, 144 of which were exporters and 132 importers. If the assumption is made that the proportion of companies of different sizes in the sample of 41 is representative of the global trade as a whole, then, once the volume of trade in different groups of species is known for each size category of company (Table 3.11), global estimates can be made (Table 3.12).

Clearly the trade estimates depend heavily on the number of companies involved in wholesale trade of marine aquarium organisms and are conservative because only the number of

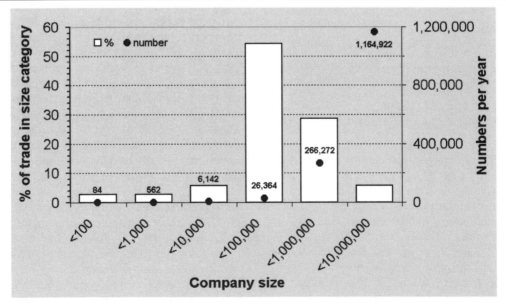

Fig. 3.3   The relative size of wholesale companies trading in aquarium species of marine fish. Company size is ranked according to a logarithmic scale of the number of organisms sold annually (in this example fish, but similar calculations were carried out for sales of corals and other invertebrates). The median number of fish sold annually for each company size category is plotted on the second y-axis. Therefore, companies that sell between one and ten million fish annually sell a median of 1,164,922 fish per year, and each year companies in this size category account for 6% of the global trade in marine aquarium fish (first y-ax-

Table 3.11 The mean and median annual sales in numbers of fish, invertebrates, and corals for differently sized companies (very small to very large)

Company size	Annual sales	Fish			Invertebrates			Corals		
		Mean	Median	N	Mean	Median	N	Mean	Median	N
Very small	< 100	84	84	1	0	0	0	688	678	0
Small	< 1,000	562	562	1	716	716	3	3,044	2,655	4
Medium small	< 10,000	3,811	3,811	1	4,461	3,527	8	26,535	22,440	11
Medium large	< 100,000	41,135	25,233	21	26,396	27,439	10	144,300	119,260	19
Large	< 1,000,000	348,299	287,783	9	244,895	252,867	7	0	0	1
Very large	<10,000,000	2,144,135	1,303,317	3	1,293,762	1,293,762	1	0	0	0
Totals				36			29			35

Source: Calculated from the Global Marine Aquarium Database.

Note: Sales data were provided by 41 wholesale exporters and importers for a variety of years from 1988 to 2001. In many cases data for periods less than a complete calendar year were received so annual sales were calculated from mean monthly sales multiplied by twelve. Company size categories are defined in the legend to Figure 3.3. N = number of companies in each size category trading in fish, invertebrates or corals. Both mean and median annual sales were calculated, but medians were used in further calculations (Table 3.12).

known companies has been used. Some companies will inevitably have been overlooked, but no attempt has been made to estimate the actual number of operational companies. If the proportion of larger companies has been underestimated, then the global trade in each group of organisms will be underestimated. This is unlikely, however, because larger companies have larger profiles within the industry and the resources to invest in marketing tools such as websites and other forms of advertising. It is more likely, therefore, that it is the proportion of smaller companies that has been underestimated.

Fish The best estimate of the global trade in fish is 27 million per year (Table 3.12). This figure refines the previous estimate made by Wood (2001) of between 14 and 30 million fish annually. The most traded species of fish is *Chromis viridis*, the blue-green damselfish (Table 3.13).

Corals The best estimate of the global trade in corals is 16 million per year (Table 3.12).

Table 3.12 Estimated global annual sales in numbers of marine aquarium fish, invertebrates, and corals

Company size	Fish		Invertebrates		Corals	
	Sales (p.a.)	Percent	Sales (p.a.)	Percent	Sales (p.a.)	Percent
Very small	322	3	0	0	0	0
Small	2,154	3	10,222	10	41,873	11
Medium small	14,607	3	134,284	28	973,233	31
Medium large	2,031,257	58	1,305,694	34	8,934,278	54
Large	9,928,514	25	8,423,070	24	0	3
Very large	14,988,148	8	6,156,525	3	0	0
Totals	26,965,001		16,029,795		9,949,384	

Source: Calculated from the Global Marine Aquarium Database.

Note: p.a. = per annum. The totals have been calculated by assuming a global total of 276 exporters and importers and by applying a multiplier derived from the proportion of companies in each size category to the annual trade in that category. For example, 58% of 276 companies (160) are assumed to be medium large, each trading 25,233 fish annually (median, Table 3.11). The total must be halved to compensate for double counting between exporters and importers, to give an estimated total annual trade of 2 million fish for companies of this size.

This figure refines the previous estimate made by Green and Shirley (1999) of 35 million corals annually. The most traded species of coral is *Trachyphyllia geoffroyi* (Table 3.13).

Invertebrates The best estimate of the global trade in invertebrates other than corals is 10 million per year (Table 3.12). No previous estimates of the trade in invertebrates other than corals have been made, though they are extremely common in home aquariums, and both the number of species on traders' lists and the quantity of specimens in pet shops intuitively suggest that this is an important sector. The most traded invertebrate species is *Tectus niloticus* (Table 3.13).

Future Data Needs for the Global Marine Aquarium Trade

Other than efforts to compensate for the weaknesses that currently exist in the data, GMAD will be expanded during 2002–2003 to include data on cultivated organisms and organisms certified under the standards developed by the Marine Aquarium Council, recognizing the importance of tracking these species in relation to global trade. People concerned for and about the aquarium trade are already using information derived from GMAD as a benchmark. If the quality of decision making among both traders and policy makers improves in line with the quality of the data, then the principal objective of this project will have been fulfilled.

Acknowledgments

The development of GMAD has been a true team effort. The following people have contributed at various stages. At UNEP-WCMC: Lucy Conway, Rachel Donnelly, Sarah Carpenter, Florence Jean, Anna Morton, Jenny Kettle, Elena Bykova, Jodie Humphreys, Michelle Taylor, Mark Taylor, Ed Green, Mark Spalding, Phil Fox. At MAC: Paul Holthus, Peter Scott, Rezal Kusumaatmadja, Gayatri Lilley, Aquilino Alvarez, Esaroma Ledua, David Vossler. Major financial support from the David and Lucile Packard Foundation is gratefully acknowledged, as is some contributing funding from the Bloomberg Foundation.

References

Froese, Rainer, and Daniel Pauly, Editors. 2002. FishBase. World Wide Web electronic publication. www.fishbase.org, 27 March 2002.

Global Coral Reef Monitoring Network (GCRMN). 2000. *Status of coral reefs of the world: 2000.* Edited by C.R. Wikinson. Australian Institute of Marine Science, p 363.

Green, Edmund P., and H. Hendry. 1999. Is CITES monitoring the international trade in corals effectively? Coral Reefs, 18: 403–407.

Green, Edmund P., and Francis Shirley. 1999. *The Global Trade in Coral.* World Conservation Monitoring Centre, World Conservation Press, Cambridge, U.K. 70 pp. ISBN 1-899628-13-4.

Hawkins, J., and C.M. Roberts. 2000. The threatened status of restricted-range coral reef fish species. Animal Conservation, 3: 81–88.

McManus, Roger. 2001. U.S. efforts to protect do-

Table 3.13 The 10 most traded species of fish, corals, and other invertebrates as a percentage of the global trade in each of these three groups of organisms

Fish species	Percent	Coral species	Percent	Invertebrate species	Percent
Chromis viridis	5.8	*Trachyphillia geoffroyi*	10.2	*Tectus niloticus*	14.3
Chrysiptera cyanea	4.5	*Catalaphyllia jardinei*	8.5	*Stenopus hispidus*	10.6
Zebrasoma flavescens	4.4	*Heliofungia actiniformis*	6.8	*Stenopus hispidus*	8.9
Amphiprion ocellaris	4.2	*Euphyllia glabrescens*	5.9	*Radianthus malu*	8.5
Chrysiptera hemicyanea	3.7	*Goniopora minor*	5.7	*Rhynchocinetes uritai*	7.2
Dascyllus trimaculatus	3.1	*Goniopora lobata*	5.4	*Condylactis passiflora*	6.0
Dascyllus aruanus	3.0	*Euphyllia divisa*	5.2	*Lima scabra*	5.2
Dascyllus melanurus	2.7	*Goniopora stokesi*	4.9	*Stoichactis kenti*	5.1
Pomacentrus australis	2.1	*Sabellidae pavona*	23.3	*Hetrodactyla hemprichii*	4.7
Chrysiptera parasema	1.9	*Euphyllia cristata*	2.8	*Lysmata amboinensis*	4.5

Source: Calculated from the Global Marine Aquarium Database.

mestic and international coral reefs: trade in the larger context. Proceedings of a symposium on: Global Trade and Consumer Choices: Coral Reefs in Crisis held at the 2001 Annual Meeting of the American Association for the Advancement of Science, San Francisco, USA, February 19, 2001, pp 11–14.

Moore, F., and Barbara Best. 2001. Coral reef crisis: causes and consequences. Proceedings of a symposium on: Global Trade and Consumer Choices: Coral Reefs in Crisis held at the 2001 Annual Meeting of the American Association for the Advancement of Science, San Francisco, USA, February 19, 2001, pp 5–10.

UNEP-WCMC. 2002. The Species Conservation Database: http://www.unep-wcmc.org/species/dbases/fauna/index.htm

United States Coral Reef Task Force. 2000. The National Action Plan to Conserve Coral Reefs, March 2, 2000, Washington, D.C., p 41.

Wood, Elizabeth M. 2001. Collection of coral reef fish for aquaria: global trade, conservation issues and management strategies. Marine Conservation Society, U.K. 80 pp.

4

World Trade in Ornamental Species

Katia Olivier

Introduction

Ornamental fish today represent a growing sector of international fish trade because of the increasing popularity of aquariums in households in many parts of the world. The importance of the ornamental fish trade, however, goes far beyond its mere share in international trade. The sector is an important source of income for rural, coastal, and insular communities in developing countries and is frequently a welcome provider of employment opportunities and export revenues.

At the same time, the issue of sustainability of harvest and culture practices has become as relevant for ornamental fish as it is for food fish, and the listing criteria of the Convention on International Trade in Endangered Species of Wild Fauna and Flora (CITES 2002)[1] and environmental labeling have been put firmly on the sector's agenda. Growing consumer concern for the environment has also resulted in numerous responses from the ornamentals industry, including the establishment of the Marine Aquarium Council. In addition, several ecological labeling schemes are now in place both for ornamental fish and corals.

The importance of accurate trade statistics that permit correct analysis of trends and developments cannot be overestimated. It is a common perception in the ornamentals sector that international trade figures frequently are underreported, with the principal reason being the noninclusion of small shipments or misclassification in the food fish category. In particular, this leads to problems in estimating correct unit values, as weights reported are frequently gross weights including packaging. Other times, quantities are given as the number of fish. The ornamentals sector should, therefore, through the national associations, encourage governments to upgrade their reporting capabilities for the ornamentals trade.

For the same reason, this chapter will discuss marine ornamentals in detail only where the data allow it. In the main, freshwater and marine species will be discussed together, as most international trade statistics make no distinction between the two categories.

International trade statistics of the United Nations Food and Agriculture Organization (FAO 2001) are based on statistics received from FAO member countries. The reporting process as well as the compilation of aggregate numbers is time consuming, and there is necessarily a time lag between the receipt of data from individual countries and the compilation of global statistics by FAO. FAO's statistics for ornamental fish show a strong growth in trade, with a peak in 1997, a decline in traded value in 1998 and 1999, and new growth registered in 2000. More recent national statistics for 2001 confirm the break in the negative trend, with import and export values rising again. This positive development has brought new optimism to the sector, promising increased benefits for all, including operators and suppliers in developing countries.

This chapter is based on a study by the FAO Fishery Department's Fish Utilization and Marketing Service carried out by Katia Olivier under the supervision of Dr. Audun Lem. The main study is available from FAO as GLOBEFISH Research Programme Report number 67.

Executive Summary

Distribution channels for ornamental fish form a complex and highly dynamic system. As in most other sectors, there is a trend toward bypassing intermediaries and more vertical integration.

At present Asia contributes more than 50% of the world supply of ornamental fish. Singapore is by far the largest exporter, followed by Hong Kong, Indonesia, Malaysia, and the Czech Republic. Since the 1980s the value of world exports has been increased markedly, from US$40 million to almost US$200 million, but it is still less than 1% of the total international fish trade. U.S. exports have fallen drastically in recent years.

Fish keeping is a hobby that is practiced mainly in industrialized countries. The main importers are the United States, Japan, and Europe (particularly Germany, France, and the United Kingdom). The value of world imports has increased markedly too, from US$50 million to US$250 million over the past two decades.[2]

In international trade, freshwater species represent about 90% of market value compared with 10% for marine species. The author estimates that, on average, 90% of freshwater species are farmed compared with 10% collected in the wild. For marine species the great majority, almost 99%, are collected in the wild and just 1% are farmed. World turnover of ornamental fish aquaculture is estimated at about US$200 million. The species that dominate the market are all freshwater species, although marine species are becoming more popular. Many experts also believe that marine aquariums with live coral reefs are the trend for the twenty-first century.

Estimates in the mid-1990s of the total value of the wholesale trade in ornamental fish were about US$900 million (Bassleer 1994), with retail trade of live animals for aquariums valued at about US$3 billion (Davenport 1996). Since the 1980s these figures have been increasing, and fish keeping today is a hobby that has become institutionalized as a standard consumer commodity for millions of people around the world.

At the same time, supply problems are increasingly common. Such problems can be due to climatic phenomena and seasonal patterns or to the degeneration of some species in breeding. Problems of supply can also be caused by excessive collection of wild species and to environmentally destructive collection methods. Cumulative mortality rates in the chain can be more than 75%, and in the fish keeper's aquarium, some 50% of fish may die within six months. With such high losses it is only natural that the issue of more sustainable development of the sector is now being raised.

The Distribution Network

The Main Players, Objectives, and Activities

The Fishers or Collectors Collectors are typically small-scale fishers from tropical countries who work alone or in small groups (often as family units). They usually work with artisanal equipment, and in many areas of the Amazon, collection is even carried out by hand.

The fishers/collectors are the first link in the chain. Although a few are grouped into associations, most of them work for a wholesaler or an exporter. They are paid for the number of fish they collect—very often a derisory sum. In Indonesia, for example, the fisher is paid US$0.10 for an *Amphiprion percula*, which is worth up to US$12 at the retail level in an importing country (Hemdal 1984). At the same time, competition is sometimes so strong between collectors that it is not unusual (in the Amazon, for example) to see them defend their concession with a gun.

The Breeder The great majority of breeding/rearing businesses are based in Asia, especially in Singapore, but there is a substantial number in North America, particularly in Florida, and also in South America. Other countries, such as the Czech Republic and Israel, have also developed breeding businesses.

Different types of breeders exist. In Asia, especially in Singapore, fish-rearing businesses are family firms. Increasingly, they are grouped in agrotechnology parks developed by government agencies. But much more rudimentary businesses also exist. In the Philippines, for example, people rear fish in urban slums, where sanitary conditions can be quite poor. In Guyana, rearing basins are sometimes holes in the soil lined with waterproof canvas and floating plants on the surface.

In the United States, investment has been made in high-capacity industrial breeding operations. Europe, on the other hand, has few commercial rearing businesses. Here groups of amateurs may supply local retailers and individuals, and some professional breeders sell to wholesalers, retailers, or directly to the public. In most countries, rearing facilities must obtain a license or permit from the Ministry of the Environment or other responsible government authority, which then sends biologists and/or veterinary officers to check the breeder's sanitary conditions.

The Wholesaler Wholesalers in exporting countries buy from collectors, breeders, and importers and then sell to exporters. Wholesalers in importing countries buy from importers and sell to retailers. Traditional wholesalers are disappearing from the scene, mainly because of the introduction of transshipping and the development of the electronic market.

The Exporter Exporters buy fish either from wholesalers or directly from collectors and breeders. In some cases, they may employ their own collectors. They then put the fish in quarantine in order to detect potential diseases and begin to acclimatize the fish to life in captivity. The quarantine period can last from a few days to a few months, according to the health of the fish and according to the professional practices of the exporter. Infrastructure needs can be substantial, requiring warehouses full of containers, aquariums, tanks, or even outdoor concrete basins.

Finally, exporters must obtain a special license for each consignment to allow it to leave its country of origin. The consignment must be checked by the veterinary services to obtain a phytosanitary certificate and must then be declared to customs before shipment.

The Importer Importers buy from exporters but may simultaneously be involved in both activities. They receive consignments covered by a phytosanitary certificate and, like the exporters, must then put the fish in quarantine and continue to acclimatize them to life in captivity, in particular to the feeding cycle and to the condition of the water. They thus have to ensure in-depth technical monitoring of their stock. They sell their products to wholesalers or directly to retailers, or re-export to other countries. Countries like the United States, at the global level, or the Netherlands, at the European level, are trading hubs that re-export a large part of their imports to other countries. In the same way, Singapore imports a major part of the Asian production and then bulks it together in order to re-export worldwide.

The Transshipper During the 1970s and 1980s, wholesalers began to face competition from another import system known as "transshipping." This system radically changed the ornamental fish sales network worldwide. The transshipper groups orders together to reduce freight costs. The transshipper receives the consignment at the airport, declares it, and then redistributes it countrywide or to neighboring countries, without acclimatizing the fish. This service obviously incurs costs, but retailers can buy fish at perhaps half the price the wholesaler would charge.

Despite this competition, wholesalers retain an important clientele, because shopkeepers can import directly through a transshipper only if they are able to abide by certain rules and regulations. They must obtain an authorization from the veterinary services, and they must have appropriate commercial premises for acclimatization. Acclimatization also requires qualified staff with knowledge of potential diseases and their treatment, equipment, and stock management. This represents a substantial investment for a retail shop. Because acclimatization is becoming an increasingly sensitive technical task, in particular for marine fish species that are more and more popular, the investment is not risk free. When fish are transshipped, all the risks are borne by the retailers; not only do they bear the costs of mortality on arrival, they also receive a product of generally reduced quality, which affects the cost of acclimatization. In consequence, the apparent benefits of the cheaper import prices offered by the transshipper are substantially reduced.

Transshippers are responding to the criticism, however. Their future depends on increased quality and better services. Some transshippers already practice "consolidating," which is the second generation of transshipping that has been introduced in the United States.

Consolidating means providing a more complete service to the retailer: taking responsibility for the fish for a period of 48 hours after delivery, giving refunds if there is mortality, and checking the origin of the fish.

The Retailer   Retailers can buy fish directly from importers or wholesalers, or they can order from a transshipper or transship themselves. If they order through a transshipper, they buy the fish at better prices but then have to acclimatize them, which means having the necessary equipment and obtaining an operating license and equipment certification. If retailers buy stock from a wholesaler who acclimatizes, the fish will then be fit for sale. Licensed retail shops, which can be specialized ornamental fish shops, pet shops, and garden centers, are all regularly inspected by the veterinary services.

Other Important Players

Governments   The governments of many exporting countries are directly involved in ornamental fish production or trade. Some offer financial assistance, while others set up improved management schemes or enact better trade regulations. In Singapore, for example, the responsible government agency, the Primary Production Department, has developed agrotechnology parks. In some countries the government sets fishing quotas, prohibits collection from certain sites or of certain species, or prohibits certain catching methods (e.g. cyanide, dynamite). Each country has different laws, so it is essential for all interested parties to contact the Ministry of Agriculture or the Ministry of the Environment or other appropriate government authority to obtain the latest legislation covering protected species and quotas, the regulations covering sanitary standards, and conditions for import and export and general trading in ornamental fish. In Europe, for example, traders must apply to the Ministry of the Environment to obtain technical certification to sell live animals. They must also declare all imports and exports to the Ministry of Finance.

CITES   The Convention on International Trade in Endangered Species of Flora and Fauna (CITES) had by January 1, 2002 been ratified by 157 countries. The aim is to regulate in-ternational trade in the species identified therein. The species covered by the convention are classed in three annexes, according to the threat of extinction that they bear. Annex I identifies immediately endangered species, and international trade in those species is totally prohibited. Annex II itemizes species that risk becoming endangered within a short period of time. In order to avoid exploitation that would threaten their survival, international trade in these species is strictly regulated through licenses or permits. Annex III lists species that are endangered within the territory of one or more countries. For these species, specific measures aim to prevent or reduce their exploitation. Where fish are concerned, there are several species in annex I and many more in annex II, but only a few are "ornamental." Among molluscs, there is a species classified in annex I that particularly interests the fish-keeping market—*Tridacnidae* (the giant clam). Finally, in annex II, there are various corals: blue coral, organ pipe coral, black corals, and fire coral.[3]

Persons in possession of items listed in CITES must be able to justify, at any time, that they are held in accordance with the relevant terms and conditions. A CITES import license, from the Ministry of Environment, must be presented to the specialized customs service. This document is given only after presentation of a CITES export license from the country of origin. For example, numerous requests are made to obtain authorizations to import and export coral. If, on arrival, products subject to CITES regulations do not have the necessary license, they are confiscated pending receipt of the license. If none is received, the products are returned to their country of origin. The increase in the number of offences noted by customs authorities concerning trade in fauna and flora testifies to the importance of the traffic.

Other Players   Other players with interest in CITES issues but not parties to CITES are international bodies such as the Food and Agriculture Organization of the United Nations (FAO). Others are international nongovernmental organizations for the protection of the environment, such as the World Wildlife Fund (WWF), in particular the British WWF. Moreover, those involved in the sector in a particular country can form associations or syndicates like the Interprofessional Syndicate of Manufacturers and Distributors of Products for Pets (PRODAF) in France, the Pet Industry Joint Advisory Council (PIJAC), and the Marine Aquarium Council (MAC) in the United States. In the exporting countries there is also a growing number of associations, cooperatives, and syndicates.

Airline Companies International airline companies transport fish from the exporting to the importing country, which in turn may re-export the consignment. In the price of the fish, the freight charges sometimes represent more than half—maybe as much as two thirds—of the landed price for the importer. The airlines are thus key players, because the better the quality of the transport, the better the final quality of the fish. The long-haul transport of ornamental fish has become an important business for the airlines and a vital link in the distribution chain.

Overview of the Supply Situation

Main Exporting Countries and Market Share

More than 100 countries are now involved in exporting ornamental fish, with 11 countries exporting about three fourths of the total (Table 4.1).

Singapore is the world's largest exporter of

Table 4.1 Value and share of world trade in ornamental fish for the main exporting countries in 2000

	Value in US$million	Share of total exports (%)
Singapore	43.5	23.9
Hong Kong	17.2	9.4
Indonesia	12.8	7.0
Malaysia	11.5	6.3
Czech Republic	10.3	5.7
Japan	8.5	4.7
United States	8.3	4.6
Sri Lanka	7.7	4.3
Philippines	6.7	3.7
Israel	6.7	3.7
Peru	4.8	2.6
Others	44.2	24.2
Total	182.2	100

Source: FAO.

ornamental fish and is the trading hub for Asia. Singapore does not produce all its exports but re-exports additional species that are collected and reared in other countries in Asia. Singapore is a duty-free zone, so there are no heavy import duties to pay. It exports to more than 60 countries around the world. In value terms in decreasing order, the top nine are Japan, the United States, the United Kingdom, Germany, France, Italy, Spain, the Netherlands, and Australia. The United States and the United Kingdom traditionally occupied the first two places, but since 1994, Japan has been the main customer.

The United States was traditionally an important exporter, breeding freshwater fish industrially. It also re-exports fish from Asia and Latin America. The business is mainly concentrated in Florida, where most of the breeders are located. Some 22% of imports arrive in Miami and 39% in Los Angeles. The main export markets for the United States are Canada and Japan.

Hurt by the strong dollar, U.S. exports continue to fall, dropping to US$7 million in 2001. Canada is the major export market, with US$4 million, or 58% of the total, whereas exports to Japan, traditionally the second largest U.S. export market, have fallen from US$4 million in 1997 to a mere US$364,000 in 2001.

The Philippines and Indonesia are the main marine fish exporters and, to a much lesser extent, Thailand, Hong Kong, and Sri Lanka.

The Czech Republic has become an important exporter of reared fish. After the collapse of the communist regime, fish keeping provided an opening for entrepreneurial initiatives. A breeders' association was established, which began to export fish and import equipment. Germany became the main customer, and France, the Netherlands, and Denmark followed. More recently, the Czech Republic has begun to re-export fish that come mostly from Germany (62%).

The dominant role of Asia as an exporter is clear (Table 4.2). Europe, because of the Czech Republic, Germany, Belgium, Spain, and the Netherlands, takes second place. Most of these countries, except for the Czech Republic, only re-export. A massive 99% of European freshwater fish exports are destined for Western Europe. For marine fish, 90% are destined for Western Europe, and 10% for Asia (particularly Japan). The main exporters of freshwater fish

Table 4.2 Value and share of world exports by continent in 1999

	Value in US$1,000	Share of total exports (%)
Asia	93,014	54.9
Europe	36,682	21.6
North America	17,820	10.5
South America	10,084	6.0
Middle East	6,688	3.9
Oceania	3,044	1.8
Africa	2,124	1.3
Total	169,456	100

Source: FAO.

are Germany, Belgium, and the Netherlands, while the main exporters of marine fish are France, Spain, and Italy. Africa's market share is minimal (1%), despite considerable potential.

Evolution of World Exports and the Main Exporting Countries

Since the 1980s, the value of world exports has been increasing markedly (Figure 4.1).

All countries experienced a net increase of their exports of ornamental fish up to 1997, after which there was a decline for most countries and new growth in 2000 (Table 4.3 and Figure 4.2). Exports from Singapore and the United States are decreasing because other countries, such as Malaysia, have considerably developed this activity and now compete with the traditionally dominant ones. Furthermore, countries like the Czech Republic, Israel, and Sri Lanka have gained a foothold in export markets, and lately also Hong Kong. Japan holds a stable

share of the market because it is the specialist in breeding and exporting koi carp, and holds on to its advantage in this field.

The Main Origin of Ornamental Fish

Asia contributes more than 50% of the total supply of ornamental fish worldwide (see Table 4.2). Of this, 80% is freshwater fish produced in breeding farms, 15% is marine species collected in the wild, and 5% is freshwater fish, also collected in the wild. Singapore specializes in breeding freshwater species but also exports marine species collected mostly in other Asian countries, such as Indonesia, the Philippines, Sri Lanka, and the Maldives. Other Asian countries specialize in breeding freshwater species, such as Thailand, Hong Kong, and Japan. Countries such as Malaysia, China, and Sri Lanka have started to develop the breeding of freshwater species, which are then exported to Singapore, although Malaysia increasingly exports directly.

The United States breeds mostly freshwater species, but some marine species are also captured in the wild. It also imports most of the ornamental fish produced in South America. The majority are freshwater species collected in the wild (especially in the Amazon river), which are re-exported to Canada, Europe, and Japan. In Africa ornamental fish are mostly freshwater species collected in the natural environment (big lakes and rivers), but the region represents only a small part of international trade, despite a good potential, which is largely unexploited. In the Near East supplies from Israel are significant and consist almost entirely of reared fish.

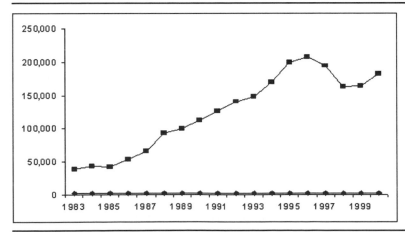

Fig. 4.1 Evolution of the value of world exports in US$1,000 1983–2000 (FAO).

Table 4.3 Evolution of exports (in US$1,000) from the main exporting countries and percentages of total

	1986	1987	1988	1989	1990	1991	1992	1993	1994	1995	1996	1997	1998	1999	2000
Singapore	21,730	25,588	29,653	32,874	40,104	40,234	44,194	45,666	52,631	59,411	58,574	52,757	43,156	42,417	43,502
Percent	40.3	38.7	31.7	32.9	35.5	32.0	31.5	30.7	31.1	29.8	28.3	27.2	25.0	25.0	23.9
Hong Kong	n.a.	n.a.	n.a.	n.a.	n.a.	n.a.	n.a.	n.a.	n.a.	n.a.	n.a.	n.a.	10,379	1,616	17,185
Percent													6.0	1.0	9.4
U.S.A.	4,884	8,493	13,071	10,365	12,512	13,261	15,337	17,467	19,203	19,943	15,522	14,717	10,609	11,007	8,289
Percent	9.1	12.8	14.0	10.4	11.1	10.5	10.9	11.7	11.3	10.0	7.5	7.6	6.2	6.5	4.6
Malaysia	172	387	397	475	1,369	1,555	2,191	4,058	4,488	6,673	9,835	8,819	8,646	10,595	11,520
Percent	0.3	0.6	0.4	0.5	1.2	1.2	1.6	2.7	2.6	3.3	4.7	4.6	5.0	6.3	6.3
Czech Rep.	n.a.	n.a.	n.a.	n.a.	n.a.	n.a.	n.a.	n.a.	n.a.	9,055	9,748	8,927	10,489	10,316	10,273
Percent										4.5	4.7	4.6	6.1	6.1	5.7
Indonesia	1,239	1,609	4,904	7,070	4,590	5,668	7,058	8,527	8,847	9,264	7,880	2,655	1,122	10,286	12,841
Percent	2.3	2.4	5.2	7.1	4.1	4.5	5.0	5.7	5.2	4.6	3.8	1.4	0.7	6.1	7.0
Sri Lanka	n.a.	n.a.	n.a.	n.a.	1,425	1,059	1,477	861	945	3,378	4,973	7,955	7,925	7,940	7,716
Percent					1.3	0.8	1.1	0.6	0.6	1.7	2.4	4.1	4.6	4.7	4.2
Japan	2,494	2,258	5,333	5,790	6,969	6,933	7,221	7,092	7,724	8,790	9,832	9,204	7,468	7,088	8,458
Percent	4.6	3.4	5.7	5.8	6.2	5.5	5.1	4.8	4.6	4.4	4.7	4.8	4.3	4.2	4.7
Philippines	n.a.	n.a.	n.a.	n.a.	n.a.	5,530	6,198	7,404	8,178	8,808	7,705	7,315	6,403	6,475	6,737
Percent						4.4	4.4	5.0	4.8	4.4	3.7	3.8	3.7	3.8	3.7
Israel	n.a.	n.a.	3,503	3,554	4,569	4,977	5,087	5,605	7,156	8,385	8,238	8,369	7,086	6,083	5,399
Percent			3.7	3.6	4.0	4.0	3.6	3.8	4.2	4.2	4.0	4.3	4.1	3.6	3.0
Belgium	n.a.	n.a.	877	738	901	932	1,553	1,705	2,795	3,197	4,471	4,291	4,202	4,827	4,496
Percent			0.9	0.7	0.8	0.7	1.1	1.1	1.6	1.6	2.2	2.2	2.4	2.8	2.5
Colombia	1,122	1,258	1,373	1,768	1,723	5,339	3,583	4,558	4,770	4,713	4,819	4,423	3,705	4,265	3,162
Percent	2.1	1.9	1.5	1.8	1.5	4.2	2.6	3.1	2.8	2.4	2.3	2.3	2.1	2.5	1.8
Brazil	859	1,023	1,078	1,110	1,393	1,848	2,396	3,277	3,833	4,252	4,249	3,921	3,345	3,371	3,235
Percent	1.6	1.5	1.2	1.1	1.2	1.5	1.7	2.2	2.3	2.1	2.1	2.0	1.9	2.0	1.8
Others	17,497	20,543	26,700	28,004	27,489	28,468	34,272	38,467	44,940	49,784	56,823	56,383	53,780	39,658	37,488
Percent	32.5	31.1	28.5	28.0	24.3	22.6	24.4	25.9	26.5	25.0	27.4	29.1	25.2	23.4	20.6
Total	53,873	66,117	93,585	99,927	113,124	125,819	140,270	148,775	169,456	199,505	207,179	193,763	172,493	169,456	182,152

Source: FAO.

Note: n.a. = not applicable

55

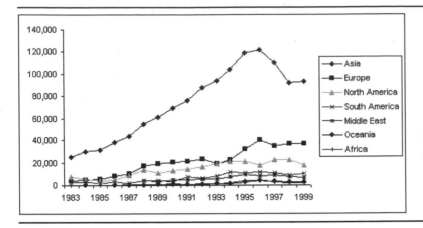

Fig. 4.2 Evolution of the value of exports by continent in US$1,000 (FAO). (Similar data for Oceania and Africa result in overlapping graph lines.)

Similarly, in Central Europe the Czech Republic specializes in breeding freshwater species but also re-exports.

In international trade, freshwater species represent about 90% in value terms, as opposed to 10% for the marine species.[4] The author estimates that globally, 90% of freshwater species are reared and 10% are collected in the wild. For marine species, 99% are collected in the natural environment and 1% are reared. In volume, the percentage for marine species would be less because of their stronger unit value. However, their market share has been increasing over the past few years. These figures suggest that in total about 20% of the value of international trade is composed of species collected in the wild.

Main Reared Fish Species and the Reared Fish Market

Collecting ornamental fish in the wild can have numerous negative effects on the environment and potentially on the resource. These effects include overexploitation, damage to habitats caused by destructive fishing methods, and even irreversible changes to natural cycles. Such problems reveal themselves gradually over time and thus will be even more significant in the future if nothing is done to resolve them. Regulatory mechanisms are currently inadequate, and research institutions do not give a high priority to the sector. High levels of mortality at each step in the marketing chain also contribute to significant waste of the resource. The development of local breeding and rearing could resolve most of these problems. It would, however, also significantly change relations between the different players and their livelihoods, resulting in a radical change in the sector as a whole.

Estimates of Volume and Value   Currently only a limited number of breeding farms raise marine species. Aqualife Research in the Bahamas was the first farm to produce marine ornamental fish commercially (*Amphiprion ocellaris*). There is Dynasty Marine Associates in Florida, and more recently the farm C-Quest has been established in Puerto Rico. There are also some *Tridacna* (giant clam) farms in the south Pacific. In addition, there are some amateur fish breeders in the United States and Europe. Other sea farms existed in the past, such as Instant Ocean and Sea World, but given the high-risk nature of marine fish farming, investors preferred to withdraw and move to more predictable and profitable activities.

The producers of marine species (fish and invertebrates) face numerous problems, most notably the difficulty of producing an animal of good quality and marketing it at a competitive price compared with animals collected in the wild. In addition, the breeding of marine species requires new and innovative technologies and much infrastructure. This implies high costs of production, equipment, and above all labor, which is particularly expensive in the United States and Europe. Currently, labor costs represent more than a third of the total cost of production for a breeding farm.

Nevertheless, the quality and life expectancy of farmed fish are higher than those collected in the wild, and this justifies the mark-up and thus the profitability of farmed marine fish. However, as long as distributors and consumers consider purchase price to be more important than quality, life expectancy, and the environmental benefits of rearing fish, the production of most marine species will remain a risky business. At the same time, the price of collected marine species will increase, because excessive collection will cause reduced availability and also because of increased protection. One can even envisage a future in which one day only reared marine fish may be sold and exported legally.

The United States of America One good example of success in ornamental fish aquaculture is undoubtedly Florida. According to PIJAC, the Florida aquaculture industry's sales in 1998 reached US$60 million, of which fish made up 80% and plants 20%. The major part of fish production was tropical freshwater fish.

PIJAC estimates the number of breeders of tropical freshwater fish at 300 in Florida and 500 in the other states. As a reference point, in the United States there are about 100 importers, 500 wholesalers, and about 1,000 retailers of ornamental fish. Even if the number of breeders in Florida is smaller than that in the other states, Florida produces and supplies 95% of the tropical ornamental fish that are sold in North America. The history of ornamental aquaculture in Florida goes back to 1926, and the breeding farms today are mostly small family-run farms or medium-sized companies. Florida produces about 300 different species of ornamental fish and a total of 500 varieties. Although the typical breeder rarely masters more than 10 species, it is important to offer a wide range of species to the wholesalers.

Several factors explain the success of this activity in Florida. First, the climate is almost ideal to the rearing of tropical fish, with only minimum investment needed to warm the water. Second, Florida has an excellent air transport service. Third, the producers are well organized in syndicates or cooperatives, which gives them weight in negotiations with the other players and more freedom of action.

Singapore Singapore also owes its success in ornamental fish culture to its tropical climate, combined with excellent market conditions. These include not only a large number of experienced enthusiasts, but also a wide range of governmental subsidies for fish rearing and also for the development of exports. Singapore's strategic position is also important, with an excellent air transport service linked to the main European markets, as well as North America, Australia, and the rest of Asia.

Ornamental fish breeding is the most important industry in Singapore's primary sector, with origins in the 1930s. Exports rose from less than US$1 million in the 1960s to US$30 million in 1982 and about US$44 million in 2000. Singapore produces around 30 species, for a total of 300 varieties. Most of them are reared in small family-run farms, usually specializing in one particular species. At present there are 70 farms.

The typical profile of an ornamental fish breeding farm has evolved considerably over the past two decades. Previously farms had several varied activities, such as raising fish and poultry and growing flowers. The modern farm is usually located in an agrotechnology park and practices only fish rearing (or occasionally the cultivation of aquatic plants). As a rule, farms rear only one species but several varieties. The techniques and know-how are handed down from generation to generation in the family-run farms, which is their key to success.

Over the past few years Singapore's share of world exports has decreased. Fish farms in Singapore are suffering from disease caused by excessive intensification of breeding in Singapore and in the other Asian countries that supply the Singapore market. Suppliers and retailers increasingly complain that fish from Singapore, having been overtreated with antibiotics, are now resistant to normal treatments and may transfer their symptoms to healthy fish. As a result, whole stocks, both at the supplier and later at the retailer level, may be contaminated. Moreover, with the gradual intensification of breeding, the brood stock has a tendency to degenerate, resulting in the disappearance of the desired species in its original form.

As a result, importers are turning to other countries, such as the United States or more recently the Czech Republic. In Asia, Hong Kong, Malaysia, and Indonesia have also increased their exports. Western importers are increasingly diversifying their sources of supply, because they are becoming stricter about the quality of the fish. Despite this, Singapore still holds the predominant place in the market, which means the quality of fish is still decreasing. This situation continues to benefit those who see in this reduced quality only a corresponding reduction in price.

A substantial increase in fish breeding in the importing countries is to be expected as long as the sanitary problems remain unsolved. This should occur as a way to offset a decrease in quality and perhaps to hedge against a possible prohibition of imports from Singapore for health and sanitary reasons. Thus, aquaculture appears to offer a solution to ensure the survival of the fish-keeping business.

The Czech Republic   The Czech Republic is an important supplier on the European market. Most fish come from breeders, although the new trend is to re-export. Starting from a low-cost position, costs are now aligned to European averages, and this is having an impact on the prices of ornamental fish. In spite of these problems, the Czech Republic still benefits from some advantages: short transportation times, good product quality, and low mortality rates.

Estimated Value of Ornamental Fish Aquaculture Production   The author estimates production values as follows: 35% for Singapore, 30% for the United States, 20% for the rest of Southeast Asia, 10% for Europe, and 5% for the rest of the world. In 1998 PIJAC set the production value of ornamental fish and plants in the United States at about US$60 million. Consequently, the value of world ornamental aquaculture can be estimated at approximately US$200 million.

Overview of the Demand Situation

Main Importing Countries and Market Shares

Fish keeping is a hobby that is practiced mainly in industrialized countries. The main importers are the United States, Japan, and Europe (Tables 4.4 and 4.5). Some 90% of imports of ornamental fish to the United States come from Asia and 10% from South America (Bassleer 1994). Imports from Indonesia and the Philippines are mainly composed of marine species.

More recent data indicate that 70% of U.S.

Table 4.4   Value of imports in the main importing countries and percentage of world imports in 2000

	Value in US$million	Share of imports (%)
United States	60.0	24.4
Japan	32.9	13.4
Germany	22.0	9.0
France	20.3	8.3
United Kingdom	20.0	8.1
Singapore	10.1	4.0
Belgium	9.6	3.9
Italy	9.6	3.9
Netherlands	8.5	3.5
Hong Kong	8.0	3.3
Canada	6.4	2.6
Others	38.1	15.5
Total	245.5	100

Source: FAO.

Table 4.5   Value and percentage of imports by continent in 1999

	Value in US$1,000	Percentage of total (%)
Europe	114,180	47.1
North America	65,756	27.1
Asia	58,446	24.1
Middle East	2,338	1.0
Africa	975	0.4
South America	433	0.2
Oceania	316	0.1
Total	242,444	100

Source: FAO

imports are from Asian countries such as Thailand, Singapore, Indonesia, the Philippines, and Malaysia. After a sharp fall in 1999, imports recovered in 2000, reaching US$60 million, but still short of the US$67 million registered in 1998.

Germany is Europe's largest market in value. According to European Union statistics (EUROSTAT), the principal exporters of freshwater ornamental fish to Germany in 2000 were from Singapore (27%), the Czech Republic (21%), Brazil (9%), Japan (7%), and Thailand (5%), with 31% coming from other countries. Its marine species imports came mainly from Indonesia (21%); the Netherlands (17%); the Philippines, the United States, and Sri Lanka (each 9%); Kenya and the Maldives (each 5%); and 25% from other countries.

The second most important European importing country in value is France. Its major suppliers of freshwater fish are Singapore (31%), the Czech Republic (13%), the Netherlands (9%), Germany (8%), Belgium (8%), Sri Lanka (5%), Thailand (4%), Israel (3%), and Indonesia (2%), with 17% from other countries. Its imports of marine fish come from Indonesia (41%), the United States (10%), the United Kingdom (9%), the Netherlands (9%), the Philippines (7%) and Sri Lanka (6%), with 16% from other sources.

Imports of freshwater fish to the United Kingdom, the third largest European importer by value, come mainly from Singapore (34%), Japan (24%), Israel (8%), the United States (7%), Malaysia (6%), Hong Kong Province of China (5%), and the Czech Republic (3%), and 13% come from other suppliers. Its marine imports are from Indonesia (19%), the United States (16%), the Maldives (12%), the Philippines (12%), Kenya (7%), Fiji (7%), Sri Lanka (6%), and Australia (5%), with 16% from other countries.

Evolution of World Imports and the Main Importing Countries

Since the 1980s, the value of world exports has seen a net increase (see Figure 4.1). In 1994 the value of imports rose dramatically (Figure 4.3) but has since declined. In 2000, a modest but most welcome increase was registered.

Table 4.6 shows the evolution in imports of ornamental fish. International demand and trade are influenced by a number of factors, including the economic situation in the importing country and the popularity of fish keeping as a hobby in competition with other activities.

The evolution of imports by continent shows Europe, Asia, and North America dominating, with Europe showing particularly strong growth (Figure 4.4). On the other hand, imports into South America, Oceania, the Near East, and Africa are virtually nonexistent, and only the richest countries on these continents import ornamental fish (e.g., Australia, Brazil, Israel, and South Africa).

Species and Prices

The number of species known as ornamental is estimated at about 1,600, of which 750 are freshwater and the remainder marine species. Marine species represent only 10% to 20% of the total volume, however. Consequently, freshwater species dominate the market, although the market for marine fish is expanding.

Exporters charge different prices in different countries, according to production costs, stock quality, and local inflation levels. In the same way, importers may raise their prices according to the cost of mortality levels on receipt of their consignments. Some species are rare in certain regions for a variety of reasons that push up prices (e.g., climatic problems, rarefaction of a species). The mark-up coefficient between the wholesale and retail price can vary from 2 to 10, and sometimes even more for marine species because the losses at each stage are

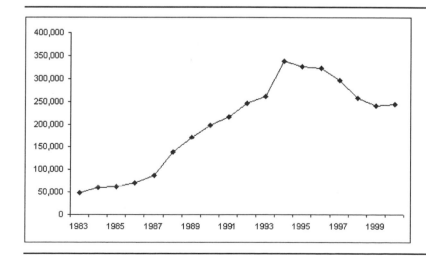

Fig. 4.3 Evolution of world import value 1983–2000 in US$1,000 (FAO).

Table 4.6 Evolution of the value of imports (in US$1,000) in the main importing countries and percentage of total imports

	1986	1987	1988	1989	1990	1991	1992	1993	1994	1995	1996	1997	1998	1999	2000
United States	26,503	29,459	35,666	57,972	61,539	56,654	63,899	68,528	69,685	80,128	77,860	72,937	67,309	57,359	60,008
Percent	37.7	33.7	25.7	33.9	31.2	26.2	25.9	26.3	20.6	24.5	24.1	24.6	26.1	23.7	24.4
Japan	9,074	10,430	14,888	18,673	23,148	31,562	42,850	54,324	67,318	78,841	73,941	52,821	39,340	35,525	32,873
Percent	12.9	11.9	10.7	10.9	11.7	14.6	17.4	20.8	19.9	24.1	22.9	17.8	15.3	14.7	13.4
Germany	10,546	13,121	15,279	15,787	18,185	19,555	22,612	21,531	23,336	27,000	27,452	26,455	24,759	22,515	21,954
Percent	15.0	15.0	11.0	9.2	9.2	9.0	9.2	8.3	6.9	8.3	8.5	8.9	9.6	9.3	9.0
France	1,306	6,479	10,201	11,509	13,726	15,446	16,054	16,347	32,735	21,536	23,306	22,679	21,143	20,516	20,291
Percent	1.9	7.4	7.4	6.7	6.9	7.1	6.5	6.3	9.7	6.6	7.2	7.7	8.2	8.5	8.3
United Kingdom	n.a.	n.a.	12,556	13,908	18,214	21,627	21,125	18,979	41,049	19,403	19,872	21,330	20,113	20,104	19,954
Percent			9.1	8.1	9.2	10.0	8.6	7.3	12.2	5.9	6.2	7.2	7.8	8.3	8.1
Belgium	n.a.	n.a.	3,812	3,943	5,315	6,122	7,261	7,039	14,533	10,101	10,443	9,984	10,123	11,800	9,610
Percent			2.8	2.3	2.7	2.8	2.9	2.7	4.3	3.1	3.2	3.4	3.9	4.9	3.9
Italy	n.a.	n.a.	4,081	4,780	5,845	7,246	8,604	8,172	17,714	10,117	10,825	10,564	9,943	9,664	9,564
Percent			2.9	2.8	3.0	3.3	3.5	3.1	5.2	3.1	3.4	3.6	3.9	4.0	3.9
Singapore	4,495	5,169	6,148	6,554	7,710	10,802	11,830	14,814	13,007	15,051	14,228	14,558	8,975	9,589	10,107
Percent	6.4	5.9	4.4	3.8	3.9	5.0	4.8	5.7	3.9	4.6	4.4	4.9	3.5	4.0	4.0
Netherlands	3,133	3,722	5,052	6,015	7,143	7,871	8,497	7,396	8,052	9,480	10,160	11,421	11,723	9,210	8,538
Percent	4.5	4.3	3.6	3.5	3.6	3.6	3.4	2.8	2.4	2.9	3.1	3.9	4.6	3.8	3.5
Hong Kong	6,470	7,553	11,053	8,638	8,024	7,808	8,519	7,552	7,781	7,730	8,453	7,820	5,592	6,242	7,972
Percent	9.2	8.6	8.0	5.0	4.1	3.6	3.5	2.9	2.3	2.4	2.6	2.6	2.2	2.6	3.3
Canada	n.a.	n.a.	5,073	5,568	5,408	4,466	4,899	4,911	5,264	5,573	4,451	5,418	5,750	6,186	6,411
Percent			3.7	3.3	2.7	2.1	2.0	1.9	1.6	1.7	1.4	1.8	2.2	2.6	2.6
Others	8,837	11,499	14,725	17,724	23,270	27,168	30,694	31,373	37,151	42,268	41,999	40,320	32,835	33,734	38,237
Percent	12.6	13.2	10.6	10.4	11.8	12.6	12.4	12.0	11.0	12.9	13.0	13.6	12.7	13.9	15.6
Total	70,364	87,432	138,534	171,071	197,527	216,327	246,844	260,966	337,625	327,228	322,990	296,307	257,605	242,444	245,519

Source: FAO
Note: n.a. = not applicable

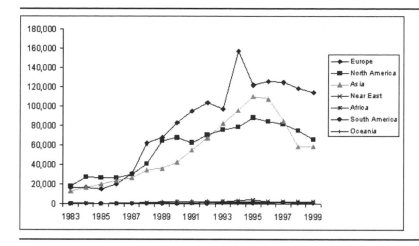

Fig. 4.4 Evolution of the value of imports by continent in US$1,000 (FAO). (Similar data for the Near East, Africa, South America, and Oceania result in overlapping graph lines.)

greater. Also, for marine species, prices fluctuate much more than for freshwater species, so the price differences may be even greater. The selling prices of a wholesaler who acclimatizes are about double those of a transshipper.

The price of an ornamental fish is considerably higher than the price of fish destined for human consumption. On average, there is a ratio of 1 to 100 between prices of fish for human consumption and those for aquariums.

Estimated Wholesale and Retail Sales

World wholesale trade in ornamental fish is worth about US$900 million (Bassleer 1994). Total world retail trade, including live fish, associated equipment, and accessories is estimated at US$7 billion, and the retail trade in live products at US$3 billion (Davenport 1996).

National statistics in most markets showed a significant increase in import values during most of the 1990s (FAO 2001). At the same time, surveys in countries such as France indicate that the turnover in the wholesaling and retailing sector was steadily increasing (TMO 1998). As a hobby, however, it competes with other hobbies for the money spent by consumers, and over the past few years, the growth of the Internet and home computer use has been seen as a challenge to the sector.

Trends in Supply and Demand

Sustainability of Supply

Supply problems are increasingly common. They are often due to climatic phenomena and seasonal patterns, including monsoons, fires, storms, hurricanes, and typhoons that occur in Asia, South America, and the Caribbean.

Supply problems can also be due to the degeneration of species in some breeding farms, mainly in Asia, where the customer criteria are not fully respected. In some cases, however, species degeneration has created new demand, and some breeders have specialized in genetic mutation. This can lead to dwarfism, rachitis, or physical deformation and even cause diseases. Sometimes the growing scarcity of a species is linked to the trade itself and to human intervention in the natural environment. There are also problems of overintensive collecting in the wild.

Demand—The Fish-Keeping Hobby Is Changing

The majority of domestic aquariums are filled with freshwater. A freshwater aquarium is easier to maintain and costs less, but today marine aquariums are becoming more and more popular. In the United States and in Europe, prices are becoming more accessible, and this sector is expanding rapidly. The proliferation of public aquariums is also an important trend and serves as an excellent promotional vehicle for the sector.

In the past few years, the emphasis has moved toward the technical and economic development of marine fish keeping. More and more enthusiasts want to create reef aquariums, notably to keep invertebrates, but above all they create "mesocosms": dynamic systems of coral reefs, where fish, corals, shellfish, molluscs, and plants are represented. The animals are living as if they were in a natural environment, with numerous fish species laying their eggs and even some of the corals reproducing. According to many professionals, the marine aquarium with reefs is the trend for the twenty-first century.

Fish keeping as a hobby is continuing to develop and will continue to gain a higher profile. It frequently develops as a hobby in a time of crisis and as a result of the pressures of the urban life. It is often cited as a haven from the aggressiveness of the modern world.

Problems in the Ornamental Fish Trade: A Solution?

Many of those involved in the ornamental fish trade lament a lack of professionalism and credibility in the sector. Lack of qualified staff in the distribution chain leads to quality problems. At the retail level, giving correct information and advice to the consumer becomes much more difficult. As a result, the consumer may start to see the fish as a simple disposable item. Fish bought in garden centers, for example, frequently have diseases that have not been treated and which then contaminate all the fish in an aquarium. Other recurrent problems are insufficient veterinary control, incompetence on the part of many retailers, a certain disregard for order specifications placed through some transshippers, frequent nonacclimatization of animals on arrival, undersized storage tanks with poor installations, unfiltered water, and generally poor management.

The majority of those involved in the sector are also concerned about the quality of the supply. Fish quality is irregular, above all in freshwater species. There are disease problems in entire stocks, which then contaminate the others, compounded by problems of poor quality fish arriving via transshippers without having been acclimatized. As the number of middlemen increases, the quality of the fish decreases. Some problems are due to bureaucracy and customs procedures that are slow and may incite some importers to import their stock illegally. In addition, the delays, risks, and the cost of transport, which sometimes increase the price of fish tenfold, cause significant problems.

Environmental problems are also serious. These include the methods used to collect marine species, inadequate breeding and rearing facilities, overexploitation of natural resources, and excessive demand for certain species, together with problems of coastal pollution and water quality. Operators lament too little scientific research into biological problems such as stress and mortality and into the reproduction cycle of certain species.

Most of these problems are exacerbated by poor organization of the sector. Better communication among all parties is required in order to improve transparency and to ensure a more general awareness of the risks for the sector, its problems, and ways to resolve them.

In recent years, however, the industry has started to take action and to confront the problems of organization, of image, and of sustainability of supply. In many countries, industry associations have become better organized and taken a proactive stand on many of the issues, including communication to the public and to the sector itself of the importance of sustainable collection, breeding, and handling practices. Likewise, the new growth in international trade and in demand for ornamental fish registered in 2000 and 2001 has created new optimism among industry operators, as has the fresh enthusiasm among amateur aquarium owners.

Notes

1. CITES lists are available at www.cites.org.
2. In trade statistics, import values are always higher than export values, as import values include freight costs to destination.
3. The European Community has adopted more restrictive measures. The list of species for which trade is totally prohibited (annex I of the convention) has been extended, and some species that are in annexes II and III benefit from tighter protection.
4. International trade statistics as well as those compiled by FAO do not, at the moment, differentiate between freshwater and marine species. Figures for the respective share of marine and freshwater species in international trade are therefore estimates only.

References

Bassleer, Gerald. 1994. The international trade in aquarium/ornamental fish. *Infofish International* No. 5, pp 15–18.

CITES 2002. www.cites.org

Davenport, Keith E. 1996. Characteristics of the current international trade in ornamental fish, with special reference to the European Union. *OIE Revue Scientifique et Technique*, Vol. 15 No. 2. pp 436–443.

EUROSTAT, Statistical Office of the European Commission CD-ROM, Eurostat's Intra-Extra Trade Statistics.

FAO 2001. Fishery Statistics, Commodities, 1999. Vol. 89. Rome, Italy.

Florida Agricultural Statistics Service, Orlando, Florida.

Hemdal, Jay. 1984 In defense of current marine fish prices. *Freshwater and Marine Aquarium*, Vol. 7, No. 9, pp 56–58.

PIJAC, Pet Industry Joint Advisory Council, www.petsforum.com

TMO 1998 "French people and fishkeeping," *Animal Distribution*, special issue. February.

5

The Consumption of Marine Ornamental Fish in the United States: A Description from U.S. Import Data

Cristina M. Balboa

Introduction

Coral reefs are among the most diverse biological systems on earth. According to some estimates, coral reefs supply hundreds of billions of dollars in goods and services to humans in the form of shoreline protection, aesthetic beauty, recreation and tourism, and as a source of food, pharma, and other revenues. The economic value of the ornamental fish trade represents a deceptively minimal use of coral reef goods. Its impact on coral reefs and the communities that depend on them far exceeds its economic reach. Global trade in marine ornamental fish is estimated to include more than 1,200 species from 45 countries (Bruckner 2001). Estimates value the global trade at US$28 to US$44 million annually (Wood 2001). The marine ornamental trade as an extractive industry can have both positive and negative effects on the resources and the communities it involves. Performed sustainably, this trade can offer livelihood for coastal communities while using the resource in a way to ensure it continues to be productive for generations. At the same time, collecting fish and entering them into trade creates several concerns. Destructive fishing techniques, such as the use of cyanide or poison, have been documented throughout Southeast Asia (Barber and Pratt 1997). Often the trade targets the juveniles in a species that, in turn, causes stunting among the number of eventual adults within a community. Many of the fish targeted for trade are herbivores that regulate the algae on the reefs. Without this balance the algae overgrow and kill corals and inhibit coral settlement. The handling of the fish post-harvest often results in high mortality of coral reef fish. Rare fish often fetch a higher price at market, creating perverse

incentives for conservation. These are just a few of the potential problems that trade in marine ornamental species could inflict upon the resource (USCRTF 2000).

U.S. Role in the Trade

Estimates show the United States consumes anywhere from 50% to 60% of all marine ornamental fish globally (USCRTF 2000). Eight to 11% of U.S. households keep marine fish as pets (Tomey 1997). Given the role of the United States as the top consumer nation of the products of this industry, the United States must also play a leadership role to address the degradation of coral reef ecosystems resulting from this trade (USCRTF 2000). As a response to this knowledge, Executive Order (#13089) for the Protection of Coral Reefs charged the U.S. Coral Reef Task Force (USCRTF) with analyzing and addressing the role of the United States in the international trade of coral reef species (USCRTF 2000). To date, however, there is no baseline against which to measure the U.S. role in this trade. The data are difficult to collect and analyze in a meaningful manner. The quality of the data can be high, and although it is not a large trade in volume compared with other fisheries, the quantity of data is overwhelming. This chapter provides an overview of the U.S. consumption of ornamental fish, especially marine ornamental fish, to start the process of creating this needed baseline.

The goal of this chapter is not to give an exhaustive baseline of the U.S. consumption of marine ornamental fish. Too few data are analyzed to create this robust description. It is designed to examine the process of creating a U.S.

baseline and to show the kind of data available and how they can be used. On an aggregate level, data are provided on the value of freshwater and marine ornamental fish from 1996 to 1999. On a detailed, species level, a typical month's ornamental fish imports into the United States are presented and analyzed.

Methods

Previous Studies

Since the 1970s there have been several studies to determine either the global or the U.S. consumption of ornamental fish. With each study building on the findings of previous ones, the data become more refined and relevant to those involved in the trade.

An analysis that examined the global trade of the combined value of all freshwater and marine ornamental fish was published by the Food and Agriculture Organization of the United Nations (Conroy 1975). Data for the study were provided by both importing and exporting countries and varied in quality and metrics. Data supplied by the United States indicated that most of the ornamental fish imported into the United States in 1972–1973 came from Latin America through ports in Florida (Conroy 1975). The sources and species composition of trade into the United States have changed greatly as will be demonstrated later.

The United Nations Conference on Trade and Development (UNCTAD) published a 1979 market study on tropical aquarium fish in the United States and European consumer countries (UNCTAD 1979). This study gave an overview of the participants and processes in the trade. Using import and export trade statistics, the UNCTAD provided value estimates of the trade for freshwater and marine fish in aggregate.

In 1985, research was published that analyzed the same data set used in this chapter to examine U.S. imports of ornamental fish in October 1971 (Ramsey 1985). The Ramsey study concentrated on freshwater fish. Because it indicated October to be an average month of imports, October was chosen for the analysis in this chapter in order to compare current data with historical data.

In 1993, another study was published, which followed on Ramsey's research and analyzed the same data sources for October 1992 (Chap-man et al. 1993). This study also examined the characteristics of aggregated imports of ornamental fish from 1982 to 1992, and like Ramsey's work, concentrated on freshwater ornamental fish.

The Marine Conservation Society published a comprehensive overview of the marine ornamental trade data on international trade, the collection and transport of fishes, and the conservation and management issues tied with the trade (Wood 2001). Trade data used were extracted from the statistical databases of importing and exporting countries. Years after the first call to report import and export statistics at a more detailed level, Wood found that reported data still aggregated the freshwater and marine trades, sometimes with other types of marine organisms. The Convention on International Trade in Endangered Species (CITES) requires data to be collected on certain marine ornamental organisms. Because no coral reef fish species are listed in this convention, however, no data are collected on coral reef fish.[1]

Currently the World Conservation Monitoring Centre, in collaboration with the Marine Aquarium Council, is collecting and collating data on an ongoing basis for its Global Marine Aquarium Database (GMAD). Data are provided on a voluntary basis directly from industry members. Because the industry tracks its shipments with species-level data, the data collected by GMAD are provided in significant detail. These data being collected on a global level, combined with the data in this chapter, have the potential to provide a fairly complete and detailed view of the trade.

Data Sources

This chapter uses two types of data: U.S. Fish and Wildlife Law Enforcement Management Information System (LEMIS) data and data from the Customs Declaration packets (Form 3-177) filed by importers of live animals and plants to U.S. Fish and Wildlife.

LEMIS aggregates imports by shipment, with each shipment identified by its control number. Each control number is linked to information on the value of the shipment, the wildlife description code describing the data, the quantity shipped and in what unit (kilogram, individual fish, etc.), country of origin, country of re-export, source (wild or cultured), the date received and the date cleared from customs, whether the shipment is alive or dead, and the port of entry into the United States. These data are electronically formatted and easily accessible. However, the data are aggregated, which diminishes their relevance. For example, tropical marine fish are categorized "tropical fish," aggregated and not distinguished from freshwater fish. There are categories in the database to determine if the imported fish are alive or dead, cultured or wild-harvested, but this information is aggregated, inaccurately portraying shipments with mixed live and dead, wild-harvest, and cultured organisms. Often shipments contain more than just fish. In these cases, marine mollusca, echinodermata, and coral are also aggregated into the data. These data are helpful only in understanding the economic value of the trade, not the quantity. Quantities listed in this database are listed by weight and number. Because no standard method exists of converting one value to the other, any information in this category is of no comparative value. The values included in this database include the cost of livestock, insurance, and freight (c.i.f. value). Understanding the limits of these data, we can make a general analysis of the quantity of imports, the value of imports, and the countries of origin on a daily to an annual basis.

Customs declaration data comes from Form 3-177, which all importers are required to file for all imported shipments of tropical organisms. In addition to Form 3-177, these declaration packets almost always contain the importer's actual invoice showing full detail of the shipment, including quantity, value, and country of origin for each individual species. Unfortunately, as useful as this level of detail is, these declaration data are not in an easily accessible form. As the shipments come into the United States, Customs commercial declaration forms must accompany them. Customs officers determine whether another agency needs to inspect the shipment. If the importer hasn't already called the Fish and Wildlife Service (USFWS) for clearance, the customs officer then calls USFWS in to clear the shipment. USFWS inspects the paperwork (Form 3-177) and determines whether or not to physically inspect the shipment. Approximately 25% of the shipments are physically inspected. Once the shipment is cleared, the USFWS officer codes the data on

Form 3-177 and sends the form to USFWS headquarters, where a limited amount of data are entered into the computer system and the paper forms are filed (Einsweiler, U.S. Fish and Wildlife Service, personal communication 2001). These records, which contain the complete, species-level information this chapter examines, are kept for 5 years.

Methods for Data Analysis

The Freedom of Information Act (FOIA) was used to request information from USFWS for use in this analysis.[2] First, the LEMIS data for tropical fish from 1996 to 2000 were requested. At the time of the data request, the last two months of 2000 data were not complete. Thus, this analysis concentrates on 1996–1999. Once the LEMIS data were obtained, declaration packets for all shipments received in October 2000 were requested. October 2000 was used as an exemplary month of data, in order to compare the data with the data collected in October 1992 and October 1972. This chapter will focus primarily on the October 2000 data. For detailed comparisons with other data collected, see other publications by the author. In each month of 1996–2000, the number of shipments imported ranged from 1,017 to 1,619. This chapter examines 1,185 shipments during October 2000, which represent approximately 8% of the total shipments in 2000. Each customs declaration packet contains from 3 to 30 pages of inventory. Once these data were requested via FOIA, an Access Database was used to enter the data, and scientific names were assigned to common names and clarified where necessary. Standard scientific name data sources were used (Froese and Pauly 2002). Regional labels were also assigned to the shipment data. The costs reflected in this set of data include some packing but do not include all freight, insurance, and taxes. This process, including entering all the data, took place over the course of several months.

Results

LEMIS Data

Between January 1996 and December 1999, 62,310 shipments of tropical organisms entered the United States. These shipments were valued at US$192.4 million, or an average of US$48 million per year. Each year, the value of imports declined between 5% and 12% (Figures 5.1, 5.2, and 5.3). This is consistent with the decline in prices of marine ornamentals over the past 10 years (Biffar 1997), in addition to the steady decline of shipments of fish to the United States [with a slight (2%) increase in 1999]. From the UNCTAD study in 1979, it is apparent that the industry doubled in just 5 years in the 1970s. U.S. imports of ornamental fish increased from US$9.5 million in 1973 to US$23.9 million in 1977. Comparing current data with historical data, we see a steady growth of the industry until 1996, when a downward trend began.

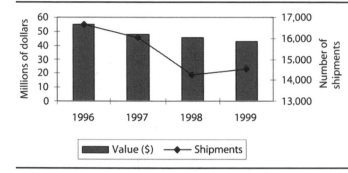

Fig. 5.1 U.S. imports of live tropical ornamentals based on U.S. import data from USFW. Dollars are US$

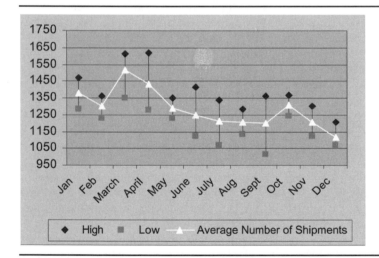

Fig. 5.2 Average number of shipments to the United States, by month, from 1996 to 1999.

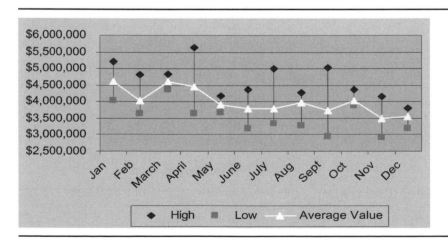

Fig. 5.3 Average value of U.S. imports of tropical ornamentals by month, from 1996 to 1999. Dollars are US$.

The 1993 study used the same LEMIS data from 1982 to 1992 (Chapman et al. 1993). From 1984 to 1987, imports were relatively stable, ranging from US$25.7 million to US$29.2 million. Imports from 1989 to 1992 increased substantially, with total annual value ranging from US$36.1 million to US$41.1 million (Figure 5.4).

Regional Breakdown of Imports Of the shipments to the United States, on average 50% came from Southeast Asia, 17% from South America, 10% from Central America and the Caribbean, 6% from Oceania, 5% from East Asia, with the rest of the continents less than 5%.

On average over the 4 years, 67% of the value of tropical fish imports into the United States came from Southeast Asia. South America supplied 13% of imports, and Central America and the Caribbean supplied 5% of imports, while the other regions all supplied less than 5%. Comparing percentage of shipments to percent-

age of value, Southeast Asia and Australia and the Pacific export higher-value shipments to the United States than other regions (Figure 5.5).

Top 10 Source Countries From 1996 to 1999, shipments were received from 147 countries, with only 66 countries having exports to the United States every year. The top 10 countries in number of shipments exported to the United States were the same in every year: Philippines, Indonesia, Thailand, Singapore, Colombia, Brazil, Hong Kong, Peru, Sri Lanka, and Trinidad and Tobago (Table 5.1). Although many of the top 10 countries swapped ranks with each other over the years, this is the average order from highest to lowest number of exports. The United States typically receives 68% of its annual imports from these top 10 countries.

On average for this period, the top 10 countries by value are Singapore, Thailand, Indonesia, Philippines, Hong Kong, Colombia, Brazil, Peru, Malaysia, and Sri Lanka (Table 5.2). Sri

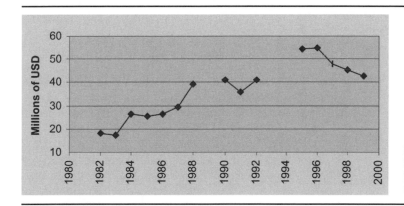

Fig. 5.4 Annual value of U.S. imports of live tropical organisms, 1982 to 1999. Dollars are US$.

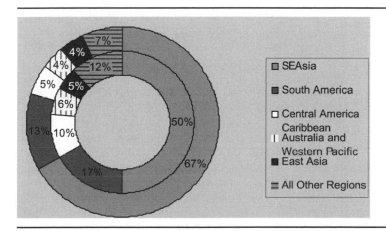

Fig. 5.5 Regional sources of U.S. imports of live tropical ornamentals by average percentage of shipments (inner circle) and average percentage of value (outer circle) from 1996 to 1999.

Table 5.1 Top 10 country sources of U.S. imports of live tropical ornamentals by average number of shipments 1996–1999

Country	Average annual number of shipments	Percent of average annual shipments
Philippines	2,366	15
Indonesia	2,040	13
Thailand	1,172	7
Singapore	1,035	7
Colombia	941	6
Brazil	795	5
Hong Kong	761	5
Peru	593	4
Sri Lanka	518	3
Trinidad and Tobago	477	3
Top 10 Total	15,645	68

Source: USFW LEMIS Data.

Table 5.2 Top 10 country sources of U.S. imports of live tropical ornamentals by average value 1996-1999

Country	Average annual value (US$)	Percent of average annual value
Singapore	8,605,651	18
Thailand	8,589,200	18
Indonesia	4,969,910	10
Philippines	4,450,147	9
Hong Kong	4,286,473	9
Colombia	2,178,867	5
Brazil	1,959,165	4
Peru	1,670,610	3
Malaysia	1,168,019	2
Sri Lanka	921,593	2
Top 10 total	38,799,634	81

Source: USFW forms 3-177 import data.

Lanka was replaced in the top 10 twice in the 4 years, once by Japan (1996) and once by Saudi Arabia (1998). These 10 countries supplied about 81% of the annual value of tropical organisms to the United States.

Ports of Entry   Imports of tropical organisms come through one of 38 official ports of entry in the United States. On average, 41% of all tropical organism shipments from 1996 to 1999 entered the country through Los Angeles, 21% through Miami, and 16% through New York City. The other ports each received no more than 5% of the shipments. In this same period, on average, Los Angeles was the port of entry for 54% of the value of tropical organisms imported into the United States, while Miami received 18% of the value and New York City received 15%. The rest of the cities received no more than 5% each of the value of shipments (Table 5.3).

Form 3-177 Data

Customs declaration data provide a much more detailed picture of imports of tropical organisms into the United States. This chapter is based on October 2000 import data from Forms 3-177 and their accompanying invoices. In October 2000, the United States received imports from 47 countries in 10 regions. LEMIS data indicate that live tropical organisms imported into the United States in October 2000 were valued at US$4.1 million. Analysis of the customs declaration data shows the actual value of live finfish, both freshwater and marine, as declared by the importer on a per fish basis. Shipping and costs associated with transport were not included in these values, since these costs are added to the invoices after the individual pricing of the fish. Including only finfish, which have values associated with them on a per fish basis, the total amount imported into the United States that month was just over US$3 million.

For this analysis, data were disaggregated into several categories. They are freshwater fish, marine fish, echinodermata, mollusca, corals, arthropods, and so on, with the focus primarily on freshwater fish and marine fish. The analysis examines imports into the United States by both quantity of fish imported (actual number of fish) and the value of fish imported (minus shipping costs).

Division of Data: Freshwater versus Marine
The United States imported more than 11 million ornamental fish in October 2000. Ninety-three percent of those were freshwater fish, while only 7% were marine fish. Freshwater fish were the overwhelming majority in number of fish imported (Figure 5.6). Looking at the value of fish imported, however, we see these percentages change. By value, freshwater fish were 69% (US$2.1 million) and marine fish

Table 5.3 Port of entry into the United States with average annual number of shipments and average annual value (US$) for live tropical ornamental imports

Port of entry	Average shipment	Percent of average shipments	Average value	Percent of average value
Agana, GU	56	0	39,302	0
Alcan, AK	0	0	0	0
Anchorage, AL	0	0	0	0
Atlanta, GA	144	1	127,379	0
Baltimore, MD	6	0	4,018	0
Blaine, WA	12	0	13,571	0
Boston, MA	23	0	17,121	0
Buffalo/Niagara Falls, NY	142	1	83,646	0
Calais, ME	2	0	6,188	0
Chicago, IL	**754**	**5**	**1,772,144**	**5**
Dallas/Fort Worth, TX	44	0	61,525	0
Detroit, MI	127	1	95,109	0
El Paso, TX	1	0	6,693	0
Golden, CO	1	0	0	0
Highgate Springs, VT	0	0	0	0
Honolulu, HI	585	4	565,481	1
Houlton, ME	0	0	0	0
Houston, TX	1	0	1,895	0
Laredo, TX	1	0	0	0
Los Angeles, CA	**6,357**	**41**	**20,815,382**	**54**
Lukeville, AZ	0	0	0	0
Miami, FL	**3,233**	**21**	**6,929,935**	**18**
Minneapolis/St. Paul, MN	53	0	41,695	0
New Orleans, LA	3	0	4,019	0
New York, NY	**2,449**	**16**	**5,939,577**	**15**
Newark, NJ	223	1	181,029	0
Nogales, AZ	1	0	2,429	0
Pembina, ND	0	0	0	0
Philadelphia, PA	1	0	1,696	0
Port Huron, MI	13	0	12,140	0
Portland, OR	9	0	4,367	0
San Diego/San Ysidro, CA	4	0	1,906	0
San Francisco, CA	801	5	1,231,483	3
Sault Sainte Marie, MI	2	0	0	0
Seattle, WA	98	1	83,211	0
Sweetgrass, MT	0	0	20	0
Tampa, FL	250	2	652,332	2
Washington-Dulles, VA	8	0	10,254	0
Other cities	1	0	0	8

Source: USFW LEMIS Data.

were 31% (US$957,792) of all live fish imports that month (Figure 5.7). This illustrates that although marine ornamental fish are not imported in the same numbers as freshwater ornamental fish, they are higher in value. In 1973, it was estimated that marine ornamental fish made up only 1% of all ornamental fish imports into the United States (Conroy 1975). By comparing data from October 1971, 1992, and 2000, we see that marine ornamental fish now represent a growing percentage of both the quantity and value of this trade.

Freshwater Fish To give context to the marine fish data, this section will present freshwater fish data for October 2000. There were 809 freshwater fish species from 81 families traded. Freshwater fish came from 36 countries and 10 regions. However, 84% of all freshwater fish (in number of fish traded) imported into the United States came from Southeast Asia. The region that supplied the next highest level was South America, with only 4%. Seventy-seven percent of the value of live freshwater fish imported came from Southeast Asia. East Asia is the next

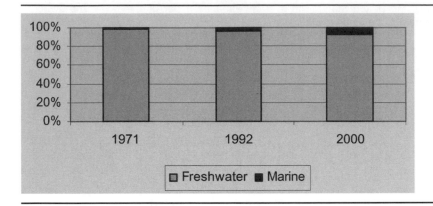

Fig. 5.6 U.S. freshwater and marine imports in percent of fish in October 1971, 1992, and 2000.

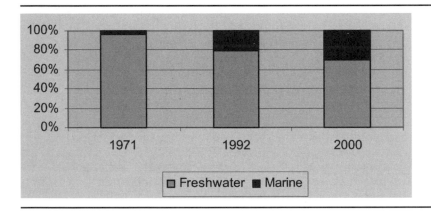

Fig. 5.7 U.S. freshwater and marine imports in percent of value in October 1971, 1992, and 2000.

highest exporter to the United States, comprising 11% of the imported value. The rest of the regions each contribute 3% or less of the total import value (Figure 5.8).

Ten families represent 93% of all freshwater fish imported (in order of quantity of fish, starting with the highest): Poeciliidae (livebearers), Cyprinidae (minnows or carps), Characidae (characins), Belontiidae (gouramies), Cichlidae (Cichlids), Gyrinocheilidae (algae eaters), Cobitidae (loaches), Loricariidae (armored catfishes), Ambassidae (glass perchlets, Asiatic glassfishes), and Callichthyidae (Callichthyid armored catfishes). Poeciliidae makes up 25% of all freshwater fish traded. The top 10 species traded make up 57% of all freshwater fish imported into the United States in October 2000: *Paracheirodon innesi* (neon tetra), *Betta splendens* (Siamese fighting fish), *Xiphophorus maculatus* (southern platyfish), *Poecilia reticulata* (guppy), *Tanichthys albonubes* (white cloud mountain minnow), *Carassius auratus* (goldfish), *Poecilia velifera* (sail-fin molly), *Danio*

rerio (zebra danio), *Gyrinocheilus aymonieri* (Chinese algae eater), and *Puntius tetrazona* (Sumatra barb). The top twenty species traded represent 70% of all freshwater fish imported into the United States.

Ten families represent 89% of the total value of freshwater fish imported (in order of value starting with most valuable): Poeciliidae (livebearers), Cyprinidae (minnows or carps), Cichlidae (cichlids), Belontiidae (gouramies), Characidae (characins), Cobitidae (loaches), Loricariidae (armored catfishes), Gyrinocheilidae (algae eaters), Callichthyidae (Callichthyid armored catfishes), and Ambassidae (glass perchlets, Asiatic glassfishes). The top species represent 51% of the total value of freshwater fish imported (in order of value starting with most valuable): *Betta splendens* (Siamese fighting fish), *Poecilia reticulata* (guppy), *Carassius auratus* (goldfish), *Xiphophorus maculates* (southern platyfish), *Cyprinus carpio carpio* (common carp), *Astronotus ocellatus* (oscar), *Paracheirodon innesi* (neon tetra), *Poecilia*

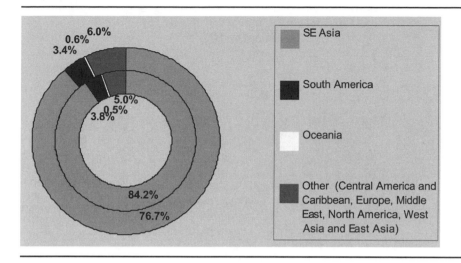

Fig. 5.8 Regional distribution of freshwater quantity (inner circle) and value (outer circle), October 2000.

velifera (sail-fin molly), *Botia macracanthus* (clown loach), and *Symphysodon discus* (red discus). The top 20 species traded contribute 67% of the total import value.

An examination of the average price of the freshwater fish traded makes it evident that a discrepancy exists between what is reported in the declaration packets and actual market prices for these fish. For example, on the basis of the average price of the freshwater fish traded, the most expensive fish imported was *Altolamprologus calvus* at US$270 per fish, an extremely high price for this Cichlid. However, the price is based on one invoice from Zambia containing four fish. This demonstrates a problem in working with such a small data set—8% of the total year's data. The next most expensive freshwater fish in this month was *Benthochromis* priced at US$65. Although expensive, this is plausible. The least expensive freshwater fish imported was a *Poecilis* at an average US$0.01 per fish.

Marine Fish There were 1,038 marine fish species from 95 families traded. Marine fish came from 38 countries in 9 regions. Sixty-nine percent of the value of live marine fish imported originated in Southeast Asia, while 19% originated in Australia and the Pacific. The Middle East supplied 4% of the marine fish, while all other regions each supplied 3% or less. Eighty-five percent of the quantity of marine fish imported came from Southeast Asia (Figure 5.9).

The top ten families in quantity of marine fish traded represent 85% of all marine fish

traded this month (in order of quantity traded): Pomacentridae (damselfish), Pomacanthidae (angelfish), Labridae (wrasses), Acanthuridae (surgeonfishes, tangs, unicornfishes), Gobiidae (gobies), Chaetodontidae (butterfly fishes), Callionymidae (dragonets), Ariidae (sea catfishes), Scorpaenidae (scorpionfishes or rockfishes), and Apogonidae (cardinalfishes). Pomacentridae alone, the most numerous of the families, makes up 53% of all marine fish imported this month. The next most traded family, Pomacanthidae, makes up only 6% of all marine fish traded. The top 10 species of marine fish imported represent 36% of all marine fish traded: *Dascyllus trimaculatus* (threespot dascyllus), *Chrysiptera cyanea* (sapphire devil), *Dascyllus aruanus* (whitetail dascyllus), *Chrysiptera parasema* (goldtail demoiselle), *Chromis viridis* (blue-green damselfish), *Dascyllus melanurus* (blacktail humbug), *Amphiprion ocellaris* (clown anemonefish), *Chrysiptera hemicyanea* (azure demoiselle), *Daschyllus albisella* (Hawaiian dascyllus), and *Arius seemanni* (tete sea catfish). The top 20 species made up 50% of all marine fish traded. Fourteen of the top 20 species traded were from the family Pomacentridae.

The 10 most valuable families in imports made up 80% of all marine fish import value (in order of value, starting with most valuable): Pomacanthidae (angelfishes), Pomacentridae (damselfishes), Acanthuridae (surgeonfishes, tangs, and unicornfishes), Labridae (wrasses), Chaetodontidae (butterfly fishes), Balistidae (triggerfishes), Gobiidae (gobies), Scor-

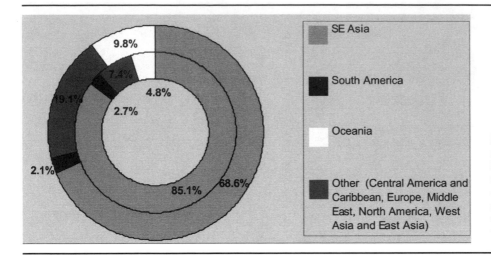

Fig. 5.9 U.S. regional distribution of marine quantity (inner circle) and value (outer circle), October 2000.

paenidae (scorpionfishes or rockfishes), Serranidae (sea basses, groupers, and fairy basslets), and Apogonidae (cardinalfishes). The ten most valuable species of marinefish imported contributed 28% of the total value of marinefish: *Paracanthurus hepatus* (palette surgeonfish), *Centropyge loriculus* (flaming angelfish), *Pomacanthus imperator* (emperor angelfish), *Chrysiptera cyanea* (sapphire devil), *Balistoides conspicillum* (clown triggerfish), *Pterapogon kauderni* (Banggai cardinalfish), *Zebrasoma xanthurum* (purple tang), *Pterois volitans* (red lionfish), *Pomacanthus xanthometopon* (yellow-face angelfish), and *Amphiprion percula* (orange clownfish). The 20 most valuable marine fish species make up 41% of all marine fish value.

The most expensive marine fish traded in October 2000, on the basis of average price, is the *Carcharhinus melanopterus* or black tip reef shark. Its price was US$350, based on one fish imported. The least expensive marine fish imported was the *Paragobiodon echinocephalus* (redhead goby) at US$0.07 per fish. Although the average price of marine fish comes from a larger range of prices than the freshwater fish traded, the most commonly traded family (Pomacentridae) of marine fish also includes 5 of the 10 lowest-priced fish. The second most traded fish (Pomacanthidae) has two species among the 10 most expensive marine fish. Most fishes were priced on average below US$2.00 (Figure 5.10).

Recommendations

Merely a glimpse has been given here of the marine ornamental imports into the United

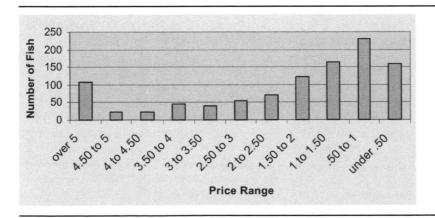

Fig. 5.10 Distribution of marine fish imported to the United States over average price ranges (US$), October 2000.

States. It represents about 8% of all the imports for the year 2000. It is not a random sample of data, so extrapolating October 2000's data to the rest of the year is not advised. This one month's worth of data took an extraordinary amount of time to collect and sort but is illustrative of the kind of data that can be collected on this trade.

In order for these data to be more useful and more easily accessible, they need to be collected at a more detailed level than is currently available. Dividing the category into marine fish, freshwater fish, and other marine organisms would go a long way toward making the LEMIS data more relevant. This would allow shipments to be aggregated while disaggregating the data for collection.

The USFWS should continue to investigate the possibility of electronic data collection and reporting. One deterrent to creating an electronic data collection system is the shear quantity of species imported. With almost 2000 species traded, one can imagine that scientific and common names would be difficult to consistently report. This study, however, has created a database that cross-references common and scientific names of fishes. USFWS is encouraged to use this database template as a basis for creating its own voluntary electronic reporting template. Although not every exporter will have access to a computer to file these electronic reports, it is estimated that at least half of the Forms 3-177 collected were accompanied by computer-generated invoices. With the exporters of at least half the shipments capable of electronic reporting, the reduction in data collection time could be significant.

Policy Implications of These Data

Aside from the above recommendations, these data could have direct impacts on many areas of the ornamental fish trade. Aquaculture is often referred to as a potential relief to the wild harvest of these species. Until it is known what is traded and in what quantities, the aquaculture industry cannot know what effect it has on the health of the wild resources. Currently, only about 1% of commercially traded marine ornamentals are captive bred (Bruckner 2001). A recent survey of hobbyists, however, shows that the majority of marine ornamental hobbyists engage in some sort of culture of marine organisms and/or buy cultured organisms whenever

possible (Moe 2001). One hundred sixty-eight species have been successfully cultured, and another 102 species are currently under research (Moe 2001). The culture of marine organisms is a growing industry. Complete data on the trade could help shape the research agenda for aquaculture by determining what species are consumed and in what quantity.

One of the concerns in the marine ornamental industry is that it is a supply-driven industry. Suppliers often end up harvesting what is available, regardless of the demand for the product. With constant and up-to-date access to these data, the industry could review and predict trends in the consumption of marine ornamental fish to help them better service the demand. Consumers could also see trends in consumption and link the fishes' area of origin with biodiversity and sustainability reports. With this information, consumers would be better poised to make educated decisions on the fishes they demand and keep.

On the supply side, such information could make the global trade transparent to otherwise isolated communities. These data, coupled with a knowledge of transportation costs and market trends, could help fishing communities negotiate fair prices for their product across the board. The data could also give the ornamental industry the information necessary to set the prices for their product and anticipate price fluctuations as a result of market floods, and so on.

Conservation groups could use these data to direct their efforts. By reviewing trade trends in species and product origin, they could better determine where their efforts would have the most influence. Aware of emerging supply areas, these groups could perform the necessary baseline assessments before the industry grows full-blown in those areas.

Policy makers could use this information to direct aid policy. By examining the species and product origin data, they could link this information with other assessments to make informed decisions on the protection of species. They could also direct aid projects to create alternative livelihoods in areas where the demand for ornamental fish outweighs the supply.

Conclusions

The marine ornamental trade size is disproportionate to its effects on coral reef resources. As the leading consumer of marine ornamental

ROSEWARNE
LEARNING CENTRE

fish, the United States has a great opportunity to lead the trade toward sustainability. Data to describe the trends of this trade are available but difficult to access and analyze. By collecting data more systematically and analyzing it on a regular basis, USFWS could contribute to the sustainability of the trade in a unique and important way. With the proper monitoring of these data, USFWS could contribute to USCRTF's goal of ensuring that "the international trade in coral reef species for use in U.S. aquariums does not threaten the sustainability of coral reef species and ecosystems" (USCRTF 2000).

Notes

1. The Convention on International Trade in Endangered Species (CITES) was created in 1973 to ensure that the trade in wild animals and plants does not threaten their survival. At the heart of CITES are three appendixes. Appendix I lists species threatened by extinction. These species may be traded only in special circumstances. Appendix II species are all species that may become threatened with extinction unless trade is subject to strict regulation in order to avoid utilization incompatible with their survival. Trade in these species is allowed as long as it is not detrimental to the survival of the species. Appendix III lists species that are protected in at least one country, which has asked other CITES parties for assistance in controlling the trade. Currently, no ornamental fish are listed on any CITES appendix. For more information, see www.cites.org.

2. The U.S. Freedom of Information Act (FOIA), 5 U.S.C. 552, was created to provide the right of access to federal agency records, except for certain records protected by special exclusions such as national security. It is based on the principle of openness in government. By specifically citing the FOIA in a written letter, which describes the records sought and the willingness to pay fees, individuals or public or private organizations can request and receive information from federal agencies.

References

Barber, Charles V., and Vaughan Pratt. 1997. Sullied seas: strategies for combating cyanide fishing in southeast Asia and Beyond. Washington DC, World Resources Institute and International Marinelife Alliance.

Biffar, M. 1997. The worldwide trade in ornamental fish: current status, trends and problems. *Bulletin of European Association of Fish Pathologists* 17(6): 201–204.

Bruckner, Andrew W. 2001. Tracking the trade in ornamental coral reef organisms: The importance of CITES and its limitations. *Aquarium Sciences and Conservation* 3(1–3): 79–94.

Chapman, Frank, Sharon Fitz-Coy, Eric Thunberg, Jeffrey T. Rodrick, Charles M. Adams, and Michel Andre. 1993. An analysis of the United States of America international trade in ornamental fish. Washington DC, United States Department of Agriculture: 55.

Conroy, D. A. 1975. An evaluation of the present state of world trade in ornamental fish. Rome, Food and Agriculture Organization of the United Nations: 126.

Froese, Rainer, and Daniel Pauly, Editors. 2002. FishBase. World Wide Web electronic publication. www.fishbase.org, 04 June 2002.

Moe, Martin A. 2001. Culture of marine ornamentals: For love, for money and for science. Presented to the Marine Ornamentals 2001 Conference, Orlando, Florida.

Ramsey, J. 1985. Sampling aquarium fishes imported by the United States. *J Alabama Academy of Science* 56(4): 220–245.

Tomey, W. A. 1997. Review of developments in the world ornamental fish trade: update, trends and future prospects. Sustainable Aquaculture: Proceedings of INFOFISH-AQUATECH 96, International Conference on Aquaculture, Kuala Lumpur, Malaysia, INFOFISH.

United States Coral Reef Task Force (USCRTF). 2000. International trade in coral and coral reef species: the role of the United States. Report of the Trade Subgroup of the International Working Group to the USCRTF: Washington DC VI. Annex I list of all freshwater ornamental fish imported to the United States in October 2000.

UNCTAD. 1979. International trade in tropical aquarium fish. Geneva, International Trade Centre, UNCTAD.

Wood, Elizabeth 2001. Global advances in conservation and management of marine ornamental resources. *Aquarium Sciences and Conservation* 3(1–3): 65–77.

6

The U.S. Wholesale Marine Ornamental Market: Trade, Landings, and Market Opinions

Sherry L. Larkin

Introduction

The United States plays a pivotal role in the market for live aquatic species that are used for ornamental purposes. Historically this market has been dominated by freshwater fish species. Marine species (fish and invertebrates) are becoming more popular as "mini-reef" systems become more technologically and economically feasible (Larkin and Degner 2001). In addition, an increase in the number of airline flights to remote areas has increased access to a larger number of marine species, which has fostered industry growth. In the United States, aquarium tanks are believed to be one of the most popular hobbies. According to *Pet Product News* more money is spent at the retail level on the aquarium hobby than any other pets; product sales were reported at nearly US$1.2 billion in 1998 (Hellwig 1999). Moreover, the United States is the primary market for aquarium species worldwide, and trade statistics show that the U.S. trade deficit in live ornamentals is increasing.

In order to characterize the U.S. market, a comprehensive research project was initiated in 1998. Sponsored by the Florida Sea Grant College Program, the project had three goals: (1) to summarize the Florida landings statistics that had been collected since 1990, (2) to obtain U.S. trade statistics and distinguish by location and product type (i.e., freshwater versus saltwater, fish versus invertebrates) to the extent possible, and (3) to determine the existing market channels of U.S. wholesalers and their opinions regarding the market. Although the results of each objective have been published in individual technical reports (Adams et al. 2001, Larkin et al. 2001a, Larkin et al. 2001b) the results

have yet to be synthesized. This analysis integrates the key information from each report and draws conclusions about the future of the wholesale market for live marine ornamentals in the United States.

This chapter begins with a summary of the background of the industry, including landings and trade statistics and the characteristics of the survey respondents. Observed differences among survey respondents by their location and size are then presented. The results of open-ended questions designed to ascertain opinions regarding the industry are also summarized. These questions touched on four main subjects, namely: advantages and disadvantages of Florida-caught species, explanations for trends in observed landings in Florida, factors that may be limiting the sales of Florida species, and anticipated short-term changes in the wholesale market. In addition, because of the conversational nature of the interview process, several other opinions and topics were discussed during the survey. This anecdotal evidence is also summarized.

Background

This section summarizes data on the market for live aquatic ornamental species in the United States and the market for marine species in particular. Historically, trade data have not distinguished between marine and freshwater species. The U.S. Fish and Wildlife Service (USFWS), which maintains records on all traded species (primarily for monitoring endangered species), included a variable that identifies the water environment for each shipment beginning in January 2001. Because saltwater and freshwater shipments are not distinguishable in the trade data through 2000, sample data were used to draw inferences regarding the size of the marine component. In addition to trade statistics, U.S. landings data (primarily from Florida) are included to show the number and quantities of species collected for commercial purposes from U.S. waters.

U.S. Trade Statistics

There are two established sources of trade data for live ornamental aquatic species in the United States, namely the USFWS and the Bureau of Census (i.e., Customs data). Customs data are easily traceable across countries because the reporting system uses harmonized tariff codes. Customs data do not, however, include information from all shipments since those with a declared value below a certain threshold (which increased from US$1,250 to US$2,000 in 1998) are exempt from reporting. Because aquatic species need to be shipped in relatively small quantities as a result of the need to include water, it is likely that a large portion of shipments would not appear in the Customs data set. An examination of manifests submitted to the USFWS (via the Law Enforcement Management Information System) for a 6-week period in October and November of 1999 revealed that 44% of all shipments were valued at less than US$1,250 (Adams et al. 2001). Given that the Customs data are incomplete and that the USFWS data also include quantities traded, this analysis only summarizes the USFWS data.

According to the USFWS, the traded value of live tropical fish and invertebrates (codes TROP and OLIN, respectively) totaled nearly US$850 million in the United States in 1998 (Adams et al. 2001). The majority of the value of the trade represented imports. Thus, the United States is a net importer, with a trade deficit of approximately US$478 million in 1998. These figures represent a 9% increase in total trade value compared with 1994 (when U.S. trade was valued at nearly US$773 million). The trade deficit increased 80% (from US$266 million) in the 5 years beginning in 1994, however, indicating that the value of imports into the United States has increased substantially more than the value of exports.

The USFWS data also include trade volume measured in the number of each species included in the shipments. Between 1994 and 1998 the trade volume increased from 2.6 billion to 2.8 billion (8%). However, the number of invertebrate species traded increased from 19 million to 35 million (84%) during the same 5-year period. Thus, the invertebrate category is substantially smaller but is experiencing tremendous growth. This category is also likely to include more marine species, since "mini-reef" marine aquariums are becoming increasingly popular. The growth in invertebrate landings could be reflecting the increase in marine invertebrate organisms that are needed in reef-based aquariums.

When comparing the aggregate 1998 trade value with the number of specimens traded, an average dockside price of approximately US$0.30 per specimen is obtained. Given that the observed retail value of live aquatic species can reach several hundred dollars, it is worthwhile noting that the tropical fish statistics (TROP) contain a relatively large volume of low-valued specimens. Although these trade statistics allow us to observe trends in the overall U.S. market, they do not reflect the market situation of any particular species. Market differences would be expected in regions that also supply specimens because of differences in stock abundance and, thus, relative prices.

When comparing the value of live aquatic ornamental species traded in the United States by region, geographic differences and trends emerge. From 1994 to 1998 the value of species imported and exported from Florida (the primary collecting region in the United States) increased 490% and 95% to approximately US$59 million and US$115 million, respectively. In 1997 imports and exports reached highs of US$135 million and US$129 million, respectively, such that the 1994 to 1998 percentages may be conservative figures. In 1998 Florida accounted for only 8% of the total value of U.S. imports but 63% of the value of domestic exports. No other state accounted for a significant share of exports. California received the majority of imports. These statistics suggest that Florida plays a significant role in the domestic aquarium market, especially for exports of live specimens.

U.S. Landings Statistics

The role of Florida as a major exporter, despite the relatively large trade deficit, is explained by the commercial marine life industry (which is primarily a collection industry located in the Florida Keys) and freshwater fish farms (primarily located near Tampa). Until 1996 the marine life collection industry also included live rock, which is used for the establishment of reef-based aquariums. The live rock industry is now managed with aquaculture lease sites in the Gulf of Mexico and around the Florida Keys. Collection of native live rock from nonleased lands is no longer allowed.

The state of Florida requires all collectors to have a "marine life endorsement" to collect commercial quantities (i.e., more than 20 fish or invertebrates per day and one gallon of plants). In order to buy from collectors and resell, a business needs a wholesale or retail dealer's license. Many collectors also have a wholesale dealer's license. Since 1990, the year that data collection and licensing began for this industry in Florida, the number of active collectors and dealers reached 229 and 114, respectively. In 1998, however, there were only 128 endorsed collectors with reported landings (i.e., "active" collectors) and 66 licensed wholesale dealers with reported purchases. Many active collectors participate in the industry on only a part-time basis (e.g., during the summer when water temperatures are higher) and sell to other collectors who have a dealer's license. Few individuals have only a dealer's license (Larkin et al. 2001a).

Since 1990, 318 marine species (181 fish and 137 invertebrates) have been collected and landed for commercial purposes in Florida. The total value of fish landings has ranged from US$766,900 in 1990 to over US$1.6 million in 1994. Annual fish landings reached a peak of 425,800 specimens, for an average value of US$3.79 per fish. The value of invertebrate landings (live rock, plants, anemones, sponges, mollusks, shrimps, sand dollars, etc.) also peaked in 1994 at US$2.66 million (Table 6.1). Landings of invertebrates cannot be aggregated because the measurement units differ (e.g., pounds of live rock, gallons of plants, number of sand dollars). From 1990 to 1998, annual landings of fish and invertebrates have been valued at more than US$2.7 million dockside.

Despite the relatively large number of species harvested in Florida, the value of landings is dominated by a few species and species groups. For example, the angelfish complex accounts for 54% of total fish landings in terms of value from 1990 through 1998. On average, 71,793 angelfish are landed per year and sold to licensed wholesale dealers in Florida for US$7.60 each (i.e., the dockside value of angelfish totals approximately US$544,000 per year). Collectively, the top 10 species groups (of the 67 total groups, which include all 181 species) account for 84%, on average, of the value of landings (Table 6.2).

Table 6.1 Florida commercial marine life landings, 1990–1998

	Fish			Invertebrate	
Year	Number	Value (US$)	Average price (US$)	Value (US$)	Total value (US$)
1990	245,401	766,868	3.12	635,950	1,401,818
1991	291,311	986,885	3.39	1,357,720	2,344,605
1992	393,497	971,115	2.47	2,061,135	3,032,250
1993	355,017	1,283,871	3.62	2,282,590	3,566,461
1994	425,781	1,612,597	3.79	2,660,887	4,273,484
1995	259,387	944,172	3.64	2,528,508	3,472,680
1996	205,832	832,603	4.05	1,773,081	2,605,684
1997	278,105	903,923	3.25	1,134,274	2,038,197
1998	201,212	759,363	3.77	1,136,385	1,895,748

Source: Larkin et al. (2001b, p. 17).

Table 6.2 Average annual Florida commercial marine life landings and value by species type and sorted by value, 1990–1998

Top species groups	Average annual landings[a]	Landings change (percent)	Average price (US$/unit)	Price change (percent)	Annual value (US$)
Fish					
Angelfish	71,793	−31.6	7.60	44.5	543,546
Hogfish	9,911	−13.1	7.55	13.6	75,189
Damselfish	26,408	−34.0	1.33	−10.5	35,152
Jawfish	12,901	−6.8	2.42	17.4	32,651
Wrasse	19,735	−42.4	1.64	13.5	32,113
Butterflyfish	11,029	−48.3	2.86	26.4	30,431
Seahorse	48,426	184.4	0.77	−29.2	26,515
Parrotfish	5,308	−39.5	4.87	97.9	25,905
Surgeonfish	7,317	18.3	3.09	3.9	22,351
Drum	9,230	−43.0	2.11	15.3	19,429
Invertebrates					
Live Rock	623,279	−63	1.38	91	837,491
Snail	373,587	791	0.40	−45	140,261
Anemone	275,812	−26	0.57	30	125,372
Crab	236,674	755	0.57	−62	101,539
Starfish	205,012	1,824	0.39	−89	81,078
Gorgonian	28,736	129	2.29	22	76,116
Sand Dollar	438,850	203	0.14	−33	60,332
Urchin	36,823	29	1.14	234	42,884
Sponge	17,534	1	2.40	80	41,063
Live Sand[b]	42,876	—	—	—	34,185

Source: Larkin et al. (2001b, p. 19 and p. 25).

[a] Landings are number of specimens for all but live rock and live sand, which are measured in pounds.

[b] Live sand collection did not begin until 1994.

Even though angelfish have continued to account for the majority of fish landings in Florida (in volume and value), landings of these species have fallen nearly 32% since 1990. Conversely, landings of seahorses, which average 48,426 annually, have increased more than 184% (annual landings have ranged from 5,969 to 110,948). The primary species harvested in Florida is the dwarf seahorse (Figure 6.1). The corresponding average prices for these species have responded accordingly; the average angelfish price increased 44%, while the average price of a seahorse fell 29%. These trends indicate that the market price has changed in response to changing supplies. Overall, 7 of the top 10 species groups experienced a reduction in landings (ranging from 7% to 48%) and a corresponding price increase (ranging from 13% to 98%) from 1990 to 1998.

Table 6.2 also contains statistics for the invertebrate species. Live rock (Figure 6.2) dominated the invertebrate species by accounting for approximately 49% of the total value of

Fig. 6.1 The dwarf seahorse (*Hippocampus zosterae*), pictured here with some encrustation, is the primary seahorse species landed in Florida; landings reached 98,779 in 1994. (Photo credit: Jeffrey Jeffords, http://divegallery.com) (See also color plates.)

Fig. 6.2 Live rock from the Florida Keys (left) and Gulf of Mexico (right). (Photo credit: Triton Marine, Inc., http://liverocks.com) (See also color plates.)

landings from 1990 to 1998, even though non-cultured (i.e., native rock) landings were prohibited in 1997 and 1998. As with the fish species, the top 10 invertebrate groups accounted for the vast majority of landings, more than 89% in terms of value. In general, invertebrates are low valued as indicated by the lower unit prices. Over the 1990 to 1998 period, dramatic changes have taken place in the quantities and types of invertebrates landed. For example, landings of starfish, snails, and crabs increased 1,824%, 791%, and 755%, respectively; the increased landings, in turn, resulted in lowering the average dockside prices by 45% to 89%. Gorgonians and urchins experienced both increased landings and increased prices, most likely because these species are prime candidates for the increasingly popular reef-based aquariums (Larkin et al. 2001a).

The only other state that has commercial landings of live marine ornamentals is Hawaii. The Department of Land and Natural Resources collects information (i.e., the number caught and sold, total value, and fishing time and location) on 30 "aquarium fish" species, which also includes invertebrates. Data on these species appear to be included in two categories, Reef Fish and Other Miscellaneous, for statistical reporting. A review of the 1999 data revealed that 95,060 surgeonfish/tangs were sold, at an average price of US$1.33 each. Sales of squirrelfish and parrotfish were nearly identical, at 37,639 and 32,993, respectively, at average prices of

US$3.13 and US$2.34 each. Wrasse sales totaled 6,574 specimens at US$3.45 each. Collectively, these four fish species groups accounted for US$344,534 in dockside revenues for commercial collectors in Hawaii in 1999. Landings of parrotfish and the surgeonfish/tangs experienced the largest changes since 1990, increasing 37% and 126%, respectively.

Landings statistics from Florida and Hawaii show an increase in the marine ornamental industry as a whole and a change in the market. A larger number of invertebrates, which are needed in reef tanks, are harvested in Florida. Hawaii has experienced an increase in the harvest of its fish species, which are generally considered to be more attractive than those from the Atlantic and Caribbean (Larkin and Degner 2001). Compared with traditional fish-only tanks, a reef tank contains more invertebrates and fewer fish. Thus, Hawaii may have a comparative advantage in the supply of colorful fish.

U.S. Wholesaler Survey

The U.S. market for live marine ornamentals at the wholesale level can be described and quantified using data from a telephone survey conducted in 1999. One objective of the study was to distinguish the opinions and characteristics of firms located in Florida. This is because the commercial collection of live marine species is subject to an increasing regulatory environment. Results of that survey supported the distinction of marine life wholesalers by region and also by annual sales (Larkin and Degner 2001).

Target respondents were identified from industry organizations (e.g., the American Marinelife Dealers Association), business directories (e.g., the *Pet Supplies Marketing Directory*), and the state of Florida (i.e., wholesaler–dealer license holders). There were 174 target respondents. All were wholesalers, but not necessarily exclusively, of live ornamental marine species. Nearly half were located in Florida (90 of the 174). Because of the relatively small size of this industry at the wholesale level, all firms were invited to participate in the study.

A total of 54 firms (34 in Florida) were declared "inactive," that is, their telephone had been disconnected or they did not buy or sell any marine species during the previous year. These firms were removed from the sample. Of the remaining 120 firms, 52 completed the survey (43%). Respondents were located throughout the continental United States (i.e., no completed surveys were received from firms in Hawaii or Alaska). Significant resources were devoted to finishing all surveys, including contacting different people within a firm to answer separate sections. In Florida, 25 firms completed the survey, which is 45% of all active wholesalers. In the other states, 27 firms finished the survey, which is 42% of all active wholesalers.

Survey Results

Respondent Characteristics

Firms located in Florida differ considerably from their counterparts in other states within the continental United States. On average, Florida firms are smaller (in the number of employees and holding space) and derive more income from the sale of marine species, especially invertebrates. Firms outside Florida keep larger inventories of fish and sell more dry goods. Table 6.3 presents the average characteristics of marine life wholesalers in Florida and those of firms located in the remaining states (exclusive of Hawaii and Alaska, i.e., the "other states").

Overall, respondents in this survey were

Table 6.3 Average characteristics and standard deviations (in parentheses) of U.S. wholesalers responding to the survey

	Florida	Other states
Experience in	16.8	19.2
years	(12.0)	(11.6)
Holding space in	22,140	78,723
total gallons	(48,803)	(191,427)
Employees in number	1.9	17.7
of FTEs	(2.1)	(25.9)
Value of marine fish	$223,444	$270,312
sold in 1998[a]	(419,555)	(265,580)
Estimated annual sales	$788,203	$1,044,987
of live species[a]	(1,773,706)	(1,380,464)
Share of total sales	11%	25%
from dry goods	(0.10)	(0.27)

[a]Averages are based on 21 Florida firms and 24 firms from other states that provided sales figures. One extremely large Florida firm is excluded in order to obtain a better representation.

well established. Florida respondents averaged 17 years of experience in the ornamental fish industry, and respondents located in other states averaged 19 years' experience. Firms in Florida have less holding space, with an average of 22,000 gallons versus 79,000 gallons for firms outside Florida. There is a high degree of variability in holding space among firms. Total holding space for firms in Florida ranged from zero gallons for collectors who sell to dealers immediately to 200,000 gallons for firms that maintain an inventory and ship worldwide. Florida firms also tend to have far fewer full-time employees, less than two, compared with the average of nearly 18 for their out-of-state counterparts. The majority of Florida wholesalers are owner-operated and perhaps use family members to a greater extent in the business (Larkin et al. 2001a).

The composition of sales also differs by location of the marine life wholesaler (Florida versus other states). The average value of marine fish sold per firm in 1998 is slightly lower for Florida wholesalers than for wholesalers located outside of Florida (i.e., US$223,444 versus US$270,312, respectively). Florida wholesalers allocate nearly twice as much of their inventory to invertebrate species as do wholesalers in the rest of the country. Invertebrates account for 50% of the value of inventories in Florida but only 27% of the value of inventories for out-of-state firms. On average, all marine species (fish, invertebrates, live rock, and live sand) make up 59% of the total value of live inventories of Florida firms but only 44% of inventories of firms located in other states. When considering the average percentage of each firm's inventory allocated to marine species, these shares increase to 90% for Florida firms and 64% for their out-of-state counterparts. On a share basis, Florida firms generated comparatively more revenue from invertebrates, while firms located outside Florida add considerable revenue through the sale of freshwater species and dry goods. Annual sales of all live aquarium species are approximated to be US$788,203 for firms located in Florida versus US $1,044,987 for firms in the rest of the nation. Fifty-two percent of wholesalers located outside Florida sell dry goods. For these firms, sales of dry goods account for 25% of their total sales. Only 22% of wholesalers located in Florida sell dry goods, which accounts for 11% of their total sales.

The sales statistics imply that Florida wholesalers (on average) earn more revenue from marine fish and invertebrates, while wholesalers in other states earn relatively more from freshwater species and dry goods. Given the average sales figures, this suggests that larger firms devote relatively more of their inventory to freshwater species. According to the standard deviations, however, a high degree of variation exists in the sales figures for Florida firms. The larger standard deviations for average Florida statistics serve as a reminder that these are generalizations across widely differing firms. For example, annual sales of marine fish among Florida wholesalers ranged from US$3,000 to US$3 million. These statistics suggest that it may be appropriate to differentiate behavior by firm size as well as by geographic location.

Supply by Location

Significant differences exist regarding supply sources of marine life of wholesalers in Florida versus wholesalers in other states. Florida firms as a group typically rely more on marine species, especially invertebrates, rather than freshwater species. As might be expected, Florida firms rely on domestic supplies more than imports, and they collect and or culture their own inventory far more than their out-of-

state counterparts. In fact, the Florida firms in the survey use domestic sources for fish and invertebrates for 67% of inventories by value compared with 42% for firms located elsewhere. The islands of the Caribbean Sea and Atlantic Ocean are the primary sources of imports for Florida firms, while islands in the Pacific, especially Indonesia, are the primary sources for firms in other states.

Sixty-five percent (by value) of marine life inventories held by Florida firms were obtained from self-collection or culture. An additional 19% come from other local collectors. The remaining 16% of domestic supply comes from other wholesalers, nearly two thirds of which are located in Florida. Thus, more than 90% of the value of domestic supplies held by Florida wholesalers originates from marine resource stocks in Florida. On the other hand, the primary source of supply for wholesalers in other states is other wholesalers. Thus, there is an additional intermediate market segment for the majority of retailers buying from wholesalers outside Florida. Sixty-four percent of inventories of firms located in other states come from other wholesalers, while 31% are provided by other collectors. Because only 24% of these other wholesalers are based in Florida, less than half of the domestically supplied marine life sold by wholesalers located in other states originates in Florida. Inventory supply sources based on total inventory value are presented in Table 6.4.

Distribution by Wholesaler Location

According to the survey respondents, marine life wholesalers in the United States tend to focus their sales on the domestic market (Figure 6.3). This confirms the observation that the United States is the primary final outlet for aquarium species. Florida wholesalers export 28% of their product (the majority to Europe and Canada), whereas other wholesalers export only 3%. The Florida and non-Florida wholesalers also differ in their reliance on alternative domestic market outlets. Florida wholesalers sell primarily to other wholesalers or exporters (57%) and depend, to a lesser extent, on retail pet shops (22%). Wholesalers based outside Florida sell primarily directly to retail pet shops (80%) but also supply other wholesalers (12%). Thus, Florida wholesalers tend to supply other

Table 6.4 Average supply sources by wholesaler location

	Florida	Other states
	%	%
Total supply		
Freshwater	41	56
Marine	59	44
Marine supply		
Live rock and sand	6	14
Invertebrates	50	27
Fish	44	59
Invertebrates and fish supply		
Foreign	33	58
Domestic	67	42
Domestic supply		
Self collect or culture	65	5
Other collectors	19	31
Wholesalers	16	64
Wholesaler location		
Florida	62	24
West Coast	12	42
Midwest	0	9
East Coast	26	25

suppliers, while wholesalers from other states primarily supply the retail market. In other words, Florida marine life wholesalers are located closer to the collectors in the marketing chain.

Florida wholesalers sell a relatively large share of marine life to other Florida firms (27%), although most of their product is sent to the Midwest (46%) (Figure 6.4). By comparison, wholesalers in the rest of the country send most of their product to the East Coast (42%) and Midwest (36%). The West Coast receives relatively little marine life in general (14% and 16% for Florida and non-Florida firms, respectively), perhaps because Hawaiian wholesalers are poorly represented in the sample.

Distribution by Wholesaler Size

To determine variations in supply and distribution patterns on the basis of firm size, the sample was divided into "large" firms, with 1998 revenue above the median value of US$ 290,000, and "small" firms, with 1998 revenue below US$290,000. As determined earlier, smaller wholesalers tend to differ from larger wholesalers by focusing on marine species rather than freshwater species. Marine species made up 77% of small firms' inventories (by value) compared with 47% for large firms. Firm size does not seem to affect the composition of marine inventories (fish made up a slightly smaller fraction of small firms' inventories, 48% versus 53%), nor does it affect the likelihood of using foreign sources (47% of small firms' inventories are imported versus 48% for large firms). Self-collecting or culturing is the predominant source of domestic marine supply regardless of firm size. However, small firms tend to rely more on other collectors. Large firms, on the other hand, primarily use other

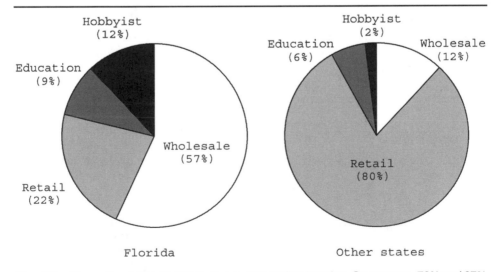

Fig. 6.3 Domestic wholesale distribution outlets by firm location. On average, 72% and 97% of sales by firms in Florida and the other states, respectively, remain in the United States.

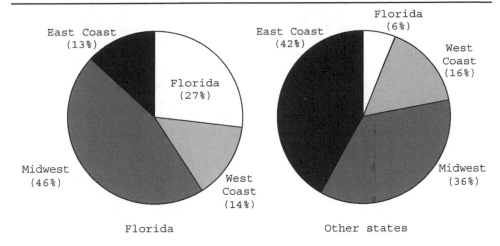

Fig. 6.4 Domestic wholesale distribution areas by firm location. On average, 72% and 97% of sales by firms in Florida and the other states, respectively, remain in the United States.

wholesalers. The shares of these supply sources by the size of wholesalers are presented in Table 6.5.

Firm size does, however, affect the pattern of distribution of marine life between foreign and domestic outlets. For example, small firms (those with 1998 revenues less than $290,000) export only 1% of their inventory on average. Conversely, exports account for 17% of the sales of large firms. The distribution of domestic sales between outlets is, however, relatively robust for firm size; the value of sales to educational facilities (research centers and aquari-

Table 6.5 Average supply sources by wholesaler size

	Large	Small
	%	%
Total supply		
Freshwater	47	23
Marine	53	77
Marine supply		
Live rock and sand	10	12
Invertebrates	37	40
Fish	53	48
Invertebrates and fish supply		
Foreign	48	47
Domestic	52	53
Domestic supply		
Self-collect or culture	42	38
Other collectors	20	36
Wholesalers	33	26

ums) or direct to consumers (e.g., via the Internet) accounted for just 11% and 13%, respectively, for small and large firms, on average. Thirty-five percent of small firms' sales are to wholesalers compared with 30% for large firms. Similarly, both groups sell slightly more than half their marine life to retail pet stores (54% for small firms and 57% for large firms).

Wholesalers' Perceptions of Florida's Marine Life Industry

Respondents were asked a series of open-ended opinion questions. The questions considered here concern the advantages and disadvantages of Florida's marine life products versus imports, factors that limit the sale of Florida species, and reasons for the shift from fish to invertebrates. The different opinions emphasized by each group illustrate the structural differences between wholesalers based in Florida and wholesalers located in other states (with the exception of Hawaii and Alaska). A summary of responses to additional questions is available (Larkin et al. 2001a).

When asked what the advantages are of Florida species compared with identical imports, the most common response was that Florida products are generally of higher quality. This response accounted for 42% and 50% of all responses by firms in Florida and other states, respectively. Quality was defined by respondents in terms of survival rates and coloration (e.g., high survival and bright colors would characterize a high-quality specimen). Survival rates depend not only on the initial health of fish but also the number packed per box and travel time. In this respect, survivability is a measure of the skill of the collector as well as the durability of the fish. The findings regarding quality also help explain why good collector relations was the advantage stated second most frequently (38% of responses) by wholesalers from other states. By developing good relations with Florida collectors, these firms feel they not only receive more information about specific species but also better care in packaging.

For Florida firms the advantage stated second most often (17% of responses) is the lower cost at which the product can be obtained. This is a particularly interesting response in that, for wholesalers from other states, the most frequently cited disadvantage is the generally higher price for Florida products (35% of responses). This highlights the role of Florida wholesale firms that also collect and thereby do not observe a purchase price. Conversely wholesalers in other states sell primarily to retailers and, therefore, are most concerned about prices paid to other wholesalers (i.e., their profit margin). Often, many species available from Florida's waters are also available elsewhere in the Caribbean, where they can be collected with lower labor costs. These imports can often be priced lower than Florida products, even with transportation costs.

Opinions were also divergent between Florida wholesalers and those located in other states regarding the disadvantages of Florida products. For Florida firms, the biggest disadvantage is that the supply volume is limited (32% of responses). Several respondents complained of being unable to deliver the total quantities requested of them and having to buy supplies from other dealers. Related disadvantages are that supplies are seasonal, and collectors can be unprofessional or unscrupulous (cited by 10% to 13% of respondents, respectively). Because collectors often are part-time workers, supplies can decline when biological or weather conditions make procurement more difficult throughout the year. The skill of the collector can clearly determine the quality (and profitability) of the product.

When wholesalers were asked which factors most limit the sale of Florida species, the answers again varied by region. Whereas there was some overlap with the disadvantages mentioned above, Florida firms overwhelmingly selected factors related to governmental policies. The most frequently mentioned factor (accounting for 20% of responses) was "red tape" or inappropriate and or ineffective regulations. Complaints included inappropriate bag and size limits, unenforceable regulations, overzealous and uninformed law enforcement officers, and too much paperwork. The second most frequently mentioned concern (accounting for 15% of responses by Florida wholesalers), and for many the most important limiting factor, is the availability of cheaper Caribbean imports and that the quantity of those imports is unrestricted. Lower labor costs abroad are thought to underlie this finding, although a lack of appropriate resource management could also be a factor. In total, 10 factors were mentioned.

There are also perceived problems with unlicensed collectors and dealers in Florida. Although this was not considered of primary importance, it was the third most frequently mentioned factor by Florida firms and the fourth most cited factor by firms located in other states (accounting for 13% and 11%, respectively, of all responses). Given that a large number of Florida wholesalers buys from part-time collectors, and the importance that wholesale firms in other states assign to good collector relations, there is considerable concern over the damage that unprofessional or unscrupulous collectors can do to the industry. Examples of misbehavior include shipments with low survival rates, incomplete or incorrect orders, and basic deception. Also, some collectors may not report all their landings and may have high mortality rates. The other limiting factors mentioned by Florida firms included poor environmental quality (primarily water quality and visibility), poor weather, lack of species diversity, and increased competition from other wholesalers.

Wholesalers not located in Florida, in contrast, seemed to focus on factors that limit the profitability of sales to retailers (which are similar to the disadvantages of Florida-caught products summarized earlier). The most frequently cited factor limiting the sales of Florida products (accounting for 38% of responses by wholesalers in other states) was a lack of adequate species selection or diversity, followed by the higher prices of Florida species relative to imports (24%). Third was the poor availability of Florida products, both seasonally and in terms of overall quantity.

The third question asked for explanations concerning the decline in fish landings and the increase in landings of invertebrates in Florida that have occurred since 1994. Once again, explanations for these changes varied according to the location of the wholesaler. For Florida firms, two of the most important reasons for the decline in fish landings are based on supply considerations: a reduction in water quality and a decline in the number of part-time divers (29% and 17%, respectively). Two other reasons mentioned frequently are a decline in the demand for fish and increased competition from imports (17% each). Firms located outside Florida believe that demand-side factors explain the decline in fish landings. For example, nearly two thirds of responses related to either a decline in the demand for fish (38%) or poor economic conditions, arising from a decline in demand for fish, for small firms (24%) as reasons for reduced landings of fish. Other explanations included an increase in competition from imports and an increase in regulations governing the industry in Florida (each response accounted for 14% of the total number of responses).

Supply and demand considerations continue to be seen in explanations for increased landings of invertebrates. For Florida firms, the ease of collecting invertebrates with very little gear was one of the most frequent answers (30% of all responses). For nearly half of the out-of-state firms (48%), the explanation lies in the increasing popularity of reef-based tanks. Respondents across the United States agreed that improvements in the knowledge of invertebrate care and corresponding technological advances are one of the most important reasons for increased landings of invertebrates (receiving 30% of responses by firms in Florida and 22% of responses by firms in other states). Many respondents noted that invertebrates are necessary in a healthy reef tank, that they are more tolerant of variable tank conditions, and that the variety of living species makes these tanks more interesting and thus more appealing to hobbyists.

Anecdotal Evidence

Because the surveys were conducted over the telephone and several questions were open-ended, many respondents offered additional comments on topics not explicitly included in the survey. A complete list of these topics is available (Larkin et al. 2001a). In this section, those topics that relate to demand and supply conditions in the market and findings that are nonquantifiable are presented.

First, the market for marine ornamentals is extremely competitive in the sense that many buyers and sellers each offer a relatively homogeneous product, and daily supply conditions can immediately change the market price. In Florida, the supply response occurs predominantly on a weekly basis when collectors fax (or post on the Internet) supply lists and current prices to prospective customers. Price discounts are routine for species that are in great supply, and conversely, significant premiums are established for rare species. Thus, supply and demand market forces operate at the species level. In terms of marketing channels, the Internet is increasingly being used in this industry. Unlike many other products, part of the appeal of the Internet is that many commercial sites also offer substantial amounts of information pertaining to the proper care for each species and for tank systems overall. Many Internet sites also include diagnostic information for common fish and invertebrate ailments.

Second, cultured marine species are perceived as a threat to collectors. This is an interesting result, given that many collectors specialize in the harvest of certain species and will trade with other specialists in order to increase the diversity of their holdings. They believe this diversity is necessary in order to compete with larger retail chains. If diversity is important to buyers, then culturing will not be a significant threat, given the costs and current status of the industry (i.e., only a dozen of approximately 1,000 species are cultured commercially).

Third, overharvesting of fish species in Florida is considered unlikely by collectors because of the difficulty associated with capture (versus invertebrate species that can be collected on foot in shallow areas). Although it is true that preventing new entrants can provide some measure of stock protection, any particular species would not be protected if prices were to increase sufficiently. This may be a concern for certain fish species where retail prices can be more than 10 times that reported at the dockside level. Although burdensome for resource managers, the most effective measures may be trip limits on specific species. Such an approach has been successful in Florida. One of the key factors that affects the success of species-specific regulations is the knowledge level of the enforcement officers, which would be more difficult to maintain as the number of such species increases. Thus, species-specific landings quotas may not prove to be effective long-run management tools.

Finally, many collectors use global positioning systems (GPS) to document productive areas and return for continued supply. These collectors are behaving as culturists by inventorying their supply in the ocean. This approach allows them to harvest only when the product is needed and when it has reached its optimal size (to achieve the highest price). This production behavior suggests that a management plan based on the allocation of private lease sites could be workable and more efficient, especially given the relatively low number of active collectors in Florida. Such a system would be similar to that currently used in Australia. Other factors suggest that a private rights-based management system could be the most effective and efficient long-run management tool for the commercial collection industry in Florida. One factor is that many members of the Florida collection industry have been active for decades. These industry pioneers have initiated the implementation of virtually all of the regulations that exist today. Their demonstrated commitment to the sustainability of the resource stocks and the industry provides the appropriate foundation on which a successful private rights system can be established. In addition, such a system could allow enforcement to move onshore and reduce enforcement costs and effectiveness.

Discussion

The U.S. market for live marine ornamental species destined for public and private aquariums is part of the larger aquarium trade that includes freshwater species. Florida is the largest domestic supplier of marine and freshwater species from wild-caught collection and culture facilities, respectively. Although the freshwater component of the industry is substantially larger, the marine component is capturing market share. The establishment of reef-based systems is growing in particular. The increased popularity of such systems, which rely predominantly on invertebrate species, is attributed to recent technological advances (e.g., proper lighting systems) and improved knowledge about the care of reef-based organisms. This knowledge is increasingly being transferred via the Internet, which has become an important tool in the market.

Increasing popularity of reef-based systems in the mid-1990s also corresponds with increased landings of collected species in Florida and Hawaii. In Florida, landings of invertebrate species have increased, while landings of fish have declined. Fish landings have, however, remained an important component of revenues, as prices have increased in response to lower landings. In Hawaii, which harvests relatively few species, landings have increased for fish. This is not entirely unexpected, as reef-based systems contain fewer fish but these fish may tend to be of the more colorful varieties, such as those collected from Pacific waters.

Trade patterns suggest that the United States, which is likely the world's largest market, will continue to increase its demand for live aquarium species. This trend is evidenced by the increasing trade deficit. Despite the overall deficit, the state of Florida exports specimens to Canada and Europe. Florida also accounts for the majority of exports from the United States. A 1999 survey found that firms located in Florida tend to specialize in marine species and are smaller than wholesale firms located in other states (with the potential exception of Hawaii and Alaska, which were not represented).

Florida wholesalers will likely be unaffected by the shift in market demand away from marine fish. This is because hobbyists demand a variety of species that can be offered only by natural systems. In addition, the initiation of the Marine Aquarium Council (MAC) certification system (which verifies sustainability of the resource and use of harvest, handling, and transport techniques to minimize losses) could have the most benefit for collection firms in Florida. This is because a statewide management plan already exists for the industry and because many wholesalers already specialize in providing high-quality products and service. Whether or not such a system will be adopted will depend on the costs of becoming certified and maintaining certification and the demand for certified specimens relative to noncertified specimens. In addition, there will likely be differences in the adoption rates of the MAC certification system by firm size since small firms would incur relatively larger costs to become certified. Ironically, according to the survey respondents, these small firms may be precisely those that would be most interested in the program because many provide quality and service to create a niche market.

The MAC certification program, in part, ensures customers (hobbyists) that the collection industry is sustainable and the resource stocks will not be adversely affected in the long run. This program is not without cost to participating firms, which must pay an independent certification authority and the MAC organization. Currently, many firms offer survival guarantees on their specimens. This guarantee is an alternative to obtaining formal certification because the seller is essentially saying that the specimen is healthy and will remain healthy after the buyer arrives home. The cost of the guarantee would be determined by the probability that the fish will die and the replacement value. Such a strategy could be cheaper than MAC certification, which also aims to lower mortality rates.

In summary, the domestic wholesale marine life industry is varied, dynamic, and competitive. It is not currently threatened by increased culture activities and may benefit from a recently proposed certification system. Firms that also collect, which are primarily located in Florida, face a different set of constraints (i.e., labor costs, regulations, market share) than do wholesale firms in other states. Florida firms al-

so tend to specialize in marine species, which account for the majority of U.S. exports. To remain competitive in the increasingly global market, all firms will need to ensure the sustainability of the resources and be cognizant of changes in retail demand that can have immediate effects on the market.

Acknowledgments

This chapter was supported by the Florida Sea Grant College Program, project R/LR-A-29, and the National Sea Grant College Program of the U.S. Department of Commerce, National Oceanic and Atmospheric Administration, under Grant No. NA76RG-0120.

References

Adams, Charles M., Sherry L. Larkin, Donna J. Lee, Robert L. Degner, and J. Walter Milon. 2001. International trade in live, ornamental "fish" in the U.S. and Florida. Florida Sea Grant Technical Paper No. 113, University of Florida, Gainesville, Florida.

Hellwig, Greg. 1999. 26th annual *Pet Product News Buying Guide Directory* state of the industry report. *Pet Product News Buying Guide Directory* 53(4): 5–11.

Larkin, Sherry L., and Robert L. Degner. 2001. The U.S. wholesale market for marine ornamentals. *Aquarium Sciences and Conservation* 3(1/3): 13–24.

Larkin, Sherry L., Robert L. Degner, Charles M. Adams, Donna J. Lee, and J. Walter Milon. 2001a. 1999 U.S. tropical fish wholesalers survey: results and implications. Florida Sea Grant Technical Paper No. 112, University of Florida, Gainesville, Florida.

Larkin, Sherry L., Charles M. Adams, Robert L. Degner, Donna J. Lee, and J. Walter Milon. 2001b. An economic profile of Florida's marine life industry. Florida Sea Grant Technical Paper No. 111, University of Florida, Gainesville, Florida.

PART **II**

Progress and Current Trends in Marine Ornamentals

B. *Health Management*

7

Disease Diagnosis in Ornamental Marine Fish: A Retrospective Analysis of 129 Cases

Ruth Francis-Floyd and RuthEllen Klinger

Abstract

Techniques used to accurately diagnose fish disease are increasingly sophisticated, and dramatic advances have occurred over the past 5 to 10 years. The scope of diagnostic work available for fish has expanded significantly in technical terms and with regard to the range of species evaluated. Work with marine ornamental species lags behind many groups of freshwater fish. The high value of individual fish and the scarcity of populations of production animals are deterrents to the sacrifice of animals for diagnostic purposes. There has been a general lack of quality diagnostic support for the marine ornamental industry. This will change as the industry matures and the captive production of animals increases. The goal of this chapter is to summarize diagnostic techniques that are available for evaluation of fish and to summarize findings from 129 cases involving marine ornamental species that were presented to the fish disease diagnostic laboratory at the University of Florida over the past 14 years.

Introduction to Diagnostic Methods

Clinical History

Clinical history is the collection of data that establishes conditions present before a disease outbreak, including mortality patterns and behavioral observations. Several references contain a summary of pertinent points that should be discussed in collection of a clinical history (Table 7.1) (Stoskopf 1988; Noga 1996).

Written records are far more valuable than recollections and should include the following: (a) date the animal was received, (b) source, (c) condition on arrival, (d) housing and nutrition, (e) appetite and other relevant behavior, and (f) data on mortality and treatment. A disease case should always be assessed with the entire population in mind. With marine ornamental species this may indeed be a single animal of value, although the population should include all animals housed in the same water system.

Water Quality and Environmental Issues

Experienced fish disease diagnosticians know the importance of water quality and should never overlook collection of the most routine data as part of any disease investigation. Experienced hobbyists, however, may assume that water quality is acceptable and fail to collect needed data. The importance of collecting water quality data at the time the disease epizootic is in progress cannot be overemphasized. Stoskopf presents an excellent overview of environmental requirements for most marine ornamental fish (Stoskopf 1993) (Table 7.2).

Most marine ornamental species prefer salinity near 33 ppt and tropical temperatures in the mid to high 70s (degrees Fahrenheit). The pH of marine systems is higher than most freshwater systems (usually 7.9–8.4). Because of these higher pH values, ammonia tends to be more toxic in marine systems, as a greater percentage of total ammonia may be in the toxic un-ionized form. In addition, some marine fish are sensitive to nitrite. In channel catfish production, salt is used to decrease the percentage of hemoglobin molecules that are transformed to methemoglobin by the presence of nitrite

Table 7.1 When a client contacts a diagnostic laboratory, he or she will be asked a routine set of questions to help identify the problem. The following is a selection of questions a fish health specialist will ask and the client should be prepared to answer:

General:
  What is the size and design of the system involved?
  How old is the system?
  What are the species and numbers of each species in that system?
  What are their sizes and ages?
  Which species are in trouble? Which are not?
  Have there been recent additions? Which species and when?
  When was the abnormal behavior or death first noticed?
  Number of sick fish per day? Number of mortalities per day?
  Have there been problems in this system before?
  Have there been problems with this or these species before?

Behavioral changes:
  What are the fish doing (e.g., are they flashing, is their breathing rate increased, are they lethargic)?
  What is the position of the fish in the water column (at surface, vertical, lying on the bottom, near the aerator)?
  Are the fish eating? If not, when did they stop?

Physical changes:
  What is the fish's body condition (e.g., thin, bloated)?
  Are one or both eyes normal, sunken in, or popped out?
  Are the fins clamped down, frayed, or bloody?
  Are the gills discolored, bloody, or frayed?
  Are there lesions or growths on the fish?
  What else looks abnormal on the fish?

Routine procedures:
  What type and size of feed is fed?
  How much and how often is fed per day?
  Has there been any change in feeding or system maintenance recently?
  When was the last water change? How much was changed?

Previous treatments:
  When was the last treatment?
  What was the treatment(s) and dosage(s)?

Table 7.2 Water chemistry parameters acceptable for most marine fish

Parameter	Range
Ammonia	0.0–0.05 ppm
Calcium hardness	60–80 ppm
Carbonate hardness	5.35–6.54 mEq/liter
	15–18 dKH (German)
Conductivity	51,000–53,000 microsiemens
Copper (routine)	0.00–0.05 ppm
(therapy)	0.13–0.20 ppm
Iron	0.1–0.5 ppm
Nitrate	Up to 20.0 ppm
Nitrite	0.00–0.1 ppm
Oxygen, dissolved	5.0 to saturation
pH	7.9–8.4
Phosphate	0.1–0.2 ppm total phosphate
Redox potential	350–390 mv
Salinity	27.5–32 ppt
Specific gravity	1.022–1.025
Temperature	23–26°C
	74–79°F

Source: Stoskopf, Michael K. 1993.

(NO_2) (Tomasso et al. 1979). The literature for nitrite toxicity and environmental disease in freshwater fish has been reviewed (Schwedler et al. 1985). The use of salt as a "treatment" for brown blood disease in freshwater fish has led to the mistaken conclusion that NO_2 is not a concern in marine systems. This is not true, but few controlled studies have assessed NO_2 toxicity to marine fish. Species-specific sensitivities to NO_2 toxicity are recognized in freshwater fish. It is reasonable to assume that similar species-specific sensitivities are also present in marine fish. Nitrate (NO_3) is much more likely to accumulate in marine systems than in freshwater systems because fewer living plants are maintained in marine systems, and water changes are usually less frequent and involve a comparatively smaller volume. Nitrate is quite toxic to marine invertebrates and can also be detrimental to some fish species. Few data exist on specific tolerances of different species, although anecdotal evidence suggests that surgeonfish may be more sensitive than some other reef species.

If water from a public water system is used to create salt water for marine tanks, the aquarist must be concerned about toxicity from chlorine or chloramine. Chloramine (a combination of ammonia and chlorine) is used by some public water suppliers because it is more stable than chlorine. Aquarists should be aware that dechlorination of water that has been treated with chloramine will result in the release of ammonia. Excess ammonia can be eliminated by a conditioned biofilter if one is available. If source water contains chloramines, however, it may be advisable to use a product that removes ammonia after removing the chlorine as opposed to a standard dechlorinator.

A toxin that is frequently introduced into marine tanks by aquarists is copper. Copper has been used for years as a parasiticide in marine systems. Copper chemistry is extremely complex and has been reviewed (Cardeilihac and Whitaker 1988). As the data presented below demonstrate, improper use of copper is a cause of mortality in marine systems, particularly for the home aquarist. Invertebrates are extremely sensitive to copper and will be killed by the introduction of any copper to the system. Although copper is an effective parasiticide with a long history of use in marine fish, the safety margin is quite small, and it is easy to overtreat the system and kill the fish. The safe use of copper is more difficult when chelated compounds are used because they can interfere with tests that measure free copper, making interpretation difficult. Chelated products may confuse the aquarist by suggesting dosages of 2.0 mg/L. The best recommendation for the novice is to find an alternative to copper. The experienced aquarist should precisely follow the label directions of the compound selected.

Physical Examination

The first step of physical examination is careful inspection of the appearance of the fish through the glass of the aquarium. This should include an assessment of behavior, body condition, finnage, and any lesions that are visible. Inspection of the eyes of marine fish is very important, as some parasites attach to the cornea and are most easily detected when viewing the fish through the glass. If the fish is removed and handled, anesthesia should be considered to facilitate handling and minimize potential injury to the fish and its handlers.

Anesthesia

The most common method for anesthetizing fish is the use of methane tricaine sulfonate

(MS-222). Induction with MS-222 can usually be accomplished by using a concentration of 75 to 125 mg/L. Once the desired plane of anesthesia has been achieved, a maintenance concentration of 50 to 75 mg/L is desirable. Current anesthetic techniques for fish have been reviewed and are available (Stetter 2001).

Once the fish is anesthetized it can be handled. A weight should be taken each time an animal is examined and entered into the medical record of the individual. The length of the fish should also be measured and recorded. The most commonly used measurement of length is total length, a measure of fish length from the tip of the rostrum to the tip of the tail fin. Electronic pit tags are inexpensive and can be used to "label" individual fish, which allows a medical record to be maintained on that animal. This is a greater level of management than has historically been practiced, but the method is in use for some zoological collections and research animals.

Biopsy Techniques

Routine biopsies of gills, skin, and fins are used to assess parasitic infestations of fish. These techniques are easily learned, provide a great deal of information, and are rarely harmful to the fish. Rare complications from gill biopsy may involve hemorrhage. Bleeding usually is insignificant and stops quickly. In extremely rare cases blood may not clot properly. In the experience of the authors, this involved extremely sick fish. Bleeding can be stopped by using silver nitrate tipped swabs that are used to control excessive hemorrhage when the toenails of dogs or cats are clipped. Biopsy techniques have been described (Noga 1996).

Fecal samples should be collected and examined by direct smear with a light microscope. Fish often defecate during the onset of anesthesia. This is an excellent opportunity to collect a sample. Old fecal material from the bottom of a tank is not a good quality sample because a range of commensal organisms is often present and can confuse the diagnosis or interpretation of findings.

Ulcers or other skin lesions should be examined. Skin or mucus samples should be collected from the margin of a lesion and examined with a light microscope. Bacterial culture of open wounds is strongly recommended and discussed in greater detail below.

Blood Collection

Although the clinical value of blood collected from fish is arguably less than that taken from birds or mammals, it should be considered during examination of larger specimens (≥ 100 g). Blood collection protocols and introductory hematology of fish have been reviewed (Campbell and Murru 1990). Blood smears can be examined for blood parasites or other anomalies. Efforts to culture bacteria from fish blood have been reported (Klinger et al. 2001).

Imaging

Imaging techniques, particularly radiography and ultrasound, have evolved as important diagnostic aids for nonlethal examination of individual fish. These techniques allow visualization of internal organs, including the swim bladder, and have proven extremely useful to clinicians. Radiology techniques and clinical interpretation for fish have also been reviewed (Love and Lewbart 1997). Ultrasound and imaging techniques have been used successfully to diagnose and manage traumatic injury in koi (Bakal et al. 1998) and these techniques can be easily applied to clinical medicine for ornamental marine fish species.

Necropsy Techniques

If a fish dies and is to be necropsied, it should be refrigerated, kept on ice, and submitted to a diagnostic lab as soon as possible. A fish that is found dead should be assessed for decomposition before submission. In general, an acceptable sample will have clear eyes, red gills, and not smell bad. Decomposition occurs very quickly in fish. This is especially true with small animals housed in warm water such as the tropical temperatures of most marine tanks. Under such conditions a sample will have diagnostic value for a very short time (perhaps less than an hour) after death. Once refrigerated or placed on ice, the sample may be good for 24–48 hours. The smaller the animal the shorter is the window of opportunity for necropsy of a specimen.

During the necropsy process, gross visual examination of all organs is expected. In addition, microscopic examination of selected tissues using wet mounts or cytology is strongly recommended. This examination may result in identification of parasites in tissue, as well as potentially significant microscopic lesions such as granulomas. Microbial cultures should be taken before examination and dissection of individual organs to minimize contamination. Lesions that become evident during the examination of tissues can be cultured, but it may be necessary to sterilize the area first if contamination occurs during the examination process.

Microbiology

Enriched blood agar (5%) is an excellent all-purpose medium for use in marine fish. Salt supplementation is recommended if the agar is not enriched with blood. Incubation at room temperature is adequate for most samples. Additional media that are used routinely for fish diagnostic work include Ordal's for *Cytophaga*, Lowenstein's for *Mycobacteria*, Sabouraud's for fungal agents, and Mueller-Hinton for sensitivity testing. Microbiology techniques for isolation and identification of fish pathogens have been reviewed (Elliott 1994; Shotts 1994).

Specialized Techniques

Vast arrays of specialized diagnostic techniques can be pursued if a case investigation warrants the investment. These include histopathology, toxicology, electron microscopy, viral isolation, and an assortment of molecular tests. These tests are usually ordered by a diagnostician after the preliminary necropsy and case assessment. There is much less experience with ornamental marine fish species than with many freshwater species for many of these techniques.

Diagnostic Case Load Involving Marine Ornamentals

Since 1987 there have been 129 disease cases involving marine ornamental fish species presented to University of Florida's fish disease laboratory in Gainesville, slightly less than

10% of all submissions. Most fish disease investigations involve both food and ornamental species of freshwater fish. However, the importance of marine ornamental fish to Florida's tropical fish industry is well appreciated. One of the difficulties in pursuing diagnostic evaluations of marine ornamental fish is that a case often involves a single animal rather than a population. It is much easier to achieve an accurate diagnosis and take steps to solve a problem when working with a population. The use of nonlethal techniques for diagnosing fish disease is increasing and does improve the chance of solving a case that involves a single animal. The situation is complicated significantly for many marine ornamental species since the single animals are normally quite small.

Distribution of Cases

Cases were distributed by species and source. The clientele (source of animals) were divided into five groups identified as pet owners, retailers, collectors, aquaculturists, or public display. Of 129 cases, 23 involved pet animals, 24 were submitted by retailers, 26 were from collectors, 21 cases involved cultured fish, and 35 cases were from public aquariums.

All cases that were from aquaculture facilities involved some species of clownfish. Pet animals were diverse and included clownfish (2), tangs (2), angelfish (7), damselfish (2), snapper (1), and one elasmobranch. Fish submitted from retailers were limited to three groups: tangs (9), angelfish (6), and damselfish (4). Fish submitted by collectors included tangs (9), angelfish (2), groupers (6), grunts (4), and elasmobranchs (4). Fish submitted by public aquariums were very broadly distributed and included: clownfish (1), tangs (3), angelfish (4), damselfish (1), groupers (4), snapper (1), puffers (2), grunts (2), seahorses (2), and elasmobranch (1). In addition to the most common groups of fish listed above, other animals were submitted that did not fit into one of these broad categories. These represented a single submission not easily assigned to a general category and were not specified. Examples of these "uncategorized" fish include moray eel and batfish.

Analysis of the species distribution data presented above indicates that marine angelfish (15%) and tangs (14%) were the groups most commonly submitted for diagnostic evaluation (n = cases). Clownfish were the third most common submission, representing 10% of the total cases evaluated. Groupers were fourth, representing 8% of the total cases, damselfish were fifth, representing 5% of the total, and grunts and elasmobranchs tied for sixth, representing 4.6% each of total submissions. Other fish with more than one submission were snappers, puffers, and seahorses, with each representing 1.5% each of total submissions.

Environmental Diseases

Environmental diseases include water quality problems and toxicities, particularly copper toxicity. Environmental factors were the primary cause of mortality in approximately 9% of the total cases examined.

Copper Toxicity Copper toxicity was the most significant cause of environmental disease and was identified as a cause of mortality in four cases. Three of the cases involved retail stores using chelated copper products in recirculating systems. As mentioned above, copper chemistry in salt water is extremely complex. The safe use of chelated copper products is very difficult and not generally recommended except for the most experienced aquarist. The other case of copper toxicity involved a pet owner who had applied an over-the-counter copper product to a home aquarium.

Water Quality Problems Water quality problems identified in disease cases included nitrite toxicity, low dissolved oxygen, and inappropriate temperature. Surprisingly, ammonia toxicity was not recorded. Low dissolved oxygen was reported as a cause of mortality in one outdoor commercial exhibit and in systems maintained by two collectors. Low dissolved oxygen was not recognized as a problem in home aquariums, retail stores, or aquaculture facilities. Temperature concerns were also a problem at holding facilities maintained by collectors. In contrast, nitrite was a problem in recirculating systems in retail shops (n = 2). Although most marine aquarists know that ammonia levels should be monitored, a surprising number do not appreciate the importance of monitoring nitrite.

Parasitic Diseases

The most common parasitic organisms found were *Amyloodinium*, *Cryptocaryon*, *Uronema*, monogenetic trematodes, and *Brooklynella* and *Trichodina*. Findings regarding each of these are mentioned briefly below.

Amyloodinium   Although *Amyloodinium* was found most frequently on cultured clown fish (8 of the total 15 cases, or 53%), it was also found on fish from collectors (13%), retailers (20%), and commercial exhibits (13%). *Amyloodinium* was not a factor in any case involving a pet animal. This finding suggests that fish infected with *Amyloodinium* are either treated effectively or die before reaching the consumer. There is no effective treatment for *Amyloodinium* approved by the Food and Drug Administration (FDA). Freshwater dips and excellent sanitation and quarantine practices are effective under some circumstances (Vermeer, personal communication 2002). The most effective treatment for *Amyloodinium* in captive marine ornamental fish is chloroquin (10 mg/L, prolonged bath). Neither the FDA nor the Environmental Protection Agency in the United States has approved this treatment. Use of this compound in food animals or open systems is completely irresponsible and should never be attempted.

Cryptocaryon   *Cryptocaryon* is a ciliated protozoan that is capable of causing catastrophic mortality when fish are crowded or maintained in a closed system. Ten cases of *Cryptocaryon* were evident in the 129 cases that were investigated. Infected fish were from collectors (33%), pet owners (33%), retailers (20%), aquaculturists (10%), and public aquariums (10%). The fact that fish from all clientele groups were found to be infected with *Cryptocaryon* reinforces the observation that the parasite is a common problem and that it is not always easily identified by the layman. *Cryptocaryon* is sometimes called "white spot disease" or "salt water Ich" because of its apparent similarity to the freshwater parasite *Ichthyophthirius multifiliis*. The parasite has a complex life cycle and is easily transmitted. Control of *Cryptocaryon* is usually achieved with prolonged (+/- 3 weeks) use of copper (0.15 to 0.20 mg/L) in marine aquariums. As demonstrated by data presented earlier in this chapter, copper toxicity is an important cause of mortality for captive marine fish, especially at the retail and pet owner levels of the hobby. With this concern in mind, aquarists should proceed cautiously when considering the use of copper in marine aquariums. Formalin (25 mg/L) may be a reasonable alternative in some situations.

Uronema   *Uronema* is a ciliate that can be quite pathogenic to marine fish because the organism can invade tissue. When it becomes embedded in host tissue, the organism is refractory to chemical treatment. *Uronema* was identified in 11 cases, involving pet fish (3), fish from retailers (3), and fish from collectors (3). The two remaining cases were from fish submitted by a public aquarium and a fish submitted by an aquaculture facility.

Monogenetic Trematodes   Monogenetic trematodes are recognized as a common and important cause of fish disease in public aquariums (Stetter et al. 1999). That observation was supported with data from these cases. Of nine cases in which monogenetic trematodes were identified, five involved fish submitted by public aquariums. Of the remaining four cases, two were submitted by collectors and two were submitted by retailers. Control of monogeneans can be challenging because the organisms that are most problematic in marine systems, *Neobenidinea* and *Benedinia*, are egg layers. The eggs are quite resistant to chemical treatment, and the life cycle is not understood well enough to predict when they will hatch. The most effective control of monogenia in public marine aquariums has been achieved using praziquantel at 2 mg/L (Stetter et al. 1999). The chemical is very expensive and is not approved for this use. If there is effluent from the system it must not be used. Praziquantel must not be used in food fish production. Formalin (25 mg/L) may be moderately effective under some circumstances and is labeled by the FDA as a parasiticide for fish. Nonchemical control may be attempted by using excellent quarantine, sanitation, and freshwater dips. Freshwater dips alone will not be effective because the eggs may remain on fish tissue.

Brooklynella* and *Trichodina   *Brooklynella* was found in two cases involving fish submitted by collectors. The ciliated protozoan can be difficult to control, especially in recently collected or transported animals that are likely to be immunocompromised. The parasite is frequently controlled by copper; however, formalin (25 mg/L) or freshwater dips may be effective in some situations. *Trichodina* was also found in one case submitted by a collector and in one case submitted by a retailer. *Trichodina* is a ciliate that usually responds well to copper, formalin, or freshwater dips. Heavy infestations with *Trichodina* are often associated with high organic loads and crowding in freshwater systems. There is inadequate information to make similar deductions regarding infestations of marine fish, although the possibility should be kept in mind.

Bacterial Diseases

Gram negative bacteria make up the majority of systemic infections in fish, and those described below are ubiquitous in the environment. Of the 129 diagnostic cases, 45% had a bacterial component, but only approximately 25% of the time was the bacterial infection implicated as the leading cause of disease.

Vibrio **spp.** Vibriosis or "saltwater furunculosis" is the most common bacterial infection described in the five clientele categories in 28 total cases. Seven species were identified—*Vibrio damsela, V. carchariae, V. anguillarum, V. ordallii, V. nereis, V. cholerae* (non-01), and *V. alginolyticus*. Vibriosis spreads rapidly in overcrowded systems, making it a potential problem in commercial production systems. However, only one case of vibriosis came from a commercial facility. The majority of cases were from commercial exhibits (9) and retail (9) facilities.

Aeromonas **spp. and** *Pseudomonas* **spp.** Twelve cases of *Aeromonas* spp. (21%) and 11 cases of *Pseudomonas* spp. (19%) were reported. Both of these bacterial diseases are rarely a primary cause of disease but can become systemic quickly if diagnosis and treatment of the initial etiologic agent is ignored. *Aeromonas salmonicida* was implicated in two cases. One was from a commercial exhibit and one from the retail level. These bacteria (which are unusual in marine systems) can create acute mortalities making quick response critical.

Aeromonads were observed in all five clientele categories, with the highest occurrence in retail (5). Pseudomonads were reported in commercial exhibits (4), collectors (3), and retail (4), but were not listed in culture facilities or in pet fish.

Edwardsiella tarda *Edwardsiella tarda*, normally a pathogen of freshwater fish, has been reported in marine fish that live in organically rich waters (Kusuda et al. 1977). There has been one report of *E. tarda* isolated from coral reef fish (grunts and squirrelfish) at a commercial aquarium (Reimschussel et al. 1993). One pet fish case involved a keyhole angelfish (*Centropyge tibicen*) that was diagnosed with *E. tarda* infection.

Streptococcus **sp.** Streptococcal infections of fish are not common but can result in acute mortalities when they do occur. They have been isolated from a variety of ornamental freshwater fish and aquaculture saltwater species. One pet case involving a puffer fish (species not identified) was diagnosed with a streptococcal infection.

Myxobacteria A common environmental inhabitant of marine and freshwater systems, myxobacteria, which includes *Flexibacter* spp., can be a cause of significant disease and mortality, especially after fish have been handled. Overcrowding of animals and poor husbandry can provide the right conditions for myxobacteria to multiply in the system. A high bacterial load overwhelms the fish and results in invasion of damaged tissue, causing disease for which treatment may be required. Proper handling of fish, and treatment of water with ultraviolet light and/or ozone, can decrease the incidence of myxobacteria and other opportunistic bacterial infections. One case from a culture facility and two cases from exhibit aquariums were diagnosed with *Flexibacter* spp.

Mycobacterium **spp.** *Mycobacterium* spp. are acid-fast bacteria that create a wasting condition in fish leading to low chronic mortalities. Because there is no treatment for mycobacteriosis, and because the bacteria can thrive under aquaculture conditions, devastating losses can occur in culture facilities. *Mycobacterium* spp. was diagnosed in one culture facility case and one commercial exhibit. Mycobacteria has zoonotic potential, and personnel handling infected fish or systems should be educated about the risk.

Future Directions

The obvious benefit of a retrospective study is the opportunity it offers to look into the future and to predict trends and develop research strategies. It seems clear from the review of the 129 clinical cases summarized in this chapter that most captive ornamental marine fish suffer from identifiable and treatable diseases. Efforts to improve diagnostic support to the industry, with emphasis on access to quality service, should result in decreased mortality rates on an industrywide basis. This service can be provided by universities, private laboratories, and trained veterinary practitioners. Given modern economic considerations, a prime constraint is the ability or willingness of clientele to pay for the service. Professional fish disease diagnosticians should be able to save small business operators enough money to more than compensate for the cost of their service. These savings will be manifest in decreased mortality, decreased investment in chemical treatments, and decreased man-hours spent treating fish. Pet owners are encouraged to solicit the assistance of trained veterinarians in meeting the health care needs of their aquatic pets. Most veterinary schools offer elective courses in fish medicine. Students who take these courses often seek employment in clinics that specialize in the treatment of exotic pets. Fish owners should ask their veterinarian about their experience and interests in fish medicine or their knowledge of other clinicians in the community who have that interest. Rapid clinical advances in fish medicine are expected to continue, and the ornamental marine fish industry should be poised to take advantage of the situation.

References

Bakal, Robert S., Love, Nancy E., Lewbart, Greg A., Berry, C.R. 1998. Imaging a spinal fracture in a kohaku koi (*Cyprinus carpio*): techniques and case report. Vet Radiol Ultrasound 39(4):318–321.

Campbell, Terry, and Murru, Frank. 1990. An introduction to fish hematology. Comp Cont Educ Pract Vet 12(4):525–531.

Cardeilihac, Paul T., and Whitaker, Brent R. 1988. Copper treatments: uses and precautions. In *Tropical Fish Medicine*, edited by M.K. Stoskopf, Vet Clin No Amer: Sm Anim Pract 18(2):435–448.

Elliott, Diane G. 1994. General procedures for bacteriology. In *Bluebook: Suggested Procedures for the Detection and Identification of Certain Fish and Shellfish Pathogens*, edited by J.C. Thoesen, Fish Health Section, American Fisheries Society, Bethesda, Maryland. Bacteriology: Section I. 11 pp.

Klinger, RuthEllen, Francis-Floyd, Ruth, Riggs, Allen. 2001. A non-lethal approach to diagnosing bacterial disease. Proc Int Assn Aquat Anim Med 32:92.

Kusuda, R., Itami, T., Munekiyo, M., and Nakajima, H. 1977. Characteristics of an *Edwardsiella* sp. from an epizootic of cultured crimson sea breams. Bull Jap Soc Sci Fish, Tokyo. 42:271–275.

Love, Nancy E., and Lewbart, Greg A. 1997. Pet fish radiography: technique and case history reports. Vet Radiol Ultrasound 38(1):24–29.

Noga, Ed J. 1996. *Fish Disease: Diagnosis and Treatment*. St. Louis: Mosby Yearbook.

Reimschussel, Renate, Whitaker, Brent, Arnold, Jill. 1993. *Edwardsiella tarda* infection in coral reef fish at the National Aquarium in Baltimore. Proc Int Assn Aquat Anim Med 24:36–39.

Schwedler, Tom E., Tucker, Craig S., Beleau, Marshall H. 1985. Non-infectious diseases. In *Channel Catfish Culture*, edited by C.S. Tucker, pp. 497–541. Amsterdam: Elsevier Science Publishers.

Shotts, Emmett B. 1994. Flowchart for the Presumptive Identification of Selected Bacteria from Fishes, In *Bluebook: Selected Procedures for the Detection and Identification of Certain Fin Fish and Shellfish Pathogens*, edited by J.C. Thoesen, Fish Health Section, American Fisheries Society, Bacteriology: Section II. 5 pp.

Stetter, Mark D. 2001. Fish and amphibian anesthesia. Vet Clin No Am: Exotic Anim Pract 4(1):69–82.

Stetter, Mark D., Davis, Jane, Capobianco, Jane, Coston, Chris. 1999. Use of praziquantel for the control of monogenetic trematodes in public marine aquariums. Proc Int Assn Aquat Anim Med 30:85–86.

Stoskopf, Michael K. 1988. Taking the history. In *Tropical Fish Medicine*, edited by M.K. Stoskopf, Vet Clin No Amer: Sm Anim Pract 18(2):283–291.

Stoskopf, Michael K. 1993. Environmental requirements of marine tropical fishes. In *Fish Medicine*, edited by M.K. Stoskopf, pp. 620. Philadelphia: W.B. Saunders Co.

Tomasso, J.R., Simco, B.A., Davis, K.B. 1979. Chloride inhibition of nitrite-induced methemoglobinemia in channel catfish (*Ictalurus punctatus*). J Fish Res Bd Can 36:1141–1144.

8

Captive Nutritional Management of Herbivorous Reef Fish Using Surgeonfish (Acanthuridae) as a Model

G. Christopher Tilghman, Ruth Francis-Floyd, and RuthEllen Klinger

Introduction

Nutrition is the basic fuel and building material for all living things. Even the simplest prokaryote extracts nutrients from its environment to survive. Many of the basic metabolic processes and their relation to various nutrients have been investigated, and much has been learned about nutritional deficiencies in fish. Some nutrient deficiencies also interact with environmental conditions to produce disease. Juvenile swordtails (*Xiphophorus* sp.) fed a diet deficient in vitamin C developed scoliosis or lordosis when water temperature was maintained over 32° C. Others fed the same diet in cooler water exhibited normal vertebral growth (Francis-Floyd et al. 2001). Other nutrients may interact with other environmental factors as well to enhance or inhibit fish health. Some of the diseases observed in captive fish are directly or indirectly linked to nutritional deficiencies (Lovell 1989; DeSilva and Anderson 1995).

Surgeonfish (Acanthuridae) are an extremely popular aquarium fish characterized by their brightly colored disc-shaped bodies. Although hardy, these fish are notoriously susceptible to a syndrome known as head and lateral line erosion syndrome (HLLES) (Blasiola 1989). Clinical lesions of affected fish include superficial erosions of the head and face that eventually progress down the lateral flank to involve the lateral line system (Blasiola 1989). Clinical disease is usually nonfatal but can result in permanent scarring of the skin surface (Fig. 8.1)

(Hemdal 1989). There is much speculation on the etiology of HLLES, including ascorbic acid deficiency (Blasiola 1989), reovirus (Varner and Lewis 1991), and copper exposure (Gardner and LaRoche 1973). Some resolution of lesions after dietary supplements of vitamin C has been reported (Blasiola 1989). A reovirus-like agent was isolated from fish displaying the initial lesions associated with HLLES. Healthy fish exposed to this agent developed similar symptoms (Varner and Lewis 1991). Copper exposure caused lesions in the mechanoreceptors of the cephalic extension of the lateral line and olfactory organs. These lesions occurred regardless of exposure time or concentration (Gardner and LaRoche 1973). It is important to note that HLLES is not seen in the oceans, which suggests that aquarium systems lack one or more parameters required for healthy acanthurids.

Great care and effort have been devoted to the replication of water chemistry, natural microflora/fauna, and lighting in model ecosystems. Little has been done to determine the nutritional requirements of captive ornamental marine fish. After environmental management, nutritional management is very important to the health of captive fish. Unfortunately, duplicating the natural diet of a wild fish can be quite difficult, if not impossible. The foraging behavior and diet preference of Atlantic acanthurids have been quantified (Tilghman et al. 2001). *Acanthurus chirurgus* was found to actively seek out sand and brown and green algae. *Acan-*

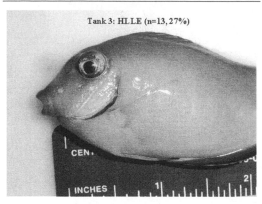

Fig. 8.1   Flake-fed ocean surgeonfish. Note the pitlike lesions along the preopercle and forward of the eye.

thurus bahianus preferred sand and green algae. *Acanthurus coeruleus* preferred red brown and green algae. A correlation of diet preference and stomach morphology was also noticed. The sand-eating fish (*A. chirurgus* and *A. bahianus*) possessed muscular, gizzardlike stomachs, while the algae-eating *A. coeruleus* was found to have a very thin-walled stomach.

The results reported in this chapter demonstrate that there is a correlation among diet, health, and growth in these fish. These data can be utilized by professional aquarists and hobbyists to improve the health and longevity of these beautiful reef fish.

Methods

Fish (n = 89) were captured in the waters near Marathon Key, Florida in June of 2001. Divers were instructed to target juvenile fish (<50 mm total length). Fish were captured using handheld nets and kept in a 300-liter flow-through holding tank until transported to the Whitney Marine Laboratory at Marineland, Florida. Fish were initially quarantined in the following manner: fresh water dip for approximately 2 minutes followed by a 10 ppm praziquantel bath for 3 hours. Many (n = 33) of these fish developed heavy cryptocaryosis within the first 14 days of the study and died.

A later (8 weeks after the start of the study) dive yielded replacement fish (n = 59), which were quarantined as before, with the addition of a 25 ppm formalin bath with water changes every 24 hours for one week. During formalin exposure, tank valves were closed and water

was vigorously aerated. This procedure proved effective, and no further infestations of any kind were observed. Fish were kept in 330-liter flow-through tanks. Water was pumped from an offshore pipe in Marineland, Florida and filtered with a sand filter. Water chemistry was tested daily. Ammonia and nitrite levels were negligible to undetectable, pH was an average of 7.8, and temperature ranged from 25.6 to 18.3°C. Lighting was provided by standard soft white florescent lamps.

Fish were marked using an injectible fluorescent elastomer (Figure 8.2). These marks were highly effective, with approximately a 90% retention rate over a 19-week period. Each fish received a unique mark so that the individual's growth and health could be tracked through the study. Fish were separated into three groups of similar numbers of species, and size class per species, and placed into three 303-liter flow-through tanks. Flow rate was 18 liters/min.

The first treatment group was fed *Ulva* sp., which was rinsed in freshwater to remove any epiphytes or epifauna. The second treatment group was fed a commercially produced extruded pellet tropical marine fish food. The third treatment group was fed a commercially produced flake food for tropical marine fish. These feeds differed in nutritional analyses in several categories (Table 8.1). Fish were fed to satiation daily. Each week, fish were anesthetized using tricaine methane sulfonate at 100 ppm, weighed, and measured. Each fish was also examined thoroughly, and health issues were noted. After being weighed and measured, fish were returned to their tanks.

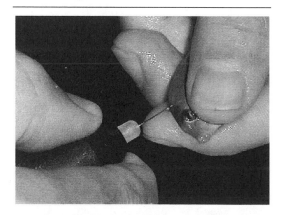

Fig. 8.2   Injecting the fluorescent elastomer mark.

Table 8.1 Nutritional analysis of the three feeds

	Ulva	Pellets	Flakes
Crude protein	18.7%	42.3%	38.8%
Crude fat	4.0%	11.8%	22.7%
Crude fiber	5.8%	1.9%	0.5%
Ash	30.3%	9.5%	5.4%
Calcium	1.4%	1.5%	1.0%
Phosphorus	0.2%	1.3%	1.0%
Magnesium	4.1%	0.3%	0.1%
Sodium	1.8%	0.7%	0.3%
Potassium	2.7%	0.8%	0.8%
Copper	17 ppm	23 ppm	9 ppm
Iron	387 ppm	364 ppm	127 ppm
Manganese	34 ppm	78 ppm	15 ppm
Zinc	396 ppm	181 ppm	90 ppm
Beta-carotene	NA	14,846 IU/lb	1,509 IU/lb
Niacin	NA	395.5 IU/lb	59.7 IU/lb
Thiamine	NA	83.3 mg/lb	5.6 mg/lb
Choline C1⁻	935 mg/lb	2738 mg/lb	964 mg/lb
Vit. A	<566 IU/lb	130,551 IU/lb	16,106 IU/lb
Vit. C	<12 mg/lb	232.7 mg/lb	91.7 mg/lb
Vit. C (polyphosphate)	<12 mg/lb	298.5 mg/lb	708.6 mg/lb
Riboflavin	4.9 mg/lb	64.6 mg/lb	5.2 mg/lb
Vitamin E	62.4 mg/lb	633.5 mg/lb	116.1 mg/lb

Note: NA indicates that parameter was not analyzed because of insufficient sample size. Vitamins A, C, and C from polyphosphate were below measurable levels for *Ulva*.

Results

Fish fed *Ulva* steadily lost weight throughout the study. Blue tangs lost the least weight (average), followed closely by the ocean surgeonfish. Doctorfish lost the most weight. Extreme wasting (cachexia) was observed in 39% of these fish, and 61% of the fish fed *Ulva* died before the end of the study.

The pellet-fed fish fared the best of the three treatments. On average, all fish steadily gained weight, with ocean surgeonfish gaining slightly more weight than the others. There were no noticeable disease signs in any of these fish.

Fish fed the flake diet also gained weight, with the exception of doctorfish during the last weeks of the study. Lesions suggestive of HLLES were observed in 27%, and 39% died before the end of the study. Exophthalmia, corneal opacity, and apparent blindness were observed in 16% of these fish.

Discussion

The feeds used in this study had very different nutritional compositions. Pellet feed had nearly 10 times the beta-carotene and eight times the vitamin A as the flake feed (Table 8.1). Beta-carotene, a yellow pigment produced by plants, is essentially two conjoined retinol (vitamin A) molecules that are separated through hydrolysis (Lovell 1989). Although the role of vitamin A in vision is well known (Lovell 1989), other functions are less well understood. It appears to be involved with epithelial cell differentiation (Lovell 1989), as well as bony and connective tissue formation (DeSilva and Anderson 1995). Vitamin A deficiencies in channel catfish (*Ictalurus punctatus*) have been known to cause exopthalmus, lens displacement, corneal thinning, and opercular deformation (Lovell 1989). The optic lesions and apparent blindness observed in the fish that were fed the flake feed may have been due to insufficient levels of vitamin A and beta-carotene. The degeneration of epithelial tissue associated with HLLES may also be related to a vitamin A deficiency. Deformation of fin spines is often observed in fish affected with HLLES. This is interesting because of vitamin A's role in bony tissue formation. Exopthalmus and HLLES were seen in fish fed the flake feed, which had far lower levels of beta-carotene and vitamin A than the pellet feed. *Ulva*-fed fish exhibited no optic lesions, blindness, or HLLES, even though the vitamin A content of that feed was negligible. This may be a result of the exceptionally high mortality rate of fish fed this feed (61%). Exopthalmia and HLLES-related lesions were not noticed until 4 to 6 weeks after the study began. *Ulva*-fed fish were often cachectic by then, which may have precluded the formation of lesions associated with HLLES.

The weight gain and loss observed in the three treatments may be attributed to the respective protein levels of the feeds. *Ulva* contained 18.7%, the pellet feed had 42.3%, and the flake feed had 38.8% protein. The relatively low level of protein in *Ulva* may have been insufficient for growth of these juvenile, fast-growing fish. The technical literature has reported that optimum levels of protein for fish growth are from 25% to 50% (Lovell 1989). Because vegetable-derived protein often lacks some essential amino acids (DeSilva and Anderson 1995), it can be less effective in feeds. This effect may have been confounded by the comparatively low energy content. *Ulva* also contained 30.3% ash. It is undesirable to have ash content above 12%, because ash is indigestible and is present at the expense of nutritive content (DeSilva and Anderson 1995). These three factors may have compounded to cause the extreme weight loss observed. Other factors can influence fish growth rates: age and size of fish, temperature, available nonprotein energy, and social interactions. The temperature of the three tanks was identical, and no competitive behavior was noted. These fish were juveniles, which typically require more energy and protein than adults of the same species. More positive weight gain may have been seen in this study if mature fish had been used.

Other nutrient levels observed may have influenced the growth and incidence of disease in this study. The relatively low levels of vitamin E in the flake feed may have allowed the oxidation of cellular membranes along the lateral line, which had begun to degenerate as a result of other causes. The flake feed also had relatively low levels of minerals (Table 8.1). Perhaps various minerals found in these feeds have unknown roles in fish metabolism and immune function. Vitamin C is known to have a positive effect on wound healing in fish. Although vitamin C levels in the flake feed were substantially higher than those of the other feeds, fish fed this feed were unable to resolve their HLLES-related lesions. The time to grow out the fish was severely limited because of the need to restock fish for the flake feed treatment. These fish may have healed if the study had continued. The water temperature of the system was beginning to fall in mid-November, so the study was terminated because it had inadequate means of temperature control. The formation of HLLES-related lesions in these fish suggests that there may be other factors involved in the manifestation of this disease.

The immune system of fish has been shown to be responsive to levels of vitamins, minerals, and other factors that exceed the minimum requirement for good growth (Landolt 1989). This may explain why fish fed the pellet feed survived the cryptocaryosis outbreak while those in the *Ulva* and flakes treatments suffered high mortalities. At least 22 minerals are essential in the diet of animals (Underwood 1977) and are also probably needed by fish as well.

Different feeding behavior and diet preferences of these fish may have been a factor in the outcome of this study as well. Acanthurids, which are generally believed to be herbivores, actually have a somewhat diverse diet. Ocean surgeonfish, in particular, were found to have items from seven phyla in their stomachs (Tilghman et al. 2001). Most surgeonfish also feed on many species of small benthic algae. Surely these diverse forage items contain various levels of nutritional components that have an impact on the health and growth of the fish.

Surgeonfish are browsers with lips and dentition for snipping off the tips and branches of algae. They also have long thin-walled digestive tracts, and some species have a sand-filled, muscular, gizzardlike stomach. The blue tang has a long thin-walled digestive tract, while the doctorfish and ocean surgeonfish have a sand-filled, muscular gizzardlike stomach (Tilghman et al. 2001). These are adaptations for foraging on relatively soft algal filaments and blades. Tangs, particularly when young, have a stringent requirement to feed almost continuously, undoubtedly because of a relatively poor utilization of their algal food (Thresher 1980). Whereas the long, thin-walled intestine is probably well adapted to absorbing the cell contents of crushed cells, it may be poorly suited to handling cellulose. This has considerable bearing on their survival in aquariums, since they must have a food source that allows almost continuous foraging (Adey and Loveland 1991). This behavior is seen in schooling acanthurids. Herbivorous fish are not known to produce cellulase or any other enzymes to digest cell wall components (Lobel 1981). They are, however, capable of digesting the materials inside plant cells, if they have developed mechanisms to break the cell walls. This can be done in two ways: trituration can occur through chewing or material may be ground up in a muscular, gizzardlike stomach. Cell walls can also be destroyed by acidic stomach secretions, typically secreted by thin-walled stomachs (Lobel 1981). Acanthurids found in the Florida Keys possess both stomach types (Tilghman et al. 2001), so it is likely that they utilize both mechanisms to break algal cell walls. This may be why blue tangs, which probably rely on acid secretions to break down the cell walls of their algal forage, lost the least weight of all the *Ulva*-fed fish. Perhaps the addition of calcareous reef sand to that tank would have allowed ocean surgeonfish and doctorfish to make better use of the algae. This observation may have important ramifications for the housing requirements of these species.

This study has shown the importance of nutrition in marine ornamental fish. Marine fish nutrition is evolving from an art to a science, but there is much to be learned. An abundance of fish species are kept in aquariums, and they have varying nutritional needs. Unfortunately, most of these unique requirements are unknown. Proper nutrition will surely prolong the lives of captive marine fish as well as enhance the experience of keeping reef aquariums to the hobbyist. Nutritional management is also a crucial step toward future efforts to successfully culture all life stages of these fish.

Acknowledgments

The authors wish to thank the following for funding, expertise, man-hours, or other support: Roy Herndon, John Than, Pablo Tepot, Terri Seron, Theresa Floyd, José Nuñez, The Whitney Marine Laboratory, Alan Riggs, Luiz Rocha, Ilze Berzins, Andy Stamper, Chuck Cichra and Jeff Hill. This project had partial support from Florida Sea Grant, project R/LR-A-30, and the National Sea Grant College Program of the U.S. Department of Commerce, National Oceanic and Atmospheric Administration, Grant No. NA76RG-0120.

References

Adey, Walter H., and Karen L. Loveland. 1991. *Dynamic aquaria. Building living ecosystems*. Academic Press Inc., San Diego, California. 643 pp.

Blasiola, George C. 1989. Description, preliminary studies and probable etiology of head and lateral line erosion HLLE of the palette tang, *Paracanthus hepatus* Lineaus 1758 and other acanthurids. *Bulletin de l'Institut Oceanographique Monaco* 5: 255–263.

DeSilva, Sena S., and Anderson, Trevor A. 1995. *Fish nutrition in aquaculture*. Chapman and Hall, London, England. 319 pp.

Francis-Floyd, Ruth, Klinger, RuthEllen, Riggs, Alan, Watson, Craig. 2001. Effect of Dietary ascorbic acid and water temperature on development of scoliosis in juvenile swordtails *Xiphophorus helleri*. Proceedings of the International Association of Aquatic Animal Medicine 32: 164.

Gardner, G.R., and G. LaRoche. 1973. Copper induced lesions in estuarine teleosts. *Journal of Fisheries Research Board of Canada* 30: 363–368.

Hemdal, Jay. 1989. Marine angelfish: color and style. Aquarium Fish Magazine. August: 15–20.

Landolt, Marsha L. 1989. The relationship between diet and the immune response of fish. *Aquaculture* 79: 193–206.

Lobel, Philip S. 1981. Trophic biology of herbivorous reef fishes: alimentary pH and digestive capabilities. *Journal of Fish Biology* 19: 365–397.

Lovell, Richard T. 1989. *Nutrition and feeding of fish*. Van Nostrand Reinhold, New York, 260 pp.

Thresher, Ronald E. 1980. *Behavior and ecology on the reef and in the aquariums*. Palmetto Publishing, St. Petersburg, Florida.

Tilghman, George C., Francis-Floyd, Ruth, and Klinger-Bowen, RuthEllen. 2001. Feeding electivity indices of surgeonfish (Acanthuridae) of the Florida Keys. *Aquarium Sciences and Conservation*. 3(1): 215–223.

Underwood, Eric J. 1977. *Trace elements in human and animal nutrition*. Academic Press, New York.

Varner Patricia W., and D.H. Lewis. 1991. Characterization of a virus associated with head and lateral line erosion syndrome in marine angelfish. *Journal of Aquatic Animal Health* 3 (3): 198–205.

PART **II**

Progress and Current Trends in Marine Ornamentals

C. *Certification*

9

The Marine Aquarium Industry and Reef Conservation

Bruce W. Bunting, Paul Holthus, and Sylvia Spalding

Introduction

Those who work to protect coral reefs and those who spend considerable time and money caring for reef animals would seemingly form a natural partnership. After all, home and public aquariums could not have beautiful tropical fish, fanciful invertebrates, and amazing corals unless coral reefs remained healthy and productive. Conversely, without the marine aquarium trade, many rural coastal villages would lack an economic incentive to conserve their reefs, and public aquariums could not showcase these tropical organisms, thereby raising public awareness and concern about coral reef ecosystems.

Unfortunately misperceptions about the potential environmental impact of the industry and a lack of communication have prevented this "symbiotic" relationship between reef conservation and the marine aquarium industry from forming. The current unnecessary presumption is often that the marine aquarium industry and reef conservation are incompatible. If the general public knows anything about the marine aquarium industry, it is usually the "bad news," such as the destructive impacts of cyanide use. Controversial headlines grab the reader's attention, but rarely do the media explain that the marine aquarium industry operators who use unsustainable practices are in the minority, and that overall, the responsible industry plays an almost inconsequential role in reef degradation.

In short, the marine aquarium hobby and industry are poorly known and understood by the conservation community, government agencies, and the general public. As a result, the way the marine aquarium trade could support long-term conservation and sustainable use of coral reefs is a message left unheard. The problem is in part nobody's fault—there are just too few credible data on the collection and trade in marine aquarium organisms and their impacts (or lack of impacts). More fundamentally, many of the different groups look at coral reefs in different ways and do not understand or appreciate how and why reefs are important to each other.

This chapter aims to enlighten the discussion by addressing the following issues:

* Why coral reefs are important to different groups;
* How humans impact reefs;
* How marine ornamentals collection and trade affect reefs (both positively and negatively); and
* How marine aquarists, the marine aquarium industry, and conservationists can work with the Marine Aquarium Council (MAC) toward a sustainable future for marine ornamentals and coral reefs.

A responsible marine aquarium industry and conscientious marine aquarists can play a key role in ensuring that both coral reefs and reef aquarium keeping are healthy and thriving for a long time to come.

Why Are Coral Reefs Important?

Coral reefs are important to a lot of people for a lot of different reasons. An important and growing interest in coral reefs is due simply to the vast biological diversity they support—much of which is not yet known (Sebens 1994). Although they occupy less than one quarter of

1% of the marine environment, coral reefs are among the most biologically rich and productive ecosystems on Earth (McAllister 1995). Tropical reefs support about 4,000 species of fish (more than a quarter of all known marine fish species), 800 species of reef-building corals, and a myriad of sponges and other invertebrates—with many species yet to be discovered (Paulay 1997). Coral reefs are often portrayed as "the rain forests of the sea" and are the focus of major conservation efforts by environment groups, governments, and international organizations.

These incredibly diverse reefs provide a wide range of direct and indirect benefits for millions of people and are estimated to be worth about US$375 billion each year (Costanza et al. 1997). The benefits include obvious goods, such as subsistence and commercial fisheries, and services, such as tourism, as well as less obvious functions, such as protecting the coast from erosion and providing a source of pharmaceutical compounds for fighting human diseases. Overall, 1 km^2 of healthy coral reef is

Fig. 9.1 Coral reefs are among the most biologically rich and productive ecosystems on Earth. © International Marinelife Alliance. (See also color plates.)

worth about US$12,000 per year in fisheries production and US$1.16 million over 25 years in fisheries, tourism, and coastal protection value (Cesar et al. 1997). Some of these benefits of coral reefs are examined below in more detail.

One-fifth of all animal protein consumed by humans comes from marine environments, and 1 km^2 of healthy coral reef can produce up to 37 metric tons of fish (Alcala 1988). Most coral reefs are located in developing countries, and much of the world's poor depend directly on reef species for protein. In fact, reefs contribute about one-quarter of the total fish catch—feeding as many as one billion people in Asia alone (Kaufman and Dayton 1997). Maintaining the health of reefs so that they can continue to support food production and the employment and income benefits of fisheries has become an important issue, especially to developing countries that have many poor people and many mouths to feed.

Reefs also support nonconsumptive commercial fisheries, such as organisms for the marine aquarium trade. The collection and export of marine ornamentals, when done responsibly, have numerous benefits, as discussed in more detail below.

Coral reefs also offer a wide range of enormously valuable services to humans, especially for tourism and recreational use. Tourism is the largest and fastest growing sector of the global economy and is largely focused on the coast, often in coral reef areas. Increasingly, travelers to tropical coasts venture into the water for snorkeling, scuba diving, and recreational fishing—all of which multiply the tourism benefits of coral reefs. For example, Florida's 220-mile long reef tract generates US$1 billion in annual fishing and tourism revenues (Lee 1996). In Australia, the Great Barrier Reef annually attracts more than 2 million visitors and generates revenues equivalent to about US$600 million (Alcock 1996). The potential for reef-based tourism is particularly important to developing countries, where most coral reefs are found.

Other benefits of coral reefs include the buffering of adjacent shorelines from storm and wave impacts and the creation of sand on many of the beautiful beaches that attract visitors for "sun, sand, and sea" vacations. The sand is the natural carbonate sediment production from reefs.

Finally, it is important to note a few less obvious coral reef benefits. In recent years scientists have begun searching the oceans for new disease cures. Coral reefs are especially promising because of their exceptionally high diversity of plants and animals. The chemicals produced by many of these organisms may contain important biochemical compounds. Up to one-half of all new cancer drug research focuses on marine organisms, and much of this targets coral reefs (Fenical 1996).

Reefs also yield a host of fishery products for human use, such as corals and shells for jewelry and tourism curios. Less well-known reef products are the materials—sand, gravel, and limestone rock—extracted for a variety of construction purposes (Salvat 1987). On a much broader scale, coral reefs reduce the impacts of global warming by incorporating carbon dioxide during photosynthesis and carbonate production.

More and more people around the world are recognizing these various benefits of coral reefs. At the same time, coral reefs are receiving considerable attention because they are being destroyed at a disturbing rate by a variety of impacts.

What Is Happening to Coral Reefs?

Humans increasingly impact coral reefs largely because more people are on the planet and are progressively becoming concentrated in the coastal zone. Overall, about 37% of the world's inhabitants—more than 2 billion people—live within 100 km of the sea (Cohen 1998). Almost half a billion people—8% of the total global population—live within 100 km of a coral reef (Bryant et al. 1998).

The increasing coastal population and growing use of the oceans are resulting in a range of impacts on coral reefs. A recent global overview of the state of coral reefs—appropriately entitled "Reefs at Risk"—documented the alarming state of the world's coral reefs (Bryant et al. 1998):

- 58% percent are potentially threatened by human activity—ranging from coastal development and destructive fishing practices to overexploitation of resources, marine pollution, and runoff from inland deforestation and farming.

Fig. 9.2 Coral reefs are negatively impacted by the growing human population, which is becoming progressively concentrated in coastal zones, such as this village in the Philippines. © Marine Aquarium Council. (See also color plates.)

- At least 11% contain high levels of reef fish biodiversity and are under high threat from human activities. These "hot spot" areas include almost all Philippine reefs, as well as many of the reefs of Indonesia, Tanzania, the Comoros, and the Caribbean.
- Coral reefs of Southeast Asia, the most species-rich on Earth, are the most threatened of any region. More than 80% are at risk, primarily from coastal development and fishing-related pressures.
- Although more than 400 marine parks and reserves contain coral reefs, most of these sites are very small—more than 150 are less than 1 km^2 in size—and at least 40 countries lack any marine protected areas for their coral reefs.

Coastal Development, Pollution, and Tourism

The growth of tropical coastal urban areas generates a range of threats to nearby coral reefs (Brown 1997). The most obvious is outright destruction from airport construction and other landfill projects on top of the reef (Maragos 1993). Dredging for harbors and extracting construction materials (sand, gravel, and limestone rock) also destroy reefs and reduce their ability to produce sand for beaches and protect the shoreline (Brown and Dunne 1988).

It is the indirect effects of development, however, that are the most pervasive and dam-

Fig. 9.3 Dredging of coral reef for construction material (sand, gravel, and limestone rock) in Micronesia. © P. Holthus. (See also color plates.)

Fig. 9.4 Coastal landfill, like that in New Caledonia shown here, is one of many land-based sources of pollution that result in the most pervasive and damaging impacts on coral reefs. © P. Holthus. (See also color plates.)

aging. In fact, the most widespread human impact on coral reefs undoubtedly comes from sediment and nutrients that are discharged into reef waters from the land (Johannes 1975; Richmond 1994).

Sediments come from shoreline construction or the clearing of forests and other vegetative cover from inland watersheds, which makes them vulnerable to erosion and flooding. When the sediments hit the reef, the plume may physically smother the corals as well as reduce light levels, thus limiting coral growth and the establishment of new corals on the reef (Holthus 1991). These sources of pollution may be right at the coast or transported by rivers into reef waters. The effect is usually the same, and it can take decades for a reef to recover, if at all (Holthus et al. 1989).

Sewage and upland sources of excess nutrients, such as agricultural runoff with fertilizer, often create alga blooms that block sunlight and reduce coral growth. Nutrient-rich runoff promotes the growth of bottom-dwelling algae that overgrows coral (Maragos et al. 1985).

Unregulated tourism can concentrate many of these land-based impacts onto coral reefs, because hotels and resorts are often located near to the reefs and may discharge sewage directly into the ocean, polluting reef waters and promoting algae growth. In addition, snorkelers and divers have destroyed corals through trampling, while boat anchors create further damage to some areas (Tilmant 1987).

Other pollution threats include mine runoff, pesticides, and industrial effluents (Johannes 1975).

Whereas oil spills make the headlines, these major accidents and other ship-based pollution are a less significant threat to coral reefs because they are rare in reef areas. When they occur, however, ship-based pollution often leaves coral reefs more vulnerable to other types of disturbances.

Fishing

Fishing is probably the most pervasive human use of coral reefs, affecting even areas far removed from the impacts of land-based human activities. Fishing has a negative impact on coral reefs when it includes overfishing or the use of destructive harvesting methods. The "fishery" for reef corals and live rock to support home "mini-reef" aquariums is also of concern because it entails the direct removal of portions of the reef habitat, and this fishery will be discussed further later in this chapter.

Overfishing The intense collection of the same fish species from limited areas may create the potential for overfishing. Many reef species—such as giant clams, sea cucumbers, sharks, lobsters, and large groupers—are now the target of specialized fisheries, and some of these reef species have been overfished to local extinction (Bryant et al. 1998; Hodgson 1999).

When overfishing occurs, it can result in shifts in fish size, population abundance, and species composition within reef communities, and the removal of key species may ultimately create changes at the ecosystem level (Roberts 1995). For example, decades of overfishing in parts of the Caribbean have reduced the number of algae-eating fish. Because of this, herbivorous sea urchins have played an increasingly important role in keeping down algal growth. In the early 1980s, huge numbers of these urchins succumbed to disease. Without grazing fish or urchin populations (and with extra algal growth from nutrient-rich pollution), algae quickly dominated the reefs, inhibiting coral settlement and sometimes overgrowing living corals. In places, hurricanes further compounded the damage, reducing coral to rubble. Once thriving reefs became low-diversity and low-productivity algal systems (Roberts 1995; Jennings and Polunin 1996).

Destructive Fishing Methods   Several methods of fishing on coral reefs are inherently destructive. These include fishing with dynamite, cyanide, or other poisonous chemicals; muroami netting (pounding reefs with weighted bags to scare fish out of crevices); and deep-water trawling (Alcala and Gomez 1987; Eldredge 1987a; Gomez et al. 1987). Because destructive methods are generally nonselective, large numbers of non–target species, along with undersized target species, may be killed in the process. Habitat destruction may also occur.

The havoc on coral reefs caused by destructive practices, especially the use of chemicals and dynamite for a variety of fisheries, has been well documented thanks to the important work of conservation organizations and reef scien-

Fig. 9.5   Coral in Indonesia destroyed by dynamite used to catch fish for consumption. © P. Holthus. (See also color plates.)

tists. Fishing with cyanide is particularly notorious and has been well publicized by conservation groups in the media.

Cyanide was first used to stun and capture aquarium fish in the 1960s in Taiwan and/or the Philippines (McAllister et al. 1999). Since the late 1970s, cyanide has also been used to capture larger live reef fish for sale to specialty restaurants in Asian cities that have large Chinese populations (Johannes and Riepen 1995). Even though cyanide fishing is nominally illegal in most countries, the high premium paid for live reef fish, weak enforcement capacities, and corruption have spread the use of the poison across the Asia-Pacific region (Barber and Pratt 1997). Because the use of cyanide is so destructive and is peculiar to the live fish trade, it will be considered in more detail later in this chapter.

Fishing on coral reefs with dynamite and other explosives is an alarmingly common method of food fishing on coral reefs and may be more of problem in this fishery than cyanide use (Mous et al. 2000).

Global Climate Change

Climate change will likely elevate sea surface temperatures in many places, cause sea levels to rise, and result in greater frequency and intensity of storms that can damage reefs. Areas of unusually high water temperatures caused by El Niño oceanographic events, and possibly due to global warming, have been linked to the bleaching of corals. Bleaching is a result of corals becoming stressed as a response to natural factors such as changes in water temperature or salinity (Brown 1997). The corals expel their zooxanthellae—the symbiotic algae that provide coral polyps with nutrients—from the coral tissue and may die.

Cumulative Effects

In many cases it is difficult to pinpoint the exact causes of the serious declines in coral reef health now occurring around the world. Degradation frequently occurs through a combination of human-caused factors that leave reefs less resistant to periodic natural disturbances, such as temperature extremes, hurricanes/cyclones, and other natural events. Healthy reefs are resilient and will recover with time. Reefs damaged by human activity may be more vulnerable to natural disturbances and take longer to recover (Bryant et al. 1998).

How Does the Marine Aquarium Industry Impact Coral Reefs?

The general assumption is that the marine aquarium industry and reef conservation are at odds with each other. The assumption primarily revolves around three issues: the use of destructive fishing methods by some collectors, the possibility of overfishing in some areas, and excessive mortality after collection in some cases. Ill-informed publicity and debate on these issues often overshadows the existence of responsible, sustainable aquarium industry operations and the role of the marine aquarium trade in maintaining reef health and creating socioeconomic benefits in coral reef areas. It is important to be fully and objectively informed about both the negative and positive impacts of the marine aquarium industry, and each will be considered below.

Negative Impacts

Destructive Fishing Destructive aquarium fishery collecting practices include the use of cyanide and other chemicals to stun and catch fish and the breaking of corals. Fishing on coral reefs with dynamite and other explosives has not been part of the marine aquarium trade.

The use of cyanide involves dissolving tablets of the chemical in a bottle of seawater and then squirting the solution at the target fish, which is usually hidden in a coral crevice. The stunned fish can then be caught, sometimes after divers pry the reef apart to collect their prey. The fish is revived in uncontaminated seawater.

It is difficult to know how many targeted fish are killed directly by the use of cyanide. The effects of the chemical also affect how well the fish survive the additional stress of handling and transport. Many cyanide-caught fish die before or soon after they have been sold, with mortality figures ranging up to 80% (Hanawa et al. 1998).

Cyanide also kills or damages corals and non–target fish and invertebrates, although only limited field research and data are available on

Fig. 9.6 The destructive practice of cyanide use continues to some extent in aquarium fish collection. It involves squirting cyanide solution in coral crevices to stun hiding fish. As a side effect, the cyanide solution harms the coral, other nearby invertebrates, and the collectors. © International Marinelife Alliance. (See also color plates.)

this (Jones and Steven 1997). In addition, cyanide use is a health risk for fishers, through accidental exposure to the poison and careless use of shoddy compressed-air diving gear by untrained divers (Barber and Pratt 1997).

Although cyanide use has now spread to many coral reef countries, considerable efforts are being made on the cyanide issue. For example, in the Philippines, the International Marinelife Alliance (IMA) has teamed up with the government to form the Cyanide Fishing Reform Program. The program, along with previous efforts by the Haribon Foundation and other groups, has trained thousands of fishers to use barrier nets, with the nets often provided by the industry. A public awareness campaign in the media and public schools is also helping to educate Filipinos about the value of coral reefs. The government has also stepped up enforcement of anticyanide fishing laws by establishing a network of cyanide detection laboratories, operated by IMA, that randomly sample fish exports at shipment points throughout the country.

How widespread cyanide use is and whether it is increasing or decreasing is the subject of considerable debate. Cyanide fishing has been reduced but not ceased in the Philippines, despite the reform efforts there. Recently released figures from the cyanide detection laboratories show a drop in the number of aquarium fish testing positive with cyanide from more than 80% in 1993 to 47% in 1996 to 20% in 1998 (Rubec et al. 2000). The monitoring of cyanide presence in the Philippines is a valuable step toward addressing cyanide use. However, there is no law or industry requirement for mandatory screening and verification of "cyanide-free" status for exports.

Some aquarium fish importers and retailers claim that their fish are "net caught only" or "cyanide free." Some of them may believe this is true to the best of their knowledge. Others may make blatant claims that cannot be substantiated to attract customers. The bottom line is no independent system exists to verify these claims, so aquarists cannot confidently know they are buying "net caught" or "cyanide-free" fish.

Overfishing   The depletion of fish stocks due to collection of marine ornamentals has often been considered an unlikely event (Randall 1987), although rare species may be an exception (Lubbock and Polunin 1975). Overfishing seemed especially improbable for abundant species with pelagic eggs, as recruitment patterns result in tremendous spatial and temporal variation in reef fish populations. This variation also complicates the ability to determine the effects of fishing effort on reef fish populations (Doherty 1991).

A growing amount of qualitative evidence of aquarium fish populations in fished areas has come from underwater surveys and observations by reef scientists and fisheries experts. The information is mixed. Some indicate some fish populations may be reduced, at least temporarily, among heavily fished species. Others report no noticeable decline in fish diversity and abundance. Clearly there is a need for improved information on fishing effort, catch, and location and more research on the effects of aquarium fishing.

The only systematic study on aquarium fish harvesting effects, undertaken in Hawaii, found declines in six of the seven most abundantly collected fishes (Tissot 1999). The study also showed no evidence of habitat destruction due to fishing practices and no increase in algae growth where herbivore populations were being collected.

Because live coral and live rock form part of the reef structure, their collection and export create additional concerns. High levels of harvesting from limited areas could potentially affect the ability of the reef to maintain itself and its ecosystem functions. The trade in all hard (stony) corals, both live and dead, is regulated under Appendix II of the Convention on International Trade in Endangered Species of Wild Flora and Fauna (CITES), which provides an important level of control and data gathering for the trade in coral species.

The most comprehensive review of the coral trade has been based on CITES data and concluded that globally the trade in coral is not a high-impact industry (Green and Shirley 1999). In Indonesia, coral harvesting for the aquarium trade is managed by a fairly broad quota system, and the harvest of coral is considered to be a relatively minor risk to the overall health of the coral reef ecosystem (Bentley 1998). Because live coral is important to the reef ecosystem, specific area-based fishery management is needed for coral harvesting and for the growing collection of live rock in some countries (Grigg 1984; Wells et al. 1994).

Determining sustainable harvest levels is critical for any fishery, including marine aquarium fish, corals, other invertebrates, and live rock. For most coral reef fisheries, this calculation is difficult because of the large number of species involved, but best efforts must be made. To ensure sustainability, monitoring programs that are independently verified should also be a part of the aquarium fishery management regime.

Post-Harvest Health and Mortality   Even when collected in an environmentally sound manner, aquarium organisms often suffer from poor husbandry and transport practices, resulting in stress, poorer health, and increased mortality. This mortality increases pressure on coral reefs because more organisms are collected to make up for those that die. Wide variations exist in the estimated levels of post-capture fish mortality—from a few percent for net-caught fish that are handled by high-quality operations to 80% or more for cyanide-caught fish.

Many variables affect post-harvest health and mortality, such as collection method, characteristics of the species involved, and the level of experience of the collector and others handling the animal. The quality of the husbandry, handling, holding and transport facilities, and practices, for example, water quality and packing densities, are particularly critical to the health of the fish. Fortunately, most of the quality controls needed to maintain the health of aquarium animals and minimize mortality are known.

Many aquarium industry operators have excellent facilities and high-quality practices, and some retailers and importers in a few countries adhere to a "code of practice." Unfortunately, no system exists to independently verify which companies have high-quality practices and facilities or to determine through which facilities an organism passes as it journeys from the collection area to the retail shop. To be effective in ensuring high quality and health in marine ornamentals, an independently verified quality control system must extend from "reef to retail."

It is also important to recognize that some marine ornamental species do not survive well in the trade or in the home aquarium. This obviously depends on the skill of the aquarist and will change with developments in technology and information. Nonetheless, unnecessary mortality could be reduced if industry wholesalers or retailers did not deal in species unable to survive in captivity, and aquarists recognized the limits to their skill.

Species Introductions   In general, species introductions—intentional and unintentional—have caused some of the most serious human impacts on natural systems, including the marine environment and coral reefs (Eldredge 1987b). Many sources other than the marine aquarium industry, such as ballast water, are considered the prime potential candidates for introducing alien species into the marine environment.

Most tropical marine aquarium organisms are sent to temperate areas that are far from the sea, limiting the possibilities for introductions to occur. To date, no populations of non-native reef animals have become established as a result of introduction caused by the marine aquarium trade or hobby. Nonetheless, the possibilities of imported marine aquarium organisms becoming established in some locations, for example, Florida, must be taken seriously and precautions should be established.

User Conflict   One of the most vocal complaints against the marine aquarium trade is that it reduces fish populations in areas important for marine tourism, for example, scuba diving, snorkeling, and glass bottom boat tours. This has particularly been an issue in Australia and Hawaii. In the latter, the government banned aquarium collecting along 30% of the west coast of the Island of Hawaii (a high tourism area and the principal site for marine aquarium collection in the state) after a study showed reduced fish populations for several species (Tissot 1999). These kinds of user conflicts will undoubtedly increase with the growth of coastal tourism, scuba diving, and marine recreation.

Positive Impacts

The collection, export, and keeping of coral reef animals have numerous benefits that are often overlooked.

Socioeconomic Benefits   Collecting and exporting marine aquarium organisms creates jobs and income in rural low-income coastal areas in developing countries that have limited resources and economic options (Holthus 1999). Collecting marine aquarium organisms provides one of the few possibilities for a sustainable local industry. For example, there are an estimated 7,000 aquarium fish collectors in the Philippines, many of them supporting families. A UNESCO report estimates the number of people in Sri Lanka directly involved in the export of reef animals is as high as 50,000 (Kenchington 1985).

Marine ornamentals are in fact one of the highest value-added products possible to harvest sustainably from coral reefs, bringing a higher economic return than most other reef uses. For example, live coral in trade is estimated to be worth about US$7,000 per tonne, while the use of harvested coral for lime production yielded only about US$60 per tonne (Green and Shirley 1999). The figures for reef fish are even more striking. Reef fish harvested for food from one island country were valued at US$6,000 per tonne. Aquarium fish from the same county realized a return of more than US$496,000 per tonne (Food and Agriculture Organization 1999).

Reef Conservation Benefits   Because of the important socioeconomic benefits the aquarium trade brings to rural, coastal communities in developing countries, fishers and their families have an incentive to ensure their reefs are healthy, protected, and productive. Consequently, collectors of marine ornamentals and their communities often become active reef stewards. They guard these valuable resources against destructive uses and sometimes create informal management systems or de facto conservation areas.

These reefs are often located in far-off reaches, beyond the capability of government to provide resource management or law enforcement. Many government agencies in developing countries will never have the staff or funds to adequately manage or police most coral reefs. Nor will there be outside investments—such as beach hotels, dive tours, or eco-tourism—for the vast majority of the world's "working" reefs. These coral reefs and the adjacent coastal communities depend on each other for their survival.

On the other hand, without sustainable uses of coral reefs—such as the responsible collection of aquarium animals and the incentives that this creates for local resource stewardship—reefs would quite probably become open to more destructive uses. This could be in the form of destructive fishing by outsiders who have no stake in the future of the local reefs. Or it could be by the local fishers themselves. Without a sustainable income from aquarium fisheries or other sustainable use, they could be forced into poverty-driven use of destructive fishing practices to get food for their families. Or, without income generation options in the rural areas, they could be forced to migrate to overpopulated urban areas, adding to social problems in their countries.

Nature Appreciation and Environmental Awareness On the other side of the world from most coral reefs, humans have become progressively removed from nature and seek to find ways to reconnect with it. For the modern, urban dweller in an industrialized nation, an aquarium—especially a reef aquarium—brings the chance to experience a complete natural system in the home in a way that no other hobby can. There is undeniably something both appealing and comforting in an aquarium, as evidenced by the continued growth of aquarium keeping and new public aquariums. In fact, researchers have found that aquarium viewing reduces blood pressure and anxiety—undoubted-

Fig. 9.7 Aquariums raise public awareness of aquatic environmental problems and help contribute to reef conservation. © R. Watt. (See also color plates.)

ly contributing to the popularity of aquariums in doctor and dentist waiting rooms (Katcher et al. 1984).

Aquarium keeping thus creates an opportunity to know and experience nature where none exists and contributes to our ability to perceive the need to conserve and protect nature. By helping to instill the love of nature in general—and of otherwise inaccessible coral reefs in particular—home aquariums contribute to the conservation of coral reefs. In fact, some authors suggest that the popular appeal of aquarium keeping could be used to raise public awareness of aquatic environmental problems, noting that some aquarium industry associations are funding conservation efforts (Andrews 1990).

Advances in Reef Science There is much not known about coral reefs. In part, this is because scientific observation or experiments with reef animals and their interaction in nature is difficult, time consuming, costly, and complicated. But rigorous, regular observation and trial-and-error experimentation of reef animals and systems is exactly what so many aquarists do best.

Many aquarium keepers have made significant contributions to reef science through their efforts to study, observe, and record occurrences in their tanks. To give just a few examples, aquarium keepers have led to important advances in the understanding of fish behavior, reproduction, feeding, and growth; the propagation and growth of corals, soft corals, and other invertebrates; and the balance of nutrients, light, and water motion needed to maintain a reef ecosystem.

The Future: Reef Conservation Plus the Aquarium Industry

Many successful industry operations and aquarists provide and maintain high-quality, healthy aquarium organisms with minimal mortality. Thus, we know that the collection and export of marine aquarium organisms can be based on quality and sustainability and achieve a balance among reef health, aquarium animal collection, and the numerous benefits described above. Most aquarists would prefer to support this kind of industry. There has been no system in place, however, to identify and document quality products and sustainable practices and allow the consumer to reward those in the industry operating on this basis.

Fortunately, the situation is rapidly changing for the better, and the "symbiotic" relationship between reef conservation and reef aquarium keeping is quickly becoming a reality. The Marine Aquarium Council (MAC) is working with aquarists, the aquarium industry, conservation organizations, government agencies, and public aquariums to implement a certification and labeling system for the marine aquarium trade. This system is the most useful means to ensure market demand and support for quality products and sustainable practices in the marine ornamentals industry.

Certification and Labeling

The demand from informed consumers for sustainable products and practices creates an incentive for industry to adopt and adhere to standards for quality and provide quality-assured, higher value-added marine organisms. The single most important market force in the marine aquarium industry is the purchasing power that the aquarists possess. Market assessments show that there is a strong demand for environmental responsibility in the marine aquarium industry and that this demand will increase rapidly when there is a comprehensive, international, independent certification system.

Although government agencies, industry, and some conservation groups have made important efforts to address the impacts of the marine aquarium trade in some areas, these have not been able to transform the industry because they have been undertaken in isolation and only addressed limited aspects of the marine aquarium industry. No single government or other party has been positioned to work with the industry's full "chain of custody," the international range of other stakeholders, the global consumer demand for marine aquarium organisms, and the trans-boundary aspect of marine conservation issues.

Certification and labeling for marine ornamentals will work in much the same way as in other industries, where we might look for the sign of quality from other accepted standards that we trust, such as the industry certification logo on electrical products, as an aquarium lamp. None of us probably knows the details of the quality standards that our fixture had to adhere to. We trust the system, the standards, the testing, the certification label, and the third-party organization behind the system. This is what was needed for the marine aquarium trade.

Marine Aquarium Council

The MAC was established as an international multi-stakeholder institution to address the situation comprehensively and achieve market-driven quality and sustainability in the marine aquarium industry. The Council began as an initiative of a cross section of organizations representing aquarium keepers, the aquarium industry, conservation groups, international organizations, government agencies, public aquariums, and scientists.

After incorporation in 1998, MAC became established as an independent, third party institution whose goal is to transform the marine aquarium industry into one that is based on quality and sustainability in the collection, culture, and commerce of marine ornamentals. As the international, multi-stakeholder certification organization for the marine aquarium industry, MAC is making this happen by:

- Developing standards for quality products and sustainable practices;
- Establishing a system to certify compliance with these standards and label the results;
- Creating consumer demand and confidence for certification and labeling;
- Raising public awareness of the role of the marine aquarium industry and hobby in conserving coral reefs and other marine ecosystems;
- Assembling and disseminating accurate data

relevant to the collection and care of ornamental marine life; and

- Encouraging responsible husbandry by the industry and hobby through education and training.

MAC is now fully established and recognized as the lead organization for developing and coordinating efforts to ensure sustainability in the international trade of marine ornamentals. MAC has made rapid progress in creating a global MAC network of partners, raising awareness of the needs and opportunities for certification and initiating certification system development.

As of late 2001, the rapidly growing MAC network already consisted of more than 2,700 individuals in 60 countries from industry, hobby, government, and environmental organizations. Network members receive regular information through the MAC website (www.aquariumcouncil.org) and the *MAC News*, a quarterly online newsletter. The broad basis for MAC support is reflected in the current membership of the MAC board, which includes:

- American Marinelife Dealers Association
- American Zoo and Aquarium Association
- Indonesia Ecolabeling Institute
- IUCN—the World Conservation Union
- Indonesia Coral, Shell and Ornamental Fish Association
- International Marinelife Alliance-Philippines
- Marine Aquarium Societies of North America
- Ornamental Aquatic Trade Association
- Ornamental Fish International
- Pet Industry Joint Advisory Council
- Philippine Tropical Fish Exporters Association
- The Nature Conservancy
- World Wildlife Fund

MAC is an international organization and seeks to work everywhere important numbers of marine aquarists or marine aquarium industry operators are found. In the future, MAC will develop into a largely self-financed system based on improved economic return from certified marine aquarium organisms, industry willingness to pay for certification, and the consumer's willingness to pay for marine ornamentals of verifiable quality that are more robust and longer lived.

Certification System Development Developing the standards of practice and certification system is at the core of MAC's efforts. A working draft of the standards was produced in mid-1999 and reviewed at workshops during the Marine Ornamentals '99 conference (Kona, Hawaii, November 1999). The standards of practice were finalized in mid-2001. After successful completion of feasibility studies in collection areas and field tests throughout the chain of custody, the MAC Board of Directors adopted the MAC Core Standards in November 2001.

The MAC Core Standards set the basic criteria to address critical and urgent issues related to sustainability, environmental impact, husbandry, transport, and so on in three areas: (1) ecosystem and fishery management; (2) collection, fishing, and holding; and (3) handling, husbandry, and transport.

The standard for Ecosystem and Fishery Management addresses management and conservation of the collection area's habitat, biodiversity, and stocks and includes requirements for:

- Defining and describing the collection area;
- Outlining the history and volume of collection in the area;
- Outlining the history of management of the collection area; and
- Outlining effects of other users, for example, environmental impacts of other fishing activities.

The standard for collection, fishing, and holding verifies that the collection, fishing, and pre-exporter handling, packaging, and transport of marine aquarium organisms ensures the ecosystem integrity of the collection area, sustainable use of the marine aquarium fishery, and optimal health of the harvested organisms. The standard includes requirements for:

- Prohibiting the use of any destructive fishing methods;
- Requiring fishers to collect only what has been ordered;
- Ensuring minimal stress of organisms during collection through the use of appropriate equipment and handling methods;

Fig. 9.8 The MAC Collection, Fishing and Holding Standard requires the use of nondestructive fishing methods, for example, the use of nets instead of cyanide. © Marine Aquarium Council. (See also color plates.)

- Verifying that all fishers/collectors can meet, or have been trained in, the standards; and
- Ensuring water quality and temperature are maintained from reef to exporter.

The standard for handling, husbandry, and transport verifies that the care, handling, packing, and transport of marine aquarium organisms will ensure the optimal health of the organisms at export, import, and retail. This includes requirements for:

- Acclimatizing all organisms;

- Prohibiting the comingling of certified and uncertified organisms;
- Ensuring that water quality must be monitored and maintained;
- Verifying that all handlers can meet, or have been trained in, the standards for handling and transport;
- Maintaining the traceability of organisms; and
- Shipments staying within a maximum allowable dead-on-arrival mortality rate.

Best Practice Guidance documents support the MAC standards by providing advice on actions that will lead to likely compliance.

In the years to come, the more comprehensive MAC Full Standards will be under development. These standards will address a more comprehensive range of issues and approaches to ensuring sustainability for the marine aquarium trade, as well as including standards for mariculture and aquaculture.

During this time, MAC will also undertake major outreach efforts to widely inform the marine aquarium industry operators and hobbyists about the MAC standards and MAC certification system and the ways they can participate. By looking for the MAC certification label on shop windows and tanks, hobbyists will be able to identify MAC-certified industry operators, facilities, and organisms.

MAC will also work with partner organizations to conduct training to ensure that the industry in developing countries (especially collectors) has the skills and capacity to supply certifiable marine ornamentals.

Figs 9.9 and **9.10** Under the MAC Handling, Husbandry and Transport Standard, species must be segregated and packed to ensure optimal health during transport. © Marine Aquarium Council. (See also color plates.)

Fig. 9.11 The MAC Certification label will enable the easy identification of MAC-certified industry operators, facilities, and organisms.

Partnerships and Politics

MAC is increasingly recognized as the global voice for sustainability in the marine ornamentals industry. MAC is also increasingly sought out by the media and is active in responding to media attention to the marine ornamentals trade that is often based on inadequate or incorrect information. To improve the capacity in this important area, MAC has formed key partnerships with media communications organizations and public aquariums. To ensure that there is consistent, comprehensive, quality information on the marine ornamentals trade, MAC has collaborated with the World Conservation Monitoring Centre (WCMC) to develop an international data recording and reporting system known as the Global Marine Aquarium Database (GMAD).

MAC is working to ensure that the governments are aware of the considerable multistakeholder momentum to develop certification

Fig. 9.12 Working with partner organizations, MAC will train collectors to ensure they have the skills and capacity to supply certifiable marine organisms. © Marine Aquarium Council. (See also color plate 13.)

for sustainability in the marine ornamentals trade and has been very active in developing government relations, particularly with the U.S. Coral Reef Task Force and the International Coral Reef Initiative. For example, in public hearings held by the U.S. government, the overwhelming majority of public input stressed the opportunity for the marine ornamentals trade to be environmentally sound and provide incentives for sustainability and the key role of MAC and certification in achieving this.

To ensure that reef habitat and stock are sustained in the long run, MAC is partnering with the Global Coral Reef Monitoring Network/ Reef Check monitoring programs to adapt and incorporate these programs into the certification system. This will ensure that the monitoring of habitat and resources that becomes part of certification is based on internationally accepted methods and expertise.

Conclusion

All stakeholder groups involved in the marine ornamentals industry can benefit from the MAC system, but, to do so, each must do its part. The authors recommend the following for aquarium keepers, the marine aquarium industry, and public aquariums:

> For those with an aquarium at home or in the office, the MAC Certification label for marine ornamentals provides a means to identify organisms that were collected and handled in a manner consistent with the long-term conservation and sustainability of coral reefs and their adjacent community. Through their purchasing power, they can reward responsible industry operators and provide others in the industry with an incentive to reform. Aquarium keepers who deal with MAC Certified facilities/operations and purchase MAC Certified organisms will benefit by having a fish or other coral reef organism that is likely to live much longer and be much healthier than an organism caught through destructive means (such as the use of cyanide).
>
> For the marine aquarium industry the MAC Certification system provides a set of internationally approved environmental and quality standards for the collection of living marine organisms. By becoming MAC Certified they are ensuring the future of coral reef ecosystems as well as their businesses. As consumers become more aware of the importance and availability of MAC-certified or-

ganisms, certified companies will reap the rewards of consumer demand and confidence in their practices and products. For conservation organizations, MAC Certification provides an avenue to further community-based reef conservation and management, including marine protected areas. When communities derive economic benefits from a resource, they become motivated to protect those resources. By attaching an economic value to the sustainable use of coral reef resources through the MAC label, the marine ornamental industry is helping create an economic incentive for effectively conserving coral reefs worldwide. Conservation organizations can partner with local communities that depend on marine aquarium resources to help them not only to embrace MAC Certification but also to protect the greater coral reef ecosystem from its myriad threats. This will protect biodiversity, while ensuring the long-term sustainable use of the marine resources these communities depend on for their livelihood.

For public aquariums, the MAC Certification system provides an important tool for educating the broader public about the plight of coral reefs and the economic and environmental benefits of a responsible marine aquarium industry. For government agencies, MAC Certification creates the enabling environment for the private sector to pursue marine conservation. Many governments that lack sufficient financial resources to effectively protect the environment now recognize the importance of such public-private partnerships to help achieve the goal of protecting the environment. MAC Certification will especially help support efforts by rural communities and businesses to protect the environment for everyone's benefit.

Everyone working together will maintain healthy populations of fish and corals and save these "rainforests of the sea" for present and future generations.

References

Alcala, Angel. 1988. Effects of marine reserves on coral fish abundance and yields of Philippines coral reefs. *Ambio* 17:194–99.

Alcala, Angel, and Edgardo Gomez. 1987. Dynamiting coral reefs for fish: A resource-destructive fishing method. In *Human Impacts on Coral Reefs: Facts and Recommendations*, edited by Bernard Salvat, 51–60. French Polynesia: Antenne Musee, EPHE.

Alcock, Don. 1996. Tourism: The key player in the ecologically sustainable development of the Great Barrier Reef. Paper presented at 1996 World Congress on Coastal and Marine Tourism, Honolulu.

Andrews, C. 1990. The ornamental fish trade and conservation. *Journal of Fish Biology* 37 (Supplement A): 53–59.

Barber, Charles, and Vaughn Pratt. 1997. *Sullied Seas: Strategies for Combating Cyanide Fishing in SE Asia and Beyond.* World Resources Institute and International Marinelife Alliance-Philippines.

Bentley, Nokome 1998. An overview of the exploitation, trade and management of corals in Indonesia. *TRAFFIC Bulletin* 17 (2): 67–78.

Brown, Barbara. 1997. Disturbances to reefs in recent times. In *Life and Death of Coral Reefs,* edited by Charles Birkeland, 370–72. New York: Chapman and Hall.

Brown, Barbara, and R. Dunne. 1988. The environmental impact of coral mining on coral reefs in the Maldives. *Environmental Conservation* 15:159–66.

Bryant, Dirk, Lauretta Burke, John McManus, and Mark Spalding. 1998. Reefs at risk: A map-based indicator of potential threats to the world's coral reefs. World Resources Institute.

Cesar, Herman, Carl Lundin, Sofia Bettencourt, and John Dixon. 1997. Indonesian coral reefs: An economic analysis of a precious but threatened resource. *Ambio* 26 (6): 345–50.

Cohen, J. 1998. Estimates of coastal populations. *Science* 278 (5341): 1211–12.

Costanza, R., R. d'Arge, R. de Groot, S. Farber, M. Grasso, B. Hannon, K. Limburg, S. Naeem, R. O'Neill, J. Paruelo, J. Raskin, P. Sutton, and M. van den Belt. 1997. The value of the world's ecosystem services and natural capital. *Nature* 387:253–60.

Doherty, Peter. 1991. Spatial and temporal patterns in recruitment. In *The Ecology of Coral Reef Fishes*, edited by P. Sale, 261–93. Academic Press.

Eldredge, Lucius. 1987a. Poison fishing on coral reefs. In *Human Impacts on Coral Reefs: Facts and Recommendations*, edited by Bernard Salvat, 61–66. French Polynesia: Antenne Musee, EPHE.

Eldredge, Lucius 1987b. Coral reef alien species. In *Human Impacts on Coral Reefs: Facts and Recommendations*, edited by Bernard Salvat, 215–28. French Polynesia: Antenne Musee, EPHE.

Fenical, W. 1996. Marine biodiversity and the medicine cabinet: The status of new drugs from marine organisms. *Oceanography* 9 (1): 23–24.

Food and Agriculture Organization (FAO). 1999. Ornamental aquatic life: What's FAO got to do with it? *FAO News & Highlights* (1 Sept.). http://www.fao.org/news/1999/990901-e.htm

Gomez, Edgardo, Angel Alcala, and Helen Yap. 1987. Other fishing methods destructive to coral. In *Human Impacts on Coral Reefs: Facts and*

Recommendations, edited by Bernard Salvat, 67–76. French Polynesia: Antenne Musee, EPHE.

Green, Edmund, and F. Shirley. 1999. *The Global Trade in Coral.* WCMC Biodiversity Series no. 9. World Conservation Monitoring Centre.

Grigg, Richard 1984. Resource management of precious corals: A review and application to shallow water reef building corals. *Marine Ecology* 5 (1): 57–74.

Hodgson, Gregor. 1999. A global assessment of human effects on coral reefs. *Marine Pollution Bulletin* 38 (5): 345–55.

Hanawa, M., L. Harris, M. Graham, A. Farrell, and L. Bendall-Young. 1998. Effects of cyanide exposure on *Dacyllus aruanus*, a tropical marine fish species: Lethality, anesthesia and physiological effects. *Aquarium Sciences and Conservation* 2:21–34.

Holthus, Paul. 1991. Effects of increased sedimentation on coral reef ecosystems. In *Workshop on Coastal Processes in the S. Pacific Island Nations, SOPAC Technical Bulletin* 7:145–54.

Holthus, Paul. 1999. Sustainable development of oceans and coasts: the role of the private sector. *UN Natural Resources Forum* 23 (2): 169–76.

Holthus, Paul, J. Maragos, and C. Evans. 1989. Coral reef recovery subsequent to a freshwater "kill" in 1965. *Pacific Science* 43 (2): 122–34.

Jennings, S., and N. Polunin. 1996. Impacts of fishing on tropical reef ecosystems. *Ambio* 25 (1): 44–46.

Johannes, Robert. 1975. Pollution and degradation of coral reef communities. In *Tropical Marine Pollution*, edited by Elizabeth Wood and Robert Johannes, 13–51. Oxford: Elsevier.

Johannes, Robert, and Michael Riepen. 1995. *Environmental, Economic and Social Implications of the Live Fish Trade in Asia and the Western Pacific.* The Nature Conservancy.

Jones, R., and A. Steven. 1997. Effects of cyanide on corals in relation to cyanide fishing on reefs. *Marine and Freshwater Research* 48:517–22.

Katcher, A., H. Segal, and A. Beck. 1984. Contemplation of an aquarium for the reduction of anxiety. In *The Pet Connection: Its Influence on Our Health and Quality of Life*, Proceedings of the Minnesota-California Conferences on the Human-Animal Bond, Center to Study Human-Animal Relationships and Environment, edited by R. Anderson, B. Hart, and A. Hart, 171–78. University of Minnesota.

Kaufman, L., and P. Dayton. 1997. Impacts of marine resources extraction on ecosystem services and sustainability. In *Nature's Services: Societal Dependence on Natural Ecosystems*, edited by G. Daily, 275. Washington, D.C.: Island Press.

Kenchington, Richard. 1985. Coral reef ecosystems: A sustainable resource. *Nature and Resources* 21(2): 18–27. Paris: UNESCO.

Lee, D. 1996. The economics of managing Florida's coral reefs. Paper presented at the 1996 World Congress on Coastal and Marine Tourism, Honolulu.

Lubbock, H., and N. Polunin. 1975. Conservation and the tropical marine aquarium trade. *Environmental Conservation* 2:229–32.

Maragos, James. 1993. Impact of coastal construction on coral reefs in the U.S. affiliated Pacific islands. *Coastal Management* 21:235–69.

Maragos, James, Christopher Evans, and Paul Holthus. 1985. Comparison of coral abundance on lagoon reef slopes: Six years before and after sewage discharge termination. In *Proceedings of the 5th International Coral Reef Congress* 4:189–94.

McAllister, Don. 1995. Status of the world ocean and its biodiversity. *Sea Wind* 9(4): 14.

McAllister, Don, N. Caho, and C. Shih. 1999. Cyanide fisheries: Where did they start? *Secretariat of the Pacific Community Live Reef Fish Information Bulletin* 5:18–21.

Mous, Peter, Lida Pet-Soede, Mark Erdmann, Herman Cesar, Yvonne Sadovy, and Jos Pet. 2000. Cyanide fishing on Indonesian coral reefs for the live food fish market: What is the problem! *Secretariat of the Pacific Community Live Reef Fish Information Bulletin* 7:20–27.

Paulay, Gustav. 1997. Diversity and distribution of reef organisms. In *Life and Death of Coral Reefs*, edited by C. Birkeland, 303–4. New York: Chapman and Hall.

Randall, John. 1987. Collecting reef fishes for aquaria. In *Human Impacts on Coral Reefs: Facts and Recommendations*, edited by Bernard Salvat, 29–39. French Polynesia: Antenne Musee, EPHE.

Richmond, Robert. 1994. Coral reef resources: Pollution's impacts. *Forum for Applied Research and Public Policy* 9, no. 1 (Spring 1994), 55–56.

Roberts, Callum. 1995. Effects of fishing on the ecosystem structure of coral reefs. *Conservation Biology* 9(5): 989–92.

Rubec, Peter, Ferdinand Cruz, Vaughn Pratt, Rick Oellers, and F. Lallo. 2000. Cyanide-free, net caught fish for the marine aquarium trade. *Secretariat of the Pacific Community Live Reef Fish Information Bulletin* 7:28–34.

Salvat, Bernard. 1987. Dredging in coral reefs. In *Human Impacts on Coral Reefs: Facts and Recommendations*, edited by Bernard Salvat, 165–84. French Polynesia: Antenne Musee, EPHE.

Sebens, K. 1994. Biodiversity of coral reefs: What are we losing and why? American Zoology 34(1): 115–33.

Tilmant, J. 1987. Impacts of recreational activities on coral reefs. In *Human Impacts on Coral Reefs: Facts and Recommendations*, edited by Bernard Salvat, 195–214. French Polynesia: Antenne Musee, EPHE.

Tissot, Brian. 1999. Adaptive management of aquarium fish collecting in Hawaii. *Secretariat of the*

Pacific Community Live Reef Fish Information Bulletin 6:16–19.

Wells, Susan, Paul Holthus, and James Maragos.

1994. Environmental guidelines for coral harvesting operations. *South Pacific Regional Environment Programme Reports and Studies* No. 75.

10

Wholesale and Retail Break-Even Prices for MAC-Certified Queen Angelfish (*Holacanthus ciliaris*)

Sherry L. Larkin, Chris de Bodisco, and Robert L. Degner

Introduction

The demand for saltwater aquarium species has increased as a result of new technologies and information available to hobbyists, especially for reef-based systems. Concurrently there is increasing concern for the sustainability of the stocks (given the relatively high mortalities that can occur) and the reefs (because of potentially destructive collection practices). The Marine Aquarium Council (MAC) is a nonprofit organization that has as its mission "to conserve coral reefs and other marine ecosystems by creating standards and educating and certifying those engaged in the collection and care of ornamental marine life—from reef to aquarium." Its goal is to provide consumers the opportunity to buy fish guaranteed to have been procured in an ecologically responsible manner, which "will create consumer demand for quality and sustainability in the collection, culture and commerce of marine ornamentals" (Holthus 2000).

The establishment of environmentally friendly certification programs, such as that proposed by MAC, is on the rise. For example, the Marine Stewardship Council (MSC) has a certification program for food fish whereby industry participants agree to ensure that the collection will not compromise the quality of the habitat or the long-run sustainability of the stock. Aside from the largely immeasurable benefits of sustaining the stock over the long run, such programs also aim to create a branded product that can enable producers to cover the cost of implementing such programs with higher prices. The MAC program is no exception: "MAC certification will strive to achieve largely self-financed sustainable use based on the improved economic return from certified marine aquarium organisms and cost recovery from industry participation in the certification" (Holthus 2001). One of the frequently asked questions of MAC is whether the certification will result in higher retail prices for marine organisms (MAC 2001a).

Detailed knowledge about the market for marine ornamentals does not exist. If the market is highly competitive, producers will not be able to raise prices. If prices cannot be raised, the benefits of participating in a certification program must come from a reduction in mortality rates. In other words, the market value of reducing losses must be equal to or greater than the capital investment needed to upgrade operations, the initial MAC certification fee, the annual MAC certification fee, the initial fee paid to an independent certifier, and the ongoing monitoring costs of the certifier to maintain the certification (MAC 2001b).

At this time the certification program is not fully operational, so there are no data on the costs or benefits of the program. In situations where market information is unavailable but potentially useful, a technique called stated preference analysis (SPA) can be used to quantify these expected values. This technique is well accepted for eliciting perceived values of nonmarket goods, including wildlife areas and environmental amenities. Similarly, SPA can be applied to participants in the market for marine ornamentals to assess the potential profitability and value of MAC-certified specimens.

Using an SPA technique, preferences were solicited from wholesalers and retailers worldwide regarding their handling of MAC-certified queen angelfish (*Holacanthus ciliaris*). This species was selected because of its relative dominance in the trade (i.e., it is well known and abundant) and relatively high value (i.e., high market price). Results are used to assess the potential profitability of handling a MAC-certified queen angelfish and the break-even price under different conditions. Several other factors that typically influence price will also be examined and compared with MAC-certification. Before discussing the experimental results, brief descriptions of the species, the potential value of certification systems, and the SPA technique are presented.

Queen Angelfish (*Holacanthus ciliaris*)

Queen angelfish are found in the tropical western Atlantic including the Southern Gulf of Mexico and Florida's Gulf coast. The species is especially common along the shallows and reefs of the Florida Keys and is most abundant around the islands between the Caribbean Sea and the Atlantic Ocean. From the family Pomacanthidae, this species can also be found around shipwrecks and other areas where it can find shelter and food. It is not considered a viable food item, nor is it harvested commercially as such; rather, it is harvested solely to supply aquarium enthusiasts.

The queen angelfish is quite distinctive (Fig. 10.1). The species has a blunt, rounded head and a continuous dorsal fin. Its brilliant blue and yellow color easily distinguishes it from all other western Atlantic angelfish species with the exception of the blue angelfish. The appearance and coloring of the juvenile queen is quite different from the adult queen. The juveniles are dark blue, and have a yellow tail and a yellow area around the pectoral fins. Also, the young queen angelfish has blue bars on the body and a dark band that runs through the eyes. As the juveniles mature, they lose these characteristics. Furthermore, their color gradually changes from dark blue to iridescent blues and yellows. They develop a blue crown on the forehead and long streamers that are highlighted with blue and orange. As an adult, the fish may measure up to 18 inches in length and weigh more than 3 pounds.

In Florida, reported annual dockside prices of queen angelfish averaged from US$11.16 to US$17.84 per fish between 1990 and 1998 (Florida Marine Trip Ticket Information System). The retail price for the queen angelfish varies according to the size of the fish but typically ranges from US$60 to US$130. A small queen in the market will typically range in size from 2 to 3 1/2 inches (5.1 to 8.9 cm), the medium from 3 1/2 to 5 inches (8.9 to 12.7 cm), and the large from 5 to 7 inches (12.7 to 17.8 cm). A mating pair of adults would command a premium because the species tends to be monogamous.

Environmental Certification Programs

The concept of developing certification standards and certified labeling for various products is not new. In 1977 Germany established the first certification seal for products proven to have positive environmental features (Wynne 1994). To date, the German ecolabeling program has certified approximately 4,000 products. Ecolabeling usage has since expanded throughout the world and across various industries (Staffin 1996). The most successful programs tend to be based on a third-party verification of compliance, which essentially results in a market-based program (i.e., if compliance is profitable, the program creates a market and demand for certified organisms, certified businesses, and certifiers).

Fig. 10.1 Queen angelfish (*Holacanthus ciliaris*). Source: Florida Keys National Marine Sanctuary, Photographer: Mike White. (See also color plates.)

Certification has the potential to be an important tool for promoting environmentally sustainable practices, if monitored successfully. Many attempts at certification programs fail because of a lack of trust from the public. In some industries, uncoordinated attempts to establish a certification standard or the inability to monitor compliance can result in a lack of consistent worldwide or regional compliance. If compliance is lacking, it becomes possible for companies to take advantage of the eco-friendly certification label without being eco-friendly. The effect can be "green washing," whereby the claims of eco-friendliness offer no meaningful benefits (Wynne 1994). Green washing in the marine ornamentals industry could have disastrous effects on conservation efforts because a lack of trust in the certification program could negatively affect the demand for the potentially higher-priced good.

The Marine Stewardship Council (MSC), as mentioned in the introduction, is another resource-based nonprofit organization that aims for sustainable marine fisheries. The MSC promotes responsible, environmentally sustainable, socially beneficial, and economically viable commercial fishing practices. The MSC initiated a third-party voluntary certification program in 1997 for the ambitious purposes of (1) maintaining and re-establishing healthy populations of targeted species, (2) maintaining the integrity of ecosystems, (3) developing effective fisheries management systems, (4) taking into account all relevant biological, technological, economic, social, environmental, and commercial fishing issues, and (5) complying with all relevant local and national laws, standards, and international pacts. Preliminary evidence on the success of MSC is mixed (Wessels 2000). The primary criticism is rooted in the potential for the certification to increase the demand for "resource-fragile" products, that is, those that are close to being overfished. If so, the certification can result in increased pressure on the resource stocks instead of offering incentives (i.e., price premiums) for enhancing sustainable harvesting practices.

Clearly, there is both sufficient precedent and potential for the MAC certification program to succeed. That success will, however, depend on the specifics of the program (as briefly discussed below), including the consideration of lessons learned from the implementation of similar programs.

MAC Certification

The Marine Aquarium Council (MAC) is an independent, nonprofit organization with headquarters in Hawaii. MAC represents the aquarium industry, hobbyists, conservation groups, and governments and thus is well positioned to gauge the industry's willingness to support a certification program. At this time MAC has initiated a third-party industrywide certification system that covers the entire chain of custody from "reef to retail." According to the Marine Aquarium Council (MAC 2001b), certification would ensure quality and sustainability in the collection, culture, and trade of marine ornamentals. MAC seeks to implement this mission by completing the following objectives:

- Establishing an independent certification process for those in the industry that meet best practice standards;
- Raising public awareness of the role of the marine aquarium industry and hobbyists in conserving coral reefs;
- Assembling and disseminating accurate data relevant to the collection and care of ornamental marine life;
- Promoting the sustainable use of coral reefs through the responsible collection of ornamental marine life;
- Ensuring the health and quality of marine life during transport; and
- Encouraging responsible husbandry by the industry and hobby through education and training.

Paul Holthus, executive director of MAC, believes that the time is ripe for these guidelines and standards because exporters, importers, wholesalers, and retailers are beginning to see that the coral reefs are dying (Bolido 2000). Coral reefs are dying, according to Holthus, as a result of collection methods that destroy the reef, thereby destroying the potential for future survival of the species as well as continued profits.

MAC sought to address all aspects of the marine aquarium industry. The initial standards are broad. In July 2001 MAC published its first issue of standards. These standards were developed from a series of discussions and revisions by an international Standards Advisory Group composed of different stakeholders from various sectors of the industry. The draft then became available for public review and was revised accordingly and then revised again by the Standards Advisory Group. The "core" standards consist of three documents: Core Ecosystem and Fishery Management International Performance Standard for the Marine Aquarium Trade (EFM); Core Collection, Fishing, and Holding International Performance Standard for the Marine Aquarium Trade (CFH); and Core Handling, Husbandry, and Transport International Performance Standard for the Marine Aquarium Trade (HHT). The scope of the EFM standards covers "the management of the marine ecosystems where fish, corals, and other marine invertebrates, and plants are harvested through non-destructive means for the marine aquarium trade and the management of the stocks of these organisms." The purpose of the CFH standards is to ensure that the collection, fishing, holding, pre-exporter handling, packing, and transporting of the organisms uphold the integrity of the collection area ecosystem, the sustainability of the fishery, and the health of the organisms harvested. The final standards, the HHT, address the holding, husbandry, packing, and transport of fish, coral, and other marine invertebrates and plants for marine aquarium trade ensuring the optimal health of these organisms. Along with these Core Standards, MAC published accompanying "Best Practice Guidance" documents to provide advice to parties seeking certification on the specific actions that they may take in order to improve their ability to follow the standards. These six documents are to be used for a series of test certifications and will be effective until they are replaced by the completed Full Standards. The Full Standards will consist of four documents, the full standards of the previous three core documents and another for Mariculture and Aquaculture Management (MAM). Best Practice Guidance documents will accompany each of the four Full Standards. The Full Standards are expected to take effect in July 2003 (MAC 2001b).

Stated Preference Technique

To estimate the value that potential participants ascribe to the MAC certification program, it is necessary to use a stated preference technique. This is because a traditional demand analysis uses market data that do not exist for MAC-certified organisms because the program has only recently been introduced. In addition, traditional demand analysis assumes that products are homogeneous, which likely does not apply to products that are shipped individually or in relatively small groups (such as high-valued marine ornamentals). In the case of ornamental fish, the products (i.e., fish) can differ with respect to such attributes as species, country of origin, method of capture, size, color, price, and/or condition. In the context of this study, a firm's overall preference for a differentiated ornamental fish is assumed to be the profit the firm can earn from its purchase and resale. These preferences are elicited by using an experimental market approach and used to determine the relative importance of various product attributes and more generally to derive a better understanding of product preferences (Meyer 1998). Such an exercise is referred to as a stated preference analysis (SPA).

Conjoint analysis is a specific experimental market technique used to generate SPA results. In a conjoint analysis respondents are asked to compare various products and either rate or rank them in order of preference. An alternative approach, known as contingent valuation, asks respondents to specify their willingness to pay or accept any changes in a specific product (i.e., changes in attributes). The SPA approach was chosen in part because an SPA treats price as just another product attribute. Including price as an attribute minimizes many of the biases that can result when respondents are asked to assign a value to a nonmarket good. In general, SPA is appropriate for primary data collection when the market is characterized by heterogeneous products and there is a high degree of market segmentation, minimal available data, and a need to forecast the acceptance of new products (Holland and Wessells 1998). All of these characteristics describe the marine ornamentals industry and the valuation of the proposed MAC certification program.

Traditional SPA models decompose a firm's preference for a hypothetical product into part-worths for each product attribute. Aggregate market studies, on the other hand, are interested in market preferences and testing differences among types of firms. By creating dummy variables for a firm's characteristics it is possible to test whether preferences vary by firm size, location, and primary function (i.e., wholesale versus retail) (Sylvia and Larkin 1995). In this study, a firm's preference for a particular hypothetical product is assumed reflected by the profitability it would generate if the firm were to buy the product for resale. By using the ordinary least squares regression technique, the part-worths of each attribute and firm's characteristic can be estimated. Testing whether the coefficients in the model are statistically significant is testing whether the attribute (or attribute levels, depending on how the characteristics are defined) affects the profitability rating across the aggregate sample of firms.

Experimental Design

The first step in creating the market experiments is to identify and define the relevant product attributes and attribute levels. For queen angelfish the following three attributes were considered: source, price, and size. There were two levels to the source attribute: collected from the wild or collected from the wild and MAC-certified (i.e., collected using sustainable practices as described in the core standards). There were three levels of the price attribute (i.e., price per fish): US$15, US$30, or US$45. There were also three sizes considered: small, medium, or large (actual sizes were defined in inches as described previously). The number of attributes and attribute levels was kept to a minimum to reduce the response burden of each participant; there were four experiments in total, one for each species.

Using the attributes and alternative levels for queen angelfish, we defined six hypothetical products. The product profiles are shown in Table 10.1. Each respondent evaluated each of the six product profiles. The evaluation was in the form of specifying the profitability of handling each (i.e., profitability of buying each for

resale). The evaluation scale ranged from –5 to +5, with zero representing the break-even point. The survey instructed respondents against assigning the same profitability rating to multiple products if possible.

Although stated preference surveys have traditionally been conducted by personal interviews or by mail, the Internet was used in this study because geographic dispersion of the target respondents precluded use of the alternatives. Note that the telephone is not a viable alternative for this type of survey because of the need to visually compare alternative products. In addition, the Internet is used extensively by this industry (firms and hobbyists), so there should be no loss in market segment from the use of this particular survey method. Further advantages of using the Internet include the ability to use color graphics (which may otherwise be prohibitively expensive), include more survey questions (as a result of the ability to respond quickly and include more questions per page), and higher completion rates. Overall, the survey was completed for a fraction of the cost associated with traditional methods. To protect the authenticity of industry members, each potential respondent was preassigned a user name and password and invited to the site by personal letter. There is no possibility that anyone not invited completed the survey because there were no invalid attempts to log in. Further, there were no multiple responses for any participant.

Survey Results and Empirical Modeling

Characteristics of Respondents

The survey data consisted of 186 responses from 31 firms. Of these firms, 19% were classified as wholesalers, 55% as retailers, and 26%

Table 10.1 Product profiles for queen angelfish experiment

Profile	Source	Price (US$)	Size
1	Collected	45.00	Small
2	Collected	15.00	Medium
3	Collected	30.00	Large
4	Collected and MAC-certified	15.00	Small
5	Collected and MAC-certified	30.00	Medium
6	Collected and MAC-certified	45.00	Large

were primarily transshippers, distributors, or service-oriented firms. This latter group will be referred to as "other" in subsequent discussions (Fig. 10.2). Aside from information on the primary function of each firm, respondents were also asked to provide information on the magnitude of sales of marine fish in the year 2000. The majority of firms (68%) had total sales of marine fish that accounted for less than US$100,000 in 2000. For 29% of the firms, marine fish sales totaled between US$100,000 and US$1 million, and just 3% had sales over US$1 million but less than US$10 million in 2000 (Fig. 10.3).

In an effort to further identify the extent that each firm is involved in the marine ornamentals market, participants were asked about other product lines, freshwater fish, and the number of marine species handled annually. The vast majority of respondents (81%) reported selling dry aquarium goods, 65% offer tank-maintenance services, and 23% sell nonaquatic products. Approximately half of the firms (52%) sell freshwater fish that have been collected, and nearly two thirds (65%) sell tank-raised freshwater fish. Nearly all firms sell both collected and tank-raised marine fish (94% and 87%, respectively). On average, each firm sells 160 species of marine fish, although this number ranges from 15 to 500 species. The average number of marine species that are tank raised is 13 but ranges from zero to 30. A relatively high percentage (81%) use the tank-raised source as a marketing tool. Note that if tank-raised specimens are purchased out of concern for the sustainability of collected substitute species, this

statistic may indicate a market demand for MAC-certified specimens.

When considering the purchasing patterns of these firms, it can be seen that just over half (55%) always buy from the same supplier, while nearly two thirds (65%) are accustomed to receiving price discounts for large-volume orders. Fifty-eight percent of the firms had purchased queen angelfish in the past year.

Most firms in the survey (58%) indicated that they have collecting, holding, and/or shipping facilities on the United States mainland. Twenty-three percent reported having facilities outside the U.S. mainland, primarily in Indonesia (16%) but also the Caribbean, Hawaii, Europe, the Middle East, and the Far East (each were reported by 10% of respondents). The majority of respondents (55%) were managers of some type (inventory, sales, or the office). Each respondent had an average of 14.4 years of experience in the industry, but individual responses ranged from 2 to 33 years.

When asked their level of familiarity with the MAC certification program, 12% of respondents were unfamiliar with the program, 29% were somewhat familiar with it, 41% were moderately familiar, and 18% were very familiar with the program. When asked how likely they were to use MAC certification within the next year, 18% were not at all likely to use the program, 41% were somewhat likely, and 41% were very likely to use the program. Those respondents who were unfamiliar with the MAC certification program were less likely to adopt it than those who were familiar with it. Half of the respondents who were unfamiliar with the MAC program were unlikely to adopt it, while the other half were only somewhat likely to adopt it. On the other hand, two thirds of those very familiar with the program were very likely to adopt it.

Market Levels

It is useful to compare responses at the various market levels because each faces different costs and demand conditions and, thus, is likely to differ in the perceived profitability of the MAC-certification program that can alter those conditions. In general, the price relationship between the two firm types (i.e., retail versus intermediate markets, including wholesale and distributors) is known; the retail prices will be no lower than for any other market segment.

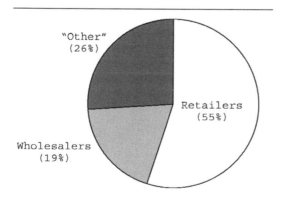

Fig. 10.2 Reported marketing functions of survey respondents.

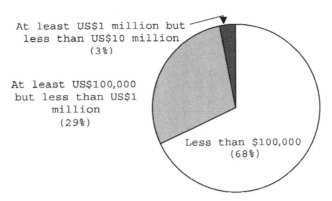

At least US$1 million but less than US$10 million (3%)

At least US$100,000 but less than US$1 million (29%)

Less than $100,000 (68%)

Fig. 10.3 Reported gross sales of respondents in 2000.

The survey responses indicate that wholesalers/other firms considered all the defined "products" less profitable than did retailers; the profitability ratings averaged 0.89 and 1.90, respectively, although respondents in both categories used the entire range of ratings (i.e., –5 to +5). Nearly half (42%) of all product choices were rated as not profitable by wholesalers. Retailers, on the other hand, considered only 12% of the products to be unprofitable (which was reflected by a zero rating). Wholesalers also are less familiar with and far less likely to use the MAC certification program than are retailers.

In terms of their product lines, wholesalers are less likely than retailers to provide tank maintenance services (43% versus 82%), to sell dry goods (57% versus 100%), and to sell non-aquatic goods (14% versus 29%). Wholesalers are more likely to collect their own marine fish (36% versus 0%). All of the wholesaler or other firms knew the source of some species (i.e., collected or tank raised) compared with only 12% of retailers. On average, retailers had higher dollar sales of marine fish in 2000. Because of these observed differences among firms based, primarily, on their market level or firm function, the empirical analysis that follows will take these differences into account.

Explanatory Variables

The product characteristics were included with four variables. The first was a dummy variable for whether the product was MAC certified (all products were assumed collected). The second variable contained the price, which ranged from US$15 to US$45 (Table 10.1). The third and fourth variables were dummy variables that were used to identify the size of the fish as either medium or large instead of small, which was the base category.

Several variables were used to account for hypothesized differences in the relative profitability ratings. Four variables were included to account for the degree of familiarity with, and degree of likelihood of using, the MAC certification system. One variable was included to account for whether the respondent had any management responsibilities under the hypothesis that it is typically the manager that makes purchasing decisions. Two dummy variables were also included to account for firm size. The first identified firms that had marine fish sales ranging from US$100,000 to US$1 million in 2000. The second identified firms that had total sales of marine fish of US$1 million to US$10 million in 2000. No respondent reported sales of marine fish above US$10 million. To account for the knowledge level of respondents, a variable was included that identified those who reported knowing the source of all their inventory (i.e., collected or cultured). Because of the observed differences between wholesalers and retailers, the model was first estimated with two dummy variables to distinguish wholesale and retail firms from "other" firms (including transshippers, distributors, and service-oriented firms, which could not be distinguished because there were too few observations). The data set was then divided into two groups: retailers and the rest (wholesalers, transshippers, distributors, and service-oriented firms), and two additional equations were estimated. Results from all three models are presented and compared.

Estimated Profitability Effects

The estimated model using all data explained 54% of the variation in the dependent variable, which was the relative profitability rating. Estimation results are summarized in Table 10.2. Fourteen of the 15 estimated coefficients were statistically significant at the 10% level or higher. All of the product characteristics were significant and had the expected signs. The "base" product is a small fish that is not MAC certified. The fish is assumed purchased by a company owner or president of a small company (in terms of the value of marine sales in 2000) who knows the source of the company's fish, is not familiar with or likely to adopt MAC certification, and does not specialize in wholesaling or retailing. If this fish were priced at US$30, the average price, the profitability rating would equal 2.93.

Because all of the variables except for price are dummy variables with values of zero or one, the estimated coefficients for each variable represent the corresponding change in profitability rating. On the basis of the product attributes, MAC certification would increase the profitability rating by 0.76, or 26%, over the profitability of the base product described in the previous paragraph. A US$10 price increase would reduce the profitability rating 0.99 points (33%), but the product would still be profitable, with a rating of 1.94. Fish size would increase profitability; the profitability rating for medium fish would increase 0.84, while that for large fish would increase 1.47 (which is 50% over the base). The higher profitability effect for a larger fish is expected because the predicted change in each variable assumes all other effects (including price) are held constant.

The variables pertaining to the familiarity with and likelihood of using the MAC certification program produced somewhat surprising results. Overall, the familiarity characteristics had larger and more highly significant effects on the profitability ratings. Those respondents who were more familiar with the program rated the profitability as much as 1.73 (59% below the 2.93 base profitability) lower. Compared with respondents who were "not at all" likely to use the MAC certification program next year, those who were somewhat likely assigned a 0.89 lower profitability rating. The lower profitability ratings can be considered a proxy for the addi-

tional costs they anticipate being associated with the program. Note that the profitability effect associated with those very likely to use MAC certification is positive; however, this coefficient was not statistically significant at the 10% level.

The remaining statistically significant variables would all work to reduce the profitability rating. If the respondent was a manager as opposed to someone holding a higher office, the profitability score would fall by 2.37, which is a relatively large effect. The hypothesis is that managers make the purchasing decisions and, therefore, would have a different rating scale. The largest single effect was associated with a firm having sales of marine fish that ranged from US$1 million to US$10 million (compared with less than US$100,000); the profitability ratings of these large firms were 4.36 below those of small firms. If the respondent did not know the product's source (cultured or collected), the profitability rating was nearly 3 points lower. The base level profitability rating, as reflected by the intercept and assumed price effect, represented "other" firms. By comparison, the profitability ratings of wholesalers and retailers were 1.27 and 1.66 lower, respectively (–43% and –57%). Thus, the "other" firms as a group are likely to be closer to the collection point. In fact, this group includes transshippers and distributors.

The statistically significant effects by firm function were expected because each market level needs to earn profits. Given the relatively large magnitude of these effects, and the likelihood that the effects of the other variables may also vary by market level, the model was re-estimated twice. The second model used data from firms in the preretail (i.e., intermediate or mid-level) of the marketing chain, that is, all firms except retailers. A dummy variable was included in this model to distinguish the wholesalers from the other firms. The third and final model used data from the retailers only. The results are also included in Table 10.2.

By estimating different regressions on the basis of market level, differences are revealed in the factors that affect profitability, although the explanatory power of each model remained approximately the same. The equation for the mid-level firms explained 52% of the variation in the assigned profitability ratings compared

Table 10.2 Estimated parameter values and standard errors (in parentheses) for each model

Variable	All firms	Mid-level	Retailers
Intercept	5.90***	6.22***	6.76***
	(0.83)	(1.32)	(1.86)
MAC-certified	0.76**	−0.64	0.86**
	(0.30)	(0.52)	(0.35)
Price	−0.099***	−0.107***	−0.093***
	(0.014)	(0.024)	(0.017)
Fish size = medium	0.84**	1.02	0.69
	(0.38)	(0.66)	(0.45)
Fish size = large	1.47***	2.09***	0.96**
	(0.38)	(0.66)	(0.45)
Somewhat familiar with MAC certification	−1.38**	−1.15	−1.18
	(0.53)	(1.25)	(0.97)
Very familiar with MAC certification	−1.73***	−1.27	−2.61**
	(0.60)	(1.01)	(1.16)
Somewhat likely to use MAC certification	−0.89*	−0.98	−0.35
	(0.50)	(1.54)	(0.82)
Very likely to use MAC certification	0.62	−0.18	1.52*
	(0.55)	(1.66)	(0.79)
Inventory, sales, or office manager	−2.37***	−2.73***	−2.50***
	(0.38)	(0.88)	(0.54)
US$100K < 2000 sales < US$1 million	−0.78**	−0.96	−0.27
	(0.38)	(1.17)	(0.62)
US$1 million = 2000 sales = US$10 million	−4.36***	−4.61***	—
	(0.98)	(1.25)	
Does not know source	−2.97***	—	−3.02***
	(0.66)		(0.62)
Primarily a wholesaler	−1.27**	−1.45*	—
	(0.54)	(0.75)	
Primarily a retailer	−1.66***	—	—
	(0.44)		
N	186	84	102
Adjusted R^2	0.54	0.52	0.50

Note: *, **, and *** indicate the variable is statistically significant at the 10%, 5%, and 1% levels, respectively.

with 54% for the overall model and 50% for the retailers. The slightly lower explanatory power was expected in the models that distinguished by market level because fewer parameters were estimated. The separate models retained their explanatory power despite having fewer statistically significant variables. Three variables retained their significance and sign across each model: price, large fish size, and whether the respondent is a manager. Some variables could not be included in the separate models because of lack of variation in the data set. For example, all of the mid-level firms knew the source of their products, and none of the retail firms had sales of marine fish that exceeded US$1 million in 2000.

In the mid-level market segment, neither MAC certification nor any of the MAC program–related variables were significant. The price effect was larger, indicating a higher level of price sensitivity. Medium-sized fish did not significantly affect the profitability rating, but large-sized fish had a statistically significant and much larger effect.

In the retail market segment, the price effect was smaller than that estimated in the mid-level and aggregate models. Large fish had a relatively smaller effect as well. In terms of the MAC certification, it would increase the profitability by 0.86, which is 22% above the base level profitability of 3.97 (using the same assumptions) in this model. This is slightly below the 26% effect estimated in the aggregate model. Conversely, large-sized fish increased profitability by only 24% at the retail level compared with 50% in the aggregate model. In terms of their familiarity with and likelihood of using MAC certification next year, only the strong responses were significant. In particular, those very familiar with the program rated the products 2.61 points less than those not familiar. Conversely, those very likely to use the program rated products 1.52 points higher. Thus, those very familiar with the program are not very likely to adopt it because of the divergent profitability effects. A comparison of the break-even prices in the subsequent section reveals this finding.

Break-Even Prices

Using the estimated coefficients from each of the three models, break-even prices were calculated. These prices were obtained by setting the estimated profitability score (i.e., left-hand side of the equation) for a given product equal to zero and solving for price. This is the maximum price that a firm could pay for the fish in order to break even and earn normal (economic) profits on its resale. The calculated break-even prices will be unique for each product type (i.e., combination of attributes) and each firm type (i.e., combination of demographic characteristics). The difference between two break-even prices would be proxy for the change in costs associated with altering the characteristics that differ between the two products and firms. To simplify the analysis of the results, break-even prices and price differences were also calculated for changes in each variable in turn. Results are summarized in Table 10.3.

To further facilitate the comparison of break-even prices and how those prices would change by changing product characteristics, the break-even price was calculated first for the base product. Recall that the base product, as previously defined, is a small queen angelfish that is not MAC certified and is purchased by an owner or president of an "other" firm that has sales of marine fish totaling less than US$100,000 in 2000. The respondent knows the source but is not familiar with or likely to use MAC certification next year. Note that price is excluded from the description because price is also solved for

assuming profitability is zero (versus solving for profitability assuming price is US$30).

The base level break-even prices ranged from US$44.46 to US$72.80 per fish, depending on the model and market segment. The break-even price for the base product from the retail equation was 5% less than that predicted by the aggregate model with the retail dummy variable. Similarly, the base level wholesale price from the mid-level model was also 5% below that predicted by the aggregate model. When comparing market levels, differences emerge between that predicted by the aggregate model (which used data from all firms) and the separate models. The aggregate model predicted profitability ratings were highest for the other firms and lowest for retailers and vice versa for break-even prices. The separate models also predict that retailers will have the highest break-even prices but that wholesalers have the lowest (versus "other" firms). This latter result may reflect the influence of service-oriented firms (as opposed to transshippers and distributors) that offer tank set-up and maintenance to business clients. In effect, service-oriented firms purchase at the wholesale level but do not rely on sales to consumers in a traditional retail pet shop. As a group, however, the break-even prices associated with the statistically significant variables of the mid-level firms are below those of the retailers as hypothesized. For the base product, the break-even price difference between these two market segments equals US$14.77; the retail break-even price is 25% above the mid-level break-even price (Table 10.3).

MAC certification would increase the break-even price from US$6 (10%) to US$9.29 (13%), depending on the model; the model that used only the retail data produced the latter, higher, effect. This result indicates that retailers perceive a MAC-certified specimen to be more profitable, and they would be willing to pay a higher price for it. Larger fish increased the break-even prices from 10% to 34%. Mid-level firms would be willing to pay US$67.53 and US$77.53 for medium and large fish, respectively. The corresponding break-even prices for retail firms increased to US$80.19 and US$83.15 (19% and 7%).

An increasing familiarity with the MAC certification program resulted in lowering the

Table 10.3 Calculated break-even prices and percentage change from the base break-even price (in parentheses) by model

Variable	All firms	Mid-level	Retailers
Base[a]	US$59.42	US$58.03	US$72.80
	(0)	(0)	(0)
MAC-certified	67.11	64.03	82.09
	(13%)	(10%)	(13%)
Fish size = medium	67.84	67.53	80.19
	(14%)	(16%)	(10%)
Fish size = large	74.23	77.53	83.15
	(25%)	(34%)	(14%)
Somewhat familiar with MAC certification	45.54	47.26	60.06
	(−23%)	(−19%)	(−18%)
Very familiar with MAC certification	41.96	46.17	44.62
	(−29%)	(−20%)	(−39%)
Somewhat likely to use MAC certification	50.44	48.92	69.03
	(−15%)	(−16%)	(−5%)
Very likely to use MAC certification	65.67	56.38	85.21
	(11%)	(−3%)	(17%)
Inventory, sales, or office manager	35.56	32.51	45.90
	(−40%)	(−44%)	(−37%)
US$100K < 2000 sales < US$1 mil	51.55	49.05	69.85
	(−13%)	(−15%)	(−4%)
US$1 mil = 2000 sales = US$10 mil	15.48	15.02	. . .
	(−74%)	(−74%)	
Does not know source	29.54	. . .	40.22
	(−50%)		(−45%)
Primarily a wholesaler	46.60	44.46	. . .
	(−22%)	(−23%)	
Primarily a retailer	76.31	. . .	. . .
	(28%)		

[a] A small fish purchased by a small "other" firm that knows the source. The fish is not MAC certified, and the owner/president is not familiar or likely to use MAC certification next year.

break-even prices by 19% to 39%, depending on the market segment. This result may indicate that once respondents become familiar with the MAC program they perceive that they will have to pay a lower price for fish in order to break even (perhaps to cover other costs associated with adoption of the program). Note also that wholesalers perceive a larger change in the break-even price and thereby anticipate the certification program will have a larger impact on their market.

An increasing likelihood that a firm will adopt the MAC certification program in the next year had mixed results on the break-even prices. Among those firms that are somewhat likely to adopt the MAC certification program, the break-even prices fell 5% to 15%, depending on the model. Conversely, if a firm is very likely to adopt the MAC certification program, break-even prices could fall 3% but also could increase as much as 17%, depending on the model. It should be noted that only the extreme changes were statistically significant. Because familiarity and likelihood of adoption may be related, break-even prices were calculated for the joint effects and are summarized in Table 10.4. Regardless of the likelihood of adoption, break-even prices fell as firms became more familiar with the program. Break-even prices declined further among those firms that were somewhat likely to adopt the program (versus those that will not). However, these price declines were tempered (and in some cases reversed) as the likelihood increased from "somewhat" to "very." In particular, among firms that are very likely to adopt MAC certification, the break-even price would increase by 17% but only among firms that were unfamiliar with the program.

Returning to Table 10.3, the effects of the remaining variables can be identified. First, if the respondent was a manager, break-even prices would fall by 37% to 44%. This represents a significantly different price and emphasizes the importance of the responsibilities of the respondent to the magnitude of the effects. Those firms that had higher sales of marine fish in 2000 reported lower break-even prices by 4% to 74%; the latter result is associated with sales in excess of US$1 million compared with less than US$100,000. This statistic may indicate the need or expectation of large firms to receive price discounts. Lastly, if a respondent did not know the source of some of his or her products, the break-even prices were between 45% and 50% lower. Hence, less informed firms cannot pay as high a price and continue to break even.

Discussion

This paper describes the results of an SPA whereby firms rated the profitability of alternative queen angelfish "products." These products were hypothetical, defined with a specific set of attributes that differed across six product choices. Results were used to quantify the effect on the relative profitability of product attributes and respondent characteristics. When all respondents were included in the model, all of the product characteristics were significant and had the expected signs. MAC certification would increase the profitability rating by 26% of the base-level profitability (assuming price was US$30). A higher-priced product would reduce the profitability rating 0.99 for each US$10 price increase. Fish size would increase prof-

itability; the profitability rating for medium fish increased 0.84, while that for large fish increased 1.47 (from 29% to 50%, respectively).

Significant differences were revealed when separate models were estimated by market level, that is mid-level (i.e., wholesalers, transshippers, service-oriented firms) and retailers. The mid-level market segment generally viewed the products as less profitable than did retailers. The break-even prices revealed the extent of cost differences, particularly in regard to the MAC certification program. Results show that the break-even price for retailers could increase as much as US$9.29 (13%) for a MAC-certified queen angelfish. This higher break-even price reflects the perception that higher prices for fish could be supported if they were MAC certified. This promising result is qualified, however, by the negative effect that familiarity with the MAC certification program has on projected break-even prices (which would decline from 18% to 39%). While the magnitude of this effect is comparatively large, it is reduced and actually becomes positive for firms that are very likely to adopt the program.

In summary, those who are very likely to become certified but are unfamiliar with the program perceive the program to be profitable and therefore are optimistic concerning the average profitability of all the products. Those who are unlikely to adopt the program rate the average profitability of the products lower and, in fact, have break-even prices that are even lower than the base.

A caveat concerning these results and their interpretation is in order. The dollar change in the base price as a result of a change in one of the factors essentially measures how much the input price of the base product would have to change in order to drive profit back to zero. Because these are linear models with no interactive terms, the dollar value of that change is independent of the base, while the relative size of the change depends on the specification of the base product. In other words, only linear effects are measured. Nonlinear or joint effects were not investigated because of the increasing complexity of the models and the need to expedite analyses of the data.

In summary, this analysis quantified the expected change in break-even price for a MAC-certified queen angelfish. It also examined how

Table 10.4 The base level break-even price and price change by familiarity and adoption likelihood by market segment

Likelihood of adoption	MAC familiarity		
	Not	Somewhat	Very
Wholesalers:			
Not likely	US$44.46	−24%	−27%
Likely	−21%	−45%	−47%
Very likely	−4%	−28%	−30%
Retailers:			
Not likely	US$72.80	−18%	−39%
Likely	−5%	−23%	−44%
Very likely	17%	0%	−22%

the break-even price would be affected by different attributes of the product and the firm. Results confirm that the perceived break-even price is affected by many characteristics, especially the primary function of the firm. Results can be used to provide a measure of the anticipated costs and benefits of MAC certification at an individual firm level. Results can also be useful to the MAC organization in its attempts to promote the program. Estimates can eventually be compared with actual costs, and firms can then decide whether carrying MAC-certified specimens would increase their profitability.

References

Bolido, Linda B. 2000. Fish collectors can help protect coral reefs. *Inquirer News Service*, November 2000.

Holland, Daniel, and Catherine R. Wessells. 1998. Predicting consumer preferences for fresh salmon: The influence of safety inspection and production method attributes. *Agricultural and Resource Economics Review* 27(1): 1–14.

Holthus, Paul. 2000. The Marine Aquarium Council: Progress towards sustainability. *Ornamental Fish International (OFI) Journal* 30(2)

Holthus, Paul. 2001. Sustainable use case study: The Marine Aquarium Council and environmental certification for the marine aquarium trade. Marine Aquarium Council, Honolulu, Hawaii. 17 pp.

Marine Aquarium Council (MAC). 2001a. Certification for the marine aquarium trade: Frequently asked questions. Marine Aquarium Council, Honolulu, Hawaii. 8 pp.

Marine Aquarium Council (MAC). 2001b. Core ecosystem and fishery management international performance standards for the marine aquarium trade. *MAC Newsletter* (1): July 1, 2001.

Meyer, L. 1998. Predictive accuracy of conjoint models for data collected by means of world wide web survey. M.S. thesis, Katholieke Universiteit Leuven, Belgium.

Staffin, Elliott B. 1996. Trade barrier or trade boom?: A critical evaluation of environmental labeling and its role in the "greening" of world trade. *Columbia Journal of Environmental Law* (21): 205.

Sylvia, Gilbert, and Sherry L. Larkin. 1995. Firm-level intermediate demand for Pacific whiting products: A multi-attribute, multi-sector analysis. *Canadian Journal of Agricultural Economics* (43): 501–518.

Wessels, C. 2000. Enlisting the consumer in sustainable fisheries production. University of Rhode Island, *Maritimes* 42(3), 3 pp.

Wynne, Roger D. 1994. The emperor's new eco-logos?: A critical review of the scientific certification systems environmental report card and the green seal certification mark programs. *Virginia Environmental Law Journal* (14): 51–60.

Progress and Current Trends in Marine Ornamentals

D. *Management*

11

Community-Based Management of Coral Reefs: An Essential Requisite for Certification of Marine Aquarium Products Harvested from Reefs under Customary Marine Tenure

Austin Bowden-Kerby

Introduction and Overview

The widespread decline of coral reefs is well documented globally, and to a large extent this decline is directly attributed to overexploitation and damage by the very human populations that depend on reefs for their sustenance (Brown and Howard 1985; Grigg and Dollar 1990; Maragos 1992; Lundin and Linden 1993; Connell 1997; Jackson 1997). Conservation of coral reefs on a global scale will thus require basic changes in human-reef interactions at the level of the reef user. Managing a resource involves regulating the behavior of the people whose activities affect that resource (Johannes 1981). Managing the behavior of reef users is often assumed to be the realm of governments; in most developing countries, however, manpower and financial limitations seriously restrict what can be done. For reef areas under customary marine tenure, management and conservation would be more effective if placed in the hands of the reef-owning communities (Gawel 1984; King and Lambeth 2000), or at least as a joint community-government responsibility. Although increased awareness and improved stewardship of reefs among reef fishers may seem beyond the realm of the marine aquarium trades, unless such a transformation among fishing communities is forthcoming, the industry ultimately will be threatened. This chapter outlines the means by which the marine ornamental industry and the "green" certification process can be effec-

tively linked into the process of community-based coral reef management in areas under customary marine tenure, for the benefit of coral reef conservation and the marine ornamentals industry.

Coral Reef Health and Resource Sustainability as an Industry Requirement

Coral reef decline not only threatens the well-being and survival of reef-dependent fishing communities (Gomez 1997; McManus 1997; Moffat et al. 1998) but also represents a serious threat to all reef-based industries: tourism, commercial fisheries, and marine ornamentals. Unregulated and unmanaged, the marine ornamentals industry, like most other extractive fisheries, has been implicated in contributing to coral reef decline: both the ornamental fish trades (Lubbock and Polunin 1975; Randall 1987; Pyle 1993) and the coral trades (Grigg 1976; Ross 1984; Wilkinson 1993; Green and Shirley 1999). However, the extent to which the various marine ornamental trades contribute to reef decline (or not) is not well documented, as the trades at present are mostly unmonitored (Pyle 1993). One aspect of the trades is certain: the high visibility of marine ornamental products makes the trades a magnet for criticism, and this high visibility tends to blow the trades out of proportion to their actual importance.

In spite of all the potential negatives and unknowns, the marine aquarium hobby in itself offers a powerful educational tool. The marine aquarium trades also offer a potential economic incentive for the management and restoration of coral reefs. If these benefits are to be realized, however, best-practice standards must be developed and rigorously implemented, as is proposed by the Marine Aquarium Council (MAC).

Even if standards for sustainable marine aquarium trades are developed and implemented, if the wider problem of coral reef decline is not reversed, it will continue to threaten the future of the industry. A reversal in coral reef decline is needed, and this will require multiple management strategies and approaches at the international, national, and local levels. The challenge before the Marine Aquarium Council today is to collaborate with governments and the industry to develop "green" certification strategies and standards that help stimulate conservation of the wider reef system, including reef areas controlled by participating communities but outside direct industry operations. This more integrated and collaborative approach will help ensure long-term viability of the industry through improved ecosystem management, while ensuring the sustainable harvest of a particular aquarium species or product from a particular reef area or country.

Site Certification and Sustainable Management of Communal Fishing Grounds

Although there will be various levels in the process of green certification of the marine ornamental trades (collecting standards, holding and shipment standards, company certification, retail sales standards, etc.), this chapter explores the management and certification of the trades at base level, the level of the collecting site. Site certification will require the development of conservation and environmental monitoring plans, as well as a process by which sustainable methods are developed and collectors are trained and certified.

The procedures for developing and implementing management plans for specific coral reef areas will vary depending on who controls the resources of the site. For reefs under customary marine tenure, where the indigenous reef users are recognized as the owners or part owners of the reef system and its resources (Christy 1982), the management planning process must include direct involvement by the resource-owning communities and final approval by their traditional chiefs.

Dealing with community-level issues, facilitating planning, and obtaining the required permissions for commercial activities within communally owned waters requires skills and processes beyond the industry, and in many cases beyond the skills of government fisheries officers. Recognized processes for community-based planning for the management of communal resources have recently been established by nongovernmental organizations (NGOs) (IIRR 1998; Ecowoman 2000) and are summarized in this chapter. These proven methods help to provide a context whereby the industry and government, assisted by NGOs, can better interact with reef-owning communities and in conformity with the requirements of traditional societies, lessening industry pressures that at present sometimes cause resource use conflicts and disunity in communally owned collecting sites.

Developing Minimum Standards for Collecting and Collectors

In addition to developing site-specific management plans for the effective conservation of reefs associated with commercial extraction, minimum standards for collecting and handling of organisms must also be developed and implemented, regardless of site. These standards should be based on reef ecology rather than "bean-counting" (i.e. permitting the collection of a particular number of organisms from a particular reef area), as is most often the current practice. Ecologically sustainable development (ESD) criteria (Kessler et al. 1992) need to be upheld, as the reefs are valuable for more than strictly economic reasons, carrying out important functions such as beach sand generation, coastal erosion prevention, climate regulation, nutrient storage, and maintenance of biodiversity (King 1995).

Even good fisheries management by species targeted by the aquarium trades by itself cannot guarantee sustainable yields of these species if the environment deteriorates as a result of land-based threats or overfishing of species important to the ecological functioning of coral reefs. Trying to manage a species or fishery when the overall environment is under threat is like "trying to improve the stopping power of a vehicle with poor brakes by renewing its tyres" (King 1995). Under ESD guidelines, the management of a particular marine ornamentals trade would be regarded as a smaller subset of coral reef ecosystem management.

The total ecosystem management approach involves managing all human activities that threaten coral reefs. Such a holistic approach must inevitably address the interests of all stakeholders: aquarium and curio-collecting companies, tourism and resorts, subsistence village fishers, commercial market fishers, village chiefs, government, conservation groups, and the like. These stakeholders often compete for the same resources and appear to have conflicting interests; however, the interest of all should be in the long-term health of the reef system. Secondary stakeholders must be involved as well. Land-based threats to coral reefs often involve such secondary stakeholders as subsistence and commercial farmers, mining and logging companies, construction firms, and municipal waste managers, and these groups must also be involved in ecologically sustainable development plans, to include conformity to regulations as set by the management plans. Such secondary groups may receive nothing of direct benefit from coral reef management, but their activities can undermine otherwise sound marine resource management plans. Obviously, legal backing for the management plans is needed for effective management.

Collection methods for particular species or groups of organisms that are ecologically sustainable will form an integral subset of ESD standards. An outline of such ecological standards is proposed toward the end of this chapter. A training program for certifying "master collectors" from the industry and "reef wardens" from the communities might also be helpful. Such a system would instill detailed knowledge of coral reef ecology and best-practice standards and would allow reef systems to be sustainably harvested and protected, not only by the industry but by the resource owners as well.

Rural Fishing Communities, Coral Reef Decline, and Sustainable Management

Coral Breakage, Rubble Creation, and Inhibited Reef Recovery

Rural fishing communities are implicated in routine practices that break and kill corals, leading to serious coral reef decline (Wilkinson 1998). Among these problems are blast fishing (Saila et al. 1993; McManus 1997; Nzali et al. 1998), fishing net and anchor damage (Rogers et al. 1988), dredging and sand mining (Clark and Edwards 1995), and coral harvesting for lime production (Venkataramanujam et al. 1981; Brown and Dunne 1988; Dulvy et al. 1995; Berg et al. 1998), for use as building materials (Brown and Dunne 1988; Dulvy et al. 1995), and for commercial sale as curios or for the aquarium trade (Wells 1981; Oliver and McGinnity 1985; Franklin et al. 1998; Green and Shirley 1999). All of these destructive practices may convert rocky reef substrata into unconsolidated rubble beds that may not recover coral populations on biological time scales (Lindahl 1998; Riegl and Luke 1998; Bowden-Kerby 2001a,b). Even though coral larvac may settle readily on these unconsolidated fragments, tiny juvenile corals are killed as the rubble rotates over time (Bowden-Kerby 2001a,b).

In perspective, the curio and aquarium trades appear to be relatively minor threats to coral reefs (Green and Shirley 1999), even though the harvest of aquarium fish may sometimes break corals, and the coral trades may damage or break considerably more corals than are actually harvested. The live rock trade may actually pose a greater threat, but this relatively new trade has not been well studied (Pyle 1993). Live rock harvesting by communities in Fiji often converts rocky substrata into unconsolidated rubble (personal observations), and this activity may prevent reef areas from recovering over the long term. Currently, local collectors and company operators understand that the corals are already dead, so they believe taking them is fine, apparently having no idea as to the long-term negative impacts of converting rocky substrate into lower-profile rubble and sand. Coral, invertebrate, and fish populations will inevitably be affected as recruitment and shelter habitat are removed. Wave impacts and associated beach erosion may also increase, as the dead coral "bommies" being mined likely serve as important wave interceptors during storms (Bowden-Kerby 1999a).

The question is whether the live rock harvest and other ornamentals trades can be sustainably managed so that damage is reduced to low-level short-term impacts and the trades become certifiably nondestructive. If this is not possible, particular wild products or species may have to be either discontinued or replaced by sustainable aquaculture. From the discussion above it becomes clear that sustainability of the trades will require certification of collection methods, training of collectors, zoning of collection sites, and monitoring of the work so that feedback can occur to enable methods to be improved.

Overfishing and Ecological Imbalances Leading to Coral Reef Decline

Overfishing on coral reefs in itself can lead to coral reef decline, causing basic shifts in ecological functioning, resulting in decreased coral cover and lower biodiversity (Hughes 1989, 1994; Done 1992; Jackson 1997; Szmant 1997). Overfishing of herbivorous fish in particular has been implicated in coral reef decline and nonrecovery, as unchecked algal growth can cover the substrata, smothering corals and inhibiting larval settlement (Lewis 1986; Hughes 1994; Connell 1997; Jackson 1997; Rogers et al. 1988; Szmant 1997). A lack of predatory fish has been implicated in an overabundance of organisms that can lead to coral death and reef decline: eroding sea urchins (Bak 1990; McClanahan 1994, 1997a; McClanahan et al. 1996), coral-eating gastropods *Drupella* and *Coralliophila* (McClanahan 1997b; Szmant 1997), coral-killing crown of thorns *Acanthaster* starfish (Birkeland and Lucas 1990; Newman 1998), and coral-killing *Stegastes* damselfish (Szmant 1997).

Involving Reef-Owning Communities to Reverse Coral Reef Decline

Because rural fishing communities are a primary force of destruction to coral reefs on a global scale (Wilkinson 1998), their involvement in the management and conservation of coral reefs will be an essential part of reversing coral reef decline (Gomez 1997; McClanahan 1997a; McManus 1997; Moffat et al. 1998; Biodiversity Conservation Network 1999; Kallie et al. 1999; World Bank 1999), particularly for those areas under customary marine tenure. Increased resources in support of community-based coral reef management are needed, and the marine ornamentals industry can potentially help provide the resources and economic incentives needed to encourage this process. The green certification process thus represents an encouraging development from the standpoint of mainstreaming coral reef conservation, with community, government, and industry involvement in the process. This process should really help stimulate a transformation of fisheries management from a sporadically implemented and mainly top-down governmental process into a more encompassing community-based process, facilitated by NGOs and government and supported (but not controlled) by the marine ornamentals industry.

Customary Marine Tenure and Implications for Management

Customary Marine Tenure in the Pacific Islands

Of the planet's 284,300 square kilometers of coral reefs (Spalding et al. 2001), roughly 70% to 80% are located in developing countries (Bowden-Kerby 2001b). Many of these reefs continue to be owned or controlled by indigenous fishing communities rather than by national or state governments, with specific communities or clans having exclusive fishing rights to particular reefs (Johannes 1981; Christy 1982). Whereas customary rights over marine resources do not exist in the Caribbean, most of the reefs in the Pacific Islands have at least some form of customary control (Gawel 1984; Johannes 1981; Munro and Fakahau 1993; Ruddle et al. 1992; Hviding 1990; Hviding and Ruddle 1991; Wright 1990; Sims 1990).

In Fiji customary marine tenure areas, the *qoliqoli,* are communally owned by kinship groups living on adjacent land areas (Veitayaki 1990; Cooke 1994; Fong 1994; van der Meerten 1996; South and Veitayaki 1998; Anderson et al. 1999; Cooke et al. 2000). Other communally owned marine tenure areas of the Pacific area that have been described in the literature are the Hawaiian *ahupua'a* (Meller and Horowitz 1987), the *tabinau* of Yap (Lingenfelter 1975; Schneider 1984), areas of Morovo Lagoon in the Solomon Islands (Hviding 1990), and marine tenure areas customarily owned by the aboriginal Yolngu tribe of North Australia (Davis 1985).

Customary tenure systems have eroded in these and many other areas as a result of colonization, commercialization, and population increases (Wright 1990; Dahl 1986; Johannes 1981, 1988; Ruddle 1988, 1993). When the colonial powers began occupying the Pacific Islands, mostly in the late 1800s, the various colonial administrations applied European law to the occupied island territories, replacing customary tenure systems with a Western legal framework of state ownership of subtidal lands (Cordell 1991). The people, however, were often unaware of the written laws, and customary ownership systems continue to be practiced in many rural and remote island areas up to the present. In the postcolonial era, the colonial laws often still stand, frequently in conflict with customary practices and mentalities (Cordell 1991; personal observations in Yap, Palau, and Chuuk Micronesia, Solomon Islands, and Fiji).

Fiji has formal constitutional recognition for customary rights over marine tenure, even though the government retains ownership (Veitayaki 1990; Cooke 1994; Fong 1994; van der Meerten 1996; South and Veitayaki 1998; Anderson et al., 2000). In the Solomon Islands and Papua New Guinea, customary ownership is recognized in policy, even if not by law (Munro and Fakahau 1993). Tonga is the only country that does not have any form of community tenure in the South Pacific region, having open access fisheries (Munro and Fakahau 1993). This is surprising because Tonga is the only country of the region that was not colonized by Europeans. In the Cook Islands, customary tenure was replaced with control by town councils (Munro and Fakahau 1993), but a degree of customary control has been re-established in recent years, associated with the establishment of traditional Ra'ui no-fishing reef closures (Dorice Reed, King Nui Council of chiefs, personal communication 1999). In many islands such as in Samoa (Munro and Fakahau 1993), customary rights have eroded, perhaps

owing to their being in opposition to government policy and law. However, recent management measures by the government of Samoa in coral reef management in a recent Australian government-funded "AUSAID" project have involved communities heavily and have served to strengthen community ownership (Walter Vermeulen, OLSS, personal communication 1999).

Customary marine tenure continues to erode in many Pacific areas, particularly near major towns, and these areas are now heavily fished with minimal control (Johannes 1977, 1978; Marriott 1984). It has been proposed by several prominent coral reef scientists and managers that customary marine tenure should be re-established and strengthened regionally (Wright 1990; Dahl 1986; Johannes 1988; Ruddle 1988; Ruddle et al.1992), as areas under community control tend to be better managed than those under governmental control. Community-based management by customary resource owners takes advantage of local knowledge, promotes self-confidence and increased understanding, lessens conflicts, and increases compliance with regulations, increasing success and lowering the cost of management and enforcement (Munro and Fakahau 1993; Johannes 1978).

Whether fully operational or in a weakened condition, for areas traditionally under customary marine tenure, commercial collection activities of all types must involve broad-based community approval rather than approval by a few individuals (the present practice in Fiji), otherwise conflict may arise between those benefiting and those who feel exploited (Johannes 1981). Community approval of commercial activities in jointly owned waters, however, is not enough; effective community-based management plans and regular monitoring of commercial activities and collecting sites need to be operational in order to ensure sustainability and protection of the resource base.

Co-Management Fiji Style: Customary Rights Modified by Colonial Law

The Republic of Fiji is sixth among reef-owning nations globally: Fiji's reefs cover 10,020 square kilometers, or about 3.52% of the planet's total (Spalding et. al. 2001). Currently in

Fiji all subtidal lands are legally owned exclusively by the national government. However, Chapter 158 of the Fiji Fisheries Act of 1974 gives exclusive subsistence fishing rights to the customary resource owners, named as the "custodians" (see South and Veitayaki 1998 for a more detailed discussion of seabed ownership in Fiji and its history).

The present "custodial" marine tenure system is a modified traditional system established by the British colonial authorities, apparently in the interest of simplifying control over the seabed and perhaps the licensing of commercial fishing activities. Whereas ownership of native land in Fiji is at the *matanqali*, or extended family level, marine ownership is legally recognized only at the *yavusa* (clan) or *vanua* (tribe) levels (Nayacakalou 1971; Ravuvu 1983). The system has several variations within Fiji but generally involves lumping together numerous smaller village-level daily usage zones known as"*kanakana*" into larger tribe-based marine tenure units referred to as "*qoliqoli.*" Currently 411 *qoliqoli* units are surveyed and mapped in Fiji's waters and thus recognized by the government (Malakai Tuiloa, Fiji Fisheries Division, personal communication 2001). Most of these *qoliqoli* areas are composed of several unmapped *kanakana* areas. The more remote reefs of the *qoliqoli* are rarely used for fishing by the community, and these areas are normally outside the boundary of any specific *kanakana*, usually being under the control of a high chief.

In Fiji, the highest chief of each coastal tribe, or *vanua,* is regarded as the "custodial chief." Each custodial chief is given exclusive authority to speak on behalf of the family (*matagali* or *tokatoka)* or clan (*yavusa*) reef-owning units that he or she represents. The Fisheries Department issues licenses for commercial fishing on the basis of this approval. The custodian has the authority to unilaterally establish *tabu* no-fishing areas within the *qoliqoli* for a prescribed time upon the death of a high-ranking member of traditional society or for purposes of conservation, to ban the taking of specific species, or to ban particular fishing gear. However, custodianship, in spite of the name, does not carry any specific monitoring or conservation obligations. Licenses for commercial fishing in the *qoliqoli* are often granted to non-Fijians, sometimes in conflict with subsistence usage.

Fees for licenses are set by the Fisheries Department and paid directly to government, conditioned on formal endorsement by the custodian. In order to seek approval by the custodial chief for commercial activities within a specific *qoliqoli*, an individual or company must first approach the custodian with a traditional *sevusevu* (presentation ceremony) of dried kava roots (*Piper methysticum*). A system of "good will payments" has also evolved, whereby considerable gifts of liquor, cigarettes, cash, and other items are given to the custodian by the marine ornamentals company requesting the permit, determined on the basis of negotiations. These "gifts" are paid directly to the custodian, who is not obligated to share these fees, although redistribution often does occur.

Fees for a license to engage in commercial fishing or the ornamentals trades in a particular *qoliqoli,* are granted for one year, renewable in January. Goodwill payments normally accompany the annual renewal of the license. These payments often range into the thousands of dollars. Government fisheries officers can be called on by the custodial chief for advice before granting licenses for fishing or collection of corals, ornamental fish, or live rock within the *qoliqoli*, but this apparently rarely takes place. Although the government authorities are required to maintain records of licenses, they do not determine the number of commercial operators allowed in a particular *qoliqoli*, having no veto power over first issuance or annual extension of licenses.

On the government side of the co-management equation, the government establishes various laws relating to fishing at the national level (species prohibitions, net limits, size limits, etc.), for the common good. Unfortunately, these government conservation laws and prohibitions are not well known to the custodians or the body of customary resource owners, and therefore these laws are not yet applied in much of Fiji. Government fisheries officers have the power to arrest violators of national fisheries laws. One aspect of the 1974 law applied throughout Fiji is the policy allowing for government arbitration between developers and the customary resource owners, with "compensation" paid to the community for damage to the *qoliqoli* resources as the result of mangrove filling, bridge building, or the like (Gawel 1984). The government of Fiji is in the process of training individuals from fishing communities as "fish wardens," whereby the trainees serve as unpaid educators and legally recognized enforcers of the government conservation laws. Most of the coastal villages in Fiji have yet to go through the fish warden program, as there are fewer than 20 Government Fisheries Officers to cover all of Fiji.

Politics and Potential Changes in Fijian Ownership Rights

The present Qarase government of Fiji, elected after the 2000 coup, is proceeding to introduce legislation into the Parliament that will transfer all vestiges of governmental control over the marine zone back to the traditional resource owners (Calamia, as cited in *PASIFIKA* 1999); however, it is not clear what impact this will have on tourism and commercial marine activities in Fiji. Resort owners fear that they will have to lease the waters their guests currently use free of charge. The distribution of licensure fees may also become openly controversial. The original reef-owning *tokatoka* and *matagali* units often yearn for a return to the precolonial tenure system and control over fishing rights in the *kanakana* and a fair share in any fees obtained from these waters. The original *kanakana* boundaries are not mapped, however, as they are not formally recognized by the Fiji government in the Fisheries Act. If such legislation is to take effect, *kanakana* boundaries will need to be set and new procedures for their use set up, and the role of the Fisheries Division will have to be redefined as well. The present management system and a potential new system are contrasted in Figures 11.1 and 11.2.

Strengthening the Local Management Infrastructure

Past attempts to manage coral reefs have mostly neglected the resource users (Gawel 1984; IIRR 1998; Calamia cited in *PASIFIKA* 1999; Ecowoman 2000). Common experience indicates that when the people fishing on the reefs are actively involved in making the management plans that affect them, the plans tend to be considerably more effective. This is because the entire body of resource owners acts together to police their own reef areas (Johannes 1981, 1988) and because private censure by friends and relatives is more effective in traditional Pacific societies than legal measures or police action (Gawel 1984).

The Community-Based Management Process

The Role of NGOs as Facilitators of the PLA Process

Community-based coastal resources management, the process by which communities take on the primary management role (IIRR 1998; King and Lambeth 2000), is not something that the majority of communities have much experience with, and thus the process will require a high degree of facilitation. Perhaps the most complete tool kit for community-based management is the three-volume booklet set published by the International Institute of Rural Reconstruction (IIRR 1998). The four major aspects of the community management process are resource tenure clarification, capacity building for management, environmental conservation, and sustainable livelihood development.

A team of trained facilitators can best carry a community through the sensitive and challenging process of participatory management planning. The participatory workshop process is most often referred to as "participatory learning and action" (PLA) and is closely related to the development tool known as "participatory rural appraisal" (PRA). Although governmental officers and people from the community can potentially be trained as PLA workshop facilitators, currently NGOs dominate this facilitation role. Figure 11.3 diagrams the participatory process and role of the facilitating agency.

Whereas communal decision making has been happening for untold generations in village settings, higher-level decisions are mostly made by traditional chiefs, either on the basis of first-hand knowledge or on the recommendations of advisors or older community members. Decision making in traditional societies, therefore, tends to be male dominated and excludes the youth (Ecowoman 2000). In order for communities to develop workable fisheries regulations and management plans, it is important for all fishers to be included, and often these are the women and younger men. One of the jobs of the facilitating NGO is to ensure a culturally appropriate means for the participa-

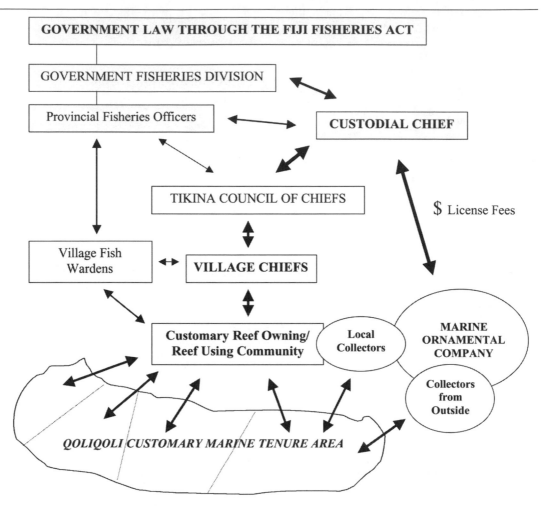

Fig. 11.1 The present custodial co-management system as practiced in Fiji. Smaller *kanakana* reef units (drawn) are not recognized, and the original resource owners do not gain directly from license fees for commercial activities in the *qoliqoli* even though such activities may be in direct conflict with subsistence fishing.

tion and input of all stakeholders. Understandably, if the process of community-based management is not carried out with sensitivity, it has the potential to undermine the traditional administrative structures of the community, and if this happens the process will undoubtedly collapse.

Governments tend to be top-down in their mode of operation, and fisheries extension officers are usually not trained to facilitate the process of bottom-up communal decision making. To expect marine ornamentals companies to facilitate the community-based management process on their own would be even more ab-

surd, and to consider this option would indicate a lack of understanding for the collective consultation process and all that is required. Training of workshop facilitators and capacity building in PLA techniques is basic to this process of participatory decision making.

Incorporating Traditional and Local Knowledge

The facilitating NGO must not only work within the local administrative structures but must encourage the expression of local knowledge in order to build the foundation of community

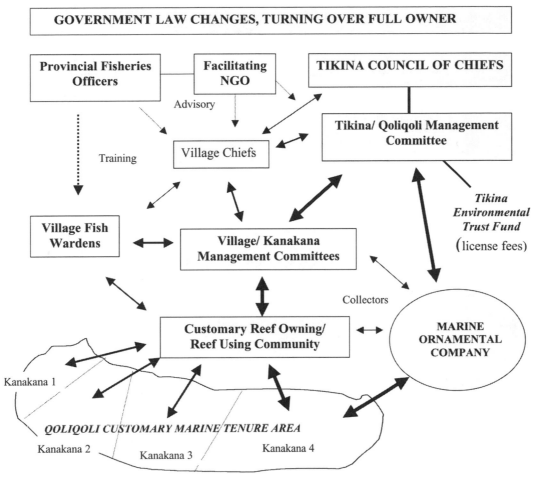

Fig. 11.2 Proposed marine resource management system for Fiji. The ultimate power rests with the District (*Tikina*) Council of Chiefs, with management planning and enforcement in the hands of *tikina* and village committees. Fees for commercial activities go into a trust fund earmarked for the environmental improvement of the district and to pay for management costs. NGOs and Government Fisheries Officers facilitate the process and serve as advisors and trainers.

awareness needed for management planning. Knowledge about such things as fish behavior, fishing methods, spawning seasons, spawning aggregations, and tides and moon phases and their influences may be important factors to consider when developing local management plans. Although traditional knowledge is an important aspect of management, management is an effective means for preserving traditional knowledge and skills, as fish behavior and other traditional information become irrelevant when all the fish are gone.

Whereas the day to day experience of fishers on the reef provides a strong foundation on

which to build an increased understanding of coral reef ecology and fisheries management, the awareness-building process must also include potential management options (reviewed in King and Lambeth 2000), as well as the biology behind the workings of each option. This awareness curriculum in fisheries management, developed by the NGO on the basis of the local needs, is best worked into the sessions as short presentations, presented if possible by government fisheries extension officers, who are briefed by the NGO beforehand. This is also a useful way of increasing governmental visibility in the meetings.

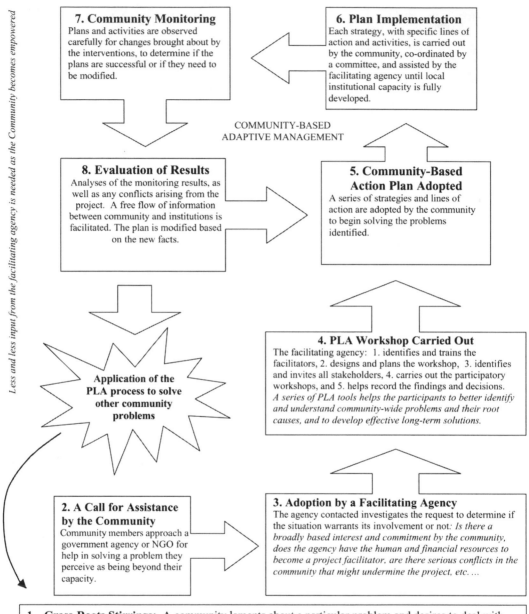

THE PARTICIPATORY LEARNING AND ACTION (PLA) PROCESS
A bottom-up approach to social and economic development   Austin Bowden-Kerby, 2002

Less and less input from the facilitating agency is needed as the Community becomes empowered

7. Community Monitoring
Plans and activities are observed carefully for changes brought about by the interventions, to determine if the plans are successful or if they need to be modified.

6. Plan Implementation
Each strategy, with specific lines of action and activities, is carried out by the community, co-ordinated by a committee, and assisted by the facilitating agency until local institutional capacity is fully developed.

COMMUNITY-BASED
ADAPTIVE MANAGEMENT

8. Evaluation of Results
Analyses of the monitoring results, as well as any conflicts arising from the project. A free flow of information between community and institutions is facilitated. The plan is modified based on the new facts.

5. Community-Based Action Plan Adopted
A series of strategies and lines of action are adopted by the community to begin solving the problems identified.

Application of the PLA process to solve other community problems

4. PLA Workshop Carried Out
The facilitating agency: 1. identifies and trains the facilitators, 2. designs and plans the workshop, 3. identifies and invites all stakeholders, 4. carries out the participatory workshops, and 5. helps record the findings and decisions. *A series of PLA tools helps the participants to better identify and understand community-wide problems and their root causes, and to develop effective long-term solutions.*

2. A Call for Assistance by the Community
Community members approach a government agency or NGO for help in solving a problem they perceive as being beyond their capacity.

3. Adoption by a Facilitating Agency
The agency contacted investigates the request to determine if the situation warrants its involvement or not: *Is there a broadly based interest and commitment by the community, does the agency have the human and financial resources to become a project facilitator, are there serious conflicts in the community that might undermine the project, etc. ...*

1. **Grass-Roots Stirrings:** A community laments about a particular problem and desires to deal with the root causes and to improve the situation.

Fig. 11.3   Flowchart of participatory processes of resource management planning. (Adapted by the author, 2002, from unpublished work of Dr. Hugh Govan)

The process leading to community-based coral reef management usually starts with fishers becoming acutely aware of declining fisheries resources and the associated negative impacts on community life and well being, with a longing to discover how to turn things around. It is at this point that community members often approach the government or an NGO.

Facilitation requires qualified personnel and financial resources, however, and the facilitating agency may or may not be able to take on the role of facilitator. If the decision is made to take on the role of facilitator, and before entering a community to begin the process, the facilitating agency should develop an exit strategy and plan for building up local capacity for management (IIRR 1998). One workable exit strategy is to help the community link with other agencies and government departments that have resources to help carry out the local management plans. This approach can also help strengthen the project overall, as some environmental problems may be too complex for communities to address on their own, and scientific or other input from the government or facilitating NGO may be required (King and Lambeth 2000).

Stakeholder Analysis and Involvement

One of the most critical steps in the resource management process is to identify and include all of the stakeholders that have the power to either make or break a resource management plan by either their inclusion or exclusion. For areas under customary marine tenure, the most basic stakeholder is that of the resource owner. It is this group in particular that must contribute actively and universally to the management plans and decisions. Women are in many places major fishers of shallow waters, and their voice must be heard and incorporated into the management plans. All plans developed by the community should be considered tentative until the formal approval and wholehearted support of the traditional leaders is obtained. These chiefs must be respected by the process and not given the impression that their traditional power is being undermined. The government also may feel threatened by the process if the various departments are excluded, as might any NGO working on conservation or resource management in the area.

Before facilitating management planning, a "stakeholder analysis" should be made by the management team (Table 11.1). After they have been identified, all stakeholders should be invited to participate at some point in the management planning process, regardless of the extra time required. Government and NGO personnel,

who will serve as workshop facilitators, should be identified and trained. The human and financial resources required for workshops to support the participatory process leading to the development of local resource management plans need to be secured, and a support and follow-up system needs to be developed, to build local capacity for monitoring and carrying out the plans. Several planning sessions at different venues can help to ensure that all stakeholders are involved, but care must be taken to ensure that the reef-owning community remains in the forefront.

The benefits of and sacrifices required for management need to be spread justly among the communal resource owners, and having all present for the planning sessions tends to ensure this. Compromises will need to be made by the overall group, such as decreased fishing effort, bans on commercial fishing, restrictions on gear, or the closure of areas to fishing as marine protected areas (MPAs), but these measures should not overburden one particular group over another. During the management planning process, new administrative structures may need to be created to carry out the management plans and to ensure the involvement and support of important groups such as women fishers. Existing local administration and chiefs should also be given a supporting role, with specific duties of follow-up as required.

Participatory Problem Identification and Problem-Solving Workshops

The PLA process takes place mainly during participatory workshops, where specific exercises and tools, such as resource mapping, time lines, and root-cause analyses, are carried out (Chatterton and Means 1996; IIRR 1998; Ecowoman 2000; Kuhn 2000). Full involvement by the community is encouraged, which may take 2 to 3 days per village, with thoughtful contemplation about environmental and related social problems. To encourage full participation, the workshop participants are often subdivided into "breakout groups" of four to five people each. Dominant people are intentionally segregated into the same group, as are overly shy people into theirs. Each group has a facilitator assigned to ensure that all topics and exercises are fully understood and are covered adequately. If staffing is insufficient, a facilitator can float among two or three groups to keep

Table 11.1   Stakeholder analysis for a customary marine tenure area: FSP-Fiji's Coral Gardens Initiative site, Cuvu District, Nadroga Province, Fiji

Stakeholders	Process stages			
	Initiation	Management workshops	Implementation	Ongoing support
CUSTOMARY OWNERS:				
Cuvu Tikina Community				
Chiefs of the 8 villages	4	5	5	5
Tikina Council of Chiefs	5	2	1	2
Tikina Environment	–	5	5	5
Komiti				
4 Kanakana Advisory Groups	–	–	5	5
8 village women's groups	0	5	5	5
8 village youth groups	0	3	5	4
subsistence fishers	1	5	5	5
commercial fishers	0	3	3	3
village churches	0	0	1	3
resident IndoFijians	0	0	2	2
FSP Fiji				
Facilitating agency	5	5	5	4
Fijian Shangri-La Resort	3	3	3	3
Government Departments				
Fisheries	3	5	4	4
Environment	2	4	3	3
Fijian Affairs	2	4	3	3
Provincial Office	4	5	4	3
Native Fisheries Com.	1	5	3	2
Agriculture	0	1	2	1
Forestry	0	1	2	1
Rural Development	0	2	3	2
Town and Country Plan.	0	1	2	1
Health	0	2	2	1
Tourism	2	4	3	3
Public Works	0	1	2	1
Education	0	1	2	3
Lands	0	2	2	1
Verata Vou Project				
Older "Model" Site	0	5	2	1
NGO Collaborators				
USP, WWF, OISCA, SPACHEE, etc.	1	3	4	2
Donors				
NZODA, Packard, MacArthur, etc.	5 passive	5 passive	5 passive	2 passive

Stakeholders are ranked based on their relative involvement in the management process: 0 = no involvement or support needed, 5 = intensive involvement or an active support role required. (Adapted by the author, 2002, from unpublished work of Dr. Hugh Govan)

things on track and moving. The appointed recorder records all information from each subgroup in poster form, with markers on newsprint, for later presentation to the larger group. This group work is best facilitated and conducted in the local language.

Several hands-on environmental mapping and assessment exercises are available for use in the workshops (IIRR 1998), each designed to stimulate thinking and to record changes in the environment and natural resources over time: resource abundance, average size of resource organisms, resource seasonality, harvest methods, destructive fishing, pollution problems, and so on. Once the major problems and challenges have been identified by the various subgroups, a root-cause analysis exercise is conducted for each problem, sometimes called a "problem tree." Once the underlying causes of problems are identified, brainstorming sessions on potential solutions are carried out. These solutions form the basis of the community resource and fisheries management plans.

Community-Based and Scientific Monitoring

Environmental monitoring of both conservation areas and collection sites needs to be carried out to observe potential changes brought about by management or by commercial harvesting activities (IIRR 1998). Community involvement in monitoring of its reefs for important food species in conservation and open fishing areas is also beneficial, to measure progress and indicate resource recovery. Community-based monitoring may be important in maintaining the community's confidence and support for the management plans. However, scientific monitoring will also be required in collection sites and aquaculture sites. So little is actually known about best-practice collection and culture methods and their potential negative impacts that basic data and scientific analysis of these data are required to reveal trends.

If possible, collectors should be involved as participants in scientific data gathering and should be educated as to the potential negative impacts of particular practices, as well as seeking improvements in best-practice methods through experimental trials. Monitoring will provide information that will allow for substantiation of and improvements in the standards.

The industry will have to shoulder the financial burden of scientific monitoring of collection sites, with the monitoring carried out by trained marine scientists or fisheries staff, with additional manpower obtained from the community, some of whom may be active in the industry as collectors. As best-practice standards become scientifically confirmed, universally understood, and applied, scientific monitoring will likely require less time and fewer resources. The fisheries wardens and other members of the community should be educated in best-practice standards for collection, to help enforce the standards in the communal waters.

The Cuvu Experience as a Model

The Cuvu project began as a small pilot in January 1999, after a request from the Cuvu community and with funding from the Pacific Development and Conservation Trust (New Zealand) and the Fijian Shangri-La Resort. On the receipt of funding from the Packard and MacArthur Foundations and New Zealand Overseas Development Assistance (NZODA) in June 2000, the project was expanded into a full 3-year project.

The Cuvu site, located on Fiji's Coral Coast, is a fringing reef site that has considerable negative terrestrial influences. The facilitating agency in the "Coral Gardens" project is the Foundation for the Peoples of the South Pacific (FSP). In 1999, nearly 2 years before approval of the management plan, the Tikina Council of Chiefs formed a *Cuvu Tikina Environment Komiti* to oversee and help facilitate the management planning and implementation process: workshops, scientific monitoring studies, and management interventions. The eight coastal villages of Cuvu Tikina (district) went through the entire PLA process with FSP-Fiji facilitating, using the participatory workshop methods described in this chapter. The management plans developed by the reef-owning communities were discussed in depth by the *Tikina Environment Komiti*, given final approval by the Tikina Council, and then endorsed by the Paramount Chief *Na Ka Levu,* the honorable Ratu Sakiusa Gavidi on July 4, 2001.

The management plan sets aside three no-fishing *Tabu* areas on the reefs as well as one mangrove no-fishing *Tabu* area for a period of at least 3 years, subject to annual review (Fig. 11.4). The comprehensive management plans set aside some 50% of the Tikina's reefs as well as some adjacent mangrove and seagrass habitats as no-fishing areas. A ban on destructive fishing methods, minimum size limits, and other measures are essential components of the community-based management plans as well. After plan adoption, the Komiti arranged for the Government Fisheries Officers to train 16 men from the communities as fish wardens, educating them in the laws and giving them the power to arrest violators of national laws. The Environment Komiti is currently involved in problem solving, enforcement, invertebrate restocking of MPAs, restoration plans, and negotiation with the Fiji National Government to assist in particular areas beyond their means. The MPA adjacent the Fijian Shangri-La Resort is being investigated for potential elevation to a permanent no-fishing-zoned *Yanuca Marine Park*, involving the resort. This will include increased usage by guests and increased benefits to the communities, with an *Environmental Trust Fund* being set up to channel usage fees from guests into resources for carrying out the management and environmental restoration plans throughout Cuvu Tikina.

FSP has begun training village youth in reef monitoring and in hands-on habitat enhancement and reef restoration activities to accelerate the resource recovery processes on reefs: crown-of-thorns starfish removal, creation of giant clam and beche-de-mer sea cucumber spawning aggregations, mangrove replanting, and so on. Project incentives in the developmental stages involve ecotourism within the MPAs, associated with the adjacent Shangri-La Fijian Resort and commercial coral and *Tridacna* clam aquaculture.

In early 2002, the Cuvu site was selected by United Nations Environment Program (UNEP) as one of the International Coral Reef Action Network (ICRAN) "model sites for coral reef conservation" in the Pacific region. Through ICRAN funding, the lessons learned will be applied to sites throughout the region, beginning

with reactivating work begun by the author in the Solomon Islands (outlined below).

Coral Reef Husbandry by Reef-Owning Communities

No-Fishing MPAs as a Community-Appropriate Management Strategy

Traditional coral reef fisheries management systems (Johannes 1978; Polunin 1984; Ruddle and Johannes 1990) have in many places been weakened or have died out (Johannes 1978). However, a cultural foundation still exists in many areas on which to build a sustainable resource development model for use by subsistence fishing communities (Dight and Scherl 1997; Gomez 1997; Biodiversity Conservation Network 1999). No-fishing MPAs, recently acclaimed as a scientific breakthrough in fisheries management (Randall 1987; King 1995; Littler and Littler 1997; Allard 1999; Gleason 1999; Hixon et al. 1999), are in fact part of a traditional Pacific Island resource management practice of *Kapu* or *Tabu* (Allard, 1999; Bowden-Kerby 2002), of which the Pacific peoples should be rightly proud. In bygone days, no-fishing areas were associated with sacred sites or with burial areas on reefs

CUVU TIKINA MANAGEMENT PLAN
Approved by the CUVU Tikina Council, July 2001

Fig. 11.4 Cuvu Tikina management plan established by the Cuvu community and chiefs to conserve and restore the environmental and associated fisheries resources.

(Bowden-Kerby 2002). In Fiji, however, fishers still readily grasp and support the closure of a reef area for a specific time period of several months after the death of a high chief. This is done for respect, although coincidentally when the area is opened, fish and shellfish are more abundantly available for the customary feast marking the end of the mourning period. With the existing understanding in Fiji of how reef closure allows for resource recovery, the establishment of more permanent no-fishing *Tabu* areas is perhaps more culturally acceptable than in areas that do not have such a management history.

Once empowered with additional scientific knowledge of how effective MPAs are and the many reasons why they work, the customary fishing rights owners more readily close reef and mangrove areas to fishing for longer periods than customary, with the potential of setting up permanent MPAs.

In several areas where reefs continue to be under customary marine tenure, the re-establishment of locally managed MPAs or *Tabu* areas is a promising development in the area of coral reef management (Bohnsack 1993; Roberts and Polunin 1993; Gomez 1997; McClanahan 1997a; Kallie et al. 1999; Biodiversity Conservation Network 1999; World Bank 1999). Establishing no-fishing and no-harvesting MPAs has been recommended for inclusion as a management tool for a sustainable and certified marine ornamentals industry as well (Randall 1987; Pyle 1993).

Low-Tech Measures to Restore Coral Reefs and Re-Establish Reef Resources

If newly established MPAs have few living coral habitats or are dominated by recruitment-inhibiting substrata, the full recovery of living coral habitats and associated fish populations could be delayed for many years. Unfortunately, if an MPA takes too long to respond positively to closure, its long-term success could be jeopardized, and the community might lose interest and reject MPAs as an effective fisheries management tool (World Bank 1999; Bowden-Kerby 2001b). Degraded and chronically impacted reefs that do not recover naturally may require low-tech interventions in order to restore coral populations and fisheries resources,

and such restoration methods are beginning to be developed and tested.

Once the reef areas are effectively protected, restocking severely overfished species of invertebrates may help to accelerate the recovery of brood stocks within MPA areas and may also help increase community involvement and support for no-fishing MPAs. Restocking reefs with giant *Tridacna* clams (Munro 1989) *Trochus* shells (Nash 1993), and *Pinctada* pearl oysters (Sims 1993) has been either carried out or suggested. Restocking MPAs with populations of these and other important invertebrates such as turban snails (*Turbo*), spider conchs (*Lambis*), cockles *(Anadara),* and valuable species of sea cucumbers *(Holothuria,* etc.) may be a viable resource restoration option, recreating breeding populations of key species on reefs where they were formerly abundant. Restocking of all of the above species with the exception of pearl oysters is currently being tested within three MPAs in Fiji (Bowden-Kerby unpublished).

As simple and low-cost restoration methods are proven and become available, communities and reef managers will increasingly have the option of intervening to restore degraded, non-recovering coral reef areas and reef resources, particularly within MPAs.

Fishing Regulations Appropriate for Inclusion in Community-Based Management

In addition to no-fishing MPAs, measures such as size limits for harvested species, bans on overfished species, gear restrictions (mesh size limits, SCUBA fishing bans, etc.), temporary reef closures and harvesting seasons, bans on destructive fishing, and reef zoning (reviewed in Alcala and White 1984) may be integral to the fisheries management plan developed by a particular community. The zoning of restricted use areas may also be warranted to avoid conflicts among subsistence fishing, commercial fishing, tourism, and the marine trades. Communities should be encouraged to establish subsistence fishing areas free of the effects of commercial fishing. Zones might also be established for collecting marine ornamentals or for commercial aquaculture of various types. Special tourism areas may also be appropriate, particularly if they overlap with no-fishing marine protected areas (MPAs).

No-fishing MPAs are a viable and highly beneficial conservation and resource management strategy recently gaining acceptance globally (Randall 1987; King 1995; Littler and Littler 1997; Ruddle 1998; Allard 1999; Gleason 1999; Hixon et al. 1999). No-fishing MPAs have been recommended as an important means to sustainably manage the marine ornamentals industry (Randall 1987; Pyle 1993). In reef areas under customary marine tenure, the establishment of locally managed MPAs is a promising development in the area of community-based reef management (Bohnsack 1993; Roberts and Polunin 1993; Gomez 1997; Mc-Clanahan 1997a; Kallie et al. 1999; Biodiversity Conservation Network 1999; World Bank 1999). Unfortunately, if newly established MPAs have few living coral habitats or are dominated by recruitment-inhibiting substrata, the full recovery of living coral habitats and associated fish populations could be delayed for many years. If a locally managed MPA takes too long to respond to closure, its long-term success could be jeopardized and the community might lose interest and reject MPAs as an effective fisheries management tool (World Bank 1999).

Degraded and chronically impacted reefs that do not recover naturally may require interventions in order to restore coral populations and fisheries resources. Methods to restore degraded reef areas are beginning to be developed and tested. Where reefs have been converted into shifting rubble so that coral recruitment is inhibited, coral has been transplanted to accelerate the recovery of coral populations (Bowden-Kerby 1997, 2001; Lindahl 1998; Fox et al. 1999). Other restoration strategies to re-establish an ecological balance more favorable to coral survival have been successfully demonstrated: removing sea urchins from reefs where excessive urchin grazing prevents coral recruitment (McClanahan et al. 1996; McClanahan and Muthiga 1999), removing overabundant macroalgae, which inhibits recruitment (Naim et al. 1997; McClanahan et al. 1999), and removing coral-killing crown of thorns starfish from reefs that have chronic outbreaks (Wachenfeld et al. 1998; Kallie et al. 1999).

A sustainable coral reef conservation and resource development model should include no-fishing *Tabu* MPAs and other conservation measures as above, fisheries habitat enhancement and coral transplantation on damaged reef areas, reintroducing commercially valuable invertebrate species within MPAs to serve as spawning aggregations, and the aquaculture of coral reef species of high commercial value currently being harvested from the wild, such as ornamental corals (Fig. 11.5) (Bowden-Kerby 1999b, 2001b).

Community Involvement in the Marine Ornamentals Trades

The Trades as Incentives for Coral Reef Conservation and Management

Community-based management is proving effective in reversing the prevailing trend of coral reef degradation in several sites in the region, and the marine aquarium industry is viewed as a vehicle to stimulate a wider application of effective grass-roots management of coral reefs. The requirement of effective community-based management as a basis for site certification would provide impetus and, it is hoped, resources for the processes required to develop and carry out such management plans. The marine aquarium industry would thus begin working more closely with governmental departments and with nongovernmental community development organizations toward the common goal of community-based marine resource conservation and environmental restoration, with the sustainable marine aquarium trades serving as a direct incentive for conservation within customary marine tenure areas. The goal would be to empower resource-owning communities to take full responsibility for the wise utilization of their own marine resources and, in accordance with existing traditional and governmental structures, stimulating the establishment of no-fishing MPAs and other conservation measures in all areas where the marine aquarium industry operates.

Perhaps the greatest threat to the success of community-based no-take MPAs is the delay in resource recovery. The closure of reefs to fishing activities deprives communities the use of portions of their fishing grounds. As discussed earlier, waiting many years for a recovery of the fisheries resources and reef system can erode support for the project and could cause the collapse of local management plans. Economic in-

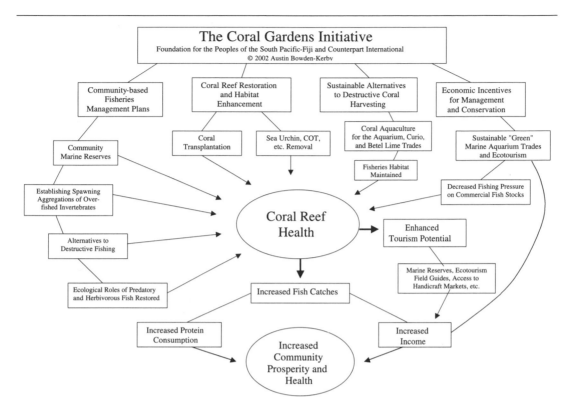

Fig. 11.5 The Coral Gardens Initiative coral reef management and resource restoration model, currently being implemented by Counterpart International in association with Foundation for the Peoples of the South Pacific international.

centives such as involvement in the marine ornamentals trades, aquaculture, or ecotourism, if tied to sacrificial management plans, would help ensure the success of community-based management and thus would contribute significantly to coral reef conservation. On the other hand, if such benefits are not tied to management, they can eventually become negative pressures on the environment.

The Negative Impact of Industry Pressures When Out of a Conservation Context

If a collecting company, after paying the license fee to the custodial chief, comes into a particular *qoliqoli* without encouraging resource management plans, without consulting the community, brings in outsiders to do the harvesting, hires community members on the basis of chiefly ties rather than *kanakana* ownership, conflict will arise. If the collecting activities in-

terfere with subsistence fishing, cause damage to the reefs, or are on reefs used for tourism (Suzuki 1989), even more conflict can be expected. Another problem is the competition between companies for the same reef area (Lubbock and Polunin 1975). Competition can result in a lack of incentives for conservation of the resource, as the companies rush to collect the most valuable organisms in competition with each other and with no clear way of assigning blame when corals are carelessly smashed in the rush to harvest ornamental fish or corals. These problems have all occurred at some point in the history of the industry in Fiji.

Community-Based Coral Aquaculture Tied Directly to Coral Reef Management

Coral aquaculture in South Pacific Island communities has recently reached the level of successful production (i.e., thousands of coral colonies) (Bowden-Kerby 1999b). Low-tech field culture methods for scleractinian and other stony corals for the aquarium and curio coral trades have been tested successfully in Fiji, the Solomon Islands, and Puerto Rico (Bowden-Kerby 1999b). These trials have met with success in all locations and are in the second phase of refinement. The fact that corals can be grown successfully has been demonstrated; however, coral aquaculture must also be nested in a community-based management approach. In the Solomon Islands, in the first relatively major commercialization of field-cultured corals, the women of Marau Sound were trained in coral aquaculture in 1997 (Bowden-Kerby 1999b). This International Center for Living Aquatic Resource Management (ICLARM)-sponsored workshop trained 30 women in the culture of scleractinian reef corals for sale to a local aquarium coral exporter. Corals were grown from 3 to 5 cm fragments of wild "seed" for the aquarium trade in 4 to 7 months. The coral fragments were planted onto small concrete disk bases. The fragments were not glued onto the bases but relied on natural self-attachment and overgrowth processes. To hold the coral fragments securely to the bases so that self-attachment could occur, both were woven securely onto wire mesh with 30-pound test monofilament line. Simple "culture tables" were constructed to elevate the planted mesh frames over the substrate, made of bent iron bars held together with baling wire of the type ICLARM uses for *Tridacna* clam culture. Initially coral fragments were collected as seed from "mother colonies," corals located on the reef that had desirable characteristics of color and form; however, the first generation of cultured corals should be grown into mother colonies in order to conform to CITES regulations regarding "captive-bred" organisms, getting away from the wild stock for all but selection purposes. Research will also be required to identify coral strains (genotypes) with superior characteristics (color, form, growth rates, survivability, etc.), and these colonies should be cultured into mother colonies, to be used in 2 to 3 years as coral seed sources, bypassing the need to rely heavily on wild-harvested seed. Corals grown by these "coral gardens" methods are easily identifiable as properly cultured corals by overgrowth onto a base, as well as by the presence of deeply ingrown monofilament line. The sug-

gestion by some that alizarin dye be used as an indicator that corals have been cultured for a long enough period to qualify as aquacultured corals is considered redundant (Bowden-Kerby 1999c).

Personal experience has shown that the aquaculture of corals does not guarantee sustainable or environmentally benign methods. The coral culture venture at Marau was not part of an awareness-raising or wider conservation effort, nor was there any follow-up or monitoring program. In an effort to plant corals as an income-generating venture, the coral farming women of Marau Sound planted some 30,000 corals within 3 months of a single 2-day workshop, causing moderate reef damage at several sites (Bowden-Kerby 1999b). Civil strife brought the coral farming activity in Marau to a halt in 1999. With the Fiji Coral Gardens site being selected as a model site by ICRAN in 2002, the increased funding has allowed the work in the Solomon Islands to be reactivated. The complete community-based management planning approach is now being applied, through Solomon Islands Development Trust (SIDT) and Foundation for the Peoples of the South Pacific International (FSPI). The corals at Marau Sound, abandoned in 1999, have continued to grow on the culture tables and have become quite large (>30 cm in height). Once the community planning process is completed, these cultured corals will serve as "mother colonies" to produce seed stock for new farms in Marau, Malaita, and Gela, to be located in or near no-fishing MPAs as an incentive for community-based management. Corals grown from this stock will in theory be CITES exempt, if they can be certified as second generation farmed corals.

The demand for stony reef corals is high, and this demand is being met almost exclusively by wild harvest. The aquarium trade removes several hundred thousand juvenile corals per year from South Pacific reefs. The curio coral trade also could absorb cultured corals, although several years of growth may be needed before salable products are available. However, trying to meet this demand by greenhouse culture of corals in cold climates would be energy intensive and would potentially burn large amounts of fossil fuels, producing a great amount of CO_2 gas over time. If such were the case, greenhouse culture of corals might not be considered a "green" practice, particularly from the standpoint of Pacific Island communities that are particularly susceptible to global warming and to the threat of rising sea levels. The exploitation of coral reef biodiversity by industry by culturing captive corals in greenhouses effectively bypasses the indigenous owners of the parent material and thus may be considered a violation of certain United Nations conventions on indigenous rights as well (Bowden-Kerby 1999b; Cullet, 2001).

In perspective, traditional coral harvesting dwarfs the ornamental coral trades. In the Solomons, Papua New Guinea, Indonesia, and other islands, live corals are harvested for the production of betel lime and for use as fill material, for example, and these activities have a major negative impact on many reef areas. Branching *Acropora* corals have been eliminated from extensive lagoonal areas to produce lime (Bowden-Kerby 1999b; 2001b). Branching corals are important shelter habitat, and reef fish have become rare where these corals have been destroyed. Traditional coral harvesting and destructive fishing practices also need to come under sustainable management, or alternatives need to be found and demonstrated to the fishers. The participatory process described in this chapter is appropriate to address these problems of reef decline as well.

The Foundation for the Peoples of the South Pacific International and its U.S. affiliate, Counterpart International, embarked the *Coral Gardens Initiative* in 1999 (Fig. 11.5). This program encourages a sustainable marine ornamentals industry and involves indigenous communities in the culture of corals as an incentive to the establishment of marine reserves, low-tech coral reef restoration, and the development of community-based resource management plans (Bowden-Kerby 1999e, 2001b). Commercial coral aquaculture for the aquarium and ornamental coral trades and for the production of betel lime is being introduced in project communities and will help offset the initial economic burden of discontinuing the wild coral harvest, establishing no-fishing areas, and application of new fishing regulations. The initial work has demonstrated that communities and traditional leaders are responsive to the participatory methods, becoming more environmentally aware as they eagerly participate in the coral aquaculture, coral reef restoration, and

conservation aspects of the project (Bowden-Kerby 1999b,c,d).

Minimizing the Environmental Impact of the Marine Ornamentals Trades

Empowering Collectors to Understand the Ecological Basis of "Green" Methods

The essential knowledge base required of collectors should be taught by neutral third-party instructors and based on reef ecology and natural recovery processes. The first requirement, before commercial activities are certified, should likely be ensuring that local management plans and monitoring are in force.

Sustainable live rock extraction is another potential activity, now that the management plans and community-based monitoring are in force. The focus of the live rock harvest, if it is to progress beyond the experimental stage, would be to target areas of the reef flat that dry during extreme low tides and to create tidal pools of higher biodiversity and enhanced fisheries habitat. Live rock would thus be a by-product of habitat enhancement rather than an end in itself.

Recommendations for Ecologically Sustainable Development of Reefs under Customary Marine Tenure

The Fijian communities and their chiefs must take full responsibility for monitoring and regulating what goes on in their own waters, as these customary owners are the ones who stand to lose or to gain in the long term. Programs must be created to empower the custodial system with the knowledge and capability for improved self-monitoring and regulation. The Industry must also work to ensure sustainable practices and must monitor its own collectors in the field on a regular basis, upholding *qoliqoli* management plans and regulations. The government's role might be to ensure effective community-based management is in place before a permit is granted or renewed, with supervision in part delegated to NGOs working directly with the communities. Inventing an elaborate system of fisheries conservation officers and fines for violations is not recommended as an effective or workable system. The authority of government may best be exercised by ensuring that collectors and companies have undergone training in best-practice methods, restricting the issuing of permits to waters with active community-based management plans, and collecting and channeling fees toward sustainable management.

Because of the realities of collective ownership of the marine resources, any sort of commercial harvesting within a customary tenure area, regardless of the country, would ideally be permitted only in areas with an established resource management plan and with established processes of adaptive management, likely involving a management committee operating within the particular cultural and governmental framework and involving the fishers and resource users. Community-based management plans would most likely have areas reserved exclusively for subsistence fishing, areas for special use such as for tourism or aquaculture development, effective no-fishing marine reserves (*Tabu* areas), as well as the commercial fishing and collecting areas. A system of zoning such as this would help prevent user-based conflicts. Resource management plans might also include reef restoration activities and resource recovery plans for overfished species.

Wherever the aquarium and curio trades operate, the community needs to be capable of a certain amount of self-regulation and self-monitoring. Therefore, the wider community should have a high level of coral reef awareness, and a concerted educational program targeting fishers and resource users should be undertaken. In order to train, educate, and support communities in marine resource management, government departments, educational agencies, and NGOs must be given the mandate and financial resources to create and carry out programs in community-based awareness, management, and monitoring.

Communities would benefit if they were more involved as partners to the industry, gainfully employed and justly compensated. As communal resources are being used for profit, a portion of benefits and profits should go to a community fund. With the exception of individuals who have specific technical skills unavailable in the communities, community members should be employed as collectors to avoid conflicts and to encourage improved reef steward-

ship. Collectors should also be trained in sustainable, best-practice methods (see Table 11.1). Only one company should operate in a given coral reef area to ensure accountability and sustainable practices. The larger tenure area might be broken into smaller subsections for licensing, but 10 km of healthy reef area should perhaps be considered a minimum harvest area for the fish and coral trades by a single company, to prevent overexploitation. Several of the smaller reef areas might also need to be combined in order to support a single company. No single company should be allowed to purchase more licenses than required for its operations for purposes of excluding other companies. The trades must not be allowed to conflict with or detract from the tourism industry or subsistence fishing activities.

Through a system of government-imposed fees on the industry, the community-based management approach could be financed. This money would be specifically earmarked for sustainable management and monitoring of the industry, community education, the training for collectors, marine conservation projects, and for relevant research, and deposited into a "Fiji Coral Reef Trust Fund." Because of the volume of the current trades in Fiji, the fees need not be excessive. An initial suggestion for these environmental taxes are 3% to 5% of wholesale value. On the basis of present prices for Fiji live rock, this tax would amount to 30 cents per kilogram (5% of Fj $6 per kilogram). Aquacultured corals could be exempt from taxation if they were part of a community-based aquaculture project.

A process by which NGOs and other organizations could develop proposals and apply to the fund would fuel local efforts directed toward industry sustainability, community-based management, and coral reef conservation. With the required resources, the industry would be well on its way toward stimulating community awareness, improved coral reef management, a system of community-based marine reserves, restored coral reefs, and compatibility between the tourism and marine industries, as well as improved international opinion and "green" certification for Fiji-based industries. Having a pool of funds available locally would greatly help local NGOs implement effective conservation, monitoring, and management training programs.

Specific Recommendations for Ecologically Sustainable Collection and Management

The Live Rock Trade

On the basis of casual field observations at the major live rock collecting sites on Fiji's Coral Coast, it appears that sustainable management of the live rock trades is possible. Currently (2002) no guidelines exist for collectors in Fiji, and apparently no research or monitoring has been done to determine best practices in which to train these collectors. Casual observation of sites indicates that while the coral making up the bulk of live rock is indeed dead, it provides a substratum for recolonization by corals and other organisms, and thus for recovery of coral populations over time. However, if live rock harvesting converts this hard rock substratum into shifting rubble and sand, natural reef recovery processes are prevented (Bowden-Kerby 1999a).

Research should be conducted to investigate changes in coral reef biodiversity in collecting areas, as well as changes in wave energy and currents. The climate change implications of the live rock trade should be seriously considered and studied, as storm wave energy and currents may be affected by modifications to the reef flat. Live rock mining could potentially help create tide pools on the outer reef flat, and such tide pools should be studied for potential biodiversity enhancement and wave energy dissipation.

With no real data on which to base best practices, for the time being, a list of principles based on the ecological situation of the coral coast fringing reefs is provided, as follows:

1. Reef flat mining for live rock should not compromise the basic habitat structure of the reef.
2. Collection should not destroy reef flat tidal pool environments.
3. Collection or activities associated with collection should not convert rocky habitats into rubble or sand environments that do not recover.
4. Collection should be focused on the outer reef flat/reef crest zone, where regeneration is rapid, avoiding the dead coral "bommies" surrounded by sand of the inner reef flat.
5. Care must be taken not to break through elevated "algal rim" structures on the reef made of live rock, which serve as dams and keep the reef from drying out, trapping seawater on the reef flat during low tides.

Aquarium Fish Collection

Tabu no-fishing marine reserve areas should be maintained within the *qoliqoli* waters that include good habitat for highly desirable species of aquarium fish. Active research should be undertaken in developing new methodologies and equipment for collecting fish without damaging coral cover. Advances should be shared within the industry. An effort to plant corals or other habitats for larval attraction in lagoonal areas as aquarium fish habitat should be encouraged. For collection in calm lagoonal areas, unattached colonies can be picked up and the fish removed, with the coral replaced to its original position.

1. For collecting probes, only lightweight, hollow aluminum or fiberglass rods should be used in fish collection; heavy iron bars, which break corals easily, should be prohibited.
2. Less than 10% of coral branches should be broken or damaged when fish are collected. Intentionally breaking up the colony should be prohibited. When minor breakage occurs, the branches should be wedged tightly within the original colony or so that their survival will be more likely or transplanted to a reef restoration site.
3. All dislodged branches or dislodged coral colonies in higher-energy areas must be replanted in coral reef restoration sites maintained by the collectors and the community.
4. For aggregate coral-loving fish, coral colonies should not be dislodged from the reef. Sustainable techniques must be developed.
5. To prevent overharvesting of the rarer species, once catch per unit effort of a particular species drops by 50%, the industry should discontinue harvesting these species for at least one full year.

Aquarium Coral Collection

Sustainable coral aquaculture should be undertaken as an alternative to wild coral harvesting for the aquarium and curio coral trades. Potentially coral farming and selection of seed stock could produce a higher-quality, higher-value product, with lower mortality and superior characteristics desired by the consumer. Coral farming should not be underestimated as a potential economic incentive for community-based coral reef conservation and reef restoration projects. Ultimately, a complete phaseout of the trade in wild corals may be possible; however, in the meantime the following principles based on reef ecology are suggested as guidelines for sustainable harvest of corals:

1. Collecting of corals should not be permitted on damaged, stressed, or recovering reefs, or on reef areas with less than 30% coral cover.
2. Juvenile corals with no other corals within a 50-cm radius must not be harvested, as they indicate a situation of low local coral cover.
3. Slow-growing genera (massive corals, brain corals, etc.) should be collected exclusively from extreme shallows, where their ability to grow into adult corals is prevented, or from areas where fast-growing corals are overgrowing and thus killing them.
4. Broken or unusable corals should be replanted into reef restoration sites, not left in areas where their survival is unlikely.
5. Rare species and unusual color morphs and growth forms should be propagated both on the reef at restoration sites and in coral farms to preserve biodiversity and to maintain and improve the quality of the industry.

Curio Coral Collection

1. Communities involved in the curio coral harvest must maintain reef restoration sites, whereby for every coral harvested, others will be replanted in damaged or rubble areas proven to be suitable for their growth. This will utilize the inevitably broken corals, replanting them where they have a good chance of growing and enhancing habitat.
2. Curio coral collecting should be confined to areas of healthy reefs, with coral cover greater than 50%. No more than 20% of the natural coral cover should be removed from a particular reef area. No more than half of the cover of a particular species should be removed from a localized reef area, to prevent the species desired by the trade from

becoming rare where it is naturally abundant.
3. Coral colonies should be thinned from dense areas of coral growth. Isolated coral colonies (>1 m away from other corals) should not be collected. The primary target of the industry should be corals in competitive overgrowth relationships with other corals, which ultimately result in the death of the slower-growing species. Table-forming corals, important fish habitat species, may allow corals to survive underneath and thus may not automatically fall into this category.
4. Slow-growing "massive" corals or rare corals should be removed only from crowded or competitive overgrowth situations or from extreme shallows where they will ultimately die because of frequent natural disturbances.
5. Communities or individuals involved in the collection of curio corals must begin to engage in activities that protect the reef, such as removing crown-of-thorns starfish from the *qoliqoli* waters wherever they reach epidemic proportions, preventing mass-mortality to the corals.
6. Coral aquaculture should be the focus for the future of the curio coral industry, with a planned phaseout of much of the wild harvest over time.

References

Alcala, Angel C., and Alan T. White. 1984. Options for management. Pages 31–40 In Coral reef monitoring handbook. (R.A. Kenchington and B.E.T. Hudson, eds.) UNESCO Regional Office for Science and Technology, Jakarta. 281 pp.

Allard, P.J. 1999. A decision analysis approach to managing effects of eutrophication on coral reefs in Barbados. Abstract in: Proc. Int. Conf. on Scientific Aspects of Coral Reef Assessment, Monitoring and Restoration, NCRI. April 14–16, 1999, Ft. Lauderdale, Florida, p 45.

Anderson, J.A., P. Townsley, T. Tavusa, F. Poni, F. Osbourne, E. Qalo, A. Botilagi, S. Wakatibau, T. Wakatibau, and Tulala. 2000. Fiji Country Report, Volume 2b *in* J.A. Anderson and C.C. Mees. The Performance of Customary Marine Tenure in the management of community fishery resources in Melanesia. Final Technical Report to the UK Department for International Development, MRAG Ltd, London, July 1999. 125 pp.

Bak, Rolf P. 1990. Patterns of echinoid bioerosion in two Pacific coral reef lagoons. *Mar. Ecol. Prog. Ser.* 66:267–272

Berg, H., M.C. Ohman, S. Troeng, and O. Linden. 1998. Environmental economics of coral reef destruction in Sri Lanka. *Ambio* 27(8):627–634.

Biodiversity Conservation Network. 1999. Final stories from the field. Evaluating linkages between business, the environment and local communities. N. Salafsky, ed. BCN Biodiversity Support Program, Washington DC. 219 pp.

Birkeland, Chuck, and J.S. Lucas. 1990. *Acanthaster planci: Major management problem of coral reefs.* CRC Press, Boca Raton, Florida. 257pp.

Bohnsack, James A. 1993. Marine reserves: They enhance fisheries, reduce conflicts, and protect resources. *Oceanus* 36:63–71.

Bowden-Kerby, Austin. 1997. Coral transplantation in sheltered habitats using unattached fragments and cultured colonies. *Proc. 8th Int. Coral Reef Symp.* 2:2063–2068.

Bowden-Kerby, Austin 1999a. Can destructive reef flat mining practices of the "live rock" trade be harnessed for biodiversity enhancement and climate change mitigation? Abstract in Marine Ornamentals '99, UH Sea Grant. November 16–19, Kona, Hawaii.

Bowden-Kerby, Austin 1999b. Coral aquaculture by Pacific Island communities. *World Aquaculture'99.* The World Aquaculture Society. 26 April–2 May 1999, Sydney, p 91.

Bowden-Kerby, Austin 1999c. Working towards sustainable and environmentally sound "green" methods of coral aquaculture. Abstract in Marine Ornamentals '99, UH Sea Grant. November 16–19, Kona, Hawaii.

Bowden-Kerby, Austin 1999d. The Coral Gardens Initiative: Coral aquaculture as an incentive for community-based coral reef conservation and restoration. Abstract in Marine Ornamentals '99, UH Sea Grant. November 16–19, Kona, Hawaii.

Bowden-Kerby, Austin 1999e. The Community Coral Reef Initiative: Coral reef restoration in rural Pacific Island settings. Abstract in Proc. Int. Conf. on Scientific Aspects of Coral Reef Assessment, Monitoring and Restoration, NCRI. April 14–16, 1999, Ft. Lauderdale, Florida. p 60.

Bowden-Kerby, Austin 2001a. Low-tech coral reef restoration methods modeled after natural fragmentation processes. *Bulletin of Marine Science.* 69(2):915–931.

Bowden-Kerby, W.A. 2001b. Coral transplantation modeled after natural fragmentation processes: Low-tech tools for coral reef restoration and management. Ph.D. Thesis, University of Puerto Rico at Mayaguez. 195 pp.

Bowden-Kerby, Austin 2002. The original MPAs. MPA News 3(6):6.

Brown, B.E., and R.P. Dunne. 1988. The environmental impact of coral mining on coral reefs in the Maldives. *Environ. Conserv.* 15:159–166.

Brown, B.E., and L.S. Howard. 1985. Assessing the effects of stress on coral reefs. In *Advances in Marine Biology.* I.H.S. Baxter, F.S. Russell, and M. Yonge. eds. Academic Press, New York, 22:1–55.

Chatterton, P., and K. Means. 1996. Community resource conservation and development—a toolkit for community-based conservation and sustainable development in the Pacific. WWF South Pacific Program.

Christy, F.T. 1982. Territorial use rights in marine fisheries: Definitions and conditions. FAO Fish. Tech. Paper #227. 10 pp.

Clark, S., and A.J. Edwards. 1995. Coral transplantation as an aid to reef rehabilitation: Evaluation of a case study in the Maldive Islands. *Coral Reefs.* 14:201–213.

Connell, J.H. 1997. Disturbance and recovery of coral assemblages. Coral Reefs Supplement. 16:S101–S113.

Cooke, A. 1994. *The Qoliqoli of Fiji: Management of Resources in Traditional Fishing Grounds.* Unpublished master's thesis. Department of Marine Sciences and Coastal Management, University of Newcastle upon Tyne, United Kingdom.

Cooke, A., N.V.C. Polunin, and K. Moce. 2000. Comparative assessment of stakeholder management in traditional Fijian fishing-grounds. *Environmental Conservation* 27:291–299

Cordell, J. 1991. *Cultural Survival Quarterly* 15(2):3–10.

Cullet, Philipe. 2001. Property rights regimes over biological resources. International Environmental Law Research Centre. http://www.ielrc.org/Content/A01041P.pdf. 17 pp.

Dahl, Christopher 1986. Traditional marine tenure: A basis for artisanal fisheries management. *Marine Policy* 12:40–48.

Davis, S. 1985. Traditional Management of the Littoral Zone among the Yolngu of North Australia. In K. Ruddle and R.E. Johannes, eds. *The Traditional Knowledge and Management of Coastal Systems in Asia and the Pacific.* Jakarta, UNESCO-ROSTSEA, pp 101–124.

Dight, I.J, and L.M. Scherl. 1997. The International Coral reef Initiative (ICRI): Global priorities for the conservation and management of coral reefs and the need for partnerships. Proc. 8th Int. Coral Reef Symp. 1:135–142.

Done, Terry J. 1992. Phase shifts in coral reef communities and their ecological significance. *Hydrobiologia* 247:121–132.

Dulvy, N.K., D. Stanwell-Smith, W.R.T. Darwall, and C.J. Horill. 1995. Coral mining at Mafia Island, Tanzania: A management dilemma. *Ambio* 24:358–365.

Ecowoman. 2000. Participatory learning and action: A trainers guide for the South Pacific. Ecowoman Project, South Pacific Action Committee for Human Ecology and Environment (SPACHEE). Suva, Fiji. 57 pp.

Fong, G.M. 1994. Case study of a traditional marine management system: Sasa Village, Macuata Province, Fiji. Project RAS/92/T05 Case studies on traditional marine management systems in the South Pacific. Field report 94/1. FAO/FFA. Rome. 85 pp.

Fox, Helen E., R.L. Caldwell, and J.S. Pet. 1999. Enhancing coral reef recovery after destructive fishing practices in Indonesia. Abstract in Proc. Int. Conf. on Scientific Aspects of Coral Reef Assessment, Monitoring and Restoration, NCRI. April 14–16, 1999, Ft. Lauderdale, Florida. p 88.

Franklin, H., C.A. Muhando, and U. Lindahl. 1998. Coral culturing and temporal recruitment patterns in Zanzibar, Tanzania. *Ambio* 27(8):651–655.

Gawel, Michael 1984. Involving users in management planning In Coral reef management handbook. R.A. Kenchington and B. Hudson, eds. UNESCO, Jakarta. pp 99–109.

Gleason, M.G. 1999. The importance of algal-grazer interactions in early growth and survivorship of sexual recruits and transplanted juvenile corals. Proc. Int. Conf. on Scientific Aspects of Coral Reef Assessment, Monitoring and Restoration, NCRI. April 14–16, 1999, Ft. Lauderdale, Florida. p 93.

Gomez, E.D. 1997. Reef management in developing countries: The Philippines as a case study. Proc. 8th Int. Coral Reef Symp. 1:123–128.

Green, E., and F. Shirley. 1999. The global trade in coral. WCMC Biodiversity Series No. 9, World Conservation Monitoring Centre. World Conservation Press, Cambridge, 70 pp.

Grigg, Richard W. 1976. Fishery management of precious and stony corals in Hawaii. Sea Grant Technical Report, NUIHI Sea Grant TR-77-03, University of Hawaii. 48 pp.

Grigg, R.W., and Dollar, S.J. 1990. Natural and anthropogenic disturbance on coral reefs. In *Coral Reefs*, Z. Dubinsky, ed. Elsevier Sci. Publ, Amsterdam, pp 439–452.

Hixon, Mark A., M.A. Carr, and J. Beets. 1999. Coral reef restoration: Potential uses of artificial reefs. Abstract in Proc. Int. Conf. on Scientific Aspects of Coral Reef Assessment, Monitoring and Restoration, NCRI. April 14–16, 1999, Ft. Lauderdale, Florida, p 104.

Hughes, T.P. 1989. Community structure and diversity of coral reefs: The role of history. *Ecology* 70:275–279.

Hughes, T.P. 1994. Catastrophes, phase shifts and large-scale degradation of a Caribbean coral reef. *Science* 265:1547–1551.

Hviding, E. 1990. Keeping the sea: Aspects of marine tenure in Marovo Lagoon, Solomon Islands. In K. Ruddle and R.E. Johannes, eds. *Traditional Marine Resource Management in the Pacific Basin: An Anthology.* Jakarta, UNESCO-ROST-SEA, pp.1–43.

Hviding, E., and K. Ruddle. 1991. A Regional Assessment of the Potential Role of Customary Marine Tenure (CMT) Systems in Contemporary Fisheries Management in the South Pacific. FFA Report 91/71. South Pacific Forum Fisheries Agency. Honiara.

IIRR 1998. Participatory methods in community-based coastal resource management. International Institute of Rural Reconstruction, Silang, Cavite, Philippines (3 volume set). 291 pp.

Jackson, Jeremy B.C. 1997. Reefs since Columbus. Proc. 8th Int. Coral Reef Symp. 1:97–106.

Johannes, Robert E. 1977. Traditional law of the sea in Micronesia. *Micronesica* 13:121–127.

Johannes, Robert E. 1978. Traditional marine conservation methods in Oceania and their demise. *Annual Review of Ecology and Systematics* 9:349–364.

Johannes, Robert E. 1981. *Words of the lagoon: fishing and marine lore in the Palau District of Micronesia.* Berkeley, Calif.: University California Press. 245 pp.

Johannes, Robert E. 1988. Criteria for determining the value of traditional marine tenure systems in the context of contemporary marine resource management in Oceania. Workshop on customary tenure, traditional resource management, and nature conservation. Noumea, New Caledonia SPREP.

Kallie, J., U. Faasili, and A. Taua. 1999. An assessment of community-based management of subsistence fisheries in Samoa. Workshop on Aspects of Coastal Fisheries Management, 30 June–2 July, USP, Suva, Fiji. 12 pp.

Kessler, W.B., H. Salwasser, C.W. Cartwright, and J.A. Caplan. 1992. New perspectives for sustainable natural resource development. *Ecological Applications* 2:221–225.

King, M. 1995. *Fisheries biology, assessment, and management.* Oxford: Fishing News Books, Blackwell Science Ltd. 341 pp.

King, M., and L. Lambeth. 2000. Fisheries management by communities: A manual on promoting the management of subsistence fisheries by Pacific Island communities. Secretariat of the Pacific Community. Noumea, New Caledonia. 87pp.

Kuhn, B. 2000. Participatory rural appraisal—ten years after: Did it fulfill the expectations? Development and Cooperation Journal, Germany.

Lewis, S.M. 1986. The role of herbivorous fishes in the organization of a Caribbean reef community. *Ecol. Monogr.* 56:183–200.

Lindahl, U. 1998. Low-tech restoration of degraded coral reefs through transplantation of staghorn corals. *Ambio* 27(8):645–650.

Lingenfelter, S.G. 1975. *Yap: Political Leadership and Culture Change in an Island Society.* Honolulu: University of Hawaii Press.

Littler, M.M., and D.S. Littler. 1997. Disease-induced mass-mortality of crustose coralline algae on coral reefs provides rationale for the conservation of herbivorous fish stocks. Proc. 8th Int. Coral Reef Symp. 1:719–724.

Lubbock, H.R., and N.V.C. Polunin. 1975. Conservation and the tropical marine aquarium trade. *Environmental Conservation* 2(3):229–232.

Lundin, C.G., and O. Linden. 1993. Coastal ecosystems: Attempts to manage a threatened resource. *Ambio* 22:438–473.

Maragos, James E. 1992. Restoring coral reefs with emphasis on Pacific reefs. In Restoring the nation's marine environment. G.W. Thayer, ed. Maryland Seagrant, College Park, Maryland, pp 141–122.

Marriott, S.P. 1984. A summary report on the South Tarawa artisanal fishery. Unpublished report, Fisheries Division, Kiribati. 21 pp.

McClanahan, Timothy R. 1994. Kenyan coral reef lagoon fish: Effects of fishing, substrate complexity and sea urchins. *Coral Reefs* 13:231–241.

McClanahan, Timothy R. 1997a. Recovery of fish populations from heavy fishing: Does time heal all? Proc. 8th Coral Reef Symp. 2:2033–2038.

McClanahan, Timothy R. 1997b. Dynamics of *Drupella cornus* populations on Kenyan coral reefs. Proc. 8th Int. Coral Reef Symp. 1:633–638.

McClanahan, T.R., and N.A. Muthiga. 1999. Sea urchin reduction as a restoration technique in a new marine park. Abstract in Proc. Int. Conf. on Scientific Aspects of Coral Reef Assessment, Monitoring and Restoration, NCRI. April 14–16, 1999, Ft. Lauderdale, Florida, p 133.

McClanahan, Timothy R., A.T. Kamukuru, N.A. Muthiga, Yebio M. Gilagabher, and D. Obura. 1996. Effect of sea urchin reductions on algae, coral and fish populations. *Conservation Biol.* 10:136–154.

McClanahan, T.R., V. Hendrick, and N.V. Polunin. 1999. Varying responses of herbivorous and invertebrate feeding fishes to macroalgal reduction: A restoration experiment. Abstract in Proc. Int. Conf. on Scientific Aspects of Coral Reef Assessment, Monitoring and Restoration, NCRI. April 14–16, 1999, Ft. Lauderdale, Florida, p 133.

McManus, John W. 1997. Tropical marine fisheries and the future of coral reefs: A brief review with emphasis on Southeast Asia. Proc. 8th Int. Coral Reef Symp. 1:129–134.

Meller, N., and R.H. Horowitz. 1987. Hawaii: Themes in land monopoly. In R. Crocombe, ed. *Land tenure in the Pacific* (3rd ed.). Suva: University of the South Pacific. 2544 pp.

Moffat, D., M.N. Ngoile, O. Linden, and J. Francis, J. 1998. The reality of the stomach: Coastal management at the local level in eastern Africa. *Ambio* 27(8):590–598.

Munro, John L. 1989. Fisheries for giant clams (Tridacnidae: Bivalvia) and prospects for stock enhancement. In *Marine invertebrate fisheries: Their assessment and management*. C.F. Caddy, ed. New York: John Wiley & Sons, pp 541–558.

Munro, John L., and S.T. Fakahau. 1993. Management of coastal fishery resources. Chapter 3, pages 55–72 In *Nearshore marine resources of the South Pacific*, A. Wright and L. Hill, eds. Institute of Pacific Studies, University of the South Pacific, Suva, Fiji. 710 pp.

Naim, O., P. Cuet, and Y. Letourneur. 1997. Experimental shift in benthic community structure. Proc. 8th Int. Coral Reef Symp. 2:1873–1878.

Nash, W.J. 1993. Trochus. Chapter 14, pages 452–495, in *Nearshore marine resources of the South Pacific*, A. Wright, and L. Hill, eds. Institute of Pacific Studies, University of the South Pacific, Suva, Fiji. 710 pp.

Nayacakalou. R.R. 1971. Fiji: Manipulating the System. In R. Crocombe, ed. *Land Tenure in the Pacific*. Melbourne: Oxford University Press.

Newman, H. 1998. A thorny issue: Crown-of-thorns controversy. *Asian Diver* 6(5):34–38.

Nzali, L.M., R.W. Johnstone, and Y.D. Mgaya. 1998. Factors affecting scleractinian coral recruitment on a nearshore reef in Tanzania. *Ambio* 27(8):717–722.

Oliver, J., and P. McGinnity. 1985. Commercial coral collecting on the Great Barrier Reef. Proc. 5th Int. Coral Reef Congress 5:563–568.

PASIFIKA. 1999. Documenting changes in customary marine tenure. PASIFIKA Newsletter, Marine Studies Programme, University of the South Pacific, Suva, Fiji 6(3–4):4–5.

Polunin, N. 1984. Do traditional marine "reserves" conserve? A view of Indonesian and New Guinean evidence. *Senri. Ethnol. Stud.* 17:267–283.

Pyle, Richard L. 1993. Marine aquarium fish. Chapter 6, pages 136–176, in *Nearshore marine resources of the South Pacific*, A. Wright, and L. Hill, eds. Institute of Pacific Studies, University of the South Pacific, Suva. 710 pp.

Randall, John E. 1987, Collecting reef fishes for aquaria. In Human impacts on coral reefs: Facts and recommendations. B. Salvat, ed. Antenne Museum E.P.H.E. pp 29–39.

Ravuvu, Asesela. A. 1983. Vaka i Taukei: The Fijian way of life. Institute of Pacific Studies, University of the South Pacific, Suva, Fiji.

Riegl, B., and K.E. Luke. 1998. Ecological parameters of dynamited reefs in the northern Red Sea and their relevance to reef rehabilitation. *Mar. Poll. Bull.* 37(8–12):488–498.

Roberts, C.M., and N.V.C. Polunin. 1993. Marine reserves: Simple solutions to managing complex fisheries? *Ambio* 22:363–368.

Rogers, Caroline S., L. McLain, and E. Zullo. 1988. Damage to coral reefs in Virgin Islands National Park and Biosphere Reserve from recreational activities. Proc. 6th Int. Coral Reef Symp. 2:405–410.

Rogers, C.S., V. Garrison, and R. Grober-Dunsmore. 1997. A fishy story about hurricanes and herbivory: Seven years of research on a reef in St. Johns, U.S. Virgin Islands. *Proc. 8th Int. Coral Reef Symp.* 1:555–560.

Ross, M.A. 1984. A quantitative study of the stony coral fishery in Cebu, Philippines. P.S.Z.N.I. *Marine Ecology* 5(1):75–91.

Ruddle, K. 1988. Social principles underlying traditional inshore fishery management systems in the Pacific Basin. *Mar. Res. Econ.* 5:351–363.

Ruddle, K. 1993. External forces and change in traditional community-based fishery management
systems in the Asia-Pacific Region. *MAST* 6
(1/2):1–37.

Ruddle, K. 1998. A modern role for traditional
coastal-marine resource management systems in
the Pacific Islands. *Ocean and Coastal Management,* Special Issue 40 (2/3).

Ruddle, K., and R.E. Johannes, eds. 1990. Traditional marine resource management in the Pacific
basin: An anthology. UNESCO/ROSTSEA, Jakarta. 410 pp.

Ruddle, K., E. Hviding, and R.E. Johannes. 1992.
Marine resources management in the context of
customary tenure. *Marine Resource Economics*
3(7):249–273.

Saila, S. B., V.L. Kocic, and J.W. McManus. 1993.
Modeling the effects of destructive fishing practices on tropical coral reefs. *Mar. Ecol. Prog.* Ser.
94:51–60.

Schneider, D.M. 1984. A critique of the study of kinship. Ann Arbor: University of Michigan Press.

Sims, N.A. 1990. Adapting traditional marine tenure
and management practices to modern fisheries
framework in the Cook Islands. In K. Ruddle and
R.E. Johannes, eds. *Traditional Marine Resource
Management in the Pacific Basin: An Anthology.*
UNESCO-ROSTSEA, Jakarta, pp 222–252.

Sims, N.A. 1993. Pearl Oysters. Chapter 12, pp
409–430 in *Nearshore marine resources of the
South Pacific,* A. Wright and L. Hill, eds. Institute
of Pacific Studies, University of the South Pacific,
Suva, Fiji. 710 pp.

South, G. Robin, and Joeli Veitayaki. 1998. Fisheries
in Fiji. In Lal. B. and T.R. Vakatora, eds., *Research
Papers of the Fiji Constitution Review. Vol. 1. Fiji
in Transition.* School of Social and Economic Development, the University of the South Pacific, pp
291–311.

Spalding, M.D., C. Ravilious, and Edmund P. Green.
2001. *World atlas of coral reefs.* Berkeley: University of California Press, 424 pp.

Suzuki, S. 1989. Users work to end Kona conflict.
University of Hawaii Sea Grant Publication
11(12):1,4.

Szmant, Alina M. 1997. Nutrient effects on coral
reefs: A hypothesis on the importance of topographic and trophic complexity to reef nutrient dynamics. Proc. 8th Int. Coral Reef Symp.
2:1527–1532.

Van der Meerten, S. 1996. Kubuna Qoliqoli: *A Study
of Community Dynamics in Co-management.* Unpublished master's thesis. Department of Marine
Sciences and Coastal Management, University of
Newcastle upon Tyne, United Kingdom. 77 pp.

Veitayaki, Joeli. 1990. *Village-Level Fishing in Fiji:
A Case Study of Qoma Island.* Unpublished master's thesis. Department of Geography, The University of the South Pacific.

Venkataramanujam, K., R. Santhanam, and N. Suku-
maran. 1981. Coral resources of Tuticorin (S. India) and methods of their conservation. In Proc.
4th Int. Coral Reef Symp. 1:259–262.

Wachenfeld, D.R., J.K. Oliver, and J.I. Morrissey.
1998. State of the Great Barrier Reef World Her-
itage Area 1998. Great Barrier Reef Marine Park
Authority, Townsville, Queensland. 139 pp.

Wells, Susan M. 1981. International trade in ornamental corals and shells. Proc. 4th Int. Coral Reef
Symp. 1:323–330.

Wilkinson, Clive. 1993. Coral Reefs are facing widespread devastation: Can we prevent this through
sustainable management practices? Proceedings
of the 7th Int. Coral Reef Symp. 1:11–21.

Wilkinson, Clive, ed. 1998. *Status of coral reefs of
the world: 1998.* Global Coral Reef Monitoring
Network. Australian Institute of Marine Science,
Queensland. 184 pp.

World Bank. 1999. Voices from the village: A comparative study of coastal resource management in
the Pacific Islands. Pacific Islands Discussion Paper Series No. 9. World Bank, Papua New Guinea
and Pacific Islands Country Management Unit. 99
pp.

Wright, A. 1990. Marine resource use in Papua New
Guinea. Can traditional concepts and contempo-
rary development be integrated? In K. Ruddle and
R.E. Johannes, eds. *Traditional Marine Resource
Management in the Pacific Basin: An Anthology.*
UNESCO-ROSTSEA, Jakarta, pp.301–321.

12

Sustainable Management Guidelines for Stony Coral Fisheries

Andrew W. Bruckner

Introduction

Coral reef ecosystems occur in more than 100 countries worldwide, most of which are developing countries whose peoples are dependent on reefs as the basis of their livelihood. Coral reefs are renowned for their high diversity and productivity, and they provide important sources of food, jobs, chemicals, medicines, revenue from tourism and fishing, and shoreline protection against tropical storms. Despite numerous economic and environmental benefits, reefs are being destroyed at an alarming rate from increasing human impacts precipitated by population growth, urbanization, and industrialization.

As human populations continue to expand in coastal areas, increasing pressure is placed on coral reefs to supply a growing local and international demand for food, ornamental organisms, and traditional medicines (Bruckner 2000). Because of their high value in international trade, many species are being collected at unsustainable rates, and fishers often use destructive fishing practices. Although the harvest of stony corals for international trade may seem to represent a small issue in terms of global volume, it can be a potentially significant extractive use at a localized scale. Coral fishers may cause considerable habitat damage, including breakage of undesirable corals and generation of rubble. Also, many of the most popular aquarium corals are uncommon, have a patchy distribution, exhibit slow growth, and are characterized by slow rates of annual recruitment and adult mortality, making them vulnerable to overexploitation and localized extinctions (Ross 1984). Repercussions of an unsustainable coral fishery may extend beyond the target species, affecting the diversity and abundance of associated invertebrates and fish and possibly triggering overgrowth of benthic habitats by macroalgae. These changes may cause any socioeconomic benefits associated with the coral fishery to sharply decline or be lost altogether.

Stony Coral Resources

Stony corals are the major reef framework constructors in tropical and subtropical environments. They provide topographic complexity, critical habitat, refuge, and feeding grounds for thousands of fish and invertebrates, including many commercially important species. Corals of commercial value can be divided into five groups: stony corals, semi-precious corals (black coral), precious corals (pink, gold, and bamboo corals), shallow-water gorgonians, and alcyonarians. This chapter focuses on stony corals, including all taxa in the order Scleractinia, organ pipe coral (*Tubipora musica*), blue coral (*Heliopora coerulea*), fire coral (*Millepora* spp.), and lace corals (*Stylaster* spp. and *Distichopora* spp.). All of the species under consideration are listed on Appendix II of the Convention on International Trade in Endangered Species of Wild Fauna and Flora (CITES). These species can be in international trade, provided that the collection is legal, the exporting country determines that the trade is not detrimental to the survival of the species in the wild and its role in the ecosystem, and shipments contain appropriate CITES permits issued by the exporting country.

Domestic Use of Coral

Stony corals are mined from reefs in East Africa, Fiji, India, Indonesia, Malaysia, Maldives, the Philippines, Sri Lanka, Vietnam, and other countries for use in road construction, buildings, jetties, seawalls, land reclamation, and as a source of lime for soil improvement, cement, ceramics, and the betel nut industry. Corals mined from construction include genera with dense calcium carbonate skeletons, especially *Porites*, *Goniastrea*, *Platygyra*, *Acropora*, and *Favia*, with limited collection of a few other species that coexist on shallow reef flats (Bentley 1998; Brown and Dunne 1988). The worldwide volume of coral extracted for construction materials is difficult to estimate, but in Bali, Indonesia, alone it amounted to 150,000 cubic meters during the 1970s and 1980s (Bentley 1998). In most countries coral is also collected for local curio, souvenir, and aquarium markets. Collection for local markets may be widespread and involve a high diversity of species, but it accounts for a relatively small proportion of the total coral harvest.

Corals in International Trade

Corals are harvested to supply international markets with souvenirs and curios, animals for home and public aquarium displays, jewelry, carvings, human bone replacement, traditional medicines, and biomedical purposes. A small component of the trade is also for scientific research and to supply captive-breeding facilities. More than 95% of all stony coral in trade originates in the Indo-Pacific (Green and Shirley 1999). Indonesia is currently the largest exporter of live coral, followed by Fiji, with smaller amounts from Vietnam, the Solomon Islands, Vanuatu, Tonga, and the Marshall Islands. Dead coral is currently supplied mainly by Fiji and Vietnam, although a number of other countries have supplied coral for curio markets over the past decade. The United States has consistently imported 60% to 80% of all live coral, 95% of the live rock, and more than half the curio coral (Bruckner 2001a). Imports of stony corals to the United States alone averaged about US$1.0 million annually from 1975 to 1980 (Wells 1981). By 1997 exporting nations generated about US$5 million from the coral trade, with a retail value of US$50 million (Green and Shirley 1999).

Corals for use as curios and souvenirs consist of a relatively small number of genera that have ornate skeletal structures. The preferred taxa include branching corals such as *Acropora*, *Pocillopora,* and *Stylophora*, mushroom corals (*Fungia* spp.), and two octocorals, *Heliopora* and *Tubipora*. Colonies, which are removed when alive, may be up to one meter or more in diameter. They are collected primarily in shallow water by snorkeling or wading, by using a hammer and small chisel for small colonies and long iron bars for large colonies. Specimens are bleached, cleaned, and dried before export, and some may be dyed. They are sold intact as raw coral skeletons or are manufactured into carvings, jewelry, inlays for picture frames, furniture, and other items. CITES trade records from 1985 to 1997 indicate that more than 900,000 kg and 16 million items of coral (coral skeletons) were in international trade (Bruckner 2001a). The trade in skeletons remained fairly constant between 1993 and 1998, but volumes have increased substantially as a result of a growing trade from Vietnam and other locations. Corals harvested for curios are a nonperishable product that can be taken from remote locations and transported to export facilities without time constraints.

Coral in trade as live specimens for aquarium displays consists of more than 50 genera that have branching, massive, and plating morphologies, most of which are traded at a small size (5–15 cm). The five most common taxa in the live trade in order of importance are *Euphyllia*, *Goniopora*, *Trachyphyllia*, *Catalaphyllia*, and *Acropora;* with the exception of *Acropora* spp., all of these corals were much less common in the trade before 1990 (Bruckner 2001a). The volume of trade in live coral has increased sharply in the past decade, from about 50,000 colonies in 1989 to nearly 800,000 colonies in 1999 (Fig. 12.1). Collection is carried out by local fishers operating out of small boats, who free dive or use surface-supplied air (i.e., hookah apparatus). Colonies are removed from depths of up to 50 m, in reef environments and soft-bottom communities. A small num-

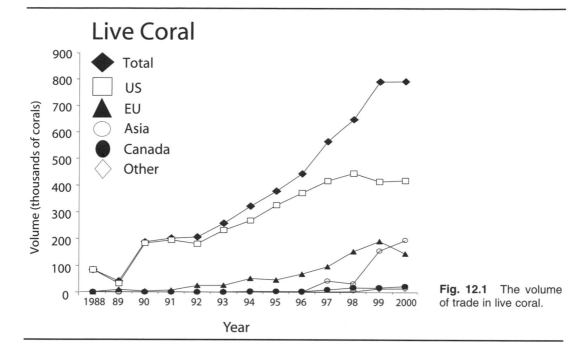

Fig. 12.1 The volume of trade in live coral.

ber of species are also collected on shallow reef flats by wading. Because only certain coral sizes and colors are preferred, fishers often cover large areas of reef or other habitats to obtain the desired animals. The harvest is highly selective, and particular environments are often targeted for single species (Bruckner 2001b). Areas of collection are generally limited to those located relatively near to airports or large holding facilities to maximize survivability.

Another component of the trade consists of material reported on CITES permits as *Scleractinia,* which includes live rock, reef substrate, and coral gravel. Live rock is any hard reef substrate (limestone skeletal material of algal or coral origin) that supports an assemblage of living marine organisms, including coralline algae. Live rock may be chipped from the reef substrate in shallow water, or it may consist of loose pieces of coral rubble collected from the reef crest or fore reef environments. The product is exported with minimal preparation and cleaning, or it is transported to a holding facility and cured through several washes to remove unsuitable organisms. Fiji has emerged as the world's largest supplier of live rock, with exports doubling or tripling each year since

1993 (Bruckner 2001a). A total global volume of more than 1.5 million kg of live rock was in trade in 1999. Reef substrate consists of small (generally up to 5–10 cm diameter) pieces of reef rock with an attached soft coral, colonial anemone, or other benthic invertebrate. The organisms of value are the attached invertebrates, but the underlying reef substrate must be removed to minimize damage to the target species. Trade in reef substrate from all exporting countries was close to 1 million pieces in 1999 (Fig. 12.2). This is likely to increase in coming years, as Indonesia established an annual export quota of 950,000 pieces of reef substrate in 2001, and it represents one of several substrate exporting countries.

Potential Environmental Impacts of Coral Collection

There are more than 650 zooxanthellate stony coral species found on tropical and subtropical reefs worldwide. Corals harvested for international trade constitute a fraction of the volume of corals that are mined for construction materials, but collection involves a much larger diversity of species and includes taxa

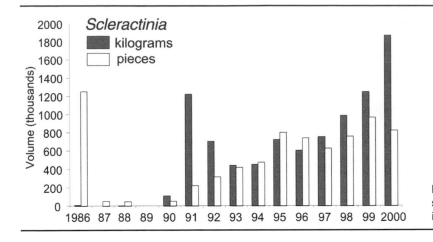

Fig. 12.2 Trade in reef substrate from all exporting countries.

with a wide variation in morphology, size, and life histories. Unsustainable coral harvest to supply international markets can contribute to changes in species composition and abundance and may lead to severe localized depletions. Breakage of corals during collection causes the generation of rubble that can cause further damage to surrounding corals through abrasion, and rubble may inhibit coral recruitment because of its unstable nature. In addition, coral collection may cause reductions of live coral cover and loss of rugosity, with cascading impacts on reef fish abundance and diversity, including species used in artisanal and export fisheries (Dulvy et al. 1995). Whereas other factors (including coral mining) have resulted in much more significant losses of corals, extraction of corals for international trade may magnify the impacts of other natural and anthropogenic stressors and possibly prevent recovery after unusual environmental disturbances.

Some of the most common corals in demand for the curio trade are ornate, small-polyp (SPS) branching corals that grow rapidly (up to 20 cm per year), are widespread, and exhibit high rates of recruitment. Even though most branching corals targeted for curio collection in the Indo-Pacific are quite successful colonizers, they are susceptible to breakage by wave action. In addition, these species are often the preferred foods of corallivores such as *Acanthaster* sea stars and *Drupella* snails. Many shallow-water reefs are dominated by species utilized as curios, and these reefs experienced widespread coral

mortality during a major bleaching event in 1998 (Wilkinson 2000). Furthermore, corals harvested for curio markets are typically sexually mature, and extensive collection may have negative impacts on reproduction. Because collection primarily involves large colonies, the removal of these corals may affect reef topography and available habitat for reef fishes and motile invertebrates.

Nine of the ten most popular live aquarium corals have a massive growth form, large colorful polyps, and prominent tentacles (LPS). Many of these are uncommon or patchily distributed and may grow slowly (<1 cm per year), but colonies often achieve a large size and can persist for centuries. The live trade is dominated by colonies that are fairly small in size and nonreproductive. Collection pressure that targets juvenile long-lived corals may lead to growth overfishing, which is of particular concern for LPS corals that recruit infrequently but have a low rate of natural mortality. In addition, many of these species occur in deep, turbid environments. If these areas are denuded of coral as a result of intense collection pressure, recruitment from other reefs may not occur if currents are unfavorable, resulting ultimately in a reef that has low coral cover and diversity.

Corals living in shallow, exposed environments are often fractured or dislodged during storms, and some colonies may die as a result of disease, predation, or other factors. Coral skeletons are rapidly colonized by algae and encrusting invertebrates and become incorporated into the reef substrate. This substrate is essential for the reef because it provides a site for future settlement and attachment of reef-building corals and other benthic organisms crucial for continued reef growth. It also provides important habitat for fish and motile invertebrates, and it contributes to the structure and total biomass of reefs. This material is collected and sold as live rock. Its removal results in the net loss of substrate that requires decades to centuries or longer to replace. In addition, many commensal (symbiotic), sessile organisms essential for the health of the reef are removed along with the rock (Maragos 1992). Intensive harvest of live rock from one location may lead to increased erosion and flooding of coastal areas. Reef degradation is of particular concern to low islands like Tonga that are most affected by rising sea level. Table 12.1 lists the consequences of unsustainable coral and live rock collection practices.

Benefits and Justification for the Coral Fishery

Some stony corals represent a valuable renewable resource. Corals have been highly valued as raw material for construction and for ornamental uses since antiquity, with the first uses dating back to 25,000 B.C. (Grigg 1989). In addition to significant cultural and economic values, a sustainably managed coral fishery can serve as a tool to promote local coral reef conservation and increase worldwide awareness of the importance of reef ecosystems.

The coral fishery is currently difficult to manage in a sustainable manner because of conflicting destructive uses, such as coral mining and blast fishing, increased nonextractive human impacts, and limited information on species in trade and the status of the resource. With proper management practices, however, it is possible to sustainably harvest some species of curio and aquarium corals. Because many of the species targeted for curios are predominantly harvested from shallow water, deep-water populations may serve as a refugium and a source for replenishment for some of these species (Lovell 2001). In areas where space is limiting, removal of dominant, fast-growing branching corals may increase coral diversity through creation of additional space and reduced competition (Ross 1984).

Existing Management Measures

In an attempt to protect coral reef resources, a number of countries have implemented severe restrictions on coral harvest. At one extreme, this involves a complete prohibition on the collection, sale, and export of coral, such as that implemented in the Philippines since the late 1970s in response to catastrophic losses of live coral from destructive fishing, overharvesting, and other impacts (Mulliken and Nash 1993). Mozambique temporarily closed its fishery to allow recovery after a severe bleaching event in 1998, with a re-evaluation of the fishery slated for 2002. Commercial harvest of stony corals has also been prohibited in most federal, state, and territorial waters in the United States because of their pivotal ecological role and importance as essential fish habitat.

Importing countries can also take broader measures to address unsustainable trade in wildlife listed under CITES. For instance, the

Table 12.1 Consequences of unsustainable coral and live rock collection practices

Impact	Description
Effect on target population	Overexploitation, localized extinctions, and reduced recruitment
Habitat impacts	Reduced coral cover, diversity, and rugosity
Effect on associated species	Loss or destruction of habitat and decreased abundance, biomass, and diversity of reef fish, invertebrates, and other species
Ecosystem impacts	Increased erosion of the reef structure, associated islands and coastal environments, with increased flooding during storm surges
Socioeconomic impacts	Conflicts with other uses and user groups, including traditional and cultural uses, fishing, and tourism

European Union allows coral imports from most countries, but it has prohibited the import of particular species from Indonesia because it feels the export quotas do not reflect sustainable harvest based on the known distribution and abundance of those species.

Prohibitions on coral harvest may be appropriate for some locations, for certain vulnerable species, to protect ecosystem biodiversity or function or to allow recovery after an unusual disturbance event. However, the closure of established coral fisheries based solely on political interests and not sound science may have severe negative implications for developing countries dependent on these resources. Closure may result in an increase in illegal commerce in stony corals. In addition, individuals whose livelihood depends on revenue generated from the coral fishery may resort to more destructive practices, including blast fishing for subsistence or cyanide fishing to supply the live reef food fish trade.

Because of the high number of species and low population numbers of most reef species, a very precautionary approach to extractive use must be taken. The fishery could be managed through the development of policies that allow operation of the industry in a competitive environment, while promoting a sound emphasis on sustainability of the resource and other fisheries resources that are likely to be impacted (Lovell 2001). Effective management measures must be carefully considered to provide the best protection for the cost and ensure that implementation of some practices will not create new ecological stress. Solutions to achieve sustainable management can be implemented through local management actions in concert with national and regional agreements and international policies. To ensure compliance by the coral fishers, management plans should be developed through consultation with all user groups. To ensure that coral resources and the habitats they come from are conserved and managed, different aspects must be considered.

The following recommendations were largely developed from discussions at the International Coral Trade Workshop (NMFS 2001), with examples of existing management approaches in Indonesia, Australia, and the United States. I have also included my own recommendations based upon independent coral reef studies, discussions with other scientists, and best management practices from other coral reef fisheries. Recommendations include:

- Provisions for licensing, training, and certifying fishers; limiting entry into the fishery; and specifications for reporting, compiling, and analyzing fishery data;
- Setting limits on the volume, size, and taxa of corals collected;
- Establishment of defined collection sites for individual collectors and cooperatives and no-take areas that include the same type of habitats used by collectors;
- Rotation of collection sites and introduction of closed seasons or periods;
- Use of acceptable gear and collection methods;
- Assessments and monitoring of collection sites and control sites to evaluate the status of the resource and harvest impacts;
- Responsibilities and requirements of collectors, middlemen, and exporters; and
- Sustainable financing schemes to ensure sufficient funds are recovered from license, collection, and export fees to pay for effective management and monitoring.

Key Considerations for Sustainable Management

A sustainable management approach for the coral fishery requires that (1) the resource is harvested in such a way that it is protected from overexploitation; (2) collection is ecologically sustainable; (3) damage to associated species is minimal; and (4) the structure and function of the reefs are not altered. Of critical importance, collectors must preserve the ecological integrity of reefs when removing coral, not simply the physical reef structure (Dustan 1999). Once the upper ecological and sustainable limits of removal are determined from sound science and monitoring, the allocations, rights, and responsibilities of collectors and exporters should be addressed. Effective management will maximize social and economic benefits, within these ecological limits, to promote equitable sharing of benefits arising from utilization of the resources.

An ecosystem-based management approach that promotes conservation and sustainable use is the most effective strategy for integrated management of marine environments and the living resources within those environments. Management is based on the application of scientific methods that focus on the structure, processes, functions, and interactions among organisms and their environment and the role that humans play in these processes. Unlike traditional, single-species fisheries management that strives to maximize yields of a target species, this approach seeks to minimize impacts to the ecosystem while maintaining ecological sustainability, and thus ensuring economic sustainability. Consideration must be given to the harvested species and all uses and impacts that may affect those species, as well as to the effect that collection has on associated species, habitat structure, and habitat quality.

Unfortunately, much of the information needed to make appropriate decisions for effective management of a coral fishery at an ecosystem level is currently unavailable. Monitoring of coral collection sites is virtually nonexistent, and baseline assessments have been completed in few sites. Many of the preferred target species for aquariums may be locally abundant but patchily distributed. Also, the favored habitat for these species is deep, turbid water (10–40 m depth), further complicating the ability to obtain distributional data. In addition to the absence of detailed information on the population dynamics and habitat requirements, the understanding of the biology of these species or their ability to recover from collection and other impacts is incomplete. As a result of these limitations, it is critical that managers adopt a precautionary approach until additional information, especially from quarterly take figures and regular monitoring, is available to support increased sustainable take.

Methods to Regulate the Coral Fishery

To achieve a sustainable fishery, regulations should be adopted for all levels of the fishery, including: (1) defined activities that can oc-

cur in the collection areas to ensure sustainable harvest; (2) activities that can be carried out by the commercial fisher; and (3) requirements that will affect the person who buys the coral from the fisher. Management approaches may vary among countries, depending on existing local and national regulations. For instance, marine tenure arrangements in Fiji, the Solomon Islands, Vanuatu, and Tonga recognize community ownership of marine resources. These communities may have the ability to prevent noncommunity participation in the fishery and also may be able to impose agreed-upon standards or protocols on community members involved in the fishery. In addition to local and national requirements, regulations must meet international standards of CITES mandated by the Appendix II listing of stony corals. Without addressing all components of the fishery, managers will not be able to provide adequate protection for the corals, associated species, and habitats or ensure maximum long-term benefits for the community.

Balancing Ecosystem Needs with Community Objectives

Effective habitat management is a critical tool for a sustainable stony coral fishery, with protective measures established for coral reefs and associated habitats—sea grass beds, mangrove communities, lagoonal habitats, and deep water soft-bottom communities utilized by certain high-value coral species such as *Trachyphyllia*, *CataElaphyllia,* and *Nemenzophyllia*. An effective tool used in Australia and other countries involves the zoning of benthic environments for specific uses. Zoning measures include ecological areas or reserves where no collection is allowed, buffer zones around these no-take marine protected areas (MPAs), and areas where specific activities can be undertaken. Coral collection sites should be separate from areas subject to other uses, such as fishing and recreational diving, to reduce conflicts among different users. Zoning can also minimize risk associated with other stresses by spatially separating collection sites. The establishment of defined collection and no-take sites also allows managers and scientists to focus their monitoring and research to specific areas and to compare collection sites with areas that are undisturbed by extractive uses. To increase capacity for compliance and enforcement, it is important to establish zones for specific uses through consultation with all user groups.

Introduction of Spatial and Temporal Closures

Sites closed to collection (no-take MPAs), rotation of collection sites, and short-term closures can be useful tools to promote the effective management of stony coral fisheries. These measures may help prevent overexploitation, allow recovery from overuse, and protect the stock at a vulnerable time in its life history or from an unusual perturbation.

No-take reserves provide numerous benefits for coral reef ecosystems. For instance, they maintain biodiversity and provide a refuge for vulnerable species, prevent habitat damage, facilitate recovery from catastrophic human and natural disturbances, and provide undisturbed spawning grounds for fishery species. Reserves help maintain adult breeding populations of fished species by allowing individuals to live longer, grow larger, and produce more offspring. As the number, density, and biomass of individuals within a reserve increases, spillover into adjacent fishing grounds will occur through emigration and larval dispersal, thereby assisting in rebuilding depleted stocks and re-establishing a natural ecosystem balance (Bohnsack et al. in press). The degree to which reserves are likely to enhance recruitment to fishing grounds is equivalent to the fraction of the total biomass of a population that they contain (Roberts and Hawkins 2000).

The extent of recruitment from no-take reserves to collection sites is dependent on (1) the size of the reserve; (2) how close it is to collection sites; (3) population dynamics, density, and life history of target coral populations within the reserve; and (4) the ability of recruits to disperse to surrounding collection sites. For most commercially important reef fish, fishery biologists recommended setting aside a minimum of 20% to 30% of representative coral reefs and associated habitats as no-take reserves (Bohnsack et al. in press). For stony corals, no-take reserves should encompass a much larger area, however, because of their unique life histories, their sessile nature, and a high demand for rare or uncommon species and juvenile life stages. First, many of the LPS corals are long-lived and slow growing and recruit sporadically, and these will take longer to become re-established after collection. Second, corals are permanently fixed to the bottom, and successful spawning relies on high population densities. Individuals need to be sufficiently close so that eggs and sperm can fuse before the sperm lose their short-lived motility. In fished areas, many species of corals, especially those that are uncommon, are likely to be much farther apart than in reserves. Thus, a small portion of the population biomass that is densely packed within a reserve can account for a large proportion of the species' reproduction and the ability to provide recruits that colonize sites open to collection. In addition, most species targeted for aquariums are removed before they reach the size of sexual maturity, which may lead to little reproduction within collection sites. As a precautionary approach, at minimum 50% of all representative habitats should be set aside as no-take areas to ensure protection for a sufficient portion of the total stock that may contribute to the recolonization in fished areas (Barbara Best, U.S. Agency for International Development, personal communication, 2002).

The concept of "rotation of collection sites" must be carefully examined scientifically and experimentally to determine whether rotation is compatible with sustainable management. Rotation of collection sites can reduce pressure on a particular area, minimize habitat damage, and allow a greater number of corals to mature and reproduce before being removed from the population. However, if collection in one area is truly sustainable for the long term, that area should continue to provide the target species at the established level indefinitely without the need for rotation. Issues about rotation are:

- If a collection site must be rotated so that it can "recover" from overexploitation or degradation as a result of collection, collection has exceeded sustainable take, and the role of collected species in the ecosystem has been diminished;

- Cycling collection areas between periods of exploitation and recovery may disrupt natural succession processes. In particular, rotation may favor those species that are more opportunistic colonizers (i.e., species that take advantage of reef patches opened up by collection) or "weedy" species that are more rapid growers; and
- The concept of rotation may undermine marine tenure rights and responsibilities. Rights-based fisheries, along with concomitant responsibilities, is the best motivator for an individual or cooperative to strive for sustainable management and long-term use of a resource. If individuals or coops are allowed to rotate out of an area into a new area, that motivation is lost. Particularly over a short time frame, an area may be overexploited, and the collectors will be allowed to move on without any consequences or responsibility for their actions.

Resource managers must also consider collection pressure and other impacts that affect coral resources when establishing collection sites and limits on collection within these sites. Temporary closures of collection sites before, during, or shortly after annual mass spawning events may enhance recruitment success. Managers should also have the authority to temporarily close a site zoned for collection in response to natural or anthropogenic disturbances that affect the harvested resource. For instance, collectors should avoid areas that have been impacted by a bleaching event, a crown-of-thorns (*Acanthaster*) sea star outbreak, or disease epizootic to allow recovery of that area. Collectors may also need to restrict collection activities during unusual environmental events, such as an *El Niño* (ENSO) event that is associated with unusually high seawater temperatures, as coral health may be compromised, and corals are unable to survive additional stresses associated with collection and handling. Such closures may last for years, with associated displacement of a community's fishery.

Coral Fishery Participants

In many countries, the people participating in the coral fishery are not identified, licensed, or limited in any way. Licensing of collectors is paramount and should include a limitation on the number of fishery participants and the amount of coral that can be collected by each fisher. The number of people that the coral fishery can support is also dependent on socioeconomic and cultural requirements of the collectors and dependent communities.

The ideal situation for a stony coral fishery is one that allows only one collector or a cooperative in each area that is zoned for collection. This rights/responsibility-based approach offers the greatest incentive for sustainable use, as individual collectors would be responsible for maintaining their collection sites.

For each new entry into the fishery, management agencies must develop and provide detailed information to fishers on the regulations for harvesting corals, including:

- Maps illustrating demarcated areas for coral collection and areas closed to harvesting;
- A list of approved species that can be collected from each collection site. If a species is not on the approved list for management and monitoring, it cannot be collected;
- Size limits and quotas for each approved taxon in that site;
- Techniques to minimize injury and mortality of target corals during collection;
- Detailed reporting requirements of fishers; and
- Safety standards for divers and reporting on diver health.

Guidance for Determining Sustainable Harvest Level

The amount of coral that can be removed from a site must be based on the carrying capacity of the ecosystem, whereas the decision to establish collection sites and the allocation of those sites depends upon the social objectives of the community. Given current limitations, allowable levels of take should be precautionary until scientific data become available that support increased sustainable collection. A precautionary approach may include "freezing" harvest levels at some average value of export for the past several years while monitoring programs are developed. An estimate of sustainable yield requires an examination of each collection area, the taxa targeted in the area, and available life-history information on those taxa (Table 12.2). The development of a database that compiles research and field monitoring data, and fishery-dependent statistics such as catch, effort and location, will facilitate this process.

Development of Species-Based Quotas

An approach often used to regulate wildlife trade involves a quota system that defines the maximum amount of collection or maximum export level. In some cases, these quotas primarily reflect the demand for a species and are based only to a limited extent on the status of the resource or its sustainable yield. Quotas should reflect the amount of sustainable harvest and not the amount of trade, with a separate quota developed for each site based on the abundance and condition of each coral species, and the ecological sustainability of that site. Other considerations include (1) the size classes within the target population and the size classes being considered for harvesting; (2) the size of the collection site, which should be small enough for reliable monitoring and management; and (3) the presence and condition of neighboring, no-collection areas that may serve as a source for larval replenishment and as control sites for monitoring. Because of the multispecies nature of the coral fishery, setting of nonspecific quotas may increase conservation problems, as fishers are likely to concentrate on the high-priced species, which are typically less common and more likely to be overex-

ploited. A separate quota must be developed for each species, determined from the coral's growth rate, the rate of instantaneous mortality, recruitment success, and other life history parameters (Grigg 1984). Finally, species that are inappropriate for the trade, or for which there is no baseline assessment and no monitoring protocol and data, should have a quota of zero. Inappropriate species include those that (1) are rare or endangered in the proposed country or region of collection; (2) have an important ecological role or provide habitat for other species; and (3) are difficult to maintain in captivity.

A quota system was first introduced in Australia in 1991, with a total allowable take established for each area zoned for collection (AFCFWG 1999). Coral areas are located in shallow environments (<6 m) dominated by *Acropora* or *Pocillopora*. Quotas were established when almost all harvested coral consisted of branching species for curios. The current management arrangement appears to be ecologically sustainable for the established species, although more comprehensive monitoring is planned (Table 12.3). Collected coral is only for domestic use, as Australia does not allow coral to be exported.

Because of an increased demand for live aquarium species in Australia, however, a number of issues have emerged regarding measuring the coral and monitoring commercial quotas. First, the current quota is based on a total allowable weight that can be removed on an annual basis, with no distinction among individual species. This approach will not be effective for live corals, as they are sold as individual pieces. Second, the aquari-

Table 12.2 Preliminary assessment for development of a sustainable stony coral fishery

Information needs	Management considerations
Type of coral utilized by the fishery	Target species, growth form, and size
Life history of target species	Growth rates, rate of natural mortality, recruitment rates, size class at maturity
Collection sites and no-take replenishment sites	Nature and size of habitat suitable for coral growth; size, depth, and zone of collection area and size and condition of adjacent replenishment areas; historical and current uses and existing natural and anthropogenic threats
Existing protection	A review of existing laws and regulations to conserve coral resources and ecosystems
Environmental impact statement	Assessment of potential collection sites prior to opening them to collection to provide initial baseline information of the standing crop (population size and density) and existing threats, impacts, and uses

Table 12.3   Provisions for sustainable management of the stony coral fishery in Australia

Regulation	Description
Type of harvest	Commercial harvest only; no recreational harvest. Dominant taxa: *Pocillopora*, *Acropora*, and *Fungia* primarily for curios, with recent increases in other taxa and live rock to supply domestic aquaria; 45–60 metric tons harvested each year from 1994 to 1997. Total allowable catch exceeds actual harvest by 60%–88% each year; harvest is 1%–2% of the standing stock.
Collection areas	50 authorized coral areas each with 200–500 m of reef front to a depth of 6 m; average size of each coral area is 25,000 m².
Quota	Total allowable catch (TAC) of 4 metric tons of coral per year for each lease area, with no distinction in amount of harvest for individual species.
Participants	14 collectors in 1983; 27–55 fishers with authority to harvest coral from 1990 to 1998. Not all areas are targeted for collection each year and not all fishers are active; each participant may acquire collection rights in up to 5 areas.

Source: Adapted from AFCFWG (1999).

um trade involves collection of many diverse taxa that often occur in areas deeper than those authorized for collection and are found at a much lower density.

Indonesia currently allows export of live coral, live rock, and reef substrate, with no export of coral skeletons or manufactured items. The proposed management plan includes the following provisions:

- An annual quota is assigned to the Indonesian Coral, Shell, and Fish Association (AKKII), whose members are the only people in Indonesia authorized to export coral. The quota for 2001 is 950,000 live corals, with separate quotas for each species divided among 10 provinces. There is a maximum size limit of 15 cm for slow-growing corals and 25 cm for fast-growing taxa. An export of 450 metric tons of live rock and 950,000 pieces of reef substrate is also allowed;
- The annual quota is based on reef accretion rates (assumed to be 10–15 mm/year); coral growth rates (linear skeletal extension rates of 2.5–30 cm/year); condition of monitoring sites [421 stations from 43 locations assessed as being in excellent condition (6%), good condition (24%), fair condition (24%), and poor condition (40%)]; and estimates of total reef area in the country (85,700 km²). Using these figures, it is assumed that a quota of 1 million corals is 0.00035% of the total reef area in good to excellent condition;
- Fishers are not nationally licensed or regulated, but permits to utilize corals may be given to Indonesian citizens living in the district surrounding the harvest area;
- Collection is supposed to occur only in sites that are in good to excellent condition and where population assessments and monitoring are undertaken. However, collection sites are not defined but must be outside conservation areas and marine tourism sites. Collection is allowed in only 10 provinces; and
- Corals will be harvested in harvest rotation systems.

Several concerns have been expressed regarding the management of the fishery in Indonesia and the calculation of the annual and species quotas. Species quotas do not appear to be based on the population dynamics or life history of the species, as some of the highest quotas are issued for rare species, while common species are traded at lower levels (Bentley 1998). Also, collection occurs in marginal reef environments and non-coral reef environments, which is contrary to assumptions used to develop the quota, and assessments or monitoring of collection sites have not been completed, with the exception of a few recent surveys (Bruckner 2001b).

In the United States, and throughout most of the Caribbean, coral collection is prohibited, as it is considered essential fish habitat and essential reef components. The United States does allow a limited harvest of black coral in state waters of Hawaii and selective harvest of precious corals in federal waters off Hawaii. The amount of harvest was determined from a mathematical model based on the classic fisheries population dynamics model of Beverton and Holt and uses data on population dynamics obtained from the collection sites and known information on species biology (Grigg 1984; Table 12.4). The black coral fishery appears to have been sustainable over a 23-year period.

Collection Guidelines

A coral fishery will benefit from the establishment of specific collection guidelines that minimize habitat impacts and maximize survivorship of targeted species. A code of practice should be established by the industry in partnership with local governments and environmental groups and must include training in appropriate collection, handling, and diver safety techniques. One method to encourage voluntary compliance is to train and certify fishers in the use of national and international standards of best practice, such as those being implemented by the Marine Aquarium Council (MAC) (see Table 12.5).

Monitoring Approaches

The monitoring of collection sites and comparable control (no-take) areas is critical, as it will provide important data on the abundance and population dynamics of the species and impacts associated with collection. Monitoring data provides managers with the ability to make sound management decisions and appropriate adjustments in response to changes in species abundance, ecosystem characteristics, or overexploitation. The monitoring pro-

tocol should be a practical and rapid technique that scientists and resource managers can use with minimal training and is easily repeatable by any individual with expertise in corals. Monitoring should be conducted with sufficient frequency to properly evaluate the impact of the trade and detect changes in the abundance or condition of taxa of interest but not so frequently that it is inefficient, too costly, or destructive. One recommendation is that monitoring should be conducted regularly over a period of 2 years, with a reassessment of existing management approaches after that time (Lovell 2001). A 2-year period will allow interpretation of the population dynamics of the target species and natural variability in abundance, obvious damage associated with collection or other impacts, and the extent of recruitment of harvested taxa.

The monitoring program should be specific enough to determine the composition, abundance, and population dynamics for each taxon in trade. Because of the high diversity of scleractinian corals found on Indo-Pacific reefs, and considerable expertise required to properly identify these species, monitoring should focus at a minimum on identification of coral genera, except for monospecific taxa, readily identifiable species, and species that are particularly rare or vulnerable to overexploitation. If the species cannot be monitored, it should not be approved for collection.

Determination of the baseline abundance of target stocks can be achieved through implementation of a variety of monitoring ap-

Table 12.4   Application of a fishery population dynamics model to the shallow-water reef-building coral *Pocillopora verrucosa* to determine sustainable yield

Data requirements	Measures of distribution and abundance, growth data (weight/size increase per unit time), rate of instantaneous mortality, recruitment rates, and age at reproductive maturity. Mortality and recruitment are determined by analysis of age frequency distribution of a portion of an unfished population.
Principle to maximize yield	Harvest must not occur until the colony reaches a specific size (age). Maximum production occurs at the point where losses due to natural mortality overtake gains from growth.
Assumptions	For simplicity, the model assumes a steady state—the yield of a single cohort over its lifespan is equal to the yield of all age classes present in a single year. Variations in annual recruitment rates can be incorporated by introducing year-specific estimates of certain population parameters.
Application	Not appropriate for the live trade, as colonies are often collected before they reach reproductive maturity and some species are collected as fragments of whole colonies.

Table 12.5 Recommended coral collection guidelines for live coral and curio coral

Parameter	Collection guidelines
Type of coral	Species with a high local abundance, fast growth, and high rates of recruitment. Focus efforts on branching species. Only collect slow-growing genera (massive corals) from areas where their ability to grow into adult colonies is prevented or from areas where fast-growing corals are overgrowing them. Remove only the number and type of species that are requested by the middleman or exporter.
Size limit	Maximum size is dependent on the biology of the species and its use; for live coral the minimum and maximum size also reflects survivability during handling and transport. Indonesia has established a maximum diameter of 15 cm from slow-growing taxa (e.g., massive corals and solitary or free-living corals) and 25 cm for fast-growing corals (e.g., branching corals) (Suharsono, 1999). Lovell (2001) recommended a maximum size of 45 cm for curio coral (*Acropora* and *Pocillopora*). MAC core standards recommend collection of corals in small size classes, with showpiece corals consisting of no more than 1% of total exports by weight.
Unit of collection	Remove whole colonies only, unless fragments can be removed from larger colonies without impact to parent colony or reef. When fragmenting a branching colony, only a small percentage should be removed, with no additional collection from the parent colony until healing or regrowth from previous collection has occurred.
Maximum harvest quota	Maximum amount of collection developed for each species within individual collection sites, based on its biology and abundance, with consideration of the desired size and reproductive status. A zero quota for rare species, ecologically important species, species for which no baseline and no monitoring data are kept and sustainability is not ensured, and species difficult to keep in captivity.
Location of collection	(1) Target areas with dense coral growth, removing species that may die as a result of competition with or shading by neighboring corals; (2) no collection of isolated corals greater than 0.5 m away from any other live coral; there is little chance of the coral being overgrown, and it is likely to reach maturity; (3) avoid reef crest environments, as aquarists cannot duplicate necessary environments, such as wave energy, and excessive mortality may occur; (4) select sites that are at a minimum distance from the holding facility or export facility to avoid long transport and unnecessary stress. Lovell (personal communication) recommends that collection be prohibited in areas with less than 30% live cover.
Amount of collection in each collection site	Remove corals in a selective manner, spreading out harvest over largest area possible, but within defined collection sites; ensure that a limited number of corals of each species are collected from the site, such that adult colonies remain and may provide a source for replenishment.
Depth of collection	Follow safe diving practices and limit collection to depths less than 20 m. Because of monitoring limitations and lack of baseline data for deeper environments, limit collection in deep water as a precautionary measure and for diver safety.
Duration of collection	Collect coral from one site for no more than 2 years, with a rotation of sites to allow recovery of the resource. Enforce periodic closure, especially after an unusual disturbance or environmental event.
Method of removal	Use only tools that remove appropriate coral species with minimal injury to the target coral, surrounding corals, and the reef structure. Collectors should use fins when collecting, as this causes less damage than walking on the reef.
Injury or breakage of nontarget corals	Place corals and other reef organisms that are inadvertently broken or dislodged during collection near the point of breakage, positioned carefully within the reef framework to stabilize the colony, with live polyps facing upward.
Handling	Minimize handling; avoid touching live polyps; maintain corals in clean seawater of ambient temperature with frequent water changes; avoid placing separate corals in direct contact, preferably one specimen per bag; minimize exposure to sunlight and air; transfer to holding facility as soon as possible.
Reporting	Collectors should maintain a logbook that shows the numbers, types, and sizes of coral collected. For each coral this should include the location, habitat, and depth of collection. Also record routine information for the collection site, including evidence of bleaching, disease, and physical damage. Management authorities should require an audit and submission of logbooks on a quarterly basis.

proaches at a range of scales. At the largest scale, countries need to determine the total area of reefs within territorial waters and the amount of that reef that will be open for harvest and other uses. Satellite and aerial remote sensing offers a quick and powerful tool for calculating such parameters and for developing maps that show the boundaries of collection sites and the total area open for collection. At a local scale, on the short-to-medium time frame, stock assessment teams or individuals need to accompany collectors to determine the extent (patch size and frequency) of habitats being exploited for target species. Suitable methodologies include manta tows and timed swims for covering large areas and, where the target species is abundant, transect approaches to obtain quantitative data on population dynamics, and free search methods to identify the presence and abundance of rare species. Because the goals are accurate density estimates and size frequency distribution data, the distribution and natural abundance of the species of interest will help determine which methods are most appropriate.

The monitoring protocol should be detailed enough to:

- Determine the status of the resource;
- Assess changes in biodiversity, standing stock, mortality, recruitment, and condition to the resource, including collection impacts and other factors contributing to coral decline or loss;
- Establish an initial quota and provide information needed to adjust the quota if a particular taxon becomes rare or overexploited; and
- Assess, quantitatively, whether a management response is warranted as a result of changes in the quarterly audits submitted by collectors.

Responsibility of Importing Countries and Consumers

The development of a sustainable management approach is primarily the responsibility of the exporting country, but it will require assistance from importing countries, the industry, and hobbyists. Hobbyists should purchase only corals that exhibit a high survivability in captivity, with a level of required care and handling that matches the expertise of the hobbyist; beginners should purchase animals that are the least challenging to maintain. Corals known to be rare or that have life history characteristics that make them vulnerable to overexploitation, and species whose collection is detrimental to the health of coral reef ecosystems, should not be purchased. Ideally, consumers should choose branching species that grow fast and can be propagated by fragmentation or corals from environmentally sound mariculture farms. Hobbyists, retailers, and suppliers should pay attention to emerging environmental issues that may affect reef health, such as a bleaching event, and avoid purchasing affected corals from the countries impacted by severe global or regional threats. When selecting appropriate corals, ask local retailers questions on the source of their coral and whether the coral was collected in a sustainable, environmentally friendly manner or from mariculture. One assurance for a quality product may be achieved through the support of ecolabeling programs like that established by MAC. Importing countries have a responsibility as well to ensure that coral reef species were taken from areas under sustainable management and were not taken with the use of destructive practices.

Conclusion

There are many basic constraints to establishing a sustainable coral fishery, especially to address live corals collected for home aquariums, offices, and restaurants. Constraints include a lack of understanding of reef processes, interdependent relationships among associated species, and regional and local connectivity among reefs and associated habitats. Data requirements for successful management rely on an understanding of the biology, life history, and habitat requirements of each species, and the population dynamics of those species within each collection site. The stony coral fishery is supported by a large diversity of corals that are collected at different sizes and in varying quantities, depending on their use, value, and demand. Environmental impacts associated with collection are difficult to measure in the short term, because of natural variation in coral reefs, limited knowledge of local retention of coral larvae, and the inability of a heavily utilized species to recolonize. Much more needs to be learned about the resource and its dynamics, including reproduction of target species, recruitment, and subsequent growth rates. Because this type of information may be prohibitively expensive to obtain, resource managers must adopt a precautionary approach and progressively work toward achieving a more thorough understanding of the resource and its sustainable yield. This can be accomplished through expanded research on the species distribution, abundance, and life history, implementation of monitoring programs in collection and control areas, establishment of extensive no-take areas, and

quarterly reports from the collectors on effort and amount of take per species (Table 12.6).

A sustainable coral fishery can be measured by long-term stability of targeted species, protection of coral reef habitat, and avoidance of detrimental cascading ecosystem impacts. A sustainable fishery will ensure social and economic security for communities dependent on the resources. One clear indication of sustainability is the ability of collectors to continue collecting corals at the same level within existing, permanent collection areas. The development of a sustainable stony coral fishery requires the financial and technical support of industry and governments in exporting and importing countries to develop appropriate management measures, commitment of local fishers to protect the resource, and demand by the hob-

Table 12.6 Summary of key points to include in a management plan for the stony coral fishery

Background information	(1) Data on the coral collection site and coral resources within that site, including the history of exploitation, a summary of pre-existing threats and existing management measures; and (2) information on the target species and proposed use.
Provisions to assess and monitor collection sites	(1) Determine the total area of coral reef habitat and the abundance and cover of the targeted coral taxa within the proposed collection area before harvesting begins (for a new entry); (2) assess the current condition of the resource (for pre-existing collection sites); (3) monitor the population dynamics of the target species and impacts of coral collection; and (4) use monitoring information to develop and modify established quotas as appropriate to continue sustainable harvest.
Zoning of coral reefs and associated habitats	Separate coral reef uses through zoning. Include: (1) areas off limit to collection; (2) defined areas for other extractive and nonextractive uses; and (2) defined collection sites with a rotational harvest that spreads out collection over a large area and minimizes the duration of collection in any given site.
Spatial and temporal closures	Establish no-take areas that include a substantial portion of all representative habitats and high densities of reproductively mature target species. Include provisions for temporary closures during key periods in the life history or during unusual environmental events.
Harvest quota	An annual quota for each species allowed in trade developed for each collection site from (1) the life history strategies of the coral; (2) its distribution and abundance and resilience based upon ecological requirements or specialization and susceptibility to threats; (3) its suitability for the trade; and (4) its status within the collection site. A list of species that can not be collected.
Collector obligations	Licensing and training for all collectors, and a system of fees to cover monitoring costs. Training should cover fishery legislation, management guidelines, size, and species that can be collected, collection and handling techniques, and reporting requirements. Logbooks that provide information on the location, amount, and type of collection should be submitted quarterly. Licenses should be revoked for failure to comply.
Collection guidelines	Specific nondestructive harvest guidelines that illustrate where and how to remove the coral to minimize impacts, proper handling and transport techniques, and training in these techniques.
Export requirements	Exporters should submit weekly, detailed information on the number and size of each species exported and the source of this coral.
Methods to minimize impacts	A management plan should include recommendations for reef enhancement in collection areas and a restoration plan for resource recovery in degraded areas.

byists and suppliers for sustainably and environmentally friendly harvested stony corals.

Acknowledgments

I am grateful to all of the participants of the International Coral Trade Workshop for background information provided on the management and trade in stony corals in their respective countries and for their input to the working groups. I am especially indebted to Eric Bornemen, Sian Breen, Ferdinand Cruz, Vicki Harriott, Gregor Hodgson, and Randall Owens for their leadership role in each working group. Technical comments on an earlier version of this manuscript by Barbara Best, John Field, and Tom Hourigan are appreciated. Support for this research was provided by the NOAA/National Marine Fisheries Service. The views expressed within this chapter are those of the author and they do not necessarily reflect the position of the National Marine Fisheries Service or the U.S. government.

References

Aquarium Fish and Coral Fisheries Working Group (AFCFWG). 1999. Queensland marine aquarium fish and coral collecting fisheries. Queensland Fisheries Management Authority Discussion Paper No. 10. 83 pp.

Bentley, Nokome. 1998. An overview of the exploitation, trade and management of corals in Indonesia. *TRAFFIC Bulletin* 17:67–78.

Bohnsack, James. A., Billy Causey, Michael P. Crosby, Roger B. Griffis, Mark A. Hixon, Thomas F. Hourigan, Karen H. Koltes, James E. Maragos, Ashley Simons, and John T. Tilman. In press. A rationale for minimum 20–30% no-take protection. *Proceedings 9th International Coral Reef Symposium.*

Brown, Barbara E., and Richard P. Dunne. 1988. The environmental impact of coral mining on coral reefs in the Maldives. *Environmental Conservation* 15:159–166.

Bruckner, Andrew W. 2000. Developing a sustainable trade in ornamental coral reef organisms. *Issues in Science and Technology* 17:63–68.

Bruckner, Andrew W. 2001a. Tracking the trade in ornamental coral reef organisms: the importance of CITES and its limitations. *Aquarium Sciences and Conservation* 3:79–94.

Bruckner, Andrew W. 2001b. Surveys of coral collection sites in the Spermonde Archipelago, South Sulawesi. In: *Review of Trade in Live Corals from Indonesia,* by Caroline Raymakers, Annex II. pp. 1–29. Belgium: TRAFFIC Europe, 17 August 2001.

Dulvy, N.K., D. Stanwell-Smith, D.R.T. Darwall, and C.J. Horrill 1995. Coral mining at Mafia Island, Tanzania: a management dilemma. *Ambio* 24:358–365.

Dustan, Phillip. 1999. Coral reefs under stress: sources of mortality in the Florida Keys. *Natural Resources Forum* 23:147–155.

Green, Edmund, and Francis Shirley. 1999. *The Global Trade in Coral.* United Kingdom: WCMC, World Conservation Press, 70 pp.

Grigg, Richard W. 1984. Resource management of precious corals: a review and application to shallow reef-building corals. *Marine Ecology* 5:57–74.

Grigg, Richard W. 1989. Precious coral fisheries of the Pacific and Mediterranean. In *Marine Invertebrate Fisheries: Their Assessment and Management*, edited by John F. Caddy, pp 637–645. San Francisco: John Wiley and Sons.

Lovell, Edward R. 2001. *Status report: collection of coral and other benthic reef organisms for the marine aquarium and curio trade in Fiji.* Fiji: World Wildlife Fund for Nature, South Pacific Program. 74 pp.

Maragos, James E. 1992. Restoring coral reefs with an emphasis on Pacific Reefs. In *Restoring the Nations Marine Environment,* edited by G.W. Thayer, pp 141–221. Silver Spring: Maryland Sea Grant Book.

Mulliken, Teresa A., and Stephen V. Nash. 1993. The recent trade in Philippine corals. *TRAFFIC Bulletin* 13:97–105.

NMFS 2001. International Coral Trade Workshop: Development of Sustainable Management Guidelines. Le Meridien, Jakarta, Indonesia. April 9–12.

Roberts, Callum M., and Julie P. Hawkins. 2000. *Fully-protected marine reserves: a guide.* Washington D.C.: WWF Endangered Species Campaign and York: Environment Department, University of York. 131 pp.

Ross, Michael A. 1984. A quantitative study of the stony coral fishery in Cebu, Philippines. *Marine Ecology* 5:75–91.

Suharsono. 1999. Condition of coral reef resources in Indonesia. *Indonesian Journal of Coastal and Marine Resources* 1:44–52.

Wells, Sue M. 1981. The coral trade in the Philippines. *TRAFFIC Bulletin* 3:50–51.

Wilkinson, Clive. 2000. The 1997–1998 mass coral bleaching and mortality event: 2 years on. In *Status of coral reefs of the world: 2000*, edited by Clive Wilkinson, pp 21–34. Cape Ferguson: Australian Institute of Science.

Fig. 6.1 The dwarf seahorse (*Hippocampus zosterae*), pictured here with some encrustation, is the primary seahorse species landed in Florida; landings reached 98,779 in 1994.

Fig. 6.2 Live rock from the Florida Keys (top) and Gulf of Mexico (bottom).

Fig. 9.1 Coral reefs are among the most biologically rich and productive ecosystems on Earth. © International Marinelife Alliance.

Fig. 9.2 Coral reefs are negatively impacted by the growing human population, which is becoming progressively concentrated in coastal zones, such as this village in the Philippines. © Marine Aquarium Council.

Fig. 9.3 Dredging of coral reef for construction material (sand, gravel, and limestone rock) in Micronesia. © P. Holthus.

Fig. 9.4 Coastal landfill, like that in New Caledonia shown here, is one of many land-based sources of pollution that result in the most pervasive and damaging impacts on coral reefs. © P. Holthus.

Fig. 9.5 Coral in Indonesia destroyed by dynamite used to catch fish for consumption. © P. Holthus.

Florida Keys National Marine Sanctuary

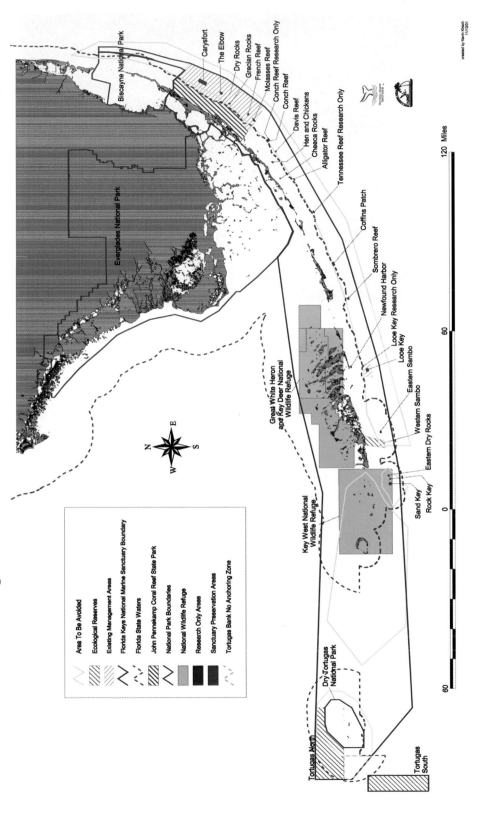

Fig. 26.1 The Florida Keys National Marine Sanctuary and marine zones used to manage the sanctuary for various uses.

PART III

The Invertebrates

A. *Live Rock Cultivation*

13

The Economics of Live Rock and Live Coral Aquaculture

John E. Parks, Robert S. Pomeroy, and Cristina M. Balboa

Abstract

Aquaculture is increasingly cited as a priority solution for reducing the harvest pressures on coral reefs associated with the marine ornamentals trade, especially in the developing Indo-Pacific. Two of the most important marine ornamental invertebrates traded internationally are live rock and live coral. The purpose of this chapter is to present results of a bioeconomic analysis of the aquaculture of live rock and live coral, based on a review of experimental and commercial production of live rock and live coral under way in Florida, Fiji, the Solomon Islands, and the Philippines. Analysis of the costs and returns of culture at different scales of production, cash flow, and price sensitivity is provided. The chapter concludes with a discussion of the benefits and constraints of culturing live rock and live coral in the United States and Indo-Pacific.

Introduction

International trade in live reef organisms for the marine aquarium industry has broad economic value and exerts considerable influence over global coral reef ecosystems. The sources of these organisms are largely concentrated in the Indo-Pacific. Overall estimates are that the Philippines and Indonesia alone contribute approximately 85% of the total volume of all live marine ornamentals that are imported into the United States (Baquero 1999). Of the remaining 15%, 5% to 10% appears to come from the Pacific Islands including Hawaii (Pyle 1993).

Two of the most important marine ornamental invertebrates traded globally are live rock and live coral. The markets for both are growing as demand increases. In recent years wide concerns have been voiced from numerous governmental and nongovernmental groups that a continued trade in wild-harvested live rock and live coral is not sustainable ecologically and should cease. Although live coral and live rock traded for the aquarium trade have not yet been listed as "endangered" under the Convention on International Trade in Endangered Species (CITES),[1] they are tracked by CITES as organisms that could potentially be threatened by the international marine ornamental trade. CITES data indicate that the trade in live coral and live rock increased by 15% to 30% each year in the 1990s (Bruckner 2001). Since 1992, Indonesia and Fiji have been the leading exporters of live rock and live coral (Bruckner 2001).

International Trade in Live Rock

The U.S. government defines live rock as "living marine organisms, or an assemblage thereof, attached to a hard substrate including dead coral or rock" (NMFS 1995). Calcareous algae, zooanthids, sponges, tunicates, bryozoa, anemones, and soft and hard corals all commonly encrust live rock, as can chitons, mollusks, serpulids, gastropods, echinoids, and even small crustaceans and fishes. Live rock commonly serves as the ecological foundation within living coral reef aquaria and is therefore heavily traded in the ornamentals market. More than 50,000 tons of live rock is thought to be currently maintained within U.S. aquaria (Green and Shirley 1999).

Before the closure of wild live rock harvest in U.S. waters during the mid-1990s, the United States supplied much live rock to its own industry. The commercial extraction of wild live rock began in Florida during the mid-1980s (Falls et al. 2000). This was followed by wild extraction and retail in Hawaii. By the mid-1990s the annual wild harvest of live rock from Florida waters alone was estimated at 300 tons per year (Falls et al. 2000). By the mid-1990s, extraction of live rock (and live coral) from U.S. state and federal waters had been prohibited.

Since the late 1990s the United States has imported nearly all (95%) of the live rock traded within the marine aquarium industry (Moore and Best 2001). Estimates suggest that millions of kilograms of live rock are imported into the United States every year, although precise figures are difficult to estimate, as live rock data are generally not reported separately from other scleractinian imports (Bruckner 2001). The total value of U.S. live rock sales during the period 1992–2000 is estimated at US$14 million, representing a volume of 2.5 million kilograms (5.5 million pounds) at an average value of US$5.60 per kilogram ($2.55/lb) live rock (Falls 2001). Fiji currently serves as the primary source of live rock to the United States (Lovell 2001; Bruckner 2000) at a rate of 4 to 5 times the weight of live rock imported from Indonesia (Falls 2001). Between 1992 and 1997, exports of live rock and coral from Fiji have doubled or tripled annually (Bruckner 2001). Since the mid-1990s, the low price of extracted live rock from Fiji has ranged between $1.25 and $2.00 per kilogram, frequently driving U.S. and international markets out of competition (Falls 2001; Heslinga 1999). In 1998, the Fijian Fisheries Department reported the export of 109,135 marine ornamentals at a value of F$349,699 (US$174,850), an average unit price of F$3.20 (US$1.60)(Fijian Fisheries Department 1998).

International Trade in Live Coral

It is known through CITES data that more than 2,000 species of coral are traded worldwide (Green and Shirley 1999). These data also indicate that 90% of all stony corals traded are sourced from only eight Indo-Pacific countries (Bruckner 2001).

The aquarium trade concentrates on scleractinian species with large flesh polyps and prominent tentacles, although small polyps are gaining popularity among hobbyists (Bruckner 2001). Nonscleractinian coral makes up only a small percent of the trade. The trade in live corals has increased tenfold since 1985 and continues to increase 20% to 30% each year (Lieberman and Field 2001). Between 1992 and 1997, more than 90% of all *Scleractinia* traded globally were imported into the United States (Bruckner 2001). Today, it is estimated that the United States imports between 56% and 80% of all live coral traded on the global market (Bruckner 2001; Green and Shirley 1999). This consistently high import consumption rate is a result of domestic prohibitions on the wild harvest of live coral and live rock due to their essential roles as fish habitat in U.S. waters (Moore and Best 2001). The next largest consumer of live coral is Germany, importing only 10% of the global production (Bruckner 2001).

Until the Philippine government enforced its ban on the export of live coral in the late 1980s, the Philippines was the top exporter of live coral (Green and Shirley 1999). Since the late 1980s, Indonesia and Fiji have become the two largest exporting countries of live coral (Bruckner 2001). Indonesia exports the largest numbers, while Fiji exports the most weight (Bruckner 2001). By 1995, 80% of nearly one million pieces of corals exported from Indonesia to the United States were shipped live (Bentley 1998). Estimates show that exporting countries generated approximately US$5 million in revenues from the live coral trade during 1997 (Green and Shirley 1999). Strong concerns continue to be raised by Indo-Pacific countries that domestic coral reefs are being consumed by the United States at unsustainable rates.

Aquaculture as a Potential Solution to Wild Harvest

As a result of growing concerns over the sustainability and ecological impacts of harvesting live coral reef organisms from the Indo-Pacific for the marine ornamentals trade, since the mid-1990s increasing interest has been expressed in exploring aquaculture as a potential solution in reducing extraction pressures associated with the live reef organism trades. The assumption is that if live marine ornamentals can be adequately and competitively cultured, it would decrease the demand for wild-caught reef organisms. Exploring culture technologies for slow-growing, sessile tropical invertebrates (e.g., coral species and the invertebrate communities that reside on live rock) is of particular interest to international decision makers, managers, and industry representatives, given U.S. consumption trends.

Consequently since the 1990s U.S. aquaculture centers (Seaman and Adams 1998; Corbin and Young 1995; National Research Council 1992) have identified ornamentals culture research as a national aquaculture priority. This priority is consistent with other international recommendations (ICRI 2001; ICLARM 2000).

Out of the Marine Ornamentals '99 Conference[2] arose the recommendation to "give highest priority to projects involving the advancement of marine ornamental aquaculture and reef preservation" (Corbin 2001). Conference recommendations were also made to encourage consumer understanding that cultured ornamentals are a more sustainable and "higher value" alternative to wild-caught live reef organisms (Corbin 2001).

Even leading entrepreneurs recognize the need to shift operations from wild harvest to propagation of live rock (Herndon and Herndon 2000) and live coral (Mcleod 2001).

Study Background

Responding to the call out of the Marine Ornamentals '99 Conference, in the year 2000 the World Resources Institute (WRI) commenced a research project designed to undertake a thorough bioeconomic assessment of the financial feasibility of culture technologies for the live reef organism trades (food fish and marine ornamentals). The goals of the project were to: (a) review and determine the potential application of such technologies as alternatives to wild capture of live reef organisms in developing country nearshore waters, and (b) if potential exists, to provide information and policy guidance for appropriate technology application within the developing Indo-Pacific island nations (i.e., from where most live reef organisms are sourced for the live reef organism trades).

A structured investigation into the techniques and feasibility of live reef organism culture was undertaken during 2000 and 2001 by the authors. As a component of this research, we undertook a review of the financial feasibil-

ity of live rock and live coral culture technologies. This chapter presents summary findings from this component of the overall research. A detailed description of the culture technologies reviewed and a complete analysis and discussion of research findings for both types of propagation, as well as other marine ornamental and live reef food fish species culture, are presented in full in the final project report (Pomeroy et al. 2002).

Methods

Interest in the culture of live rock and live coral is not new. Public aquarium staff in Hawaii and private operators in Michigan have practiced asexual propagation of corals via fragmentation since the 1960s (Delbeek 2001). During the late 1980s, the international conservation community began raising sustainability concerns regarding trade reliance on wild harvest and by the early 1990s began exploring culture technologies as both alterative income and reef rehabilitation opportunities. By the mid-1990s numerous small-scale private culture operations for profit and research commenced on the U.S. mainland ("basement" hobbyist operations) and throughout the Indo-Pacific (Delbeek 2001; Miller 1998). The majority of these, however, are not strict culture operations and rely on supplementary income from trade in wild stock to be profitable. The financial viability of strict-culture operations is poorly understood.

Data Collection

Primary data were collected through site visits to private business and research culture operations in the Indo-Pacific (Fiji, the Solomon Islands, Indonesia, and the Philippines) and the United States (Florida and Hawaii). Key informant responses were collected using structured, semi-structured, and open-ended interviews of private operators, nongovernmental organization staff, and researchers. Interview data were also collected from rural village culturists and community residents in the Solomon Islands and Fiji using focus group discussions. Interviews were designed to elucidate information regarding the life history of cultured organisms and various culture technologies being employed. Economic data were collected on site through interviews and secondary data sources, including business records, balance sheets, accounting reports, and financial plans and audits. Published and unpublished sources of enterprise case study and analysis were also reviewed where available.

Four community-based conservation project sites undertaking live coral culture through nongovernmental organization (NGO) support were also reviewed as comparative cases to that of private operators and researchers. Interviews were conducted with community representatives and NGO project staff. Because conservation projects subsidize some or most of the costs from the culture operations, enterprise data and records were used to inform and triangulate financial models but not derive them.

Study sites were selected on the basis of the following criteria: (a) they are internationally recognized by the marine ornamentals industry or marine resource management community as case examples where the state-of-the-art technology for in situ live rock and coral culture is used; (b) they have been in operation for more than 3 years; (c) they represent either a medium-scale U.S. or small-scale Indo-Pacific operational culture model; and (d) they agree to share data and be transparent in their reporting of the benefits and constraints of the technology reviewed.

In addition to the written materials provided by study sites, a limited number of relevant secondary data sources were identified through literature review of: (a) peer-reviewed journal articles and paper presentations from conference proceedings, (b) academic publications, (c) aquarium industry business and trade publica-

tions, and (d) aquarist magazine articles and other hobbyist publications.

The Two Production Models

On the basis of the data collected and the study goals, two culture production models were projected and analyzed financially: (1) a medium-sized U.S.-based producer (Fig. 13.1), and (2) a small-scale, rural village producer based in an Indo-Pacific island (IPI; Fig. 13.2). An analytical focus at these particular scales was made on the basis of the goals of the study. The medium-scale U.S. production model assumes both marine and on-land operations, whereas the small-scale IPI model assumes only nearshore marine operations.

For both models, three separate production scenarios are projected: (1) cultured live rock production (Fig. 13.3) only; (2) cultured live coral production (Fig. 13.4) only; and (3) mixed cultured live rock and live coral production. The live rock scenario is based on the "seeding" of porous, calcareous substrate (e.g., quarried limestone) that is placed on the seafloor (12 to 50 foot depth) for a sufficient length of time (2 to 7 years, depending on site placement and target coverage and quality) to allow invertebrates to recruit and become established as live rock that can then be harvested (Fig. 13.5) and held (Fig. 13.6) until sale (Falls 2001; Falls et al. 2000; Herndon and Herndon 1999; Falls et al. 1999; Antozzi 1997; Frakes and Watts 1995).

The live coral scenario is based on coral fragmentation, whereby either nubbins (pruned pieces from tips or middles) of branching par-

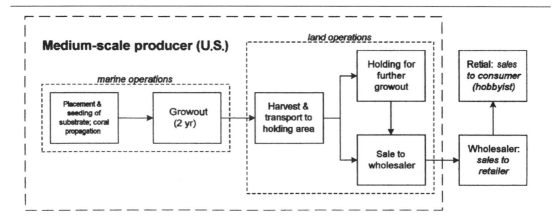

Fig. 13.1 Diagram of the production process under the projected U.S. medium-scale culture operation.

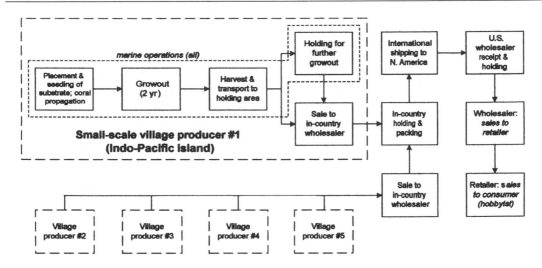

Fig. 13.2 Diagram of the production process under the projected Indo-Pacific island small-scale village culture operation. Note the multiple village producers contributing cultured product.

ent colony hard corals (Fig. 13.7) or pie-sliced segments obtained through parent colony soft coral biopsy (Fig. 13.8) are affixed to a base (substrate) using epoxy, string, wire, or mesh (see Heeger et al. 2000; Bowden-Kerby 1999a) or hung from monofilament line suspended in the water column (Bowden-Kerby 1999b) and then grown out until achieving a marketable size (often fist-sized). Other forms of asexual coral reproduction (e.g., encouraging polyp exclusion or polyp ball formation; Spotts 2001) are not considered under this scenario. The mixed live rock and coral scenario includes

Fig. 13.3 Cultured live rock produced by a Florida-based operator in the Gulf of Mexico. Such pieces are being successfully marketed as "ecofriendly" live rock in the United States. (See also color plates.)

both rock substrate placement and seeding and coral fragmentation activities.

Both the U.S. and IPI models are projected in this study as strict culture operations (no supplemental wild-harvest income) using appropriate levels of technology and selling all cultured product to a wholesaler (no retail sales).

Although the emphasis of the study is oriented toward addressing the feasibility and potentials of the IPI model in developing nations, it became clear that a comparison to a projected U.S.-based model would illustrate an alternative model of live rock and coral production from which the costs and benefits of investment in a developing country operation could become clearer. In addition, marine ornamental industry and U.S. and Southeast Asian national policy audiences both indicated their interest in a comparison of northern and southern ornamental invertebrate production operations at the outset of the study, presumably for decision-making purposes relating to the equitable and logical allocation of the necessary human and financial resources within either or both models as an alternative to wild capture operations.

Data Analysis

A financial analysis was made of primary and secondary data collected during 2001 and 2002 at WRI. This analysis allows the operational projection of the medium-scale U.S. and small-

Fig. 13.4 Cultured live hard (left) and soft (right) corals produced by Philippines- and Florida-based (respectively) operators. (See also color plates.)

scale IPI models from which the economic feasibility of both models can be examined. Essentially three questions are asked in the analysis: (1) How much capital investment must be made into the model? (2) What are the incremental costs and operating budget involved in a production cycle? (3) When does the operation become profitable? To address these three questions, three sets of economic analyses are undertaken: (1) capital investment analysis; (2) creation of an enterprise budget; and (3) cash flow projection. A sensitivity analysis is also conducted in order to identify the thresholds of

economic viability (i.e., determine how the profitability of operations can be improved). All projected expenses, income levels, and pricing structures for both production models are considered conservative (i.e., competitive and minimally required) and are based on actual market conditions at the time of study, the economics of business operations reviewed, and key informant interview responses. Additional costs beyond what would be considered legitimately necessary (e.g., ecological monitoring, travel to industry meetings and trade exhibitions) to operate a culture business were not included under

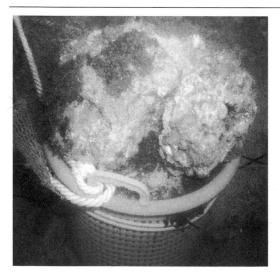

Fig. 13.5 The harvest of cultured live rock by a Florida-based operator in the Gulf of Mexico. Divers load seeded product into buckets (left) that are then hauled up into a workboat and sorted into insulated bins filled with seawater (right) for transport to an on-land wholesaler or holding facility. (See also color plates.)

Fig. 13.6 The holding facility (left) and individual holding well (right) of cultured live rock by a Florida-based operator. (See also color plates.)

either model. Finally, all equipment and materials are assumed purchased new.

It is assumed that optimal conditions are operating throughout the duration of the analysis timeframe: (1) the demand for cultured products is stable, (2) the marketability of cultured products is secure (i.e., of sufficient quality), (3) permission to access and use the coastal waters and seafloor for culture has been secured with relevant authorities (including national, provincial/state, or local actors), (4) the areas of coastal waters and seafloor where in situ culture

occurs reflect ideal abiotic and biotic site conditions (see Heeger et al. 2000 for live coral, Herndon and Herndon 1999 for live rock), (5) no natural or man-made disasters are experienced, and (6) no poaching of cultured product or broodstock occurs from the seeding, grow-out, or holding areas.

Study Limitations

There are a number of limitations to the study and two production models that must be recog-

Fig. 13.7 Pruned coral branches are affixed to substrate (left) and then held and grown out in underwater holding areas (right) or in an on-land holding facility until they have achieved a marketable size. (See also color plates.)

Fig. 13.8 Soft coral cuttings being held and grown out in an on-land holding facility until they have achieved a marketable size. (See also color plates)

nized. First is the lack of actual live rock and live coral culture businesses that were operating during the study period from which information could be sourced. Next, although such operations are increasingly being attempted, to date there is a relative scarcity of published or peer-reviewed literature on propagation techniques and business operations. As a consequence of these two constraints, the models within the study from which conclusions are drawn rely heavily on the accuracy and reliability of the

limited primary and secondary data available. This leads to three related and fundamental assumptions in the study: (1) the sample of operations visited and study undertaken is sufficient from which to project the two models presented and draw conclusions, (2) the operations reviewed together reflect the actual operating conditions (market and otherwise) at the time of the study that would likely be encountered under both models, and (3) a strict culture operation is potentially feasible in the marine ornamentals marketplace, despite all the businesses reviewed being heavily to principally reliant on supplemental income sourced from the collection and sale of wild-caught marine ornamentals.

Results

Capital investment

The first financial analysis undertaken was an estimate of the start-up funding necessary to undertake the two production models. Summary results of the projected capital investment for each scenario required under the U.S. and IPI models are displayed in Tables 13.1 and 13.2, respectively.

The projected U.S. model start-up costs for the live rock, live coral, and mixed rock and

Table 13.1   Summary of total projected capital investment costs for the live rock, live coral, and mixed rock and coral culture scenarios under the U.S. production model (all figures in US$)

Cost category	Live rock (only)	% of total capital investment	Live coral (only)	% of total capital investment	Mixed live rock and live coral	% of total capital investment
Marine operations (seeding and collection)						
• Ocean transport	190,712	(43%)	73,013	(28%)	190,712	(37%)
• Collection and holding	24,185	(05%)	807	(01%)	24,185	(05%)
• Diving equipment	12,650	(03%)	6,915	(02%)	12,650	(02%)
Subtotal	227,547	(51%)	80,735	(30%)	227,547	(44%)
Initial substrate Subtotal	34,000	(08%)	N/A		34,000	(07%)
Land operations (grow-out, holding, and shipping)						
• Holding/grow-out facility	140,178	(32%)	141,667	(54%)	212,666	(41%)
• Packing and shipping	1,480	(01%)	1,480	(01%)	2,265	(01%)
• Administration	5,395	(01%)	5,395	(02%)	5,640	(01%)
• Land transportation	32,375	(07%)	35,500	(13%)	32,375	(06%)
Subtotal	179,428	(41%)	184,052	(70%)	252,946	(49%)
Total capital investment	440,975	(100%)	264,787	(100%)	514,493	(100%)

Note: Line items for each cost category can be found in Pomeroy et al. 2002.

Table 13.2. Summary of total projected capital investment costs for the live rock, live coral, and mixed rock and coral culture scenarios under the IPI production model (all figures in US$)

Cost category	Live rock (only)	% of total capital investment	Live coral (only)	% of total capital investment	Mixed live rock and live coral	% of total capital investment
Marine operations (seeding, grow-out, and collection)						
• Ocean transport	12,834	(46%)	12,834	(62%)	12,834	(39%)
• Collection and holding	2,353	(09%)	2,585	(13%)	2,585	(08%)
• Coral grow-out stations	N/A		4,498	(22%)	4,498	(14%)
• Diving equipment	660	(02%)	660	(03%)	660	(02%)
Subtotal	15,847	(57%)	20,577	100%)	20,577	(63%)
Initial substrate Subtotal	12,000	(43%)	N/A		12,000	(37%)
Total capital investment	27,847	(100%)	20,577	(100%)	32,577	100%)

Note: Line items for each cost category can be found in Pomeroy et al. 2002.

coral culture scenarios are high, with the live coral scenario being the least capital intensive (US$264,787) and the mixed rock and coral being the most (US$514,493, or twice that of live coral). All three scenarios are highly technology-dependent as evidenced by the exorbitant capital investment needs. The U.S. live coral culture scenario is projected to be less than two-thirds (60%) of the cost of U.S. live rock culture. This is a result of reduced marine operations costs under the live coral scenario (35% of live rock costs). Live rock marine operation costs are high because of the need for a 13.7 m (45 ft) commercial workboat (43% of total costs) that will haul a sufficient volume of seed and propagated rock to achieve the target production rate. Start-up investment for live rock only is nearly (86%) that of mixed rock and coral costs. Land operation investment under the live coral scenario represents more than two-thirds (70%) of the total start-up costs, whereas under the live rock and mixed rock and coral scenarios the marine and land operations were proportionally similar (40% to 50%) with respect to total costs.

Capital investment needs for the three IPI scenarios are between 6% and 8% of the U.S. model, largely because IPI model start-up costs do not require a large commercial workboat, on-land grow-out and holding facilities, or SCUBA diving equipment. The IPI US$20,000 to US$30,000 investment required, however, is comparatively high for a rural village producer in the developing Pacific. Under all three IPI scenarios, the 7 m (23 ft) open-hull boat with

outboard engine required for ocean transport of seed rock and cultured product is the highest cost category.

Enterprise Budget

The next financial analysis completed was the creation of enterprise budgets for each of the three scenarios under the two production models. These budgets itemize the projected variable costs (costs that vary with unit output), fixed costs (unchanging capital expenditures), and income levels during a single production period of 2 years from which an end balance can be calculated (profit or loss). The 2-year production period is based on the minimum "seeding" (accretion) and grow-out time for marketable live rock (Falls 2001; Herndon 1998; Antozzi 1997), 6 to 18 month grow-back period for biopsied parent colonies, and 6 to 16 month grow-out period for nubbins to fist-sized colonies, depending on the species and culture conditions (Mcleod 2001; Herndon and Herndon 1999; Bowden-Kerby 1999a). A conservative estimate of product mortality is also factored into the budget. The summary results for projected U.S. and IPI enterprise budgets are displayed in Table 13.3.

All three scenarios under both models operate at a loss under the projected production budgets. Under the U.S. model the live coral scenario operates at the greatest loss (–US$311,197). The U.S. live rock scenario loss (–US$236,194) is three-quarters (76%) of the live coral loss. Although operating at a loss

Table 13.3 Projected summary enterprise budget for live rock, live coral, and mixed rock and coral culture operations for both U.S. and IPI models over a 2-year production cycle (all figures in US$)

Cost category	Live rock (only) U.S. model		Live rock (only) IPI model		Live coral (only) U.S. model		Live coral (only) IPI model		Mixed live rock and coral U.S. model		Mixed live rock and coral IPI model	
Variable costs												
Cost of goods	71,748	(06%)	6,000	(08%)	21,286	(03%)	432	(01%)	73,286	(05%)	6,432	(08%)
Cost of sales	66,240	(05%)	6,180	(09%)	66,240	(08%)	6,180	(09%)	96,960	(07%)	8,100	(10%)
Ocean transport	88,880	(07%)	12,540	(17%)	4,747	(01%)	12,540	(19%)	88,880	(06%)	16,060	(19%)
Land transport	30,000	(02%)	0	(00%)	30,000	(04%)	0	(00%)	30,000	(02%)	0	(00%)
Advertising/sales	10,080	(01%)	0	(00%)	10,080	(01%)	0	(00%)	10,080	(01%)	0	(00%)
Payroll	504,000	(39%)	43,776	(61%)	274,008	(33%)	43,776	(65%)	554,000	(38%)	49,920	(59%)
Supplies/repairs	126,960	(10%)	288	(00%)	126,960	(15%)	288	(00%)	151,440	(11%)	288	(00%)
Utilities	77,280	(06%)	0	(00%)	77,280	(09%)	0	(00%)	98,160	(07%)	0	(00%)
Miscellaneous	2,640	(00%)	0	(00%)	2,640	(00%)	0	(00%)	2,640	(00%)	0	(00%)
Subtotal	977,828	(76%)	68,784	(95%)	613,241	(73%)	63,216	(94%)	1,105,446	(77%)	80,800	(95%)
Fixed costs												
Insurance	83,352	(06%)	0	(00%)	65,712	(08%)	0	(00%)	88,800	(06%)	0	(00%)
Rentals	53,980	(04%)	0	(00%)	46,720	(06%)	0	(00%)	53,980	(04%)	0	(00%)
Licenses	9,200	(01%)	150	(00%)	1,100	(00%)	150	(00%)	9,200	(01%)	150	(00%)
Depreciation	54,598	(04%)	3,406	(05%)	18,116	(02%)	4,144	(06%)	66,627	(05%)	4,144	(05%)
Interest Payments	103,500	(08%)	0	(00%)	85,320	(10%)	0	(00%)	108,000	(07%)	0	(00%)
Miscellaneous	8,880	(01%)	0	(00%)	8,880	(01%)	0	(00%)	8,880	(01%)	0	(00%)
Subtotal	313,510	(24%)	3,556	(05%)	225,848	(27%)	4,294	(06%)	335,487	(23%)	4,294	(05%)
Total expenses	1,291,338	(100%)	72,340	(100%)	839,089	(100%)	67,510	(100%)	1,440,933	(100%)	85,094	(100%)
Total income	1,055,144		14,250		527,892		15,200		1,352,636		29,850	
Balance	−236,194		−58,090		−311,197		−52,310		−88,297		−55,244	

Note: Line items for each cost category can be found in Pomeroy et al. 2002. IPI, Indo-Pacific island; (%), percent of total investment.

194

(–US$88,297) the U.S. mixed live rock and coral operation performs notably better than the other two scenarios at only 28% of the live coral and 37% of the live rock losses. The improved enterprise performance under the mixed rock and coral scenario is a consequence of improved income (two-and-a-half times the live coral and more than one-and-a-third times the live rock income).

Under the IPI model the live rock scenario results in the greatest loss (–US$58,090). Variable costs account for the majority of expenses under both the U.S. and IPI models at approximately 75% and 95% of total expenses, respectively. The cost of labor is the highest expense, accounting for approximately one-third (33% to 39%) of total U.S. model expenses and approximately two-thirds (59% to 65%) of total IPI model expenses. The costs of labor were estimated on the basis of comparable pay scales under actual production models and a weighted average daily wage rate for the Pacific model.

Income estimates are based on a competitive pricing structure given operating market conditions at the time of study and based on a feasible but maximized production rate of cultured products. Price sensitivity for the U.S. model was investigated, and at a moderate price increase for live rock (i.e., a 12% to 25% increase, or US$0.50 more paid per pound, from competitive base prices of US$2 to US$4 per pound, depending on the age of the cultured rock) allows the live rock budget scenario to break even (slight profit at US$23,806), whereas at a high price structure (i.e., a 30% to 50% increase, or US$1.00 more paid per pound, from competitive base prices) the budget becomes profitable (US$283,806). Whereas the U.S. mixed live rock and coral scenario does not break even (–US$88,297), the budget is sensitive to increased pricing and profitable un-der both moderate (US$210,104) and high ($508,504) price structures for live rock (US$2.50 per pound and US$3.00 per pound) and coral (US$7 per piece and US$8 per piece) during the production cycle.

Price sensitivity under the IPI model is not responsive under both moderate and high price structures for the live rock (only), live coral (only), or mixed live rock and coral scenarios, and the model remains operating at a loss near to the original projected balance. This is a consequence of relatively low pricing structure flexibility and lower volume output at the level of the IPI small-scale village production. Reduced variable costs at predicted yield levels as a consequence of changing national economic conditions (e.g., reduced cost of living or daily wage rate) could lead to increased profitability under the IPI model.

Cash Flow

Summary results for projected U.S. cash flows over five production periods (10 years) across the three scenarios are displayed in Tables 13.4 through 13.6. Summary results for projected IPI cash flows are displayed in Tables 13.7-13.9.

Projected 10-year U.S. model cash flow indicates poor viability of the live rock and live coral scenarios before the fifth production cycle (years 9 and 10), where live rock returns a first year-end profit (US$101,806) and the live coral breaks even (US$18,735). However, both the live rock and live coral scenarios by this cycle are heavily in debt (–US$874,190 and –US$1,000,161, respectively) and would likely be predicted not economically feasible for another 10 years. The U.S. mixed live rock and coral scenario proves much more feasible, turning a first profit (US$100,145) in the third production cycle and nearly out of debt

Table 13.4 Projected cash flow for a U.S.-based live rock (only) culture operation over five production cycles (all figures US$)

	Cycle 0 (yr 0)	Cycle 1 (yr 1–2)	Cycle 2 (yr 3–4)	Cycle 3 (yr 5–6)	Cycle 4 (yr 7–8)	Cycle 5 (yr 9–10)
Gross income	0	0	1,055,144	1,211,144	1,289,144	1,393,144
Capital costs	440,975	197,150	313,510	313,510	313,510	313,510
Operating costs	0	19,288	977,828	977,828	977,828	977,828
Cycle-end balance	–440,975	–216,438	–236,194	–80,194	–2,194	101,806
Balance carried over	–440,975	–657,413	–893,607	–973,801	–975,996	–874,190

Table 13.5. Projected cash flow for a U.S.-based coral (only) culture operation over five production cycles (all figures US$)

	Cycle 0 (yr 0)	Cycle 1 (yr 1–2)	Cycle 2 (yr 3–4)	Cycle 3 (yr 5–6)	Cycle 4 (yr 7–8)	Cycle 5 (yr 9–10)
Gross income	0	0	527,892	659,865	791,838	857,824
Capital costs	264,787	197,150	225,848	225,848	225,848	225,848
Operating costs	0	19,288	613,240	613,240	613,240	613,240
Cycle-end balance	−264,787	−216,438	−311,197	−179,224	−47,251	18,735
Balance carried over	−264,787	−481,225	−792,422	−971,645	−1,018,896	−1,000,161

Table 13.6 Projected cash flow for a U.S.-based mixed live rock and live coral culture operation over five production cycles (all figures US$)

	Cycle 0 (yr 0)	Cycle 1 (yr 1–2)	Cycle 2 (yr 3–4)	Cycle 3 (yr 5–6)	Cycle 4 (yr 7–8)	Cycle 5 (yr 9–10)
Gross income	0	0	1,352,636	1,541,077	1,685,063	1,822,056
Capital costs	514,493	197,150	335,487	335,487	335,487	335,487
Operating costs	0	19,288	1,105,445	1,105,445	1,105,445	1,105,445
Cycle-end balance	−514,493	−216,438	−88,296	100,145	244,131	381,124
Balance carried over	−514,493	−730,931	−819,227	−719,083	−474,952	−93,829

Table 13.7 Projected cash flow for an IPI-based live rock (only) culture operations over five production cycles (all figures US$)

	Cycle 0 (yr 0)	Cycle 1 (yr 1–2)	Cycle 2 (yr 3–4)	Cycle 3 (yr 5–6)	Cycle 4 (yr 7–8)	Cycle 5 (yr 9–10)
Gross income	0	0	14,250	18,525	20,187	22,563
Capital costs	27,847	3,556	3,556	3,556	3,556	3,556
Operating costs	0	6,232	68,784	68,784	68,784	68,784
Cycle-end balance	−27,847	−9,788	−58,090	−53,815	−52,153	−49,777
Balance carried over	−27,847	−37,635	−95,724	−149,539	−201,691	−251,468

Table 13.8 Projected cash flow for IPI-based live coral (only) culture operations over five production cycles (all figures US$)

	Cycle 0 (yr 0)	Cycle 1 (yr 1–2)	Cycle 2 (yr 3–4)	Cycle 3 (yr 5–6)	Cycle 4 (yr 7–8)	Cycle 5 (yr 9–10)
Gross income	0	0	15,200	20,900	26,600	32,300
Capital costs	20,577	3,556	4,294	4,294	4,294	4,294
Operating costs	0	6,232	63,216	63,216	63,216	63,216
Cycle-end balance	−20,577	−9,788	−52,310	−46,610	−40,910	−35,210
Balance carried over	−20,577	−30,365	−82,674	−129,284	−170,194	−205,404

Table 13.9 Projected cash flow for IPI-based live rock and coral culture operations over five production cycles (all figures US$)

	Cycle 0 (yr 0)	Cycle 1 (yr 1–2)	Cycle 2 (yr 3–4)	Cycle 3 (yr 5–6)	Cycle 4 (yr 7–8)	Cycle 5 (yr 9–10)
Gross income	0	0	29,850	39,425	46,787	54,863
Capital costs	32,577	3,556	4,294	4,294	4,294	4,294
Operating costs	0	6,232	80,800	80,800	80,800	80,800
Cycle-end balance	−32,577	−9,788	−55,244	−45,669	−38,307	−30,231
Balance carried over	−32,577	−42,365	−97,608	−143,277	−181,584	−211,815

(−US$93,829) by the fifth cycle. Clearly the mixed live rock and coral scenario is the most economically feasible of the three scenarios based on study projections.

All three scenarios under the IPI model proved not feasible within five production cycles. By years 9 and 10, the live rock scenario is operating at the heaviest loss (−US$49,777) and the most deeply in debt (−US$251,468), whereas the mixed rock and coral experiences the least loss (−US$30,231) of the three scenarios and the live coral is the least in debt (−US$205,404). These losses and debt would not be sustainable for a small-scale village producer to maintain, and the model is therefore not economically feasible on the basis of study results.

Because of the poor viability of all scenarios under the two models, with the exception of the U.S. mixed rock and coral operation, an analysis of the budgets' sensitivity to capture of additional income was undertaken. Whereas a moderate to high price structure would not be considered competitive and therefore feasible given current market conditions, the capture of a "green" (i.e., ecologically friendly) premium on cultured products paid by the end consumer

(hobbyist) appears feasible given the "eco-friendly" marketing success experienced by U.S.-based live rock and coral culturists. Such a premium could be captured using a "sustainable product" label recognized by consumers and granted through a third-party certifier (e.g., under Marine Aquarium Council[3] labeling). To test each scenario's sensitivity to receipt of a green premium, a fee (US$0.10 on every dollar sold) was added to the cultured products, and cash flow was recalculated. Summary results for projected U.S. and IPI cash flows across the three scenarios under green premium capture are displayed in Tables 13.10 through 13.12 and 13.13 through 13.15, respectively.

The U.S. model proves responsive to capture of a moderate (10%) green premium, with the live rock scenario breaking even (US$39,406) during cycle three and turning a marginal profit (US$125,206) the next cycle. Likewise, with the green premium the U.S. live coral scenario breaks even (US$21,869) during cycle four and turns a slight profit (US$93,615) during cycle five. Under both the revised U.S. live rock and live coral scenarios, however, the operations are still in debt at the end of 10 years (−US$385,390 and −US$752,481, respective-

Table 13.10 Projected cash flow for a U.S.-based live rock (only) culture operation over five production cycles with a 10% green premium being paid on all product (all figures in US$)

	Cycle 0 (yr 0)	Cycle 1 (yr 1–2)	Cycle 2 (yr 3–4)	Cycle 3 (yr 5–6)	Cycle 4 (yr 7–8)	Cycle 5 (yr 9–10)
Gross income	0	0	1,159,144	1,330,744	1,416,544	1,530,944
Capital costs	440,975	197,150	313,510	313,510	313,510	313,510
Operating costs	0	19,288	977,828	977,828	977,828	977,828
Cycle-end balance	−440,975	−216,438	−132,194	39,406	125,206	239,606
Balance carried over	−440,975	−657,413	−789,607	−750,201	−624,996	−385,390

Table 13.11 Projected cash flow for a U.S.-based coral (only) culture operation over five 2-year intervals (1–2 coral production cycles, depending on the species) with a 10% green premium being paid on all product (all figures in US$)

	Cycle 0 (yr 0)	Cycle 1 (yr 1–2)	Cycle 2 (yr 3–4)	Cycle 3 (yr 5–6)	Cycle 4 (yr 7–8)	Cycle 5 (yr 9–10)
Gross income	0	0	573,972	717,465	860,958	932,704
Capital costs	264,787	197,150	225,848	225,848	225,848	225,848
Operating costs	0	19,288	613,240	613,240	613,240	613,240
Cycle-end balance	−264,787	−216,438	−265,117	−121,624	21,869	93,615
Balance carried over	−264,787	−481,225	−746,342	−867,965	−846,096	−752,481

Table 13.12 Projected cash flow for a U.S.-based mixed live rock and coral culture operation over five production cycles with a 10% green premium being paid on all product (all figures in US$)

	Cycle 0 (yr 0)	Cycle 1 (yr 1–2)	Cycle 2 (yr 3–4)	Cycle 3 (yr 5–6)	Cycle 4 (yr 7–8)	Cycle 5 (yr 9–10)
Gross income	0	0	1,446,130	1,689,477	1,847,023	1,997,296
Capital costs	514,493	197,150	335,487	335,487	335,487	335,487
Operating costs	0	19,288	1,105,445	1,105,445	1,105,445	1,105,445
Cycle-end balance	−514,493	−216,438	5,198	248,545	406,091	556,364
Balance carried over	−514,493	−730,931	−725,733	−477,189	−71,098	485,265

Table 13.13 Projected cash flow for IPI-based live rock (only) culture operations over five production cycles with 10% green premium paid on product (all figures in US$)

	Cycle 0 (yr 0)	Cycle 1 (yr 1–2)	Cycle 2 (yr 3–4)	Cycle 3 (yr 5–6)	Cycle 4 (yr 7–8)	Cycle 5 (yr 9–10)
Gross income	0	0	16,500	20,378	22,206	24,819
Capital costs	27,847	3,556	3,556	3,556	3,556	3,556
Operating costs	0	6,232	68,784	68,784	68,784	68,784
Cycle-end balance	−27,847	−9,788	−55,840	−51,962	−50,134	−47,521
Balance carried over	−27,847	−37,635	−93,474	−145,436	−195,569	−243,090

Table 13.14 Projected cash flow for IPI-based live coral (only) culture operations over five production cycles with 10% green premium paid on product (all figures in US$)

	Cycle 0 (yr 0)	Cycle 1 (yr 1–2)	Cycle 2 (yr 3–4)	Cycle 3 (yr 5–6)	Cycle 4 (yr 7–8)	Cycle 5 (yr 9–10)
Gross income	0	0	16,720	22,990	29,260	35,530
Capital costs	20,577	3,556	4,294	4,294	4,294	4,294
Operating costs	0	6,232	63,216	63,216	63,216	63,216
Cycle-end balance	−20,577	−9,788	−50,790	−44,520	−38,250	−31,980
Balance carried over	−20,577	−30,365	−81,154	−125,674	−163,924	−195,904

Table 13.15 Projected cash flow for IPI-based live rock and coral culture operations over five production cycles with 10% green premium paid on product (all figures in US$)

	Cycle 0 (yr 0)	Cycle 1 (yr 1–2)	Cycle 2 (yr 3–4)	Cycle 3 (yr 5–6)	Cycle 4 (yr 7–8)	Cycle 5 (yr 9–10)
Gross Income	0	0	33,220	43,368	51,466	60,349
Capital costs	32,577	3,556	4,294	4,294	4,294	4,294
Operating costs	0	6,232	80,800	80,800	80,800	80,800
Cycle-end balance	–32,577	–9,788	–51,874	–41,726	–33,628	–24,745
Balance carried over	–32,577	–42,365	–94,238	–135,964	–169,592	–194,337

ly). Fortunately this is not the case under the mixed rock and coral scenario, where the operation breaks even (US$5,198) by the second production cycle, turns a profit (US$248,545) by the third cycle (years 5 and 6), and is free of debt with net revenues of US$485,265 by the end of year 10. Under the medium-scale U.S. production model, the mixed live rock and live coral scenario with 10% green premium appears the only feasible business model for prospective strict-culture operators.

All three IPI scenarios under 10% green premium remain operating at a loss by the end of the fifth production cycle. Like the U.S. model, the most optimistic IPI scenario is mixed live rock and coral production under the 10% green premium, where the operating loss (–US$24,745) is least and debt (–US$194,337) comparably lowest during the fifth production cycle. Because of the relatively low volume of product generated, the green premium has a negligible effect on the viability of these operations over the long term and they remain unfeasible. It is important to note, however, that independent of operational profitability, village producers generate comparable to higher unit returns on cultured products sold with 10% green premium than to that of wild-caught products.

Discussion

Overall Financial Feasibility of Strict Culture

The results of this study are not particularly encouraging for live rock and coral culture entrepreneurs or resource managers and decision makers seeking sustainability relief through aquaculture promotion. Under current market conditions, a competitive pricing structure, and in the absence of a green premium and deliberate regulatory intervention restricting or limiting wild-caught live rock and coral, it does not appear overall that strict live rock and/or live coral culture operations are economically feasible or at a minimum (as with the U.S. mixed live rock and coral scenario) risk free. Under such conditions, it is clear that strict-culture operations will alone be insufficient to solve the marine ornamentals sustainability issue as an industry enterprise alternative. This conclusion has important implications for conservation and sustainable management of wild stock collection and best practices certification.

The results of this study are not unexpected and corroborate previous research on the financial feasibility of marine ornamental aquaculture (Frakes 2001; Brown 1999; Antozzi 1997; Wheeler 1996) in regard to the high start-up cost requirements, high operating costs, and comparably low returns based on the product pricing structure required to remain competitive against wild-caught product. Of the six scenarios projected, only the mixed live rock and live coral scenario under the U.S. model indicated a profit potential, and only after five production cycles (beyond 10 years). As live rock and live coral culture operations appear cost prohibitive under present conditions, short of lengthy subsidy or sustained venture capital they may become feasible as businesses only in requiring a long-term (10 to 20 years) phase-out process of wild stock capture and gradual increased fiscal reliance on propagated live rock and live coral. Although this may be regarded as sobering news to decision makers, resource managers, and conservation NGOs, the reality is that most operators in the industry are aware of the financial costs and inadequate returns constraining cultured live rock and coral in a competitive marketplace against wild-caught products.

Implications of Capital Investment Needs

Capital investment costs are high under both projected models (relative to local purchasing power) and would likely require loans or subsidies for the purchase of workboats, engines, seed rock, and on-land holding facilities (U.S. model).

Because of such high start-up costs, aquaculture has traditionally focused in developed countries, where sufficient capital investment potential exists (Tlusty 2001; King 1999). In the case of marine ornamentals, application of aquaculture in developing countries is critical, given that otherwise in situ wild stocks will continue to be depleted. However, the realities of securing the necessary capital for up-front investment in culture operations were a concern expressed by most of the Pacific operations studied (both private and NGO-led). Given the need for live rock and coral culture development in a developing country context, there are several implications for the Indo-Pacific developing island application:

1. Because of the relatively large investment needs, subsidies (likely from foreign aid or "green" investment) for the costs of boats, engines, and substrate (seed rock and/or coral bases and grow-out units) would be required to initiate village-based culture production.
2. An extension service would be necessary to provide training culture and husbandry, technology maintenance and repair, and biological impact monitoring.
3. An in-country boat buy-back program to transform excess fishing capacity and retrofit the boats for live rock and live coral culture operations could provide a feasible avenue to reduce capital investment needs.
4. Small-scale village producers operating within nearshore waters held under customary marine tenure would avoid the need for up-front seafloor lease acquisition and licensing, as well as related ongoing operating costs. Such tenurial arrangements already exist for prospective village producers in the Indo-Pacific (e.g., within Fiji or the Solomon Islands).

5. Small-scale village producers who otherwise would remain engaged in wild stock capture could easily translate their existing knowledge and familiarity with marine ornamentals into culture operations allowing for improved entrepreneurial confidence and increased probability of attracting outside subsidy opportunities.

Implications of Enterprise Budget Profitability

As study results illustrate, the culture of live rock and live coral becomes profitable under a high price structure only under the U.S. model. However, this raises two practical questions: (1) How likely is it that buyers will pay moderate or high product prices? (2) How likely is it that live rock and coral culture operators with higher prices can successfully compete in the current market against low-cost wild product? Nearly all key informants canvassed in the study responded that it would be very unlikely under both questions. The high operating costs of both models also requires that the operator pay attention to how to reduce the variable costs of production.

In addition to the financial risks associated with the budget outlays, a number of nonfinancial risks may restrict or limit financial feasibility of the culture operations. These include natural disasters (e.g., cyclones, diseases), theft of product, and diving accidents when uninsured (Herndon 1998; Antozzi 1997; Herndon and Herndon 1996).

Implications of Projected Cash Flows

Profitability results for all six scenarios were poor. The study indicates that profitability is likely achievable only after four to five production cycles when assuming a higher price structure or 10% green premium is being captured. Even when an operation turned a year-end profit during the third through fifth cycles, however, it was still operating at a significant overall debt. Subsidizing capital investment costs would allow operations to get clear of debt one to three cycles earlier and become profitable within the 10-year time frame projected. Such subsidies could be tied to grants made for culture research, coral reef conservation and rehabilitation projects, ecotourism fees, and marine protected area operations.

Because there is no income during the first production cycle, both models require an additional income source during the start-up period of culture operations. In some cases, this may need to be from supplemental wild-caught live rock and coral revenues.

The use of capital investment subsidies and supplemental income is particularly applicable for the Pacific model, which otherwise remains unprofitable even when experiencing high price structures or a green premium. It is safe to assume that strict-culture operators under the IPI model would need to subsidize culture operations for at least five to seven production cycles (10 to 14 years). It is worth noting that actual operational experience demonstrates that culture enterprises can become profitable, assuming that they are operated over a long enough time frame and are led by an individual who has firm expertise and understanding of operating conditions (Herndon 1998).

Independent of its ability to improve the economic feasibility of culture operations, the 10% green premium has an important implication for the Pacific. Results under the IPI model indicate that village-based culture operators could potentially receive higher returns than if they continued wild harvesting live rock and live coral. This finding is important because the higher returns from cultured products could potentially encourage the permanent displacement of the wild harvest effort, as opposed to merely supplementing such effort with an additional income source.

The encouraging study results regarding green premium potential are evidenced through actual U.S. operator experience where the successful "green marketing" of cultured live rock and coral as ecologically friendly has reportedly created enormous demand that producers are unable to supply. Capturing a moderate (10%) green premium through an informed and contentious consumer base may be likely and, if undertaken, would encourage the profitability and therefore expansion of strict live rock and live coral culture operations. New questions arise under this green premium context: (1) Can green labeling be used as a vehicle to raise consumer awareness of the availability of cultured product alternatives to wild stock? (2) Could such labeling lead also to increased consumer demand (preference over wild stock) for cultured products? (3) Could such labeling precede raised prices through green premium capture? These new questions underscore the need to test and promote "green labeling" opportunities for cultured marine ornamental producers through certification (e.g., Marine Aquarium Council[3]).

The Benefits of Live Rock and Coral Culture

Although live rock and coral culture may prove financially challenging to maintain, several benefits arise from its operation, beyond the promotion of a more sustainable alternative for the international marine ornamentals trade than wild extraction, that collectively may outweigh the financial costs. The first set of these benefits is related to improved livelihoods and livelihood opportunities, including: (1) salaries for commercial employment staff, (2) an important source of household income for the families of commercial staff or for rural village community residents, including income generation by women and children (Counterpart International 2000; Wheeler 1996), (3) an occupational alternative in fisheries with strong reported job satisfaction, (4) group income toward community projects from coral farm ecotourism fees (Heeger and Sotto 2000), and (5) improved fish habitat (coral reefs) resulting in replenished and/or larger food fish stocks (Herndon 1998).

A second set of benefits examines the biological and nonlivelihood social outputs from culture operations. The most commonly cited biological benefit would be coral reef rehabilitation and restoration (Heeger et al. 2000; Herndon 1998), resulting in improved ecological functioning and biodiversity maintenance. Some live rock and coral culture advocates claim that after culture operations, the in situ coral reef habitat was healthier and in better condition than before culture operations (Spears 1995; Sweeney 1992). In this regard, the costs of operational subsidy may be less than the cost of coral reef loss or degradation. Acknowledgment of such conservation incentives may provide culture operators with sources of investment capital otherwise not considered. Important social benefits not associated with livelihood are improved marine resource management, increased environmental awareness and education, and strengthened local ownership (tenure) and/or responsibility over the area being cultured.

The relative ease of operation and low-maintenance technologies used under the live rock and live coral culture operations lend themselves to developing country application. This finding is not new. Invertebrate ornamentals culture has been advocated previously (Spotts 1997; Wheeler 1996) for a small-scale tropical developing island nation level because of the low technology required coupled with quick product grow-out and high marketability. The culture of live rock and corals is also recognized as being easier to start and maintain than that required for ornamental fishes (Wheeler 1996).

Recommendations

Whereas regulatory constraints regarding the importation of wild-caught live rock and live corals into the United States would clearly work in favor of promoting cultured live rock and coral production, this scenario appears unlikely given the current U.S. political and regulatory aversion to placing government controls over the free market. In addition, import restrictions or bans would have far-reaching, negative socioeconomic impacts on developing country exporters. Assuming no import bans or limits are placed on wild-caught organisms and no price fixing of wild-caught and cultured products occurs, the following policy, economic, and management recommendations are provided to encourage a trade in cultured live rock and corals.

Policy Recommendations

To improve incentives for entrepreneurs to undertake strict live rock and live coral culture operations, it is recommended that the government achieve the following policies:

1. Where government jurisdiction exists over waters in which cultured operations take place, pass the appropriate federal, state, provincial/county, and/or local permitting policies that are integrated, nonconflicting, easy to understand and follow, reflect seed rock and coral grow-out placement needs, and encourage the entry of culture operators.

2. Decrease culture operation risk liability by providing legislation allowing culture operations to access commercial diving insurance coverage and workman's compensation and provide national insurance to businesses against catastrophic loss as a result of disasters (e.g., cyclone damage, disease, poaching).

3. Where applicable, strengthen laws by ensuring adequate monitoring and enforcement against illegal wild stock harvest activities in order to protect cultured products from being unfairly out-priced in the marketplace.

4. Maintain a stable, transparent, and noncorrupt decision-making setting within which the marine ornamentals trade is consulted and appropriately legislated.

5. Where applicable, provide a regulatory environment that advocates for the trade in cultured products over wild stock.

Economic Recommendations

Given the financial challenges facing strict live rock and live coral operations, the following economic recommendations and incentives are provided to promote the development of cultured enterprises and improve their operational viability:

1. Provide a broad, sustainable U.S. government subsidy program through the National Marine Fisheries Service and offer small business loans to marine ornamental culture start-up ventures.

2. Provide international donor (government and nongovernment) aid to developing Indo-Pacific island nations in subsidizing the start-up and operational costs of small-scale village production of cultured live rock and live coral, particularly within the leading countries of wild-caught live rock and coral exports.

3. Identify and nurture sources of "eco-friendly" venture capital and "green" private investment toward support of live rock and live coral culture operations in the United States and Indo-Pacific.

4. Provide financial incentives (e.g., sales or export tax exemption) to attract the Pacific islands investment necessary to create and maintain the in-country "base" wholesale in-

frastructure required to purchase, hold, pack, and ship (export) cultured product, as well as for formal establishment of partnerships between base wholesalers and small-scale village producers, thereby providing producers with a secure and reliable purchaser of cultured product for export.

5. Regional and national governments provide an expedited export reporting process and financially underwrite an international flight service to allow reliable export of cultured product.

6. International conservation and development NGOs should provide financial support and in-kind cost contributions to improve small-scale village operation profitability within a compatible "project" context;

7. Provide funding to support research into the science and impacts of live rock and live coral culture, including: (a) the testing of ideal site conditions and locations for live rock seeding and grow-out of propagated live rock and corals, (b) ideal vertical (water column/depth) and horizontal (reef versus seafloor) placement of propagated pieces, and (c) the ecological impacts of culture operations.

8. Where or when conditions permit, build market incentives for private sector transition from wild capture to culture of live rock and live coral products, allowing a gradual and incremental phase-out period from strict wild capture to strict-culture based on economic constraints and market conditions.

9. Improve the marketability and profitability of cultured products via green labeling and green premium capture through third-party certification, thereby increasing the viability of strict-culture operations.

Management Recommendations

Finally, we recommend that the following management actions related to aquaculture of live rock and coral be undertaken in order to improve the sustainability of the marine ornamentals trade:

1. Promote the sustainable management of cultured marine ornamentals through: (a) the requirement of and compliance with best aquaculture practices and known principles of sustainable aquaculture (New 1996; Hopkins 1996; Pillay 1996) that meet management needs and result in improved product yield and quality, (b) management support for cross-business training to transfer culture skills and technologies and share techniques of best culture practice, and (c) advancing north-south technology transfer opportunities within the industry and promoting potentials for appropriate technology transfer within a developing Indo-Pacific island country context.

2. Identify community assemblages living on live rock and coral reef species that do not do well as cultured products and develop sustainable wild harvest and management plans (Herndon and Herndon 2000) for their continued trade as marine ornamentals.

3. Undertake culture operations and interface production staff within the context of coastal ecosystems conservation and management planning and activities occurring at the local (county or village community) level (see Bowden-Kerby 1999c).

4. Encourage in situ (seafloor) area-intensive live rock and live coral culture operations to be undertaken in tandem with the declaration and maintenance of a marine protected area to enhance cultured product yield, quality, and associated benefits such as replenished fish stocks and biodiversity protection.

Conclusion

Strict-culture marine ornamental enterprises must be viewed as long-term investments, not as short-term ecological fixes. Whereas enterprise viability is difficult to ensure under either a U.S. or Indo-Pacific island model, under certain conditions cultured live reef organisms can be both ecologically and economically sustainable. A mixed live rock and live coral approach is the most promising model and likely profitable if operating under a high price structure or moderate green premium capture. To get strict-culture operations under way will require adequate start-up capital, minimized variable costs, a moderate to high price structure or green premium capture, a sufficient transition time out of wild-stock capture or other supplementary sources of income, and ideal market and operating conditions.

Although these requirements may be difficult to secure in either the short or long term, in the end the short-term cost associated with subsidizing the prospective operation may be far outweighed by the biological, social, and economic costs of coral reef ecosystems degradation, decreased fishery productivity (food and income), and biodiversity loss resulting from the wild extraction of live rock and coral for the marine ornamentals trade.

Acknowledgments

This study was funded by the Western Pacific Program of the David and Lucile Packard Foundation and the Asia-Pacific Economic Cooperation (APEC) Fisheries Working Group and its Member Economies.

This study would not have been possible without the vision, willingness, and transparency of the study partners. The authors would like to thank the following study partners: L. R. Herndon (Ocean Dreams, Inc.) and T. R. Herndon (Florida Sea Farms, Inc.); W. Vince (Solomon Islands Marine Exports) and D. Palmer and S. Gower (Aquarium Arts Ltd.); S. Wale (Solomon Islands Development Trust) and the Mala'afe community of Langa Langa Lagoon, Malatia Island, Solomon Islands; F. Durai and Leitongo community residents, Nggela Islands, Solomon Islands; A. Tawake and W. Aalbersberg (Institute of Applied Sciences, University of the South Pacific, Fiji); A. Bowden-Kerby and S. Linggi-Troost (Foundation for the Peoples of the South Pacific, Fiji); W. Smith and T. Mcleod (Pacific Aqua Farms, Fiji); and J. Gatus (International Marinelife Alliance) and the Olango Island Coral Farm community. Finally, the authors wish to express their appreciation for the personal support of the following individuals in facilitating this study: the Honorable N. Kile, Solomon Islands Minister of Fisheries; M. Sarmiento, Director of the Philippines Bureau of Fisheries and Aquatic Resources; R. B. Mieremet, Chairman of the U.S. Department of Commerce Aquaculture Task Force; the Honorable S. Kusumaatmadja, (former) Indonesian Minister of Sea Exploration and Fisheries; T. LaVina and C. V. Barber of the World Resources Institute; S. Tinkham and G. Hurry of the APEC Fisheries Working Group; J. Cato of Florida Sea Grant; and B. Cordes of the David and Lucile Packard Foundation.

Notes

1. The Convention on International Trade in Endangered Species (CITES) was created in 1973 to ensure that the trade in wild animals and plants does not threaten their survival. At the heart of CITES are three appendices. Appendix I lists species threatened by extinction. These species can be traded only in special circumstances. Appendix II species are all species that may become threatened with extinction unless trade is subject to strict regulation in order to avoid utilization incompatible with their survival. Trade in these species is allowed as long as it is not detrimental to the survival of the species. Appendix III lists species that are protected in at least one country, which has asked other CITES parties for assistance in controlling the trade. To learn more online, visit www.CITES.org.

2. The Marine Ornamentals '99 Conference was held November 16–19, 1999 in Kailua-Kona, Hawaii. It was the first international conference on marine ornamentals and was attended by nearly 350 delegates from 22 countries representing marine aquarium industry, research, conservation, hobbyist, and public decision-making sectors. The purpose of the conference was to provide an exchange of information and ideas regarding the collection, culture, and conservation of marine ornamentals. The conference resulted in identifying 20 priority recommendations to guide the sustainable development of the marine ornamentals industry into the twenty-first century.

3. The Marine Aquarium Council (MAC) is an international, not-for-profit organization that aims to conserve coral reefs and other marine ecosystems by bringing together the marine aquarium industry, hobbyists, public aquariums, conservation organizations, and government agencies to set industry standards and "best practices," which will improve the sustainability of the trade. Through third-party certification, those industry operators who adopt and follow MAC standards and practices may be awarded the MAC label on their products, identifying them to consumers as operations that are environmentally responsible and actively seeking sustainability.

References

Antozzi, William O. 1997. The developing live rock aquaculture industry. Report SERO-ECON-98-10. National Marine Fisheries Service, Southeast Regional Office, St. Petersburg, Florida.

Baquero, Jaime. 1999. Marine ornamentals trade: quality and sustainability for the Pacific region. Trade and Investment Division, South Pacific Forum Secretariat and the Marine Aquarium Council, Suva, Fiji.

Bentley, Nokome. 1998. An overview of the exploitation, trade, and management of corals in Indonesia. *TRAFFIC Bulletin* 17(2): 67-78.

Bowden-Kerby, Austin. 1999a. Working towards sustainable and environmentally sound "green" methods of coral aquaculture. Paper presentation at Marine Ornamentals '99, 16-19 November 1999, Kailua-Kona, Hawaii.

Bowden-Kerby, Austin. 1999b. Coral Gardens Initiative update, October 1999: community-based marine resource management, low-tech coral reef restoration, and sustainable coral aquaculture. Foundation for the Peoples of the South Pacific, Suva, Fiji.

Bowden-Kerby, Austin. 1999c. The Coral Gardens Initiative: coral aquaculture as an incentive for community-based reef conservation and restoration. Paper presentation at Marine Ornamentals '99, 16–19 November 1999, Kailua-Kona, Hawaii.

Brown, Christopher L. 1999. Historical perspective on marine ornamental fish aquaculture in Hawaii. Paper presentation at Marine Ornamentals '99, 16–19 November 1999, Kailua-Kona, Hawaii.

Bruckner, Andrew W. 2000. New Threat to Coral Reefs: Trade in Coral Organisms. *Issues in Science and Technology* Fall 2000.

Bruckner, Andrew W. 2001. Tracking the trade in ornamental coral reef organisms: The importance of CITES and its limitations. *Aquarium Sciences and Conservation* 3(1–3): 79–94.

Corbin, John S. 2001. Marine Ornamentals '99: conference highlights and priority recommendations. *Aquarium Sciences and Conservation* 3(1–3): 3–11.

Corbin, John S., and Leonard G. L. Young. 1995. Growing the aquarium products industry for Hawaii: Report to the 18th Legislature 1996 Regular Session. Aquaculture Development Program, State of Hawaii Department of Land and Natural Resources, Honolulu, Hawaii.

Counterpart International. 2000. Coral Gardens Initiative: restoring reefs and rural livelihoods in island communities. Counterpart International, Washington, D.C.

Delbeek, J. Charles. 2001. Coral farming: past, present and future trends. *Aquarium Sciences and Conservation* 3(1–3): 171–181.

Falls, William W., Pamela Stinnette, and J. Nicholas Ehringer. 1999. Hillsborough Community College Interdisciplinary Live Rock Project: year one annual report, prepared for the National Science Foundation. Hillsborough Community College, Tampa, Florida.

Falls, William W., Pamela Stinnette, and J. Nicholas Ehringer. 2000. Hillsborough Community College Interdisciplinary Live Rock Project: year two annual report, prepared for the National Science Foundation. Hillsborough Community College, Tampa, Florida.

Falls, William W. 2001. Florida aquacultured live rock as an alternative to imported wild harvested live rock. Paper presentation at the World Aquaculture 2001 Meeting, 21–25 January 2001, Orlando, Florida.

Fijian Fisheries Department. 1998. Annual fisheries report of Fiji. Ministry of the Environment, Suva, Fiji.

Frakes, Thomas, and M. Watts. 1995. Live rock aquaculture. *SeaScope* No 12 Spring, Aquarium Systems Inc., Mentor, Ohio.

Frakes, Thomas. 2001. "Live Rock Alchemy." In *Proceedings of the Marine Ornamentals '99, Waikoloa, Hawaii.* University of Hawaii Sea Grant College Program, Honolulu, Hawaii, pp 33–36.

Green, Edmund, and Frances Shirley. 1999. The global trade in coral. WCMC Biodiversity Series No. 9. World Conservation Monitoring Centre. World Conservation Press, Cambridge, U.K.

Heeger, Thomas, and Filipina Sotto (eds). 2000. *Coral Farming: A Tool for Reef Rehabilitation and Community Ecotourism.* German Ministry of the Environment. Manila, Philippines.

Heeger, Thomas, Filipina Sotto, Joey Gatus, Cristeta Laron, and Cccarsten Huttche. 2000. Coral farming: A tool for reef rehabilitation and community ecotourism. In *Coral Farming: A Tool for Reef Rehabilitation and Community Ecotourism.* German Ministry of the Environment. Manila, Philippines, pp 1–32.

Herndon, Teresa. 1998. Open-ocean culture of live rock: an environment-friendly approach to aquaculture with many benefits. *Today's Aquarist* 6(9): 4–7.

Herndon, L. Roy, and Teresa R. Herndon. 1996. Live rock. *Aquarium.Net*, December 1996: www.aquarium.net/1296/1296_7.shtml.

Herndon, L. Roy, and Teresa R. Herndon. 1999. Pioneering a commercial aquacultured live rock operation in Florida: an entrepreneur's perspective. Paper presentation at Marine Ornamentals '99, 16–19 November 1999, Kailua-Kona, Hawaii.

Herndon, Teresa R., and L. Roy Herndon. 2000. Business plan for Florida Sea Farms, Inc., d.b.a. Sea Critters. Dover, Florida.

Heslinga, Gerald. 1999. Culture of invertebrates, giant clams, and live rock for the marine aquarium industry. Paper presentation at Marine Ornamentals '99, 16–19 November 1999, Kailua-Kona, Hawaii.

Hopkins, J. Stephen. 1996. Aquaculture sustainability: avoiding the pitfalls of the green revolution. *World Aquaculture* 27(2): 7–9.

ICLARM (International Center for Living Aquatic Resource Management). 2000. *ICLARM Strategic Plan 2000–2020.* The International Center for Living Aquatic Resource Management, Penang, Malaysia.

ICRI (International Coral Reef Initiative). 2001. Resolution on Actions to Promote Sustainable and Equitable Practices in the International Trade in Coral Reef Species. Resolution passed at the ICRI Coordination and Planning Committee Meeting, 5–6 April 2001, Cebu City, Philippines.

King, Michael R. 1999. The role of in situ aquaculture in reaching sustainability in the marine orna-

mental industry. Paper presentation at Marine Ornamentals '99, 16–19 November 1999, Kailua-
Kona, Hawaii.

Lieberman, Susan, and John Field. 2001. "Global
Solutions to Global Trade Impacts?" In (Barbara
Best and Alan Bornbusch, eds.) *Global Trade and
Consumer Choices: Coral Reefs in Crisis.* Proceedings of Papers Presented at a Symposium held
at the 2001 Annual Meeting of the American Association for the Advancement of Science, San
Francisco, California, 19 February 2001, pp
19–24. American Association for the Advancement of Science, Washington, D.C.

Lovell, Edward R. 2001. *Status Report: Collection of
coral and other benthic reef organisms for the marine aquarium and curio trade in Fiji.* World Wide
Fund for Nature South Pacific Programme, Suva,
Fiji.

Mcleod, Tim. 2001. Tivua Aquafarm general report
July 2001. Pacific Aqua Farms. Navutu, Fiji.

Miller, Tom. 1998. "Reef Propagation Project: The
Complete Cookbook for Making Live Rock from
Cement and Other Types of Rock." *Marine Fish
Monthly,* May 1998.

Moore, Franklin, and Barbara Best. 2001. "Coral
Reef Crisis: Causes and Consequences." In (Bar-
bara Best and Alan Bornbusch, eds.) *Global Trade
and Consumer Choices: Coral Reefs in Crisis.*
Proceedings of Papers Presented at a Symposium
held at the 2001 Annual Meeting of the American
Association for the Advancement of Science, San
Francisco, California, 19 February 2001, pp. 5–10.
American Association for the Advancement of
Science, Washington, D.C.

NMFS (National Marine Fisheries Service). 1995. 50
Code of Federal Regulations 622.2, April 23,
1995. National Oceanographic and Atmospheric
Administration, Silver Spring, Maryland.

National Research Council. 1992. *Marine Aquaculture: Opportunities for Growth.* Committee on Assessment of Technology and Opportunities for
Marine Aquaculture in the United States, Commission on Engineering and Technical Systems,
National Research Council. Washington, D.C.:
National Academy Press.

New, Michael B. 1996. "Sustainable Global Aquaculture." *World Aquaculture* 27(2): 4–6.

Pillay, T.V.R. 1996. "The Challenges of Sustainable
Aquaculture." *World Aquaculture* 27(2): 7–9.

Pomeroy, Robert S., John E. Parks, and Cristina M.
Balboa. 2002. *Farming the Reef.* World Resources
Institute, Washington, D.C.

Pyle, Richard L. 1993. "Marine Aquarium Fish." In
(Andrew Wright and Lance Hill, eds.) *Nearshore
Marine Resources of the South Pacific,* pp.
135–176. Institute of Pacific Studies, University of
the South Pacific, Forum Fisheries Agency and International Centre for Ocean Development.

Seaman, William, and Charles M. Adams. 1998.
Florida marine aquaculture research and extension
issues, including the Florida Sea Grant long range
plan. Technical Paper 93. Florida Sea Grant Extension Service, University of Florida,
Gainesville, Florida.

Spears, Dennis. 1995. Live rock farmers look to future harvest. *Underwater USA* 11(32): 12–13.

Spotts, Dan G. 1997. "Progress and Pitfalls for Marine Ornamental Culture on Tropical Islands." In
*Martinique '97: Island Aquaculture and Tropical
Aquaculture.* World Aquaculture Society. European Aquaculture Society, Oostende, Belgium.

Spotts, Dan G. 2001. Progress in culturing the stony
corals *Alveopora* and *Goniopora.* Paper pesenta-
tion at the World Aquaculture 2001 Meeting,
21–25 January 2001, Orlando, Florida.

Sweeney, Louise. 1992. "Live rock" use in aquariums, elsewhere anchored in controversy. *Underwater USA* 10(30): 27.

Tlusty, Michael. 2001. The benefits and risks of
aquaculture production for the aquarium trade.
New England Aquarium, Boston, Massachusetts.

Wheeler, Jennifer A. 1996. The marine aquarium
trade: a tool for coral reef conservation. Sustainable Development and Conservation Biology Program, University of Maryland, College Park,
Maryland.

14

Aquacultured Live Rock as an Alternative to Imported Wild-Harvested Live Rock: An Update

William W. Falls, J. Nicholas Ehringer, Roy Herndon, Teresa Herndon, Michael Nichols, Sandy Nettles, Cynthia Armstrong, and Darlene Haverkamp

Abstract

Live rock is any type of rock that contains organisms attached to it that is used in marine aquariums. Traditionally live rock has been removed directly from the sea and sold as an aquarium product. In 1997, the harvest of wild live rock was banned from both state and federal waters off Florida. Since then, aquacultured live rock has been grown in a variety of locations, using an assortment of mincd rock, in an effort to supplant the wild-harvested rock. Several research projects have been conducted to determine the efficacy of aquacultured rock and are described here. Other studies on enhanced live rock and on live sand are described. Recommendations are made for future research on live rock.

Introduction: Live Rock Defined

Live rock has been described as "a living marine organism or an assemblage thereof attached to a hard substrate (including dead coral or rock, usually calcareous in nature)" (Gulf of Mexico Fishery Management Council 1994). It is also the habitat for invertebrates and larger plants. Anemones, tunicates, bryozoa, octocorals, sponges, echinoids, mollusks, tubeworms, and calcareous algae are common inhabitants on live rock obtained from Gulf Coast waters (Gulf of Mexico Fishery Management Council 1994).

As such, live rock is a collection of calcium carbonate material that contains encrusting organisms such as coralline algae, other algae, and epibenthic invertebrates (Tullock 1997; Sprung and Delbeek 1994), which is used in saltwater aquariums. The marine aquarium hobby is supported by a worldwide industry that has seen tremendous growth since the mid-1980s. In the state of Florida in particular, rapid growth and development of a commercially important new fishery has evolved via the harvesting and supply of live rock for the marine aquarium trade.

With the advent of improved filtration systems and recognition of the vital role live rock plays in a reef aquarium system, demand has grown for a steady supply of live rock. With live rock in an aquarium, a balance of nutrient levels is more easily obtained; however, the survival of live rock depends on a number of other aspects of tank set-up. Factors affecting the survival of marine life in aquariums include high intensity lighting, effective protein skimming, good water movement, and a rocky reef structure taken from the marine environment that harbors bacteria and macroscopic life, that is, live rock (Moe 1992; Tullock 1992, 1998; Sprung and Delbeek 1994). Aquarium technology has improved to the point that saltwater aquariums can be maintained successfully with an increasing variety of organisms (Sprung and Delbeek 1994; Moe 1989; Tullock 1997). Live rock has vastly improved the keeping of saltwater aquariums, in part by contributing to the control of excess nitrogen levels (Tullock 1997).

The use of live rock in aquariums has evolved from a system of trial and error to determine what live rock is and what grows on it. A few research projects have been conducted to establish which types of rock, whether natural or man-made, make the best habitat for the growth of organisms. Studies have been conducted in the marine environment, in saltwater tanks, in freshwater, and in large upland tanks hundreds of miles from the ocean.

Purpose of Live Rock

Live rock is used in marine aquariums to create a "mini-reef." Reef aquariums range in size from small tanks of less than 10 gallons, called "nano-reefs," that are currently very popular with hobbyists to massive aquariums of several hundred gallons. Both individual hobbyists and large public aquariums maintain reefs in their systems. In aquarium systems live rock is aesthetically pleasing, acts as a natural biological filter, and provides habitat for other aquatic inhabitants. Live rock is essential for the reef aquarium because it provides important habitat for motile fish and invertebrates; it provides vital substrates for the settlement and recruitment of benthic organisms; and it contributes to the structure of the reefs and to total reef biomass. Live rock acts as a natural biofilter that serves as habitat for the beneficial nitrifying bacteria (ammonia and nitrite-oxidizing bacteria), protozoans such as foraminiferans, and algae (both encrusting coralline and macro-branching).

The harvesting of live rock in Florida has evolved from removal of natural rock from the wild to the culturing of rock, both in a natural setting and in upland systems. In Florida the culture of live rock helps to protect natural reefs from destruction (it is ecofriendly). It is a sustainable aquaculture industry that takes in millions of dollars and provides a viable aquarium rock that is nearly phosphate-free (one-third the phosphate level of that of wild Fiji live rock) (Falls et al. 2001).

History of Live Rock

From the 1970s and into the 1990s, collection of live rock evolved into a commercially important fishery in waters surrounding Florida and South Pacific islands. As the demand for live rock increased, harvesting increased in Florida waters. Because so much rock was being removed, many special interest groups and environmental organizations opposed the ever-increasing harvest of natural rock from Florida waters. By 1993, the ex-vessel value of live rock collected grew to US$1,063,000 (GMFMC 1994). Collection of live rock from Gulf of Mexico coastal waters resulted in the removal of the equivalent of two patch reefs each year (Ehringer and Webb 1993). Subsequently, Amendment 2 to the Fishery Management Plan for Coral and Coral Reefs of the Gulf of Mexico and South Atlantic was developed and implemented by the National Marine Fisheries Services (NMFS). Amendment 2 allowed for closure of all wild harvest as of January 1, 1997, and allowed for the aquaculture of live rock in the Gulf of Mexico and in the south Atlantic offshore from the Florida Keys (GMFMC 1994).

A record of live rock harvesting in Florida provides a history of wild live rock collection. There were between 33 and 73 wild live rock collectors from 1990 until the federal ban in 1997, with the high in 1994. Even with the ban, illegal collection of Florida wild live rock has continued primarily because enforcement of the law is difficult. The Florida coastline consists of about 1,300 linear miles, and there are not enough Marine Patrol officers to cover it all. There is also a need for education of law enforcement agencies. The quality of Florida aquacultured live rock has improved so much that aquacultured rock is very difficult to distinguish from natural rock without removing the encrusting organisms from the rock. It is also difficult to enforce the law that restricts wild collection. Law officers must catch violators in possession of wild rock coming from Florida waters. Once it is in the pet stores, it is nearly impossible to confirm Florida wild rock is not aquacultured, wild from outside of Florida (Caribbean or South Pacific), or wild rock returned to stores from hobbyists trading old rock or getting out of the hobby.

The popularity of wild-collected live rock from Florida waters raised concerns over damage to natural reef systems that were already stressed. The rationale for the ban was a decline in reefs from overcollecting, often done with the aid of destructive chemicals and tools such as explosives and crowbars that chip the rock into small pieces. This led the Florida Department of Natural Resources (now the Florida Fish and Wildlife Conservation Commission) in 1989 to ban the harvest of wild live rock from Florida state waters, which extends 3 nautical miles offshore into the Atlantic Ocean and 9 nautical miles into the Gulf of Mexico. Before the ban, Florida live rock harvesting had reached an estimated 300 tons annually.

In Florida, the primary sources of "wild" live rock were the mounds of dead coral along the Florida Reef Tract and the limestone ledges off the Gulf Coast. The Florida Reef Tract, estimated to be 4,000 to 7,000 years old, has a base of calcium carbonate. It is speculated that the production of "new rock" on this reef is at best equal and may be falling behind that lost to natural erosion of the reef by the action of water. Around 1986, divers began to extract this material and all of the attached dependent living organisms for sale to saltwater aquarium enthusiasts as a basis to form miniature reefs.

In Florida federal waters (beyond 3 nautical miles in the Atlantic Ocean and beyond 9 nautical miles in the Gulf of Mexico) wild collection was still allowed until the federal government banned the harvesting of live rock after January 1, 1997, which terminated an estimated US$10 million per year industry in Florida. From 1990 to 2000, a total of 5,892,930 pounds of live rock was harvested from Florida waters. Of this amount, 5,388,691 pounds (91.4%) were collected from the wild and 504,239 pounds (8.6%) were aquacultured live rock.

Wild live rock collection is now banned in the Gulf of Mexico, the South Atlantic Exclusive Economic Zone (EEZ), the Caribbean (EEZ), Puerto Rico, U.S. Virgin Islands, North Carolina, Western Pacific, California, Hawaii, and Guam. Most wild live rock (95%) is imported to the United States from Fiji and Indonesia, where live rock harvesting is not banned. Imports increased dramatically after the 1997 ban went into effect. Fiji live rock is measured by weight. Data on the volume of imported live rock are at times difficult to track from U.S. customs records. Live rock is listed under the heading "Scleractinia" and is grouped with live corals that are imported into the United States and tracked by other agencies such as the U.S. Fish and Wildlife Service. Work is in progress to review and revise the method of listing and tracking imported live rock in a more precise manner. From 1996 to 1998, a significant increase in imported live rock occurred with an increase from 225,341 kg to 605,727 kg. Indonesia (by items) had a similar increase in exports to the United States from 328,624 kg in 1997 to 1,564,000 kg in 1998. This fivefold increase from Indonesia occurred a year after the Fiji increase (Robinson et al. 2000).

In June 1989, the Florida Marine Research Institute suggested that commercial aquaculture of live rock be considered as an alternative to the removal of wild live rock from Florida waters. It suggested that "affected parties are encouraged to consider culturing live rock in privately owned upland tanks or applying for aquaculture leases. The privately owned rock material would be on barren sovereign land (state or federal ocean bottom void of living organisms such as bare sandy bottom) for colonization by the living components of live rock" (Huff 1990).

Aquaculture of live rock was subsequently attempted in Florida by a variety of entrepreneurs. The beginning of aquacultured live rock in Florida state waters had a slow start, however, primarily because of the bureaucratic maze set up to establish a permit system for the culture of live rock. One early pioneer had to sue the state of Florida to get the first state permit 5 years after the application was submitted. Another early pioneer tried to get a permit from the Army Corps of Engineers but failed to obtain one. Most found it easier to seek sites and permits from the National Marine Fisheries Service in federal waters outside the jurisdiction of the state of Florida. According to the basic rules of live rock aquaculture, a site should be in an area where there is no hard bottom, no natural reef, no seagrass present, and shallow enough to support the growth of organisms on rock. Only about 50 sites in the Gulf of Mexico and the Atlantic Ocean meet such criteria, totaling 196.05 acres and they are mostly in state waters (FDACS 2002).

Initially the Florida Department of Environmental Protection was responsible for live rock aquaculture in Florida waters. Today, the Florida Department of Agriculture and Consumer Services (FDACS) regulates live rock in state waters. Live rock aquaculture participants must obtain a general permit from the Division of Aquaculture of FDACS. The Florida Marine Research Institute staff then inspects the site and sends its findings to the Division of Aquaculture for confirmation. A Saltwater Products License is also needed in order to sell the live rock when ready for harvest.

U.S. Live Rock Aquaculturists

The ornamental aquarium hobby is a leading cash crop of the U.S. aquaculture economy, with retail sales approximating $1 billion annually. Marine organisms represent 20% of U.S. aquarium retail sales (Chapman and Fitz-Coy 1997). Worldwide trade in aquarium livestock and products may exceed $7 billion annually (Andrews 1990).

In Hawaii it is illegal to conduct aquaculture operations on state ocean bottom. There is geothermal ocean water off the coast of Kona. At this location, live rock is cultured in tanks using geothermal water in a flow-through system with marine salts added. The warm enriched water allows year-round culture.

In Idaho, geothermal water is also available. Leroy and Sally Headlee rear live rock and corals in recirculating tanks with artificial seawater. They have patented an artificial rock from cement and aragonite called Aragocrete (Fig. 14.1). The Headlees introduced Florida aquacultured live rock into their grow-out tanks in 1994. They grow batches of aquacultured rock for about 2 months in a 30-gallon tank with one 18-inch fluorescent light bulb and two power heads. The Florida aquacultured rock has successfully grown macro algae and sponges on it.

In Texas state waters in the Gulf of Mexico, it is also illegal to culture live rock. Live rock is being cultured in tanks using wild rock from Texas and Fiji to seed it (Falls et al. 2001). The growth is good, and the potential for a new industry is promising.

Fiji has entered the aquacultured live rock industry. Walt Smith has developed an artificial

Fig. 14.1 Artificial substrates used for aquacultured live rock: (A) Fiji, concrete and crushed coral; (B) Idaho, Aragocrete; (C) Florida, concrete and Styrofoam beads; (D) Florida, clay. (See also color plates.)

rock with cement and crushed coral (Fig. 14.1) that is an alternative to wild-harvested rock (Walt Smith, Walt Smith International Ltd., Fiji, personal communication, 2001).

In Florida, 16 live rock aquaculturists have used several types of natural rock, including three different types of limestone rock. The three most commonly used limestones are Suwannee, Bahama, and Miami (Fig. 14.2). One requirement of a live rock aquaculturist is to provide the state or federal government were

Fig. 14.2 Three natural limestone rocks used for aquacultured live rock in Florida: (Top) Suwannee; (Middle) Bahama; and (Bottom) Miami.(See also color plates.)

a lithologic description of the rock to be used from a certified geologist plus a sample of each rock type used. Suwanee is white to light yellowish white, soft, micritic (mudstone), fossiliferous limestone composed of mollusks (gastropods and pelecypods), echinoids (sea urchins, starfish), and large foraminifera (nummulities). Bahamian is a light pale gray micrite with drussy aparry calcite filling vug and channel pore spaces; distinctly nonfossiliferous; microporous; and contains scattered grains of rounded, black, very fine phosphate, evidence of terrestrial vegetation (tree roots), and large pelecypod borings. Miami is white to yellowish granular limestone, with iron staining on the rock common and variable; it is fossiliferous, with large tubular branches of bryozoans crisscrossed within the rock, with branches up to 2 to 10 cm in length made of a matrix of oolithic carbonate sand with abundant juvenile pelecypod molds.

The Suwannee limestone is no longer used because it is too dense and not porous. Hand picked off the forest floor in the Bahamas, Bahama rock is no longer available. Miami limestone is mined near Homestead, Florida and is plentiful. Miami rock is the choice rock used by most live rock aquaculturists in Florida. The rock is primarily mined for road construction.

One of the unsubstantiated criticisms of Florida aquacultured live rock is that it contains too much phosphate. Phosphate in a reef tank will cause unwanted algae to grow, which can overgrow living corals or cause them to die and destroy the balance of the reef tank. Miami (12.5 ppm) and Bahama (5 ppm) limestone both contain less phosphate than wild-collected Fiji (37.5 ppm) live rock. Florida aquacultured live rock has one-third the phosphate level of Fiji rock (Table 14.1).

To begin the aquaculture of live rock, a large capital investment is needed. A boat in the range of 40–50 feet that can carry 5000–10,000 pounds of rock is a minimum requirement in addition to scuba equipment. Rock can average US$0.07–$0.10 per pound, plus shipping can cost as much as US$0.20 per pound. A site is needed to hold harvested rock until shipped. Typically, a small greenhouse or warehouse with recirculating tanks will suffice. It can require approximately US$200,000 to begin the culture business without any return on the investment for 3–5 years.

Table 14.1   Chemical analysis of substrates

(ppm)	Concrete/ styrofoam	Clay	Miami limestone	Suwannee limestone	Bahama limestone	Fiji	Gulf sand
Nitrogen	30	5	5	5	5	30	5
Phosphorus	50	75	12.5	37.5	5	37.5	37.5
pH	11.5	7.5	8.5	7.5	8	8	8
Potassium	200	50	50	70	45	200	200

The state government requires that seed rock be placed on the bottom site by hand. It cannot be dropped from the boat or shoveled off the barge by mechanical means. The rock must be lowered by hand in baskets to the bottom and then stacked by hand. This is quite labor intensive and time consuming. It takes several trips to carry and place 200,000 pounds of rock on a site.

A common misconception is that once the rock is on the bottom, it will grow without any intervention on the part of the aquaculturist. This is far from the truth. A live rock aquaculturist must frequent the site, monthly at least, inspect it, and often restack the rocks after storms, especially hurricanes. There are also concerns about poachers.

Live Rock Research

At the time of the 1997 ban on wild rock harvesting in Florida limited research had been conducted on live rock culture. Early live rock aquaculturists learned by trial and error. Florida Sea Grant funded Hillsborough Community College to begin the first live rock research in Florida in 1993, and the National Science Foundation built on this pilot project with a 3-year grant in 1998, which allowed Hillsborough Community College to conduct additional live rock research.

Florida Sea Grant Research

The first live rock aquaculture research was titled "Assessment of Live Rock Harvesting in Tampa Bay" (Ehringer and Webb 1993). The hypothesis was that as rocks are placed in the water, the growth of organisms should undergo successional stages much like growth on any other pioneer substrate. This is supported by an assessment of live bottom communities of Tampa Bay (Derrenbacher and Lewis 1985). Depending on environmental conditions such as salinity, temperature, and turbidity, a variety of life forms would be expected to grow on the rocks contingent upon where the rocks are placed within the bay and during which successional stage they are removed. Comparisons were made between live rock and artificial reefs.

A review of the literature indicated that more species would be found on rock near the mouth of an estuary than in the upper regions of the estuary because of variance in salinity, with consideration also being given to stability of other parameters (Mook 1980). This was also indicated to be true for Tampa Bay (Estevez 1989).

Five dominant sessile organisms exist on rock: barnacles, tunicates, hydroids, polychaetes, and mussels (Dean and Hurd 1980). The following invertebrates were known to exist in Tampa Bay: sea fans, sea whips, hydroids, anemones, tunicates, sponges, bryozoans, and hard corals (Lewis and Estevez 1988). Great seasonal variation in species abundance and diversity of organisms also existed in Tampa Bay, with the least colonization occurring in the months from September through February (Santos and Simon 1980). Therefore, the seasonal timing for the placement of rocks for growth may be of great importance.

Artificial reefs and live rock have similar requirements for placement and growth (Duedall and Simon 1991). Live rock actually becomes an artificial reef once in place. Six sites were chosen in Tampa Bay for the placement of live rock platforms (Fig. 14.3) for 6–9 weeks.

The red alga *Hypnea cervicornia* and the hydroid *Sertularella speciosa* occupied a significant amount of space on the upper surface of the rocks. Only one site close to the Skyway Bridge (Figure 14.3, Site D) was suitable for live rock culture. The Ivory bush coral *Oculina diffusa* was left on the rock in the water. At the time of this initial research it was illegal to harvest live rock with living coral on it. This is no longer illegal.

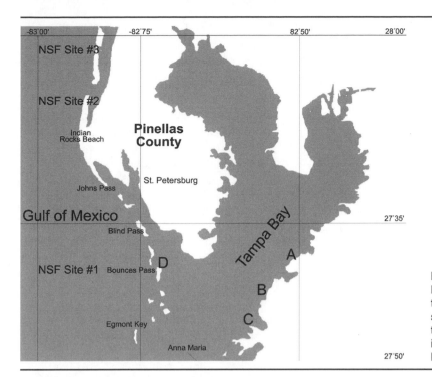

Fig. 14.3 The Tampa Bay sites (A–D) used in the 1993 Sea Grant study in comparison with the NSF study sites used in 1999 in the Gulf of Mexico.

National Science Foundation Research

In 1998, Hillsborough Community College was awarded a 3-year grant from the National Science Foundation to study growth on live rock at five different locations using a variety of rock substrates. Five locations were selected for the live rock research project (Fig. 14.4). Five different substrates (Table 14.1; Figs. 14.1 and 14.2) and two artificial substrates were chosen for placement at the five sites (Tables 14.2 and 14.3).

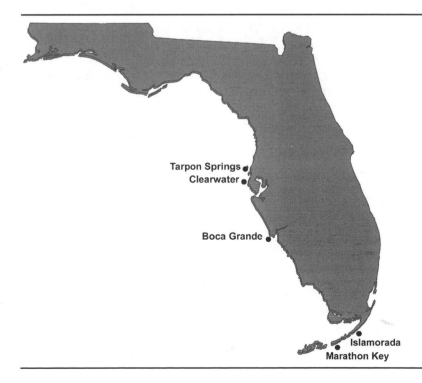

Fig. 14.4 Live rock study locations.

Table 14.2 Live rock sites and percent light transmitted at each site depth

Site location	Site depth (m)	Percent light transmitted
Boca Grande	2.03	11.9
Islamorada	6.77	52
Marathon Key	6.3	25.6
Clearwater	15.03	7.2
Tarpon Springs	8.73	10.5

Table 14.3 Live rock site depths and GPS coordinates

Site location	Approximate site depth (feet)	GPS coordinates
Boca Grande	12	N/A
Islamorada	21	N24° 51'.846 W80° 35'.914
Marathon Key	21	N24° 40'.250 W80° 58'.528
Clearwater	48	N28° 02'.762 W83° 01'.251
Tarpon Springs	31	N28° 20'.707 W82° 58'.257

This project was a multidisciplinary project that involved the art, environmental science, and aquaculture programs from three different campuses of Hillsborough Community College in Tampa, Florida. Two live rock aquaculture businesses were co-researchers with Hillsborough Community College. They were Sea Critters, of Dover, Florida, and Triton Marine, of Ozona, Florida. Both companies had federal live rock leases for submerged land in both the Gulf of Mexico and the Florida Keys.

The first live rock site was started on August 1, 1998 in Islamorada in the Florida Keys. On September 24, 1998 Hurricane George destroyed that site along with all other commercial sites in the Keys. An estimated US$700,000 of aquacultured live rock was lost to the hurricane. The rock was either buried or swept away. The live rock aquaculturists applied for federal disaster relief from the U.S. Department of Agriculture, but were denied because live rock was not recognized as part of the department's definition of ornamentals.

The Islamorada site was replanted on October 30, 1998. On November 4, 1998, Hurricane Mitch skirted the Florida Keys, destroying the Islamorada site for the second time in 2 months. The site was replanted again in March 1999, and the Marathon site was added at the same time.

Three Gulf sites were planted in August 1999. The southernmost site was abandoned after four attempts to relocate it failed. There were no hurricanes or storms in this area, nor were the other two Gulf sites disturbed. In November 2000, a third Gulf site was added in Boca Grande, Florida on a state clam farm site to determine if this inshore area was suitable for live rock culture as a supplemental crop to clams. All five sites were completed by January 2002.

Environmental Monitoring　In order to determine the environmental conditions of the live rock sites, the following parameters were measured quarterly: pH, dissolved oxygen, turbidity, phosphate, nitrate, ammonia, salinity, and light transmittance. A Hydrolab and Hach Kit were used to measure these parameters, either at the site or in a lab. Temperature was measured hourly with an Onset Temperature Sensor. The water sample collected on site was returned to Hillsborough Community College, and the Environmental Science Program performed the analysis with the Hach Kit. Data were stored on Microsoft Excel and analyzed with Microsoft Access.

The five substrates, Fiji rock, and Gulf sand were pulverized to a fine powder and analyzed for nitrogen, phosphorus, pH, and potassium (Table 14.1). A Hach Kit was used for the chemical analyses.

Research Results　The initial study of live rock centered on the recruitment of benthic organisms to the rock. There was a gradual progression of organisms attaching to the live rock, and over time, the diversity and density of these organisms increased. There was a distinct difference, however, in the rock piles created for this project compared with the live rock reefs constructed by commercial enterprises. The project rock sets were composed of approximately 7,500 pounds (1,500 pounds of each of the five substrate rock types). The 7,500-pound rock sites were 5 to 50 feet in diameter and 3 to 5 feet in height. The small size of the study sites resulted in much lower density and diversity of organisms, both benthic and pelagic, compared with the much larger (100,000 to 600,000 pounds) commercial reefs. The large reefs were proficient fish attractors, providing thousands of niches for juvenile fish. They also provided thousands of holes for crustaceans, urchins, shrimp, and worms to take refuge. An evaluation of the first 3 months of the larger reefs saw primarily juvenile grunts, snapper, and tropical ornamentals inhabiting the reef. By 6 months, a second set of juveniles had arrived and the initial group of fish had grown in size. A few small predator fish had moved in, primarily small grouper and lizardfish. Small soft corals, sea whips, had recruited to the edges of the reefs. After one year of growth, the reefs exhibited the entire trophic pyramid, with hundreds of juveniles, hundreds of subadults, and numerous adults of practically every fish common to the reef area. In the Keys reefs, dolphins were observed feeding on the abundance of fish from artificial reef to artificial reef. Large barracuda also stationed themselves on each reef after the first year. The soft corals covered the entire surface of the reefs, and scleractinian corals were first observed. Fire coral was present as small patches on most of the surface rock, and by 15 months the fire coral had grown from a flat surface up to 6 inches in height. These observations on the commercial reefs, documented by quarterly videotaping, indicate that commercial aquaculture of live rock reefs can have a large, positive impact on the local marine ecosystems.

Study Recommendations On the basis of this initial project, the following questions are recommended for further live rock research:

1. Are fish recruits to aquacultured reefs being taken from existing reefs?
2. Are live rock reefs providing niches for juvenile fish that would not survive because the natural system cannot support them, particularly in light of the loss of extensive coral growth from the existing reefs?
3. Are live rock reefs a better method of repairing reefs damaged by ship groundings than current methods? A live rock reef can be transported and placed on the damaged site in a matter of weeks, and it would already be a viable productive reef system. There would be no recovery time. The larger commercial aquaculture reefs have been

proven to withstand the hurricane damage that buried the smaller test sites, and they are constructed without having to cement or epoxy the rocks into place.
4. Are live rock reefs a better method of growing indigenous corals for transplanting to damaged reefs or for the aquarium trade?
5. Can live rock reefs be used exclusively for the production of ornamental tropicals and invertebrates to reduce the stress on natural systems and reduce damage from collection activities?
6. Can live rock reefs provide fish and invertebrate species for restocking areas damaged by pollution, taking the pressure of restocking off natural systems?

Enhanced Live Rock (Eco-Gorgonians)

The Florida gorgonians are a diverse group of octocorals that range from deeper waters of the Gulf of Mexico (30 to 40 feet deep), to specimens found in shallow water of the Florida Keys. In addition deep water specimens from the east coast of Florida have been harvested for the marine aquarium trade from waters of the state and the Exclusive Economic Zone (EEZ) for a number of years. With the closure of live rock harvest in 1997, restrictions were also applied to the harvest of octocorals to limit the amount of rock that could be taken surrounding the anchoring base. State and federal rules allow only a rocky base of one inch in any direction from the base of the gorgonian to be collected. This is to prevent removal of larger pieces of live rock that could be harvested with a gorgonian attached. It is sometimes difficult to chip a one-inch square piece of rock with the base of the gorgonian, so sometimes there is just a tiny base attached. The resulting problem is that gorgonians are sometimes harvested with little or no anchoring base, which may affect the survivability of the specimen, as well as the aesthetic appearance.

Gorgonians are an important member of coral reef environments for a variety of reasons. Gorgonian forests provide beautiful displays for viewing by divers, a source of specimens for marine life collectors, and habitat for reef fish and other organisms. Further importance of octocorals developed with the discovery of the presence of prostaglandin, an important medical compound in one species, *Plexaura homomalla* (Bayer 1973). Recently, octocorals were designated as a primary component of essential fish habitat. The designation includes essential fish habitat areas of particular concern and designations for other managed fishery species, such as grouper, snapper, and lobsters (Jeanne Wheaton, Florida Marine Research Institute, St. Petersburg, Florida, personal communication, 1999). How to balance the economic importance, as well as the ecological and environmental importance of gorgonians has become a significant question.

During the past several years, the marine aquarium hobby has seen a tremendous interest in "captive propagation" of both soft and stony corals for reef aquariums. There has been a reasonable interest in gorgonians harvested and provided by some method that is both economical and more environmentally sound than the current methods used. Harvesting small branches from individual gorgonians and attaching the clippings to small pieces of aquacultured live rock to produce "eco-gorgonians" has been attempted and now needs to be done under a controlled study to assess and determine if it is an appropriate alternative to harvesting the entire gorgonian. This method would seem to be more ecologically friendly than to continue mass harvest and is a reasonable method of sustaining the economic aspect of harvesting gorgonians, while providing for resource protection as well (Walt Japp, Florida Marine Research Institute, St. Petersburg, Florida, personal communication, 1999).

Live Sand

Along with the ongoing "evolution" of the marine aquarium hobby, one of the many changes that have greatly benefited the marine aquarium is the use of "live sand" as a substrate in the bottom of the aquarium. This concept became popular about 1993 and has since been a subject of debate. Live sand is beneficial in the nitrification processes that take place in an aquarium because of the increased surface area it provides for microbes and for the many microscopic organisms that live in the sand. The addition of live sand to marine aquariums has helped improve the ability to maintain proper water chemistry,

and thereby healthier marine life, including live rock and live corals.

A method using an elaborate structure on the bottom of aquariums that creates a platform one inch above the bottom of the aquarium and two divided layers of sand was first proposed by scientists at the Monaco Aquarium (Jaubert 1991). This device is called a "plenum" and allows for a layer of water below, a thick sand bed with a low oxygen saturation protected by a layer of screen material, and a smaller top layer open and accessible to any invertebrates in the main aquarium. Over the past several years there has been concern as to whether the "plenum" is necessary and whether there are any measurable differences between aquariums with or without a plenum-style sand bed (Sprung and Delbeek 1994).

The debate over live sand extended to discussions of where to collect the "best" sand, how the sand was handled and shipped, what kind of sand to use, and other issues. Collection of live sand from Florida waters became quite popular and additional debate over the merits of Gulf of Mexico sand versus Florida Keys sand followed. This discussion ended abruptly with the creation of the Florida Keys National Marine Sanctuary in 1998, when all collection of live sand from within the sanctuary was banned. It was perceived that collection of sand from the Florida Keys area was detrimental to the ecology of the bay. Because the desirable sand was found within those boundaries, all collection ended despite complaints from marine life collectors in the area.

Current trends with sand beds now allow or recommend the use of fine-grain sand, either live sand or dry sand (both calcium carbonate and silicate based are acceptable) to create a bed up to 6 inches deep, with or without a plenum structure. If "dry sand" is used it is inoculated with small amounts of live sand to add the desirable microscopic life and bacteria. Dry sand is being "aquacultured" in holding tanks to allow for the growth of these organisms, then sold to aquarists for their tanks. Lectures and seminars are frequently held to examine and discuss the many different life forms that live in the sand beds, and hobbyists are reporting improvements in the appearance and health of their aquariums using a deep sand bed. The trend is to use less live rock in the aquarium and more live sand.

Coral Culture

Over the past several years attempts have been made to culture live corals in captivity, in part to help decrease the demand for collecting them from coral reefs around the world. Work has been primarily at the hobbyist level and a growing amount of culturing or "propagating" by business owners who now call their facilities "coral farms."

Some of the aquacultured live rock producers in Florida have found a new market emerging for cultured corals grown attached to pieces of aquacultured live rock. This culture process takes place in land-based closed systems because of the concern that many of the desirable species are native to the Pacific Ocean. The process involves obtaining a "wild-caught" specimen that becomes the parent colony. Sometimes it is possible to use a previously propagated coral from another dealer or from a hobbyist as the parent colony, then housing that coral in a large aquarium and growing it for several months. Once the colony is large enough, it can be divided into several daughter colonies, which either can be sold or in turn be grown into more parent colonies. Propagating corals in this manner on a commercial scale can help significantly to decrease the pressure on corals reefs by decreasing the need for collecting colonies from the wild. More research needs to be done on methods for culturing corals and increasing the number of species that can be cultured.

A number of soft coral species from Florida or Caribbean waters are also desirable for propagating or culturing onto aquacultured rock. Using native species eliminates any risks of introducing nonnative species. It also allows for open or partially open systems in the appropriate regions of the state, which should enhance the growth of cultured colonies.

Conclusions

Florida live rock aquaculture activity has continued to increase since the wild harvest ban in 1997. In 2001, there were eight state and fifteen federal lease sites. In 2001, the Florida Aquaculture Association approved live rock as an official new commodity, which will give live rock aquaculturists a seat on the board of directors.

Finally, enterprising live rock aquaculturists have developed techniques to enhance aquacultured live rock by propagating and attaching various organisms to small pieces of rock. One such example is eco-gorgonians. Clippings of wild gorgonians are attached to aquacultured live rock. The wild parent gorgonian is left in the sea to continue to grow without harvesting the whole animal. The clipping will attach itself to the rock and continue to grow in a reef tank. Research continues on eco-gorgonians in Florida to develop methods to commercially grow this new environmentally friendly product that enhances the reef tank and can be used to seed damaged natural reefs. Live sand and coral propagation have emerged as new products. The future looks promising for marine ornamentals in Florida.

References

Andrews, C. 1990. The ornamental fish trade and fish conservation. *J. Fish Biology*, Vol.37(supplement A): 53–59.

Bayer, F. 1973. Colonial organization in octocorals. In: Boardman R.S., A.H. Cheetham, and W.A. Oliver (eds.), *Animal Colonies: Development and Function Through Time*. pp.69–93. Dowden, Hucthinson & Ross, Inc., Stroudsburg, Pennsylvania.

Chapman, Frank, and S. Fitz-Coy. 1997. United States of America trade in ornamental fish. *J. World Aquaculture Soc.* 28:(1)1–10.

Dean, T.A., and L.E. Hurd. 1980. Development in an estuarine fouling community: The influence of early colonists on later arrivals. *Oecologia* (Berl.) 46:295–301.

Derrenbacher, J.A., Jr., and Robin R. Lewis. 1985. Live bottom communities of Tampa Bay. In the Tampa Bay Area Scientific Symposium. pp. 385–392.

Duedall, I.W., and M.A. Simon. 1991. Artificial reefs: Emerging science and technology. *Oceanus*. 34: 94–101.

Ehringer, Jon Nicholas, and Fred J. Webb. 1993. Assessment of "Live Rock" Harvesting in Tampa Bay. Occasional Papers of the Institute of Florida Studies. No. 3. 19 pp.

Estevez, E.D. 1989. Water quality trends and issues, emphasizing Tampa Bay. In NOAA Estuary-of-the-Month Seminar Series No. 11: 65–88.

Falls, William W., Jon Nicholas Ehringer, Michael Robinson, Cynthia Armstrong, Darlene Haverkamp, Roy Herndon, Teresa Herndon, Michael Nichols, and Sandy Nettles. 2001. Florida Aquacultured Live Rock as an Alternative to Imported Wild Harvested Live Rock. Aquaculture 2001. Orlando (Oral).

Florida Department of Agriculture and Consumer Service. 2002. Division of Aquaculture. Tallahassee, Florida. http://www.FloridaAquaculture.com.

Gulf of Mexico Fishery Management Council. 1994. Amendment 2 to the Fisheries Management Plan for Coral and Coral Reefs in the Gulf of Mexico and South Atlantic (July). Tampa, Florida.

Huff, Alan.1990. Assessment of "live rock" harvesting in Florida. Southeastern Aquaculture Conference (June). Florida Marine Research Institute. St. Petersburg, Florida.

Jaubert, Jean M. 1991. United States Patent number 4,995,980.

Lewis Robin R., and E.D. Estevez. 1988. The ecology of Tampa Bay, Florida: An estuarine profile. Fish and Wildlife Service Biological Report. 85(7.18).

Moe, Martin A. 1989. The marine aquarium reference: systems and invertebrates. Green Turtle Publications, Islamorada, Florida.

Moe, Martin A. 1992. *The marine aquarium handbook: beginner to breeder*. Green Turtle Publications. Islamorada, Florida.

Mook, D. 1980. Seasonal variation in species com-position of recently settled fouling communities along an environmental gradient in the Indian River Lagoon, Florida. *Estuarine and Coastal Marine Science*. II: 573–581.

Robinson, Michael S., Bart Baca, and William W. Falls. 2000. The Marine Ornamental Trade Industry. Aquaculture America 2000.

Santos, S.L., and J.L. Simon. 1980. Response of soft bottom benthos to annual catastrophic disturbance in a south Florida estuary. *Mar. Ecol. Prog. Ser.* 3:347–355.

Sprung, Julian, and J. Charles Delbeek. 1994. *The reef aquarium—a comprehensive guide to the identification and care of tropical marine invertebrates*. Vol. One. Ricordea Publishing, Coconut Grove, Florida.

Tullock, Jeffrey L. 1992. *The reef tank owners manual*. Coralife. Harbor City, California.

Tullock, Jeffrey L. 1997. Natural reef aquariums. Microcosm. Shelbourne, Vermont.

Tullock, Jeffrey L. 1998. Cultivating live rock. Cir. No. 171:64–65. Pet Product News. (March 1998).

15

Overview of Marine Ornamental Shrimp Aquaculture

Ricardo Calado, Luís Narciso, Ricardo Araújo, and Junda Lin

Introduction

Marine shrimp are among the most popular invertebrates in the marine ornamental aquarium trade. Traded ornamental shrimp are largely concentrated in a few families: Stenopodidae, Hippolitydae, Rhynchocinetidae, Palaemonidae, Alpheidae, Hymenoceridae, and Gnathophyllidae (Table 15.1). They exhibit striking coloration and delicacy in addition to their mimetic capacity, symbiotic behavior, peculiar appearance, and rarity that helps them achieve ornamental "status." Some shrimp species can also control certain undesirable organisms, such as anemone pests and fish parasites, in reef tanks. Despite their high market value, the number of tank-raised marine ornamental shrimp is still insignificant, with the vast majority of the traded specimens being collected from the wild. Although the extent of the damage caused to coral reefs as a result of wild collection is yet to be determined, there is a worldwide concern by researchers, traders, collectors, and hobbyists on the sustainability of the marine ornamentals industry (Holthus 2001).

The aquaculture of marine ornamentals must play a key role if any conservation policy is to be successful. In spite of the fact that some bottlenecks still seem to impair the commercial rearing of ornamental shrimp, the incredible achievements in food shrimp production in the past two to three decades must not be overlooked. In a relatively short period of time, research studies successfully shortened larval duration, increased larval survival, developed standard procedures for maturation and spawning in captivity, and established programs of selective breeding for the production of disease-resistant and fast-growing penaeid shrimp (Benzie 1998).

In the past several years a growing number of research efforts have been made toward the understanding of the life history of the most traded ornamental shrimp species as well as the development of suitable rearing techniques. Although several species have been studied, research has mainly been focused on shrimp from the genera *Lysmata* and *Stenopus* (Lin 2001).

According to a recent survey, this research effort seems to reflect the wishes of most commercial breeders, who readily identified *Lysmata debelius*, *L. amboinensis*, and *Stenopus hispidus* as the most desirable shrimp species to be reared (Moe 2001). The culture of these species can guarantee their reliable supply while decreasing the pressure on natural populations.

The main objective of this chapter is to present an overview on broodstock maintenance, mating and spawning, amd larval and juvenile rearing of marine ornamental shrimp.

Broodstock Maintenance

As adults, the majority of the commonly traded shrimp are easy to keep in captivity. Certain species display amazing and often poorly understood symbiotic relationships with other marine animals. For example, alpheid shrimp, generally known as pistol shrimp, live in close association with a wide variety of other organisms (e.g., corals, anemones, sponges, and fish), the most common being with gobies. The study of the symbiotic relationship between *Alpheus djiboutensis* and the goby *Cryptocentrus cryptocentrus* revealed that chemical cues played an

Table 15.1   Most traded marine ornamental shrimp list (GMAD 2002) and their world distribution

Scientific name	Common name	World distribution
Alpheidae		
Alpheus spp.	Pistol shrimp	Atlantic and Indo-Pacific
Hippolitydae		
Lysmata amboinensis	Skunk, white banded cleaner shrimp	Indo-Pacific
Lysmata californica	Peppermint, candy-striped shrimp	Eastern Pacific
Lysmata debelius	Fire, cardinal cleaner shrimp	Indian and Pacific
Lysmata grabhami	Scarlet lady cleaner shrimp	Atlantic
Lysmata rathbunae	Peppermint shrimp	Western Atlantic
Lysmata wurdemanni	Peppermint shrimp	Western Atlantic
Saron marmoratus	Monkey, marble shrimp	Indo-Pacific
Thor amboinensis	Broken back, sexy shrimp	Atlantic and Indo-Pacific
Gnathophyllidae		
Gnathophyllum americanum	Bumblebee shrimp	Atlantic
Hymenoceridae		
Hymenocera elegans	Harlequin shrimp	Indo-Pacific
Hymenocera picta	Harlequin shrimp	Central and Eastern Pacific
Palaemonidae (Pontoniinae)		
Periclimenes brevicarpalis	White-patched anemone shrimp	Indo-Pacific
Periclimenes holthuisi	Holthuis's anemone shrimp	Indo-Pacific
Periclimenes pedersoni	Pederson's commensal shrimp	Western Atlantic
Periclimenes yucatanicus	Yucatan commensal shrimp	Western Atlantic
Rhynchocinetidae		
Rhynchocinetes durbanensis	Camel, dancing, hinge-beak shrimp	Indo-Pacific
Stenopodidae		
Stenopus cyanoscelis	Blue-legged cleaner shrimp	Western Pacific
Stenopus hispidus	White-banded cleaner, coral banded, boxing, barber pole shrimp	Atlantic and Indo-Pacific
Stenopus pyrsonotus	Ghost cleaner shrimp	Indo-Pacific
Stenopus scutellatus	Golden coral, Caribbean boxer shrimp	Western Atlantic
Stenopus tenuirostris	Blue-bodied cleaner, blue boxer shrimp	Indo-Pacific

Source: after Chace 1997 and Debelius 2001.

important role on the recognition of the fish by the shrimp (Karplus et al. 1972). A field study on two other alpheid species (*Alpheus rapax* and *A. rapacida*) also revealed the importance of tactile communication on this intriguing and fascinating symbiotic relationship (Preston 1978). Hippolytid and palaemonid shrimp also present several symbiotic associations with various invertebrates, for example, between the members of the palaemonid genus *Periclimenes* (e.g., *P. brevicarpalis*) and hippolytid genus *Thor* (e.g., *T. amboinensis*) and several species of sea anemones (Fauntin et al. 1995; Guo et al. 1996).

Some of the traded shrimp are also known for their ability to clean fish (Van Tassel et al. 1994) (Table 15.2). Although these shrimp also occur in subtropical waters, the majority are present on coral reefs, where they may establish "fish cleaning stations." Several studies have

described the mesmerizing "dance" performed by these shrimp (Sargent and Wagenbach 1975), namely the characteristic waving movement of their conspicuous white antennae. Despite the large number of studies reporting cleaner shrimp cleaning teeth and removing parasites and skin particles from several fish species, little scientific evidence has been shown to actually prove the existence of cleaning behavior (Spotte 1998).

Another fascinating behavior displayed by several of the traded shrimp species is the existence of social recognition, namely the ability to discriminate between unique individuals. These mechanisms, some of the most complex types of social recognition (Boal 1996), are believed to play a key role in the pair formation of some of the most popular ornamental shrimp species (Table 15.2). The understanding of pair formation and individual recognition has

Table 15.2 Fish cleaning behavior and social organization of the traded marine ornamental shrimp

Scientific name	Fish cleaning behavior	Social organization
Alpheidae		
Alpheus spp.	No	Solitary and in pairs (Debelius 2001)
Hippolitydae		
Lysmata amboinensis	Yes (Van Tassel et al. 1994)	In Pairs and in crowds (Debelius 2001)
Lysmata californica	Yes (Jonasson 1987)	In crowds (Debelius 2001)
Lysmata debelius	Yes (Van Tassel et al. 1994)	In pairs and in small groups (Debelius 2001)
Lysmata grabhami	Yes (Jonasson 1987)	In pairs and in groups of three (Wirtz 1995, Debelius 2001, personal obs.)
Lysmata rathbunae	Yes (Van Tassel et al. 1994)	In pairs and in crowds (Debelius 2001, Rhyne personal communication)
Lysmata wurdemanni	Yes (Van Tassel et al. 1994)	In pairs and in crowds (Debelius 2001, Rhyne personal communication)
Saron marmoratus	No	?
Thor amboinensis	No	In pairs and in small groups (Debelius 2001)
Gnathophyllidae		
Gnathophyllum americanum	No	Solitary and in pairs (Fosså and Nilsen 2000)
Hymenoceridae		
Hymenocera elegans	No	In pairs (Debelius 2001)
Hymenocera picta	No	In pairs (Seibt and Wickler 1979, Fosså and Nilsen 2000)
Palaemonidae (Pontoniinae)		
Periclimenes brevicarpalis	?	?
Periclimenes holthuisi	?	?
Periclimenes pedersoni	Yes (Jonasson 1987)	Solitary, in pairs, and in small groups (Fosså and Nilsen 2000)
Periclimenes yucatanicus	Yes (Jonasson 1987)	Solitary, in pairs, and in small groups (Fosså and Nilsen 2000)
Rhynchocinetidae		
Rhynchocinetes durbanensis	No	In large groups (Fosså and Nilsen 2000)
Stenopodidae		
Stenopus cyanoscelis	?	In pairs (Fosså and Nilsen 2000)
Stenopus hispidus	Yes (Jonasson 1987)	In pairs (Johnson 1977, Strynchuk 1990)
Stenopus pyrsonotus	Yes (Fosså and Nilsen 2000)	In pairs and in small groups (Fosså and Nilsen 2000)
Stenopus scutellatus	Yes (Fosså and Nilsen 2000)	In pairs (Debelius 2001)
Stenopus tenuirostris	?	In pairs (Emmerson et al. 1990)

helped researchers and hobbyists prevent agonistic behaviors among the conspecifics in certain species. Studies conducted on *Stenopus hispidus* revealed the existence of agonistic responses when two shrimp of the same sex were paired (Johnson 1977). These agonistic responses also existed during a very short period of time when shrimp from opposite sexes were put together but were readily followed by courtship and pair formation (Johnson 1969). The fire shrimp (*Lysmata debelius*) also displays a high level of agonistic responses when a pair of this species faces an unknown conspecific. This kind of behavior generally results in the death of the "intruder" shrimp (Simões and Jones 1999). The study of the mechanisms involved in individual recognition on *Stenopus*

hispidus, *Lysmata debelius*, and *Hymenocera picta* revealed that chemical, tactile, and visual cues all played a major role in the recognition of pair members (Johnson 1969; Seibt and Wickler 1979; Simões and Jones 1999).

The majority of the traded species are gonochorist, the individuals being born either as male or female. However, a remarkable and unique sexual system in decapod crustaceans has recently been described for the genus *Lysmata*: protandrous simultaneous hermaphroditism (PSH) (Bauer and Holt 1998; Fiedler 1998; Bauer 2000). These individuals are born as males but at a certain age become simultaneous hermaphrodites. It is puzzling that while widespread within the *Lysmata* genus, PSH is not found in any other decapod crustaceans

(Bauer 2000). In *Lysmata wurdemanni* (Fig. 15.1) at least, the ratio of male and simultaneous hermaphrodite individuals depends on group size. This is the first direct experimental demonstration of social control of sex change in decapod crustaceans (Lin and Zhang 2001a). In both caridean and stenopodidean shrimp, copulation takes place soon after the female has molted, this being the only period during which mating and egg fertilization may take place. Certain simultaneous hermaphrodite shrimp (e.g., *Lysmata amboinensis*) seem to synchronize their molts in such a fashion that both individuals can alternate sexual roles (Fiedler 1998). In all the studied simultaneous hermaphrodite species, cross-fertilization must occur because self-fertilization does not take place (Bauer and Holt 1998; Fiedler 1998).

In gonochoric and simultaneous hermaphrodite caridean shrimp, the male copulates from below, facing the female head to head (e.g., Bauer and Holt 1998; Fiedler 2000). The mating behavior of stenopodids is somehow different, the male facing the female abdomen to abdomen and head to tail (e.g., Zhang et al. 1998a). The eggs are generally extruded moments after copulation (e.g., 10 to 15 seconds in *Stenopus tenuirostris*, unpublished data) and are incubated during their embryonic development on the female's pleopods. Although eyestalk ablation, a procedure widely used in penaeid shrimp culture, has been successfully used to induce gonad development and shorten molt cycle in *Stenopus hispidus* (Zhang et al. 1997a), the vast majority of traded species will readily mate and spawn if well kept in captivity.

The number of eggs produced varies according to species and shrimp size. Inadequate broodstock maintenance, along with nutritionally unsuitable diets, may result in poor egg quality and in abnormal egg loss during the incubation period. Although several broodstock diets (e.g., enriched and unenriched fresh and frozen *Artemia* nauplii and adults, squid, mussels, clams, and polychaetes) have been tested in different shrimp species (e.g., *Lysmata debelius, L. amboinensis L. wurdemanni,* and *Stenopus scutellatus*), the nutritional suitability of each diet seemed somehow species dependent. In *Lysmata debelius*, broodstock shrimp fed with fresh mussel and polychaete produced fewer larvae (average ±s.d.= 486 ± 254) than those fed with *Artemia* nauplii (1766 ± 391) (Simões et al. 1998). Broodstock *Stenopus scutellatus* (Figure 15.2) fed with a mixture of regular frozen *Artemia* adults and hard clam (*Mercenaria mercenaria*) produced significantly more larvae (average ±s.d.= 1,589 ± 298) than those fed with frozen enriched *Artemia* adults (1,268 ± 245), frozen regular *Artemia* adults (1,006 ± 217), and frozen hard clams (806 ± 71) (Lin and Shi 2002). However, on *Lysmata amboinensis* and *L. wurdemanni,* different broodstock diets seemed to have little effect on their reproductive performance (Simões et al. 1998; Lin and Zhang 2001b). Another important issue in the assessment of egg quality that only recently started to be investigated on ornamental shrimp species is the egg's biochemical composition. Because lipids represent the most important energy source during embryonic development of most crustaceans

Fig. 15.2 The marine ornamental shrimp *Stenopus scutellatus*. (See also color plates.)

Fig. 15.1 The marine ornamental shrimp *Lysmata wurdemanni*. (See also color plates.)

(Wehrtmann and Graeve 1998), special attention must be paid to the egg lipid composition, namely its fatty acid profile. Preliminary studies on the fatty acid profile of *Lysmata seticaudata* eggs during embryonic development revealed that the embryos of smaller females presented a higher consumption rate of essential fatty acids, namely docosahaexaenoic acid (DHA) and eicosapentaenoic acid (EPA) than medium and larger females (Calado et al. 2001a). In this way a decrease of larval quality is expected and probably reflects the reduced maternal investment of smaller shrimp that are reproducing as females for the first time. Additionally, through the comparison of the biochemical composition of wild versus laboratory-spawned eggs, the suitability of broodstock diets can be evaluated and egg quality improved. The study of the egg's biochemical composition can also be a helpful tool in the assessment of the nutritional requirements of early larval stages.

Larval Rearing

In order to fulfill adequately the growing demand for ornamental shrimp species without increasing the collection effort on natural populations, the artificial production of these highly priced resources must be achieved. However, one special constraint that must be overcome is mass rearing of larvae.

Larval rearing has always been one of the most serious bottlenecks in aquaculture. Unfortunately ornamental shrimp are not an exception. Although promising results have been achieved in the larviculture of *Lysmata amboinensis* (Fletcher et al. 1995), *L. debelius* (Fletcher et al. 1995; Palmtag and Holt 2001), *L. seticaudata* (Calado et al. 2001b), *L. wurdemanni* (Zhang et al. 1998b) and *Stenopus scutellatus* (Zhang et al. 1997b), the long larval duration and/or low survival rates still impair the commercial rearing of these and other species. Most of the traded species are known or believed to have a large number of larval stages (Table 15.3), although larval descriptions of the majority of the species are still missing. These larvae are also known to display frail larval structures, such as long rostrums (e.g., *Stenopus* larvae) and long paddlelike appendages (e.g., *Lysmata* larvae, Fig. 15.3). The peculiar appendages, enormous elongated fifth pereiopods (up to twice the total larval length),

Fig. 15.3 The larva of marine ornamental *Lysmata*. (See also color plates.)

may assist in the maintenance of position and orientation in the water column (Fletcher et al. 1995; unpublished data) and feeding and defense against predation (Rufino and Jones 2001). Some of the species also have the ability to delay larval development for long periods of time when unsuitable rearing conditions are present (e.g., Wunsch 1996). This ability is generally designated as mark-time molting and can be defined as a sequence of molts displayed by a certain larval stage in which few morphological changes take place (Gore 1985). The commercial rearing of these species will be possible only if shorter larval cycles and higher survival rates to the postlarval stage are achieved.

The development of suitable larval diets certainly is one of the keys to shortening larval durations. Several larval diets have been tested (e.g., microalgae, rotifers, decapsulated cysts, newly hatched nauplii, and enriched metanauplii of *Artemia*), mostly resulting in long larval cycles (Table 15.3) and poor survival rates. The recent use of high-speed video to record the study of larval shrimp feeding kinematics (Rhyne et al. 2001) allowed researchers to evaluate the influence of prey size in the capture and manipulation effort on different larval stages. The same study documented the strong feeding response of early shrimp larval stages to inert diets. These findings will surely enable researchers to test and eventually fulfill several larval nutritional requirements through the use of prepared pellet diets.

Even though the development of suitable larval diets will certainly play a key role in the commercial rearing of ornamental shrimp, the development of suitable larval rearing systems can also promote shorter larval duration and higher survival rates. Ornamental shrimp larvae

Table 15.3   Number of larval stages and larval duration of the traded marine ornamental shrimp

Scientific name	Number of larval stages	Larval duration (days)
Alpheidae		
Alpheus spp.	9 Zoeas* (Barnich 1996)	?
Hippolitydae		
Lysmata amboinensis	?	58–140 days (Goy 1990, Fletcher et al. 1995)
Lysmata californica	?	?
Lysmata debelius	?	75–158 days (Fletcher *et al.* 1995, Palmtag and Holt 2001)
Lysmata grabhami	?	?
Lysmata rathbunae	?	25–37 days (Goy 1990, Zhang et al. 1998b)
Lysmata wurdemanni	11 (Kurata 1970)	43–110 days (Kurata 1970, Goy, 1990, D. Zhang personal communication)
Saron marmoratus	?	30 days (Kruschwitz 1967)
Thor amboinensis	Sarver 1979	Sarver 1979
Gnathophyllidae		
Gnathophyllum americanum	?	?
Hymenoceridae		
Hymenocera elegans	?	?
Hymenocera picta	?	28–56 days (Kraul 1999)
Palaemonidae (Pontoniinae)		
Periclimenes brevicarpalis	8 Zoeas* (Bourdillon-Casanova 1960)	
Periclimenes holthuisi	8 Zoeas* (Bourdillon-Casanova 1960)	
Periclimenes pedersoni	8 Zoeas* (Bourdillon-Casanova 1960)	28–40 days (Goy 1990)
Periclimenes yucatanicus	8 Zoeas* (Bourdillon-Casanova 1960)	29–45 days (Goy 1990)
Rhynchocinetidae		
Rhynchocinetes durbanensis	11 Zoeas* (Matoba and Shokita 1998)	?
Stenopodidae		
Stenopus cyanoscelis	?	?
Stenopus hispidus	?	120–210 days (Fletcher et al. 1995)
Stenopus pyrsonotus	?	?
Stenopus scutellatus	?	43–77 days (Zhang ct al. 1997b)
Stenopus tenuirostris	?	?

* Data from other species in the genus; ?, no data are available for this species.

have been reared using several types of rearing containers (e.g., beakers, bottles and plastic boxes of various volumes, aquarium tanks, and mesh baskets) (Couturier-Bhaud 1974; Crompton 1992; Riley 1994; Wilkerson 1994; Fletcher et al. 1995; Wunsch 1996; Zhang et al. 1997b; 1998b; 1998c; Kraul 1999; Rufino 1999; Calado et al. 2001b; Debelius 2001; Palmtag and Holt 2001; Simões et al. 2001). Nevertheless, all these rearing systems require time-consuming operating procedures such as daily water changes and systematic manipulation of the developing larvae, a practice that can easily induce physical damage and larval mortality. However, a preliminary study on the larval rearing of *Lysmata seticaudata* on enriched *Artemia* metanauplii in Algamac 2000® at 24 ±

1°C revealed that the use of a water recirculation rearing system based on the "planktokreisel" (Greve 1968) considerably shortened the larval duration to the postlarval stage from approximately 80 days to 27 days (Calado et al. 2001b), reduced the number of mark-time molts, and increased the survival rate from 15% to 70% (unpublished data). The main feature of this rearing system is the maintenance of larvae and food in suspension through water motion, avoiding the use of water aeration in the rearing tanks. Although the use of water aeration can also provide an adequate water circulation, this common procedure on several rearing systems may also be a source of physical damage to the frail larval stages and, consequently, induce mortality. Another important feature of this rearing system is the daily replacement of the 150 μm by 500 μm mesh screens in each rearing tank. This simple procedure allowed the daily flush of uneaten prey as well as the replacement of 24-hour-old enriched metanauplii by newly enriched ones. Because the nutritive value of enriched *Artemia* is known to decrease rapidly 24 hours after being supplied to the larvae (Navarro et al. 1999), the daily use of prey with a high nutritive value played a vital role in the rearing process. Currently, the "planktokreisel"-based system appears as the best option for the culture of marine ornamental shrimp. This avoids the traditional time-consuming water change and minimizes larval handling while providing excellent water quality and allows researchers to provide suitable diets. This rearing system also presents the advantage of allowing researchers to test on a small rearing scale (10 liter tanks) as well as on a commercial mass scale (200 liter tanks). Preliminary trials of the mass scale version of the "planktokreisel"-based system produced more than 10,000 postlarvae of the Mediterranean cleaner shrimp *Lysmata seticaudata,* with survival rates ranging from 52% to 70% (unpublished data).

Juvenile Rearing

In contrast to larvae, juvenile ornamental shrimp generally display high survival rates, for example, 99% for *Lysmata debelius* (Palmtag and Holt 2001) and 70% to 75% for *L. seticaudata* (unpublished data). Juvenile shrimp are easily fed on commercial pellets and frozen foods (such as mussel, clam, fish, shrimp, and squid meat blends with finely chopped algae) (Calado et al. 2001c; M.R. Palmtag, University of Texas at Austin Marine Science Institute, Port Aransas, Texas, personal communication, 2001). The exception to this rule seems to be the harlequin shrimp *Hymenocera*, because the juveniles of these species start feeding exclusively on sea stars a few days after metamorphosis (Fiedler 1994). In order to achieve a commercially sustainable production of this highly priced shrimp, studies are needed to find alternative preys.

The rearing system for juvenile ornamental shrimp is not an issue for gregarious species such as *Lysmata seticaudata* and *L. wurdemanni* because high densities of juveniles (e.g.,

500/m^2 for *L. seticaudsata,* unpublished data) can be easily raised with minimal mortality. However, raising juveniles of certain species that display strong agonistic behaviors toward conspecifics, such as *Lysmata debelius* and *Stenopus* spp., may require special rearing systems in order to minimize mortality rates. Another solution can be the commercialization of those juveniles before they display agonistic behavior. The majority of these species are suitable for sale to retail stores at 30 to 40 mm total length, a size that can be achieved in 3 months (for example, *Lysmata wurdemanni* [A. Rhyne, Florida Institute of Technology, Melbourne, Florida, personal communication, 2001]) to 6 months (for example, *L. debelius* [Palmtag and Holt 2001]) from the date of hatching. Because of their longevity and better adaptation to the aquarium environment, these artificially raised ornamental shrimp generally attain higher market values than wild specimens, an advantage for commercial production.

Future Directions

The popularity of marine tropical shrimp is still increasing among hobbyists. The growing effort to solve the present bottlenecks on the propagation of these species makes it likely that a considerable increase in the availability of captive raised animals for the ornamental trade will occur. Another area that needs attention is the evaluation of the rearing potential of several shrimp species from temperate and subtropical waters (e.g., *Lysmata seticaudata, Periclimenes sagittifer,* and *Stenopus spinosus*) that could fulfill the requirements of ornamental species (Calado et al. 2001c). Besides their delicacy, remarkable mimetic adaptations and associative behavior, the coloration presented by some of these crustaceans is rivaling that of tropical species.

The majority of current research studies are focused on tropical marine ornamental shrimp. However, the rearing potential of other decapod crustaceans should also be seriously considered. Several species of hermit crabs (e.g., *Calcinus* spp., *Clibanarius* spp., *Dardanus* spp., and *Paguristes* spp.), porcelain crabs (e.g., *Porcellana* spp., *Petrolisthes* spp., and *Neopetrolisthes* spp.), sponge crabs (*Dromia* spp.), boxer crabs (*Lybia* spp.), and spider crabs (e.g., *Mithrax* spp. and *Stenorhynchus* spp.) are also heavily traded species. The majority of these species have considerably fewer larval stages than those of ornamental shrimp (e.g., spider crabs have two zoea stages and one megalopa stage) and can be reared in shorter periods, for example 6–10 days for the emerald crab *Mithrax sculptus* (unpublished data). Additionally, all these species can be raised using the same "planktokreisel"-based rearing system for larval shrimp, while the juvenile crabs can be easily raised using the same procedures previously described for juvenile shrimp.

The United States, Japan, Germany, France, and the United Kingdom are the major importers of ornamental species (Lem 2001). These countries should work with the main exporters of marine ornamentals to adopt environmentally friendly and sustainable procedures for the exploration of these valuable resources and to develop the technology for the rearing of ornamental species. Many developing countries in Southeast Asia, the Caribbean, Eastern Africa, and the Red Sea regions have strong potential for ornamental species production, with excellent climate conditions and low production costs. Developing the culture technology in these countries will allow the implementation of more effective conservation programs while generating important economic incomes (Fletcher et al. 1999).

Acknowledgments

This chapter is dedicated to our deceased research colleague Fernando Simões, one of the pioneers in marine ornamental shrimp aquaculture. We would like to thank the Luso-American Foundation for Development and the Fundação para a Ciência e a Tecnologia (scholarship SFRH/BD/983/2000 and research project POCTI/BSE/43340/2001) from the Portuguese government for their financial support.

References

Barnich, Ruth. 1996. The larvae of the Crustacea: Decapoda (excl. Brachyura) in the plankton of the French Mediterranean coast (identification keys and systematic review). Ph.D. dissertation, Cuvillier Verlag, Göttingen, Germany.

Bauer, Raymond T., and G. Joan Holt. 1998. Simultaneous hermaphroditism in the marine shrimp *Lysmata wurdemanni* (Caridea: Hippolytidae): An undescribed sexual system in the decapod Crustacea. *Marine Biology* 132:223–235.

Bauer, Raymond, T. 2000. Simultaneous hermaphroditism in caridean shrimps: A unique and puzzling sexual system in the decapoda. *Journal of Crustacean Biology* 20 (Special Number):116–128.

Benzie, John A.H. 1998. Penaeid genetics and biotechnology. *Aquaculture* 164(1–4):23–47.

Boal, Jean G. 1996. Absence of social recognition in laboratory-reared cuttlefish, *Sepia officinalis* L. (Mollusca: Cephalopoda). *Animal Behaviour* 52(3):529–537.

Bourdillon-Cassanova, Laurette. 1960. Le meroplancton du Golfe de Marseille: les larves des crustacés decapodes. *Recueil des Travaux de la Station Marine d'Endoume* 30(18):1–286.

Calado, Ricardo, Sofia Morais, and Luis Narciso. 2001a. Fatty acid profile of Mediterranean cleaner shrimp (*Lysmata seticaudata*) eggs during embryonic development. Book of Abstacts, Marine Ornamentals 2001, Orlando, USA 95–97.

Calado, Ricardo, Catarina Martins, Olga Santos, and Luis Narciso 2001b. Larval development of the Mediterranean cleaner shrimp *Lysmata seticaudata* (Risso, 1816) (Caridea; Hippolytidae) fed on different diets: Costs and benefits of mark-time molting. Larvi'01 Fish and Crustacean Larviculture Symposium, European Aquaculture Society, Special Publication 30:96–99.

Calado, Ricardo, Sofia Morais, and Luis Narciso. 2001c.Temperate shrimp: Perspective use as ornamental species. Book of Abstracts, Marine Ornamentals 2001, Orlando, USA 118–119.

Chase Jr., F.A. 1997. The caridean shrimps (Crustacea:Decapoda) of the Albatross Philippine Expedition, 1907–1910, Part 7: Families Atyidae, Eugonatonotidae, Rhynchocinetidae, Bathypalaemonellidae, Processidae, and Hippolytidae. *Smithsonian Contributions to Zoology* 587:1–131.

Couturier-Bhaud, Yvone. 1974. Cycle biologique de *Lysmata seticaudata* Risso (Crustacé, Décapode). III—Étude du développement larvaire. *Vie Et Millieu* 24(3):431–442.

Crompton, W. Douglas. 1992. Laboratory culture and larval development of the peppermint shrimp, *Lysmata wurdemanni* Gibes (Caridea: Hippolytidae): Part I. Laboratory Culture. M.S. Thesis. Corpus Christi State University, Corpus Christi, Texas, USA.

Debelius, Helmut. 2001. *Crustacea Guide of the World*, 2nd ed. IKAN—Unterwasserarchiv, Frankfurt, Germany.

Emmerson, W.D., Joseph W. Goy, and S. Koslowski. 1990. On the occurrence of *Stenopus tenuirostris* De Man, 1888, in Naval waters. *South Africa Tydskrif vir Dierkunde* (4):260–261.

Fauntin, Daphne G., Chau-Chih Guo, and Jiang-Shiou Hwang. 1995. Costs and benefits of the symbiosis between the anemone shrimp *Periclimenes brevicarpalis* and its host *Entacmaea quadricolor. Marine Ecology Progress Series* 129:77–84.

Fiedler, G. Curt. 1994. The larval stages of the harlequin shrimp, *Hymenocera picta* (Dana). M.S. Thesis, University of Hawaii, Hawaii, USA.

Fiedler, G. Curt. 1998. Functional, simultaneous hermaphroditism in female-phase *Lysmata amboinensis* (Decapoda: Hippolytidae). *Pacific Science* 52:161–169.

Fiedler, G. Curt. 2000. Sex determination and reproductive biology of two caridean shrimp genera: *Hymenocera* and *Lysmata*. Ph.D. Dissertation, University of Hawaii, Hawaii, USA.

Fletcher, David J., I. Kotter, Mark Wunsch, and I. Yasir. 1995. Preliminary observations on the reproductive biology of ornamental cleaner prawns. *International Zoo Yearbook* 34:73–77.

Fletcher, David J., Elizabet Wood, and David A. Jones. 1999. Marine ornamental culture and reef conservation. Aquaculture 99 Book of Abstracts, Sydney, Australia 265.

Fosså, Svein A. and Alf J. Nielsen. 2000. *The Modern Coral Reef Aquarium,* Vol. 3, Birgit Schmettkamp Verlag, Bornheim, Germany.

GMAD. 2002. Global Marine Aquarium Database. http://www.unep-wcmc.org/marine/GMAD/.

Gore, Robert H. 1985. Molting and growth in decapod larvae. In *Crustacean Issues—Larval growth.* Vol. 2, edited by A. Wenner, pp 1–65. A Balkema Publishers, Rotherdam, Netherlands.

Goy, Joseph W. 1990. Components of reproductive effort and delay of larval metamorphosis in tropical marine shrimp (Crustacea: Decapoda: Caridea: and Stenopodidea). Ph.D. dissertation, University of Texas, Texas, U.S.A.

Greve, Wulf. 1968. The "planktonkreisel," a new device for culturing zooplankton. *Marine Biology* 1:201–203.

Guo, Chau-Chih, Jiang-Shiou Hwang, and Daphne G. Fauntin. 1996. Host selection by shrimps symbiotic with sea anemones: A field survey and experimental laboratory analysis. *Journal of Experimental Marine Biology and Ecology* 202:165–176.

Holthus, Paul. 2001. From reef to retail: Marine ornamental certification for sustainability is here. Book of Abstracts, Marine Ornamentals 2001, Orlando, USA 21–23.

Johnson, Victor R., Jr. 1969. Behavior associated with pair formation in the banded shrimp *Stenopus hispidus* (Olivier). *Pacific Science* 23:40–50.

Johnson, Victor R., Jr. 1977. Individual recognition in the banded shrimp *Stenopus hispidus* (Olivier). *Animal Behaviour* 25:418–428.

Jonasson, Mark. 1987. Fish cleaning behaviour of shrimp. *Journal of Zoology, London* 213:117–131.

Karplus, I., M. Tsurnamal, and R. Szlep. 1972. Analysis of the mutual attraction in the association of the fish *Cryptocentrus cryptocentrus* (Gobiidae) and the shrimp *Alpheus djiboutensis* (Alpheidae). *Marine Biology* 17:275–283.

Kraul, Syd. 1999. Commercial culture of the harlequin shrimp *Hymenocera picta* and other ornamental marine shrimp. Book of Abstracts, Marine Ornamentals 1999, Hawaii, USA 50.

Kruschwitz, Lois G. 1967. Aspects of the behavior and ecology of a reef shrimp. *American Zoologist* 7:204–205.

Kurata, H. 1970. Studies of the life histories of decapod crustacea of Georgia. Ph.D. dissertaion, University of Georgia, Sapelo Island, Georgia, U.S.A.

Lem, Audun. 2001. International trade in ornamental fish. Book of Abstracts, Marine Ornamentals 2001, Orlando, USA 26.

Lin, Junda. 2001. Overview of marine ornamental shrimp aquaculture. Book of Abstracts, Marine Ornamentals 2001, Orlando, USA 63–65.

Lin, Junda, and Peichang Shi. 2002. Effect of broodstock diet on reproductive performance of the golden banded coral shrimp, *Stenopus scutellatus. J. World Aquaculture Society*, 33:384–386.

Lin, Junda, and Dong Zhang. 2001a. Reproduction in a simultaneous hermaphroditic shrimp, *Lysmata wurdemanni*: Any two will do? *Marine Biology* 139:919–922.

Lin, Junda, and Dong Zhang. 2001b. Effect of broodstock diet on reproductive performance of the peppermint shrimp, *Lysmata wurdemanni. Journal of Shellfish Research* 20(1):361–363.

Matoba, Hiroe, and Shigemitsu Shokita. 1998. Larval development of the rhynchocinetid shrimp, *Rhynchocinetes conspiciocellus* Okuno and Takeda (Decapoda: Caridea: Rhynchocinetidae) reared under laboratory conditions. *Crustacean Research* 27:40–69.

Moe, Martin A., Jr. 2001. Culture of marine ornamentals: For love, for money and for science. Book of Abstracts, Marine Ornamentals 2001, Orlando, USA 27–28.

Navarro, Juan C., R. James Henderson, Lesley A. McEvoy, Michael V. Bell, and Francisco Amat. 1999. Lipid conversion during enrichment of *Artemia. Aquaculture* 174:155–166.

Palmtag, Matt R., and G. Joan Holt. 2001. Captive rearing of fire shrimp (*Lysmata debelius*). Texas Sea Grant College Program Research Report.

Preston, J. Lynn. 1978. Communication systems and social interactions in a goby-shrimp symbiosis. *Animal Behaviour* 26:791–802.

Rhyne, Andrew, Junda Lin, Ricardo Calado, and Ralph Turingan. 2001. Improvements in marine ornamental shrimp culture: High-speed video analysis of feeding kinematics in dietary study. Book of Abstracts, Marine Ornamentals 2001, Orlando, USA 29–30.

Riley, Cecilia M. 1994. Captive spawning and rearing of the peppermint shrimp (*Lysmata wurdemanni*). *Seascope* 11(4): Summer.

Rufino, Marta. 1999. Some aspects of ecology and behaviour of *Lysmata debelius* (Bruce, 1983) (Decapoda: Hippolytidae), and a review about the genus. First Degree Thesis, University of Wales,

Bangor, UK and Faculdade de Ciencias da Universidade de Lisboa, Portugal.

Rufino, Marta, and David A. Jones. 2001. Observations on the function of the fifth pereiopod in late stage larvae of *Lysmata debelius* (Decapoda, Hippolytidae). *Crustaceana* 74(9):977–990.

Sargent, R. Craig, and Gary E. Wagenbach. 1975. Cleaning behavior of the shrimp, *Periclimenes anthophilus* Holthuis and Eibl-Eibesfeldt (Crustacea: Decapoda: Natantia). *Bulletin of Marine Science* 25(4):466–472.

Sarver, D. 1979. Larval culture of the shrimp *Tor amboinensis* (De Man, 1888) with reference to its symbiosis with the anemone *Antheopsis papillosa* (Kwietniwski, 1898). *Crustaceana* Suppl. 5:176–178.

Seibt, Uta, and Wolfgang Wickler. 1979. The biological significance of the pair-bond in the shrimp *Hymenocera picta. Zeitschrift fur Tierpsychologie* 50:166–179.

Simões, Fernando, Fernando Ribeiro, and David A. Jones. 1998. The effect of diet on the reproductive performance of marine cleaner shrimps *Lysmata debelius* (Bruce, 1983) and *L. amboinensis* (de Man, 1888) (Caridea, Hippolytidae) in captivity. Aquaculture 98 Book of Abstracts, Las Vegas, Nevada, USA 497.

Simões, Fernando, and David A. Jones. 1999. Pair formation in the tropical marine cleaner shrimp *Lysmata debelius* (Crustacea, Caridea). Aquaculture 99 Book of Abstracts, Sydney, Australia 700.

Simões, Fernando, Fernando Ribeiro, and David A. Jones. 2001. Feeding early larval stages of fire shrimp *Lysmata debelius* (Caridea, Hippolytidae). *Larvi'01 Fish and Crustacean Larviculture Symposium, European Aquaculture Society, Special Publication* 30:559–562.

Spotte, Stephen 1998. "Cleaner" shrimps? *Helgolander Meeresunters* 52:59–64.

Van Tassel, James L., Alberto Brito, and Stephen A. Bortone. 1994. Cleaning behaviour among marine fishes and invertebrates in the Canary Islands. *Cybium* 18(2):117–127.

Wehrtmann, Ingo, and Martin Graeve. 1998. Lipid composition and utilization in developing eggs of two tropical marine caridean shrimps (Decapoda, Caridea, Alpheidae, Palaemonidae). *Comparative Biochemistry and Physiology Part B* 121: 457–463.

Wilkerson, Joyce D. 1994. Scarlet cleaner shrimp. *Freshwater and Marine Aquarium* 8 (August).

Wirtz, Peter. 1995. *Unterwasserführer Madeira, Kanaren, Azoren.* Delius Klasing, Stuttgart, Germany.

Wunsch, Mark. 1996. Larval development of *Lysmata amboinensis* (De Man, 1888) (Decapoda: Hippolytidae) reared in the laboratory with a note on *L. debelius* (Bruce, 1983). M.S. Thesis. University of Wales, Bangor, UK.

Zhang, Dong, Junda Lin, and R. LeRoy Creswell. 1997a. Effect of eyestalk ablation on molt cycle and reproduction in the banded shrimp *Stenopus hispidus. Journal of Shellfish Research* 16(2):363–366.

Zhang, Dong, Junda Lin, and R. LeRoy Creswell. 1997b. Larviculture and effect of food on larval survival and development in golden coral shrimp *Stenopus scutellatus. Journal of Shellfish Research* 16(2):367–369.

Zhang, Dong, Junda Lin, and R. LeRoy Creswell. 1998a. Mating behavior and spawning of the banded coral shrimp *Stenopus hispidus* in the laboratory. *Journal of Crustacean Biology* 18(3): 511–518.

Zhang, Dong, Junda Lin, and R. LeRoy Creswell. 1998b. Effects of food and temperature on survival and development in the peppermint shrimp *Lysmata wurdemanni. Journal of the World Aquaculture Society* 29(4):471–476.

Zhang, Dong, Junda Lin, and R. LeRoy Creswell. 1998c. Ingestion rate and feeding behaviour of the peppermint shrimp *Lysmata wurdemanni* on *Artemia* nauplii. *Journal of the World Aquaculture Society* 29(1):97–103.

The Invertebrates
C. *Corals*

16

Coral Culture—Possible Future Trends and Directions

Michael Arvedlund, Jamie Craggs, and Joe Pecorelli

Introduction

The culture of corals, both soft (Octocorallia) and stony (Hexacorallia), has recently witnessed an explosion in promising results, especially within asexual coral culture—so-called fragmentation or propagation. These results are the focal point in this chapter. Readers interested in studies of sexual reproduction of corals can consult a number of sources (e.g. Glynn et al. 2000, 1996, 1994, 1991; Shlesinger et al. 1998; Harrison & Wallace 1990; Richmond and Hunter 1990; Harrison et al. 1984; Fadlallah 1983). Readers interested in studies of sexual coral culture can consult other sources (e.g. Szmant et al. 2001; Rinkevich and Shafir 1998).

Asexual coral culture may be divided into several subcategories. Two useful subcategories are coral culture in situ (in the field) and coral culture ex situ (in captivity). Surprisingly, these two categories have evolved with little use of each other's results. In many cases, investigators from each category are obviously not aware of the useful results achieved from the other category. This may reduce the rate of progress of coral culture considerably—in the end it may also slow down the progress of reef restoration projects.

This chapter begins with a literature review, continues with a discussion of possible future trends and directions, especially regarding ways to improve the communication between the two categories, and finishes with a discussion of promising types of experimental coral culture.

Coral Culture in Situ

Studies of coral culture in situ have been conducted for more than three decades. This area

may (again) be divided into subcategories: for example, natural processes of coral fragmentation, field growth studies, and the area of coral transplants (fragments) as an aid to damaged reefs. Studies in the first two areas do not include any experiments in coral culture; however, the findings are vastly useful for anyone interested in aspects of coral culture. First, let's examine basic studies of coral fragmentations.

Asexual Reproduction of Corals in Nature

Asexual reproduction of corals in nature happens on a regular basis through fragmentation. According to a comprehensive review (Highsmith 1982), production of new colonies by fragmentation of established colonies is shown as an extremely important mode of reproduction and a local process of distribution among reef-building corals. Highsmith concluded that several of the most successful corals are adapted to fragment, that is, have incorporated fragmentation into their life histories. On coral reefs, fragmentation happens among branching corals but also among gorgonians, zooanthids, stony hydrozoans, alcyonaceans, and vinelike sponges, when intact colonies are broken apart, especially by bioerosion, predation, storms, and hurricanes (Smith and Hughes 1999). According to Highsmith and Smith and Hughes, several studies have documented these processes (Highsmith et al. 1980; Bak and Criens 1981; Tunnicliffe 1981; Wulff 1985; Karlson 1986; Karlson et al. 1996).

Two in-depth experimental studies of fragmentation in the branching coral *Acropora palmata* Lamarck have been conducted (Lirman 2000a, b). This species has a limited sexual recruitment, and therefore asexual reproduction by tissue fission or fragmentation may have a significant influence on its survivorship and propagation. Lirman experimentally fragmented this coral and removed these fragments for later placement on different types of substrate, for example, sand, rubble, and on top of other corals of the same species. No significant relationships were found between size and the survivorship of fragments. However, Lirman noted that the highest survivorship of *A. palmata* fragments was observed for those fragments placed on top of live *A. palmata* colonies. These fragments fused to the underlying tissue and showed no signs of loss (0% mortality) (Lirman 2000a,b). The lowest survivorship was observed for fragments on sand (58% mortality).

Another study of natural coral fragmentation was done on the Caribbean branching coral *Madracis mirabilis* Duchassaing and Michelotti (Bruno 1998). General life history theory and models of coral fragmentation predict that intra- and interspecific variation in fragment size should be positively related to survival and inversely related to dispersal. To test these predictions, Bruno examined fragmentation in the Caribbean branching coral *Madracis mirabilis*, for 60 labeled fragments for 11 months at three sites. Fragments ranged in size from 3 to 19 cm. The smallest fragments displayed the lowest survivorship (i.e., 50%). However, there was no apparent increase in survivorship with size among larger fragments.

Other valuable reports in this area also exist. They are: fragmentation in *Madracis mirabillis*, *Acropora palmata,* and *A. cervicornis* Lamarck (Bak and Criens 1981); the coral genus *Acropora* (Bothwell 1982); regeneration after experimental breakage in the solitary reef coral *Fungia granulosa* Klunziger (Chadwick and Loya 1990); fragmentation in the Hawaiian coral *Montipora verrucosa* Lamarck (Cox 1992); regeneration and growth of fragmented colonies of *Acropora formosa* Dana and *A. nasuta* Dana (Kobayashi 1984); reproduction, recruitment, and fragmentation in nine *Acropora* species from the Great Barrier Reef (*A. loripes* Brook, *A. granulosa* Milne Edwards and Haime, *A. sarmentosa* Brook, *A. longicyathus* Milne Edwards and Haime, *A. florida* Dana, *A horrida* Dana, *A. nobilis* Dana, *A. hyacinthus* Dana and *A. valida* Dana) (Wallace 1985).

Basic Field Studies of Growth in Scleractinian Corals

Basic field studies of growth in scleractinian corals are possibly a smaller field of research than coral fragmentation and coral transplantation studies. However, as with studies of natural coral fragmentation, results of growth studies may likewise help anyone interested in coral culture.

A research project in the Red Sea focused on growth rates of the hydrocoral *Millepora dichotoma* Forskal and the corals *Acropora variabilis* Klunziger and *Stylophora pistillata* Esper (Vago et al. 1997b). They found that *M. dichotoma* exhibited the slowest relative growth, whereas *S. pistillata* and *A. variabilis* were faster growers with weight-based doubling times of about 3, 1.2, and 1.3 years, respectively. In the case of *A. variabilis*, as well as in bladed colonies of *M. dichotoma*, growth rate per unit weight was size independent, whereas in *S. pistillata* a clear positive correlation was found between the relative growth rates and the initial colony buoyant weights. *M. dichotoma* and *A. variabilis* demonstrated distinct seasonal growth oscillations (2-month time lag with respect to the seasonal trend of the ambient water temperature). They concluded that the oscillations in the absolute growth rate of corals resulted from the seasonal changes in water temperature. The Vago group has conducted a number of useful studies of coral growth: for example, the use of laser measurements of coral growth (Vago et al. 1997b); nondestructive method for monitoring coral growth affected by anthropogenic and natural long-term changes (Vago et al. 1994). A recent study (Vago et al. 2001) showed that aluminium metallic substrate induces colossal biomineralization of the calcareous hydrocoral *Millepora dichotoma*.

Growth rates (i.e., extension rate; number of radial branches; skeletal mass; branch diameter) of the staghorn coral *Acropora formosa* Dana were studied at four sites on the Beacon Island platform at Houtman Abrolhos, in subtropical Western Australia (Harriott 1998). Sites were at depths of 7 to 11 m, with variable exposure to weather and swell conditions. Two sites on the western reef slope were partly exposed to the oceanic swell, and two sites in the lagoon were largely protected from wave action. The linear extension rate between 1994 and 1995 varied significantly between sites, with greater linear extension at the more protected lagoonal sites. Branch extension rate over 11.5 months ranged from a mean of 50.3 mm (range 13 to 93 mm) at a reef slope site to a mean of 76.0 mm (range 31 to 115 mm) at a sheltered lagoonal site. Growth was within the range reported for *A. formosa* from tropical sites, which is consistent with the relatively high calcification and reef accretion.

Another paper reports variation in coral growth rates with depths at Discovery Bay, Jamaica (Huston 1985). Huston concluded via a compilation of available growth data for Atlantic and Pacific corals that a strong pattern exists, with the highest growth rates a short distance below the surface (5–15 m) and a decrease with depth.

Other useful growth studies are: debates about whether normal coral growth rates on dying reefs are good indicators of reef health (Edinger et al. 2000); growth rates of eight species of scleractinian corals in the eastern pacific (Costa Rica) (Guzmán and Cortés 1989). A detailed study of growth rates in the staghorn coral *Acropora pulchra* Brook has also been conducted (Yap and Gomez 1984). They later expanded the study of this species to include experimental transplantation and observation of survival rates (Yap and Gomez 1985); and growth rates in commercially important scleractinian corals from the Philippines (Gomez et al. 1985).

Coral Transplants (Fragments) as an Aid to Damaged Reefs

Projects using coral fragmentation or whole colonies as an aid to damaged reefs have been conducted at least since 1966 (Rinkevich 2000). An example is a recent study in Africa, near Mafia Island, Tanzania (Lindahl 1998). The aim was to restore local reefs using low-tech rehabilitation through transplantation of staghorn corals (*Acropora formosa*) in shallow areas of sand and rubble. The study was designed to investigate the effects of the following three factors: (1) Attachment: Unattached corals were compared with corals tied together on strings. (2) Density: Corals were placed either densely with a cover of 30% to 40% or sparsely with a cover of 10% to 15%. (3) Site: The corals were transplanted to two sites with different sub-

strates and exposure to waves. After 23 months, the cover of the transplanted corals had increased by 51%. The colonization of other species of hard and soft corals was also promoted by the transplantation of corals. The method of tying corals together with strings before placement on the seabed was effective and could be applied for rehabilitation of degraded coral reefs that are moderately exposed to water movements.

A study of reef rehabilitation in the Maldives (Clark and Edwards 1995) reports that whole coral colonies (primarily *Acropora, Pocillopora, Porites, Favia,* and *Favites*) were transplanted and cemented in place onto three approximately 20-square-meter areas of Armorflex concrete mats on an 0.8- to 1.5-m deep reef-flat that had been severely degraded by coral mining. Growth, in situ mortality, and losses from mats due to wave action of a total of 530 transplants were monitored over 28 months. Overall survivorship of corals 28 months after transplantation was 51%. Most losses of transplants due to wave action occurred during the first 7 months when 25% were lost, with only a further 5% of colonies being subsequently lost. Within 16 months most colonies had accreted naturally to the concrete mats. Thirty-two percent of transplants, which remained attached, died, with *Acropora hyacinthus* and *Pocillopora verrucosa* Ellis and Solander having the highest mortality rates (approximately 50% mortality over 2 years) and *Porites lobata* Dana and *P. lutea* Milne Edwards and Haime the lowest (2.8% and 8.1% mortality, respectively, over 2 years). Growth rates were quite variable, with a quarter to a third of transplants showing negative growth during each intersurvey period. *Acropora hyacinthus, A. cytherea* Dana, and *A. divaricata* Dana transplants had the highest growth rates (colony mean linear radial extension 4.15–5.81 cm per year, followed by *Pocillopora verrucosa* (mean 2.51 cm per year. Faviids and poritids had the lowest growth rates. *Favia* and *Favites* showed the poorest response to transplantation, while *Acropora divaricata*, which combined a high growth rate with relatively low mortality, appeared particularly amenable to transplantation. The following was concluded: (1) species transplanted should be selected with care, as certain species are significantly more amenable than others to transplantation; (2) the choice of whether fragments or whole colonies are transplanted may profoundly influence survival; (3) considerable loss of transplants is likely from higher energy sites whatever method of attachment; (4) transplantation should, in general, be undertaken only if recovery after natural recruitment is unlikely.

Edwards and Clark were back on the public scene in 1998 with a paper debating the subject of transplantation: "The primary objectives of coral transplantation are to improve reef 'quality' in terms of live coral cover, biodiversity and topographic complexity (Edwards and Clark 1998). Stated reasons for transplanting corals have been to: (1) accelerate reef recovery after ship groundings; (2) replace corals killed by sewage, thermal effluents or other pollutants; (3) save coral communities or locally rare species threatened by pollution, land reclamation or pier construction; (4) accelerate recovery of reefs after damage by Crown-of-thorns starfish or red tides; (5) aid recovery of reefs following dynamite fishing or coral quarrying; (6) mitigate damage caused by tourists engaged in water-based recreational activities and (7) enhance the attractiveness of underwater habitat in tourism areas. Whether coral transplantation is likely to be effective from a biological standpoint depends on, among other factors, the water quality, exposure, and degree of substrate consolidation of the receiving area. Whether it is necessary (apart from cases related to reason 3 above) depends primarily on whether the receiving area is failing to recruit naturally. The potential positive and negative benefits of coral transplantation are examined in the light of the results of research on both coral transplantation and recruitment with particular reference to a 4.5 year study in the Maldives." Edwards and Clark suggest that in general, unless receiving areas are failing to recruit juvenile corals, natural recovery processes are likely to be sufficient in the medium to long term and that transplantation should be viewed as a tool of last resort. They argue that there has been too much focus on transplanting fast-growing branching corals, which in general naturally recruit well but tend to survive transplantation and relocation relatively poorly, to create short-term increases in live coral cover at the expense of slow-growing massive corals, which generally survive transplantation well but often recruit slowly. In those cases where transplantation is justified, they advocate that a reversed stance, which focuses on

early addition of slowly recruiting massive species to the recovering community, rather than a short-term and sometimes short-lived increase in coral cover, may be more appropriate in many cases.

Edwards and Clark's conclusion could be important for future studies of coral culture: Some believe that culture of massive coral species, for example, *Porites*, is of no importance to coral reef restoration (Borneman and Lowrie 2001)—it may be the opposite case.

Several other studies have been published, all of them with a focal point of using sexual or asexual coral culture for reef restoration. (Shinn 1966, 1976; Maragos 1974; Birkeland et al. 1979; Bouchon et al. 1981; Kojis and Quinn 1981, 2001; Alcala et al. 1982; Auberson 1982; Fucik et al. 1984; Yap and Gomez 1985; Plucer-Rosario and Randall 1987; Harriott and Fisk 1988a,b; Guzman 1991, 1993; Yap et al. 1992; Yates and Carlson 1993; Bowden-Kirby 1997, 2001; Hudson and Goodwin 1997; Muñoz-Chagin 1997; Vago et al. 1997a; Zeevi and Benayahu 1999; van Treeck and Schumacher 1997, 1999; Rinkevich 1995, 2000; Ammar et al. 2000; Becker & Mueller 2001; Gleason et al. 2001; Milon and Dodge 2001; Petersen & Tollrian 2001; Ortiz-Prosper et al. 2001; Sherman et al. 2001; Spieler et al. 2001).

Coral Farming in Situ

A few case stories on existing coral farms in situ deserve mentioning here, although a comprehensive review is available (Delbeek 2001). Possibly one of the first in situ farms targeting the marine ornamental market was initiated in the mid-eighties by American aquarists, namely Solomon coral farm (Paletta 1999). Local villagers were trained to run their own coral farm. Small polyp stony corals are attached with nylon line on calcium plates and left to grow in protected lagoons until ready for sale. Pictures of this farm can be found on the website: http://www.coralfarms.com/.

A coral farm in situ has recently been established in the Philippines (Heeger et al. 1999). This community-based 2-hectare coral farm is located at a small island in the Philippines. The farm was set up in 1997 with the objectives of giving an alternative livelihood for fishers who engage in destructive fishing techniques, to increase coral biomass, to rehabilitate degraded reefs with low-cost technology, and to strengthen the environmental awareness of the community. Thirty families are working in the farm. Two reef sites have been rehabilitated with 6,000 farm-grown coral fragments. Currently, the coral farm has approximately 22,000 fragments ready for deployment.

Research on coral farming in situ has also emerged, for example, the potential for coral farming in African waters (Franklin et al. 1998). The team published the results of two experiments on coral culturing and one study on temporal patterns in the coral recruitment near Zanzibar, Tanzania. The survival and growth of cultured coral fragments of different sizes were compared for several species of corals, and the effects of lesions on coral fragments were examined. Temporal patterns in recruitment were assessed through regular deployment and sampling of terracotta tiles. The results show that within 1 year, fragments as small as 1–2 cm can be reared into coral colonies that can be sold to the aquarium market or used for reef rehabilitation. Lesions had no negative effect on the growth of the fragments, while a larger initial size had a positive impact on the fragment's growth rate. Coral recruitment was seasonal, with two annual peaks in February–April and November–December.

A report on how to establish a soft coral farm in situ, including methods, strategies, and budgets was recently published (Ellis 1999). Methods are simple; cuttings a few centimeters in size are produced with the use of a razor knife. The cuttings are placed on gravel of basalt, for example, and they attach after a couple of weeks. They are ready for sale after a few months. A technical report on the recent advances in lagoon-based farming practices for eight species of commercially valuable hard and soft corals is also available by this group (Ellis and Ellis 2001).

Coral Culture ex Situ

Techniques of asexual reproduction of corals ex situ have been developed for at least the past three decades, among private aquarists (Calfo 2001, Sprung and Delbeek 1997), public aquariums (Carlson 1992), and commercial enterprises (Calfo 2001), and a few scientists have started using these results in their research (Atkinson et al. 1995, Gateño et al. 1998, An-

thony 1999, Marubini and Thake. 1999, Mueller and Becker 1999, Ferrier-Pagés et al. 2000). For reviews of coral culture ex situ, consult Delbeek 2001; Borneman and Lowrie 2001; Rinkevich and Shafir 1998.

Borneman and Lowrie's review (2001) explains in detail current techniques of affixing coral fragments and propagules. It is recommended reading (just as the other two reviews are too). However, some statements from Borneman and Lowrie are seriously questionable, for example, "growth rates of corals in captivity frequently meet or exceed those in the wild." The value of such assertions, in the absence of supportive scientific evidence, is questionable.

Ex Situ Propagation Techniques

More than 150 species of corals can currently be propagated ex situ (Borneman and Lowrie 2001); however, the technique used for each species can vary considerably. This largely depends on the coral's ability to reattach to the new substrate.

Environmental conditions for ex situ coral propagation must closely replicate those found on the reef. These conditions will vary depending on the type of reef, that is, lagoonal or exposed outer reef, and this has a profound effect on the species composition in those areas. These conditions must be understood if propagation is to be successful for each species/genus. Of paramount importance for successful propagation is water chemistry, light, and flow. The necessary ranges of essential parameters are as follows: $NH_3 = 0$, $NO^{-2} < 0.003$, $NO^{-3} < 5$, $PO^{-4} < 0.05$, $Ca > 430$ ppm, alkalinity = 2.86–3.93 dKH, specific gravity = 1.022–1.025, temperature = 24–27°C, pH = 7.9–8.6. Based upon the limits of our experience, these ranges suggest that corals can survive ex situ. It must be noted, however, that corals do not fare well in fluctuating conditions—stability is just as much a key to success. Another point to mention with the above range of essential parameters is the levels of nitrates and phosphates. Both should be kept low to replicate the oligotrophic condition typically found on reefs. Failure to do so can lead to excess growth of unwanted algae. The level of phosphates is far more important than the commonly measured nitrate levels when maintaining hermatypic corals, and levels must be kept as low as possible. The reason is due to phosphates' ability to chemically bind with free calcium ions, which inhibits calcium uptake and therefore results in decreased coral growth.

The tools needed for coral propagation are sharp scissors, wire cutters, a hacksaw, superglue, plastic mesh netting (as typically used to market citrus fruit), elastic bands, two-part epoxy putty, and a piece of rock to attach the cutting.

A few rules apply to propagation: (1) It should be done only on healthy colonies. (2) Cuttings should be placed in an area with good flow, as excess mucus production in response to the stress of propagation can reduce oxygen uptake and increase bacterial infections (Jamie Craggs personal observation).

Octocoral propagation of the genera *Sinularia, Cladiella, Lobophytum, Xenia, Sarcophyta, Nephthea, Capnella,* and *Discosoma/Rhodactis* is quite simple. Branches or pieces of the tissue are cut away by using a sharp pair of scissors. The colony will dramatically reduce in size after being cut. This is a result of the loss of hydrostatic pressure through the open wound.

Branching octocorals of the genera *Sinularia, Cladellia, Capnella, Stereonephthya,* and *Nephthea* are best propagated with the "netting technique" (Jamie Craggs personal observation). With this method of reattachment, branches greater than 1 cm in diameter are removed from the parent colony with a single clean cut. The cutting is then placed with the cut area facing the new substrate. The cutting is then covered by a small piece of netting held in place with elastic bands. The netting will secure the cutting to its new substrate but allows a continuous flow of water over the surface of the coral. This will minimize the loss of coral due to bacterial infections. Care must be taken when covering the coral. If the netting is too loose, the cutting will be swept away with the current; too tight and the cutting cannot re-expand. The speed of reattachment varies depending on the species. *Sinularia, Capnella, Stereonephthya,* and *Nephthea* can attach to their new substrate within a week, but *Cladellia* takes longer. When the cutting is attached, the elastic bands and netting can carefully be removed. Sometimes the coral can grow around the netting and in this instance simply cut away the excess netting without disturbing the coral.

When propagating *Discosomas, Amplexidiscus,* and *Rhodactis,* small pieces of tissues or sections, like pieces from a pie, can be cut from the polyp and reattached using the same method as above. *Corallimorphs* can also be propagated by making a single cut through the oral disc along the central axis. The incision should not cut the polyp in half; it merely needs to score the surface. This cut will stimulate the polyp to complete the division, and the two halves will regrow into completely separate spherical polyps. It should be noted that although the majority of coral cuttings fare best in high water flow after propagation, *Corallimorphs* by their very nature prefer calmer water and lower light intensities. Cuttings are therefore best placed in such conditions.

Octocorals that do not produce high quantities of mucus during propagation can be attached using superglue (cyanoacryle). The cut surface is simply glued to the new substrate. The authors have seen no detrimental effects through the use of superglue, and it soon dissolves in the water. Genera that can be successfully propagated in this way include *Lobophytum* and *Sacrophyta.*

The genera *Xenia* and *Cespitularia* grow via outgrowths of the stolon. An easy way of propagating these species is to place a rock next to the new growth shoot. Within a few days the new growth will attach itself to the piece of rock and the connective tissue can be cut. This process can be rather slow, especially if large numbers of cuttings are needed. For more intensive propagation a bowl can be partly filled with coral chipping (5–10 mm pieces) and submerged into the tank. Individual polyps or small groups of polyps can then be cut and placed into the bowl. The side of the bowl will prevent the cuttings being washed away in the current, and within a few days they will be attached to the substrate. The growth following this can appear slow to start with, but high quantities can be obtained in the end.

The order Scleractinia, or stony corals, represents the most important and abundant hermatypic, reef-building corals. Although survivorship of fragments can vary in situ, because of predation, substrate type (Lirman 2000a, b), and light availability, these factors can be controlled in ex situ propagation. Mortality therefore may be reduced.

In ex situ propagation, fragments are cut away from the parent colony by using wire cutters for branching (ramose) growth forms or, for more robust massive growth forms, a hacksaw. The method of reattachment depends on the speed with which the coral species can lay down new tissue.

Acropora, Montipora, and *Pocillopora* rapidly encrust on the new substrate, and superglue can be used for reattachment. Encrusting growth should appear within 1 to 2 weeks, and at this time the fragment will have naturally bonded to the substrate. When propagating ramose growth forms, the orientation of the cutting can have a profound effect on the speed of the long-term growth of the colony (Jamie Craggs, personal observation) Once the coral has encrusted over the new substrate, vertical growth begins. The time between initial fragmentation and the start of vertical growth varies with species. Around a month is normal. Cuttings that are placed horizontally grow faster, because after the period of encrusting growth, generally more than one vertical growth point appears. With cuttings that are fixed in a vertical position, growth is limited to one growth point; therefore, overall calcium deposition is comparatively reduced.

Among genera or species of corals that do not naturally encrust or encrust slowly, two-part epoxy putty can be used for reattachment. The fragment is cut and the epoxy is used to hold it in place. The putty will go hard underwater, and although some milky residue is released into the water, no detrimental effects have been observed. Genera suitable for this method of reattachment include *Seriatopora, Turbinaria, Caulastrea, Favia, Favites, Merulina, Echinopora,* and *Stylophora.* The Family Tubiporidae can also be reattached by using epoxy. Further studies in this area are available (Sprung and Delbeek 1997; Borneman and Lowrie 2001; Calfo 2001).

Possible Future Trends and Directions

Coral Culture—Exchange of Results

Advanced aquarists and public aquariums have achieved an incredible quantity of results of coral culture ex situ. These results exist mainly as popular unedited anecdotal text, without peer review, with no critical test methods and no replicates, mainly in the popular press or the Internet, or in abstracts from obscure conferences. Other au-

thors have noticed this, too (Brown 1999; Carlson 1999). Apart from the few publications cited in this chapter, nothing refereed can be found, at the time of writing this chapter, using scientific public databases as, for example, BIOSIS, Zoological Records, and the on-line web-of-science Internet service http://wos.isiglobalnet.com/.

This lack of scientific papers, books, and manuals on coral culture ex situ has of course an explanation (possibly several). The advanced aquarist is most often a talented hobbyist, from another professional field than the scientific community, and consequently untrained in the design of experimentation suitable for peer-reviewed journals. Regarding public aquariums, the missing research papers are to a large extent due to the time-consuming tasks of maintenance and management, which it takes to run these popular enterprises. Last but not least, the scientists, who possess the special skills of proper experimental protocols and scientific writing, and might have had offered their expertise to aquarists, for example, seem unaware of the many promising results from the first two groups.

However, when excellent results are not shared with the rest of the world through the peer-reviewed scientific literature, practitioners have to reinvent the wheel. If the past three promising decades of captive coral culture are labeled as phase one, the second upcoming phase of coral culture should concentrate on changing this unwanted current situation.

The following is suggested: (1) Scientists are strongly encouraged to begin communication with the advanced aquarists and public aquarium curators/technicians. (2) A service is needed offering personal guidance and help with experimental design, scientific analysis of results, and scientific writing. This could be arranged through sponsorships or funding a specialist that can help interested parties get their results published in the peer-reviewed literature of life science. (3) Local—as well as international—workshops should be established for aquarists and scientists interested in coral culture. (4) There should be better use of information exchange on the Internet (Brown 1999). The website Breeders Registry (http://www.breeders-registry.gen.ca.us/) accumulates information on the captive rearing of marine ornamentals. This website has been running and developing well the past 10 years, and all investigators of coral culture are recommended to make better use of it.

Experiments in Coral Culture—Style

Experiments in coral culture, which can later be accepted for publication in scientific journals, must follow a protocol called IMRAD (Introduction, Methods, Results, Analysis, and Discussion). It is the protocol the majority of professional researchers in the life sciences pursue. A good example of such a scientific paper (Gateño et al. 1998), which follows IMRAD, makes clear that the differences between such a paper and a paper in a high-profile hobby magazine is less than might be expected. At least in theory, many advanced aquarists (i.e., many more than currently) should be able to produce scientific papers instead of just popular papers (provided in-depth guidance has been acquired—see further in this section). Further guidance on the protocol of IMRAD is available (Zar 1996; Morris 1999; Albert 2000; Day 1989; Oehlert 2000; Dean & Voss 1999; DePoy and Gitlin 1998; Scheiner 1993). Finally, anyone who needs help, whether because of lack of knowledge or lack of time, is encouraged to contact a scientist within the field.

Experiments in Coral Culture—Types

Most studies of coral culture ex situ are currently in a phase, investigating which species can be used for fragmentation (Borneman and Lowrie 2001) and developing more techniques of attaching fragments to substrate. However, other studies such as growth under varying parameters including levels of nutrient and trace elements, water current (power, mode of current, direction of current), types of substrate,

Fig. 16.2 Set-up of the transplantation experiment. Using direct current at appropriately constructed electrodes, mineral compounds from the seawater can be precipitated on a steel matrix. The tentlike installation on the picture, which is seeded with coral fragments, serves as cathode; a titanium anode is spanned over the structure. Coral fragments ("nubbins") used for transplantation were derived from sites with recent mechanical breakage by ship groundings. (Photo: Michael Eisinger; see also color plates.)

lighting levels, and lighting types (10.000 Kelvin and 20.000 Kelvin) are yet to come. Such studies could later be combined with promising new techniques from studies in situ, such as use of electrified chicken mesh as a substrate for coral fragments (van Treeck and Schumacher 1997, 1999). Aspects of this promising new technology may be seen in Figures 16.1 through 16.11, which show the different phases of the experimental setup in shallow water in the Aqaba-bight, Egypt, the northern Red Sea. Electrified chicken mesh seems to enhance growth significantly by an electrochemical process speeding up the rate of calcification in the stony corals. In addition, using this technique, it may be possible to avoid the time-consuming task of attaching fragments to a substrate; with electrified chicken mesh, the fragments seems to attach easily to the mesh within a few weeks.

Aluminium metallic substrate induces colossal biomineralization of the calcareous hydrocoral *Millepora dichotoma* (Vago et al. 2001). What about other species? What about species in captive coral culture? What about combining this technique with all of the above mentioned?

One of the advantages of captive coral culture is the fact that many parameters can be artificially modified as opposed to field studies. However, so far nobody exploits this situation:

Fig. 16.1 Set-up of the transplantation experiment. (Photo: Michael Eisinger; see also color plates.)

Fig. 16.3 Nubbins on a staircase-like installation. After the collected fragments were stored on husbandry trays for adaptation to the new environment, they were transplanted onto the steel matrix by cutting open the grid and inserting them without using additional fixing material. All transplanted nubbins were regularly checked and measured with a caliper in order to assess mortality and axial growth rates. (Photo: Michael Eisinger; see also color plates.)

Fig. 16.4 *Acropora granulosa* nubbin 3 months after transplantation. (Photo: Michael Eisinger; see also color plates.)

for example, it would be interesting to combine super high levels of light with van Treeck's and Schumacher's electrified chicken mesh. Would this give super high growth rates? And why only use lights from the top of a tank: why not from the side? the bottom? from all sides (top, side, bottom, and back)? In combination with electrified chicken mesh or aluminium metallic substrate or both?

Just these few combinations mentioned here combined with the many different species of corals add up to hundreds of experiments just waiting to be conducted.

At the London Aquarium, the results of a recent pilot study of growing coral fragments on a plain fishing line, using *Acropora* spp. were promising (although preliminary) and should be further investigated—not only by London

Aquarium. Small fragments were tied up on a nylon fish line, hanging down from the top of the tank. The fragments attached to the line after less than one month. Such results could be useful for coral culture in situ: it may one day be possible to grow corals on-line large-scale, as has been done for decades for tropical seaweed (Sinha 1991) and in mariculture of mussels (consult Danaq Consults web page:http://www.danaq.dk for more on large-scale techniques of mariculture). Corals on-line may also solve problems with limited coastal space, as it may be possible to grow them offshore, where space is almost unlimited. There may be more than one advantage with this type of coral culture: offshore waters may enhance growth considerably because of clean water and lack of coral predators, for example parrot fishes (Family: *Scaridae*). Culture of corals on-line should furthermore be combined with van Treeck and Schumacher's techniques: why not exchange the nylon line with electrified metal line? Aluminium line? Would this speed up the growth? And this could be combined, at least ex situ, with elevated lev-

Fig. 16.5 *Acropora hemprichii* nubbin 3 months after transplantation. (Photo: Michael Eisinger; see also color plates.)

Fig. 16.7 Various nubbins 3 months after transplantation. (Photo: Helmut Schuhmacher; see also color plates.)

Fig. 16.6 *Millepora dichotoma* nubbin 3 months after transplantation. (Photo: Michael Eisinger; see also color plates.)

Fig. 16.8 Holdfast development. (Photo: Michael Eisinger; see also color plates.)

Fig. 16.9 Holdfast development. (Photo: Michael Eisinger; see also color plates.)

Fig. 16.10 Holdfast development. Almost all branched coral species (mainly acroporids) and the fire coral *Millepora dichotoma* exhibit a high regeneration potential manifested by the development of a firm proliferating foothold onto the wire matrix. Here, they were cemented onto the matrix by the electrochemical accretion. (Photo: Michael Eisinger; see also color plates.)

els of light from several directions (side, back, bottom), as well as other parameters.

Investigators should also be aware of studies reporting unusual settlement patterns, (Zeevi and Benayahu 1999), of the gorgonian coral *Acabaria biserialis*. This coral seems to be a highly successful colonizer of artificial substrata, preferring cement as a substrate on jetties in the Red Sea.

Finally, an interesting review on future concepts for domestication of colonial marine ornamentals has also been published (Rinkevich and Shafir 1998). Many excellent ideas for new research types are presented here, for example, cryopreservation of whole embryos of marine invertebrates.

Captive Coral Culture and Developing Countries

The art of coral culture may save a reef one day. Or is this just something many in the Western

Fig. 16.11 Diver checking the cable connections. (Photo: Michael Eisinger; see also color plates.)

world prefer to lure ourselves into believing? Many fishers in developing tropical countries currently depend entirely on the commercial business of collecting live reef animals for the international market. If the majority of marine ornamentals are ultimately farmed through the establishment of large-scale facilities situated in the Western world, what will these fishers then turn to for a living? Will it be something more devastating for the environment than collecting marine ornamentals, such as dynamite fishing for human consumption? Is there anything to turn to? The danger of making these relatively low-income people worse off certainly exists, with devastating effects for the local socioeconomic infrastructure. At least one sustainable solution does exist, however, and is recommended for much more future focus: Examine for instance the case stories of the Solomon coral farm (Paletta 1999) and the Heegers coral farm in the Philippines (Heeger et al. 1999). Both farms are run by local villagers who benefit greatly from their enterprise. Establishment of coral farms in situ in such areas is a strongly endorsed trend. Governmental institutions—as well as nongovernmental organizations (NGOs)—are also encouraged to consider sponsoring such future enterprises.

Conclusion

Studies of coral culture have provided numerous useful results and bring hope for successful aid to damaged reefs, for future research, as well as for sustainable large-scale farming of marine ornamentals. Scientific studies of coral culture in situ are plentiful and accessible. It would be highly advantageous if this were the same situation for studies of coral culture ex situ by the advanced aquarist and the curator/technician in public aquariums. Results from these two groups are most often obscure. It would be greatly beneficial, if these two groups in the future could find ways to share their results with each other, with scientists, and other interested parties through the peer-reviewed literature, in order to avoid many investigators reinventing the wheel. Scientists are therefore strongly encouraged to begin communications with the advanced aquarist and staff from public aquariums, since this lack in communication seems to be part of the current situation. It would then help this field secure better

chances of funding, as well as enhanced prospects of solving exceedingly difficult problems in its study area.

Most results in coral culture ex situ are currently in a phase of investigating which species can be used for fragmentation and developing techniques of attaching fragments to substrate. However, growth studies under varied parameters such as levels of nutrient and trace elements, current (power, mode of current, direction of current), types of substrate, lighting levels, and lighting types, are yet to come. Such studies could be combined with promising recent results from coral culture in situ. Combined with the many different species of corals, it would appear that this field is full of opportunity for experimental biologists.

The authors invite contact from coral culturists and other interested parties, using the electronic contact information provided.

Acknowledgments

We would like to thank Associate Professor Lis Engdahl Nielsen, The August Krogh Institute, University of Copenhagen, Denmark, and Director and Senior Consultant in Danaq Consult Ltd., Bent Højgaard, for useful comments to an early version of this manuscript.

References

Albert, Tom 2000. *Winning the publications game*, 2nd ed. Abingdon: Radcliffe Medical Press.

Alcala, A. C., E.D. Gomes, and L.C. Alcala. 1982. Survival and growth of coral transplants in central Philippines. *Kalikasan, the Philippine Journal of Biology* 11:136–147.

Ammar, Mohammed Shokry A., Ekram M. Amin, Dietmar Gundacker, and Werner E. Mueller. 2000. One rational strategy for restoration of coral reefs: Application of molecular biological tools to select sites for rehabilitation by asexual recruitment. *Marine Pollution Bulletin* 40(7):618–627.

Anthony, Kenneth. R. N. 1999. A tank system for studying benthic aquatic organisms at predictable levels of turbidity and sedimentation: Case study examining coral growth. *Limnology and Oceanography* 44(6):1415–1422.

Atkinson, M. J., Bruce Carlson, and G.L. Crow. 1995. Coral growth in high-nutrient, low-pH seawater: A case study of corals cultured at the Waikiki Aquarium, Honolulu, Hawaii. *Coral Reefs* 14:215–223.

Auberson, B. 1982. Coral Transplantation: An approach to the reestablishment of damaged reefs.

Kalikasan, the Philippine Journal of Biology 11(1):158–172.

Bak, Rolf, P. M., and Stanley R. Criens. 1981. Survival after fragmentation of colonies of *Madracis mirabilis*, *Acropora palmata* and *A. cerviconis* (Scleractinia) and the subsequent impact of a coral disease. *Proceedings of the 4th International Coral Reef Symposium, Manila* 2:221–227.

Becker, Lillian C., and Erich Mueller. 2001. The culture, transplantation and storage of *Montastrea faveolata*, *Acropora cervicornis*, and *Acropora palmata*: What we have learned so far. *Bulletin of Marine Science* 69(2):881–896.

Birkeland, C., R.H. Randall, and G. Grimm. 1979. Three methods of coral transplantation for the purpose of re-establishing a coral community in the thermal effluent area of the Tanguisson power plant. Technical report, no. 60. Guam: University of Guam.

Borneman, Eric H., and Jonathan Lowrie. 2001. Advances in captive husbandry and propagation: An easily utilized reef replenishment means from the private sector. *Bulletin of Marine Science* 69(2):897–913.

Bothwell, A.M. 1982. Fragmentation, a means of asexual reproduction and dispersal in the coral genus Aropora (Scleractinia: Astroconiidae: Acroporidae)—a preliminary report. *Proceedings of the 4th International Coral Reef Symposium, Manila* 2:137–144.

Bouchon, C., J. Jaubert, and Y. Bouchon-Navaro. 1981. Evolution of a semi-artificial reef built by transplanting coral heads. *Tethys* 10(2):173–176.

Bowden-Kerby, Austin. 2001. Low-tech coral reef restoration methods modeled after natural fragmentation processes. *Bulletin of Marine Science* 69(2):915–931.

Bowden-Kerby, Austin. 1997. Coral transplantation in sheltered habitats using fragments and cultured colonies. *Proceedings of the 8th International Coral Reef Symposium, Panama* 2:2063–2068.

Brown, Stanley. 1999. Information exchange and captive propagation. Paper read at the 1st International Conference on Marine Ornamentals, November 16–19, at Waikoloa, Hawaii.

Bruno, John F. 1998. Fragmentation in *Madracis mirabilis* (Duchassaing and Michelotti): How common is size-specific fragment survivorship in corals? *Journal of Experimental Marine Biology and Ecology* 230:169–181.

Calfo, Anthony Rosario. 2001. *Book of coral propagation*, Vol. 1, 1st ed., Monroeville, Penn.: Reading trees.

Carlson, Bruce A. 1992. The potential of Aquarium Propagation of Corals as a Conservation measure. *AAZPA/CAZPA 1992. Annual conference proceedings* 370–377.

Carlson, Bruce A. 1999. Organism response to rapid change: What aquaria tell us about nature. *American Zoologist* 39:44–55.

Chadwick, Nanette E., and Y. Loya. 1990. Regeneration after experimental breakage in the solitary reef coral *Fungia granulosa* Klunziger, 1879. *Journal of Experimental Marine Biology and Ecology* 142:221–234.

Clark, S., and A.J. Edwards. 1995. Coral transplantation as an aid to reef rehabilitation: Evaluation of a case study in the Maldive Islands. *Coral Reefs* 14:201–213.

Cox, Evelyn F. 1992. Fragmentation in the Hawaiian coral *Montipora verrucosa*. *Proceedings of the 7th International Coral Reef Symposium, Guam* 1:513–516.

Day, Robert A. 1989. *How to write and publish a scientific paper*, 3rd ed., Cambridge: Cambridge University Press.

Dean, Angela, and Daniel Voss. 1999. *Design and analysis of experiments*, 1st ed.: New York: Springer-Verlag.

Delbeek, J. Charles. 2001. Coral farming: Past, present and future trends. *Aquarium Sciences and Conservation* 3(1–3):171–181.

DePoy, Elizabeth, and N. Laura Gitlin. 1998. *Introduction to research*, 2nd ed. St. Louis: Mosby.

Edinger, E. N., G.V. Limmon, J. Widjatmoko Jompa, J.M. Heikoop, and M.J. Risk. 2000. Normal coral growth rates on dying reefs: Are coral growth rates good indicators of reef health? *Marine Pollution Bulletin* 40(5):404–425.

Edwards, A. J., and S. Clark. 1998. Coral transplantation: A useful management tool or misguided meddling? *Marine Pollution Bulletin* 37 (8–12):474–487.

Ellis, Simon 1999. *Farming soft corals for the marine trade*, 1st ed., CTSA Publication No. 140, Makapu'u Point, Waimanalo: The Oceanic Institute, Hawaii.

Ellis, Simon, and E. Ellis. 2001. Recent Advances in Lagoon-based Farming Practices for 8 Species of Commercially Valuable Hard & Soft Corals—a Technical Report. CTSA Publication No. 147. Makapu'u Point, Waimanalo: The Oceanic Institute, Hawaii.

Fadlallah, Y.H. 1983. Sexual reproduction, development, and larval biology in scleractinian corals. *Coral Reefs* 2:129–150.

Ferrier-Pagés, C., J.P. Gattuso, S. Dallot, and J. Jaubert. 2000. Effect of nutrient enrichment on growth and photosynthesis of the zooxanthellate coral *Stylophora pistillata*. *Coral Reefs* 19:103–113.

Franklin, Henrik, Christopher A. Muhando, and Ulf Lindahl. 1998. Coral culturing and temporal recruitment patterns in Zanzibar, Tanzania. *Ambio* 27(8):651–655.

Fucik, K. W., T.J. Bright, and K.S. Goodman. 1984. "Measurements of damage, recovery, and rehabilitation of coral reefs exposed to oil." In *Restoration of habitats impacted by oil spills*. J. Cairne and A.L. Buikema (eds.). pp 115–153. Boston: Butterworth.

Gateño, D., Y. Barki, and B. Rinkevich. 1998. Aquarium maintenance of reef octocorals raised from field collected larvae. *Aquarium Sciences and Conservation* 2(4):227–236.

Gleason, Daniel F., Daniel A. Brazeau, and Delicia Munfus. 2001. Can self-fertilizing coral species be used to enhance restoration of Caribbean Reefs? *Bulletin of Marine Science* 69(2):933–944.

Glynn, P.W., N.J. Gassman, C.M. Eakin, J. Cortes, D.B. Smith, and H.M. Guzman. 1991. Reef coral reproduction in the eastern Pacific—Costa-Rica, Panama, and Galapagos-Islands (Ecuador) 1. Pocilloporidae. *Marine Biology* 109(3):355–368.

Glynn, P.W., S.B. Colley, C.M. Eakin, D.B. Smith, J. Cortes, N.J. Gassman, H.M. Guzman, J.B. Delrosario, and J.S. Feingold. 1994. Reef coral reproduction in the eastern pacific—Costa-Rica, Panama, and Galapagos-Islands (Ecuador). 2. Poritidae. *Marine Biology* 118(2):191–208.

Glynn, P.W., S.B. Colley, N.J. Gassman, K. Black, J. Cortes, and J.L. Mate. 1996. Reef coral reproduction in the eastern Pacific: Costa Rica, Panama, and Galapagos Islands (Ecuador). 3. Agariciidae *Pavona gigantea* and *Gardineroseris planulata*. *Marine Biology* 125(3):579–601.

Glynn, P. W., S.B. Colley, J.H. Ting, J.L Mate, and H.M. Guzman. 2000. Reef coral reproduction in the eastern Pacific: Costa Rica, Panama and Galapagos Islands (Ecuador). IV. Agariciidae, recruitment and recovery of *Pavona varians* and *Pavona* sp.a. *Marine Biology* 136(5):785–805.

Gomez, E.D., A.C. Alcala, H.T. Yap, L.C. Alcala, and P.M. Alino. 1985. Growth studies of commercially important scleractinians. *Proceedings of the 5th International Coral Reef Congress, Tahiti* 6:199–204.

Guzmán, Héctor M. 1991. Restoration of coral reefs in Pacific Costa Rica. *Conservation Biology* 5:189–195.

Guzmán, Héctor M. 1993. *Transplanting coral to restore reefs in the eastern Pacific.* 1st ed., New York: Rolex Awards Books.

Guzmán, Héctor M., and Jorge Cortéz. 1989. Growth rates of eight species of scleractinian corals in the eastern pacific (Costa Rica). *Bulletin of Marine Science* 44(3):1186–1194.

Harriott, V.J. 1998. Growth of the staghorn coral *Acropora formosa* at Houtman Abrolhos, Western Australia. *Marine Biology* 132:319–325.

Harriott, V.J., and D.A. Fisk. 1988a. Coral Transplantation as a reef management option. *Proceedings of the 6th International Coral Reef Symposium, Australia* 2:375–378.

Harriott, V.J., and D.A. Fisk. 1988b. *Accelerated regeneration of hard corals: A manual for coral reef users and managers.* Great Barrier Reef Marine Park Authority Technical Memorandum GBRMPA-TM-16. 1st ed., Townsville: Great Barrier Reef Marine Park Authority Press.

Harrison, P. L., and C.C. Wallace 1990. Reproduction, dispersal and recruitment of scleractinian corals. In *Ecosystems of the world*, Vol. 25: Coral Reefs. New York: Elsevier, pp 133–208.

Harrison, P. L., R.C. Babcock, G.D. Bull, J.K. Oliver, C.C. Wallace, and B. L. Willis. 1984. Mass spawning in tropical reef corals. *Science* 223:1186–1189.

Heeger, T., M. Cashman, and F. Sotto. 1999. "Coral farming as alternative livelihood for sustainable natural resource management and coral reef rehabilitation." In *Proceedings of Oceanology International in the Pacific Rim* edited by Spearhead Exhibitions, pp 171–185. New Malden, Surrey: Spearhead Exhibitions Ltd.

Highsmith, R. C. 1982. Reproduction by fragmentation in corals. *Marine Ecology Progress Series* 7:207–226.

Highsmith, R.C., A.C. Riggs, and C.M. D'Antonio. 1980. Survival of hurricane-generated coral fragments and a disturbance model of reef calcification/growth rates. *Oecologia (Berl.)* 46:322–329.

Hudson, J. H., and W.B. Goodwin. 1997. Restoration and growth rate of hurricane pillar coral (*Dendrogyra cylindricus*) in the Key Largo National Marine Sanctuary, Florida. *Proceedings of the 8th International Coral Reef Symposium, Panama* 1:567–570.

Huston, Michael. 1985. Variation in growth rates with depth at Discovery Bay, Jamaica. *Coral Reefs* 4:19–25.

Karlson, Ronald H. 1986. Disturbance, colonial fragmentation, and size dependent life history variation in two coral reef cnidarians. *Marine Ecology Progress Series* 28:245–249.

Karlson, Ronald H., T.P. Hughes, and Susan.R. Karlson. 1996. Density-dependent dynamics of soft coral aggregations: The significance of clonal growth and form. *Ecology* 77(5):1592–1599.

Kobayashi, Atsushi. 1984. Regeneration and regrowth of fragmented colonies of the hermatypic corals *Acropora formosa* and *Acropora nasuta*. *Galaxea* 3:13–23.

Kojis, Barbara L., and Norman J. Quinn. 1981. "Factors to consider when transplanting hermatypic corals to accelerate regeneration of damaged coral reefs." In *Proceedings of the conference on environmental engineering*, edited by Great Barrier Reef Marine Park Authorities, pp 183–189. Townsville: GBRMPA Press.

Kojis, Barbara L., and Norman J. Quinn. 2001. The importance of regional differences in hard coral recruitment rates for determining the need for coral restoration. *Bulletin of Marine Science* 69(2):967–974.

Lindahl, Ulf. 1998. Low-tech rehabilitation of degraded coral reefs through transplantation of staghorn corals. *Ambio* 27(8):645–650.

Lirman, Diego. 2000a. Lesion regeneration in the branching coral *Acropora palmata*: Effects of colonization, colony size, lesion size, and lesion shape. *Marine Ecology Progress Series* 197:209–215.

Lirman, Diego. 2000b. Fragmentation in the branching coral *Acropora palmata* (Lamarck): growth, survivorship, and reproduction of colonies and fragments. *Journal of Experimental Marine Biology and Ecology* 251:41–57.

Maragos, J.E. 1974. *Coral transplantation: A method to create, preserve and manage coral reefs. Sea grant advisory report 74-03-COR-MAR-14.* Honolulu: University of Hawaii.

Marubini, F., and B. Thake. 1999. Bicarbonate addition promotes coral growth. *Limnology and Oceanography* 44(3):716–720.

Milon, J. Walter, and Richard E. Dodge. 2001. Applying habitat equivalency for coral reef damage assessment and restoration. *Bulletin of Marine Science* 69(2):975–988.

Morris, T.R. 1999. *Experimental design and analysis in animal sciences*, 1st ed. Wallingford, New York: CABI Publishing.

Mueller, Erich, and Lillian C. Becker. 1999. The culture, transplantation and storage of *Montastrea faveolata, Acropora cervicornis* and *A. palmata*: What we have learned so far. Paper read at International Conference on Scientific Aspects of Coral Reef Assessment, Monitoring, and Restoration, 14–16 April at Ft. Lauderdale, Florida.

Muñoz-Chagin, R. F. 1997. Coral transplantation program in the Paraiso coral reef, Cozumel Island, Mexico. *Proceedings of the 8th International Coral Reef Symposium, Panama* 1:2075–2078.

Oehlert, Gary. W. 2000. *A first course in design and analysis of experiments*, 1st ed., New York: W.H. Freeman and Company.

Ortiz-Prosper, Antonio L., Austin Bowden-Kerby, Hector Ruiz, Oscar Tirado, Alex Caban, Gerzon Sanchez, and Juan C. Crespo. 2001. Planting small artificial concrete reefs or dead coral heads. *Bulletin of Marine Science* 69(2):1047–1051.2.

Paletta, M. 1999. Coral Farming. *Sea Scope.* 16: pp. 1 and 4.

Petersen, Dirk, and Ralph Tollrian. 2001. Methods to enhance sexual recruitment for restoration of damaged reefs. *Bulletin of Marine Science* 69(2):989–1000.

Plucer-Rosario, Gyongyi, and Richard H. Randall. 1987. Preservation of rare coral species by transplantation and examination of their recruitment and growth. *Bulletin of Marine Science* 41(2):585–593.

Richmond, R.H., and Cynthia .L. Hunter. 1990. Reproduction and recruitment of corals—— Comparison among the Caribbean, the tropical pacific, and the Red Sea. *Marine Ecology Progress Series* 60(1–2):185–203.

Rinkevich, Baruch. 1995. Restoration strategies for coral reefs damaged by recreational activities: The use of sexual and asexual recruits. *Restoration Ecology* 3:241–251.

Rinkevich, Baruch. 2000. Steps towards the evaluation of coral reef restoration by using small branch fragments. *Marine Biology* 136:807–812.

Rinkevich, Baruch, and Shahan Shafir. 1998. Ex situ culture of colonial marine ornamental invertebrates: Concepts for domestication. *Aquarium Sciences and Conservation* 2(4):237–250.

Scheiner, Samuel M. 1993. *Design and analysis of ecological experiments*, 2nd ed., New York: Chapman and Hall.

Sherman, Robin L., David S. Gilliam, and Richard E. Spieler. 2001. Site-dependent differences in artificial reef function: Implications for coral reef restoration. *Bulletin of Marine Science* 69(2):1053–1056.

Shinn, E.A. 1966. Coral growth rate, an environmental indicator. *Journal of Paleontology* 40:233–242.

Shinn, E.A. 1976. Coral reef recovery in Florida and the Persian Gulf. *Environmental Geology* 1:241–254.

Shlesinger, Y., T.L. Goutlet, and Y. Loya. 1998. Reproductive patterns of scleractinian corals in the northern Red Sea. *Marine Biology* 132(4):691–701.

Sinha, V.R.P. 1991. A compendium of aquaculture technologies for developing centers 1st ed., Bombay: Oxford & IBM publishing Co. Press Ltd.

Smith, L.D., and T.P. Hughes. 1999. An experimental assessment of survival, re-attachment and fecundity of coral fragments. *Journal of Experimental Marine Biology and Ecology* 235:147–164.

Spieler, R. E., D.S. Gilliam, and R.L. Sherman. 2001. Artificial substrate and coral reef restoration: What do we need to know to know what we need. *Bulletin of Marine Science* 69(2):1013–1030.

Sprung, Julian, and J. Charles Delbeek. 1997. The Reef Aquarium: A Comprehensive Guide to the Identification and Care of Tropical Marine Invertebrates, Vol. 2. 1st ed., Florida: Ricordea Publishing.

Szmant, Alina M., Margaret W. Miller, and Tom Capo. 2001. Propagation of scleractinian corals from wild-captured gametes: Mass-culture from mass spawning. Paper read at the 2nd International Conference on Marine Ornamentals, 26 November–2 December, Lake Buena Vista, Florida, University of Florida.

Tunnicliffe, V. 1981. Breakage and propagation of the stony *Acropora cerviconis. Proceedings of the National Academy of Science USA* 76:2427–2431.

Vago, R., E. Vago, Y. Achituv, M. Benzion, and Z. Dubinsky. 1994. A nondestructive method for monitoring coral growth affected by anthropogenic and natural long-term changes. *Bulletin of Marine Science* 55(1):126–132.

Vago, R., Z. Dubinsky, A. Genin, M. Ben-Zion, and Z. Kizner. 1997a. Growth rates of three symbiotic corals in the Red Sea. *Limnology and Oceanography* 42(8):1814–1819.

Vago, R., E. Gill, and J.C. Collingwood. 1997b. Laser measurements of coral growth. *Nature* 386(6620):30–31.

Vago, R., G. Pasternak, and D. Itzhak. 2001. Aluminium metallic substrate induce colossal biomineralization of the calcareous hydrocoral Millepora dichotoma. *Journal of Materials Science Letters* 20:1049–1050.

van Treeck, Peter, and Helmuth Schumacher. 1997. Initial survival of coral nubbins transplanted by a new coral transplantation technology—Options for reef rehabilitation. *Marine Ecology Progress Series* 150:287–292.

van Treeck, Peter, and Helmuth Schumacher. 1999. Artificial reefs created by electrolysis and coral transplantation: An approach ensuring the compatibility of environmental protection and diving tourism. *Estuarine Coastal Shelf Science* 49(Supplement A AUG):75–81.

Wallace, C.C. 1985. Reproduction, recruitment and fragmentation in nine sympatric species of the coral genus *Acropora*. *Marine Biology* 88:217–233.

Wulff, J.L. 1985. Dispersal and survival of fragments of coral sponges. *Proceedings of the 5th International Coral Reef Congress, Tahiti* 5:119–124.

Yap, Helen T., and E.D. Gomez. 1984. Growth of *Acropora pulchra*. *Marine Biology* 81:209–215.

Yap, Helen T., and E.D. Gomez. 1985. Growth of *Acropora pulchra* III. Preliminary observations on the effects of transplantation and sediment on the growth and survival of transplants. *Marine Biology* 87:203–209.

Yap, Helen T., M.P. Aliño, and E.D. Gomez. 1992. Trends in growth and mortality of three coral species (Anthozoa: Scleractinia), including effects of transplantation. *Marine Ecology Progress Series* 83:91–101.

Yates, K.R., and Bruce E. Carlson. 1993. Corals in Aquaria: How to use selective collecting and innovative husbandry to promote coral preservation. *Proceedings of the 7th International Coral Reef Symposium, Guam* 2:1091–1095.

Zar, J.H. 1996. *Biostatistical analysis*. 3rd ed., Englewood Cliffs, N.J.: Prentice-Hall.

Zeevi, D. B-Y., and Y. Benayahu. 1999. The gorgonian coral *Acabaria biserialis*: Life history of a successful colonizer of artificial substrata. *Marine Biology* 135(3):473–481.

PART IV

Reef Fish

A. *Hatchery Methods*

17

Research on Culturing the Early Life Stages of Marine Ornamental Fish

G. Joan Holt

Introduction

The number of marine ornamental species that can be economically produced on commercial farms today is extremely limited. The future of marine ornamental fish farming, like marine food fish culture, depends on the ability to reliably produce eggs, raise large numbers of larvae, and transition them to juveniles. A large number of marine ornamental fish and invertebrate species have been spawned in captivity. Some species spawn naturally in large aquariums, and others have been induced to spawn by photoperiod and temperature cycles (Holt and Riley 2001) or by the use of hormones (Moe 1997). Many more species have been spawned in captivity than have been reared. The early life stages remain the critical bottleneck in the production of most marine ornamentals. Among the priority issues in hatchery technology are designing rearing systems that provide acceptable environmental conditions and identifying suitable prey for different stages during ontogenetic development.

Rearing Systems

Some major concerns in rearing the young are light, space, and water quality. The environmental conditions for raising larvae should mimic their natural planktonic habitat. This means very stable conditions of high salinity and oxygen concentrations, low nutrients, basic pH, light cycles of 12 hours light:12 hours dark, and warm temperatures. Some of these conditions are difficult to maintain once feeding is initiated because dense concentrations of live prey change the quality of the water. For example, high concentrations of algae deplete oxygen during the night and can alter the pH of the system, while large numbers of zooplankton added for food use up oxygen and add metabolites.

Creative designs are needed to maintain high-quality rearing water and at the same time avoid damaging the fragile larvae. Some systems that have overcome many of these problems use microcosms (Henny et al. 1995; Palmtag and Holt 2001), algae scrubbers, or flow-through water, but the last is highly location dependent. The author uses the microcosm approach, placing larvae and their prey (including algae) inside 10- to 18-liter baskets that are in turn placed in larger tanks (300–500 liters) connected to an external filter. Heating and aeration are carried out in the large tank, and water is slowly dripped into the baskets to replace the rearing chamber water several times a day. Mesh on the basket is small enough to retain prey. These types of rearing chambers have been used to rear a large number of fish and shrimp larvae at our lab (Table 17.1).

Feeding and Nutrition

A critical bottleneck continues to occur at first feeding, when larvae change over from internal yolk stores to exogenous feeds. Many ornamental fish (e.g. Chaetidontidae, Cirritidae, Serranidae, Labridae) spawn small pelagic eggs that hatch into small larvae with narrow mouth gapes. Rotifers and brine shrimp (*Artemia* sp) are the most widely used live food items in marine fish culture, but they are not always acceptable food. First-feeding marine larvae (<3 mm standard length) feed on a wide variety of micro-zooplankton including protozoans (tintinids, ciliates, foraminiferans), dinoflagellates, larvae of barnacles and mollusks, and copepod eggs and nauplii (Holt and Holt 2000; Riley and Holt 1993). Diatoms occur in the diets as well,

Table 17.1 Marine ornamental fish and shrimp spawned in captivity at UTMSI, their early life characteristics and reproductive strategy

Closed life cycle	
Species	Characteristics
Lined seahorse *(Hippocampus erectus)*	Egg brooder, hatch as juveniles (g)
Jackknife fish *(Equetus lanceolatus)*	Pelagic eggs (1.0) hatch at 2.7 mm (g)
Cubbyu *(Equatus umbrosus)*	Pelagic eggs (1.2), hatch at 2.8 mm (g)
Comet *(Calloplesiops altivelis)*	Attached eggs, hatch at 2.7 mm (pg)
Fire shrimp *(Lysmata debelius)*	Egg brooder, larval duration 75-158 days (sh)
Peppermint shrimp *(Lysmata wurdemanni)*	Egg brooder, larval duration 30-65 days (sh)
Spawned but life cycle not closed	
Species	Characteristics
Harlequin bass *(Halichores maculipinna)*	Pelagic eggs (0.75) hatch at 2.0 mm (sh)
Longnose hawkfish *(Oxycirrhitus typus)*	Pelagic eggs (0.75) hatch at ~ 2 mm (pg)
Pygmy Angelfish *(Centropyge argi)*	Pelagic eggs (0.73) hatch at 1.2 mm (pg)
Lemonpeel *(Centropyge flavissimus)*	Pelagic eggs (0.71) hatch at 2.3 mm (pg)
Bluehead wrasse *(Thalassoma bifasciatum)*	Pelagic eggs (0.56) hatch at 1.4 mm (pg)
Clown wrasse *(Halichores maculipinna)*	Pelagic eggs (0.59) hatch at 1.5 mm (pg)
Cuban hogfish[a] *(Bodianus pulchellus)*	Pelagic eggs (0.85) hatch at 2.2 mm (pg)
Firefish *(Nemateleotris magnifica)*	Attached eggs, hatch at 2.0 mm (pg)
Scarlet cleaner shrimp[b] *(Lysmata amboinensis)*	Egg brooder, larval duration 180+ days (sh)

Note: Egg diameters at spawning are in millemeters in parentheses; (g), gonochoristic (separate sexes); (pg), protogynous hermaphrodite (female to male sex change); (sh), simultaneous hermaphrodite.

[a] Larvae reared to 21 days.

[b] Larvae reared to 6 months.

ranging from 2% of the total items in larval guts of temperate fish to 5% in tropical reef fish. The size of micro-zooplankton in the guts of small larvae varies from 3 to 100 μm, with the majority smaller than 60 μm. Wild zooplankton has been used to rear marine ornamentals (Danilowicz and Brown 1992), but it does not always provide a consistent quantity of prey items on a regular basis. The author reared *Bodianus pulchellus* on wild zooplankton but was not able to identify the prey consumed by the larvae; predators were often introduced into the rearing chambers (Holt, unpublished). Collecting specific organisms from plankton at times or places where appropriate prey is available has proven successful.

The nutritional requirements for long-chain n-3 highly unsaturated fatty acids (HUFA) for the normal growth and development for marine fish larvae are well established (Sargent et al. 1999; Watanabe 1982). Marine copepods typically have high levels of n-3 HUFA that reflect the fatty acid composition of their diet. These HUFA are transferred to marine larvae through

feeding. There is a fundamental need for new and more nutritious prey species for rearing marine larvae that contain these essential nutrients. Although copepod eggs, nauplii, and copepedites are the natural food of young marine fish larvae, they have not generally been used extensively in aquaculture because they are difficult to culture on a continuous basis, and natural populations vary in abundance and size distribution. Various species of calanoid and harpticoid copepods have been raised in ponds and under controlled conditions, but it is thought that production is not high enough to sustain marine ornamental larviculture. Techniques have been recently described that reliably produce an intensive culture of a temperate water estuarine calanoid copepod *Gladioferens imparipes* (Payne and Rippingale 2001). They achieved 850 nauplii per liter per day for over 420 days in automated 500-liter culture chambers.

A recent breakthrough was the development of a protocol for producing resting stage or diapause eggs of the copepod *Centropages hamatus* (Marcus and Murray 2001). Initial tests of the eggs as food for marine fish larvae were encouraging. Larval comet *Calloplesiops altivelis* that were fed the copepod nauplii grew larger and survived better than larvae fed rotifers and brine shrimp (Table 17.2). Larvae grown on a mixture of wild zooplankton, rotifers, and brine shrimp were also smaller than the copepod-fed larvae. Copepod nauplii hatched from *Centropages hamatus* diapause eggs hold promise as an alternative live food for tropical ornamental fish. The next step needed is to develop a large-scale production protocol for these copepods, probably in collaboration with industry.

There has been considerable research on the fatty acid requirements of marine larvae, especially in temperate species that are of interest for aquaculture. Emphasis has been on ways to enrich live foods that do not contain adequate fatty acid profiles. It has been suggested that marine larvae require approximately 10% (of diet dry weight) as n-3 HUFA-rich phospholipids in their food, with optimum dietary levels of certain HUFA: docosahexaenoic acid (DHA) to eicosapentaenoic acid (EPA) ratios of 2:1 and DHA to aracadonic acid ratios of 10:1 (Sargent et al. 1999). Since rotifers and brine shrimp do not have this pattern of fatty acids, they can be enriched to increase the concentration of HUFA, but it is almost impossible to reach the desired HUFA ratios in brine shrimp. Many species of seahorses cannot be raised successfully on a diet of enriched brine shrimp unless they are fed copepods for the first few days of feeding (Payne and Rippengale 2000; Gardner 2001; Kucera and Holt, *Hippocampus erectus*, unpublished).

The ultimate goal for larviculture would be to develop artificial, inert diets as feed to replace live zooplankton. Several species of marine fish larvae have been successfully weaned to microdiets by cofeeding them with live prey for some length of time. Red drum *Sciaenops ocellatus* larvae have been successfully raised with only an inert diet and the algae *Isocrysis galabana* from first feeding (Lazo et al. 2000). Microdiets can be designed to provide all the nutritional requirements and energy needs of the developing larvae. So far this concept has seldom been applied to ornamental species. Two species of reef sciaenids, cubbyu (*Pareques umbrosus*) and jackknife fish (*Equetus lanceolatus*) have been weaned onto microdiets at 2 weeks by cofeeding artificial diets with ro-

Table 17.2 Standard length and standard deviation (in parentheses) of comet *Calloplesiops altivelis* larvae raised on rotifers and brine shrimp (*Artemia* sp.), copepod nauplii, or a mixture of wild zooplankton, rotifers, and brine shrimp

Age	Rotifers + brine shrimp	Copepods	Zooplankton + rotifers + brine shrimp
Day 3	3.5	3.5	3.5
Day 7	3.97 (.177)	4.79 (0.115)	—
Day 14	3.83 (—)[a]	5.05 (0.297)	4.22 (0.469)

[a] Larvae survived to day 14 in only one of the replicates.

tifers and brine shrimp. The major problem encountered in these studies was maintaining high water quality since some of the diet is uneaten and falls to the bottom of the culture chamber. With careful attention to biofiltration and occasional vacuuming and water replacement, recirculating systems can successfully be used to wean larvae. A great deal of research is needed before artificial diets will be of use in marine ornamental culture.

Uses of Cultured Ornamentals

There are many incentives for culturing marine ornamentals. Cultured organisms are positioned to play a more significant role in supplying the marine ornamental trade, but this is not a panacea. Research is needed to increase species diversity and availability and to establish a sound scientific background for culture. Open transfer of data and information would increase successful larviculture. It has been suggested that cultured animals could be useful for stock enhancement to decrease the recovery time of depleted or overfished populations (Zeimann 2001). Culture alternatives are listed in most conservation plans. Culturing fish in tropical regions by native islanders could provide alternative livelihoods as well as increased interest in conserving coral reef resources.

Moreover, culturing fish can provide valuable life history details such as size and age at spawning, measures of fecundity, morphological development, age at first feeding, stage durations, and growth rates. These life history traits are unknown for a large majority of fishes of the world, yet this information is critical for understanding fish population dynamics.

Knowledge of these vital rates is necessary for management of coral reefs and designing refuges and marine reserves. In addition, data on ontogenetic changes in larval feeding and physiological responses to environmental parameters would be invaluable for interpreting the response of coral reefs to human impacts. Thus, increased understanding of early life stages at all levels is crucial to understanding and mitigating anthropogenic impacts. Much of this information is available only through the controlled culture of marine ornamentals.

Acknowledgments

I would like to thank Hugh A. and Angela McAllister for their strong support and encouragement, and The World Wildlife Fund and Texas Sea Grant College program for funding our research on marine ornamental fish and shrimp. Charlotte Kucera, Matt Palmtag, and Cecilia Riley made important contributions to the success of the research presented herein.

References

Danilowicz, Bret S., and Christopher L. Brown. 1992. Rearing methods for two damselfish species: *Dascyllus albisella* (Gill) and *D. aruanus* (L). *Aquaculture* 106:141–149.

Gardner, Todd. 2001. The copepod/artemia tradeoff in the captive culture of *Hippocampus erectus*. Marine Ornamentals Conference, Lake Buena Vista, Florida, November 26– December 1, 2001.

Henny, D.C., G. Joan Holt, and C.M. Riley. 1995. Recirculating-water system for the culture of marine tropical fish larvae. *Progressive Fish-Culturist* 57:219–225.

Holt, G. Joan, and S.A. Holt. 2000. Vertical distribution and the role of physical processes in the feeding dynamics of two larval sciaenids *Sciaenops ocellatus* and *Cynoscion nebulosus*. *Marine Ecology Progress Series* 193:181–190.

Holt, G. Joan, and C.M. Riley. 2001. Laboratory spawning of coral reef fishes: Effects of temperature and photoperiod. Proceedings of the 28th U.S.-Japan Natural Resources Aquaculture Panel: Spawning and Maturation of Aquaculture Species. UJNR Technical Report No. 28:33–38.

Lazo, J.P., M.T. Dinis, G. Joan Holt, C. Faulk, and C.R. Arnold. 2000. Co-feeding microparticulate diets with algae: Toward eliminating the need of zooplankton at first feeding in larval red drum (*Sciaenops ocellatus*). *Aquaculture* 188:339–351.

Marcus, Nancy H., and Margaret Murray. 2001. Copepod diapause eggs: A potential source of nauplii for aquaculture. *Aquaculture* 201:107–115.

Moe, Martin A. 1997. Spawning and rearing the large angelfish *Pomacanthus* sp. *Aquarium Frontiers*. 5/6 14–24.

Palmtag, Matthew R., and G. Joan Holt. 2001. Captive rearing of fire shrimp (*Lysmata debelis*). 2001. Texas Sea Grant College Program research Report TAMU-01-201.

Payne, Michael F., and R.J. Rippingale. 2000. Rearing West Australian seahorse, *Hippocampus subelongatus*, juveniles on copepod nauplii and enriched artemia. *Aquaculture* 188:253–361.

Payne, Michael F., and R.J. Rippingale. 2001. Intensive cultivation of the calanoid copepod *Gladioferens imparipes*. *Aquaculture* 201:329–342.

Riley, C.M., and G. Joan Holt. 1993. Gut contents of larval fishes from light trap and plankton net collections at Enmedio Reef near Veracruz, Mexico. *Revista de Biologia Tropical* 41(1):53–57.

Sargent, John, Lesley McEvoy, Alicia Estevez, Gordon Bell, Michael Bell, James Henderson, and Douglas Tocher. 1999. Lipid nutrition of marine fish during early development: Current status and future directions. *Aquaculture* 170:217–229.

Watanabe T. 1982. Lipid nutrition in fish. *Comparative Biochemistry and Physiology* 73B: 3–15.

Ziemann, David A. 2001. The potential for the restoration of marine fish populations through hatchery releases. *Aquarium Sciences and Conservation* 3(1):107–117.

18

Out-of-Season Spawning of the Rainbow Shark, *Epalzeorhynchus frenatus:* Freshwater Hatchery Technology with Marine Potential

Christopher L. Brown, Brian Cole, Claudia Farfan, and Clyde S. Tamaru

Abstract

The culture of fish follows a more-or-less predictable sequence of events that can eventually lead to domestication. Aquaculture technology with freshwater ornamental fish is generally quite advanced in comparison with marine species. The examination of spawning and rearing technology in an appropriate freshwater species may provide a sense of direction for current efforts with marine ornamental fish and a forecast of what is to come.

The rainbow shark, *Epalzeorhynchus frenatus* (formerly known as *Labeo erythrurus)* is a popular freshwater ornamental fish that is subject to some seasonality in its availability. Like many of the marine ornamentals, the reproductive strategy of this species involves the scattering of a large number of small eggs. The established breeding techniques for this species depend on the selection of broodstock that are in an advanced state of ovarian development, which typically occurs in late summer/early fall in Hawaii. Because the ability to produce fry on demand would help farmers meet year-round market demand, an experiment was designed to determine whether this species could be spawned out of season. One group of fish was transferred to an environmentally controlled room and held under long photoperiod (LD 14:10) and high temperature (20–24°C) beginning in November of 1998. A second group was maintained under ambient light and temperature conditions, as controls. Routine sampling by ovarian cannulation revealed an advance by several months in the onset of ovarian maturation in the experimental group compared with controls. Four of the experimental female fish were subjected to hormonal induction of spawning in late April 1999 that resulted in 95% fertility and a 95% hatch rate. A second spawning induction was carried out on the remaining experimental fish in early May, generating fry in numbers suitable for commercial scale production. The ova of controls were still not ready for spawning at that time and did not show evidence of attainment of the ability to spawn until August, as observed previously in Hawaii. It was concluded that environmental manipulation may be a useful tool for the maturation of broodstock of this and possibly other species of ornamental fish. It is likely that these environmental and endocrine manipulations will be of value in the development of hatchery technology for ornamental marine species.

Introduction

The transition from hunter-gatherer to farmer is considered to be one of the definitive elements of civilization, although this transition has proceeded at different rates for different species that humans wish to cultivate. In the case of ornamental fish, the stepwise process leading to domestication involves the acquisition of control over survival, reproduction, rearing of young, and to some extent the genetic manipulation of the cultured species to produce new varieties or varieties better adapted to captivity.

In practical terms, this means that the mastery of feeding and husbandry necessarily precedes captive breeding, with rudimentary hatchery technology eventually followed by year-round production or breeding on demand.

Ultimately, this frequently leads to the application of selective breeding methods and the development of strains that have unique and novel characteristics. As these technical capabilities are accumulated, the reliance on wild specimens gives way to the economically viable production of and market dependence on farmed fish. The last half of the twentieth century saw this shift occurring for much of the freshwater ornamentals trade, although most marine ornamental fish species are at or near the beginning of this sequence. Similarly, the proportion of farmed freshwater fish being consumed for food is higher than the proportion of marine fish; about 80% of carp and 65% of tilapia consumed worldwide are farmed (Naylor et al. 2000). A much lower proportion of edible marine fish are farmed (approximately 10%; Naylor et al. 2000).

The closure of a fish's life cycle, or the initial success in captive breeding, is a benchmark accomplishment that is now being met in numerous marine species that are very much in demand by both the food fish and ornamentals markets. The first reports of success in rearing pygmy angelfish of the genus *Centropyge*, announced in tandem at the Marine Ornamentals 2001 conference, exemplify breeding accomplishments in this area (F. Baensch, Reef Culture Technology, personal communication 2001; C. Laidley, Oceanic Institute, personal communication 2001). Once mastery of spawning and larval rearing are established, attention typically turns to technical refinements leading to culture on a mass scale and culture on demand, as opposed to production only within the constraints of the natural patterns of spawning seasonality.

The purpose of this chapter is to explore technical developments in a representative freshwater species that has been chosen as an appropriate model. Like many of the more challenging marine ornamental culture prospects, spawning and rearing the rainbow shark in captivity requires some effort and attention to detail.

The rainbow shark *Epalzeorhynchus frenatus* (formerly known as *Labeo erythrurus)* is a species appreciated by aquarists for its aesthetically pleasing appearance and generally peaceful behavior in freshwater community aquariums. Although reliable statistics documenting the market share held by any particular species are not available (F. Chapman, University of Florida, personal communication 1998), the rainbow shark and the related red-tail shark, *Labeo bicolor* are both valued commodities in the aquarium trade. Because demand for these moderately priced fish is predictably high, ornamental fish farmers have had an interest in their propagation for years (Axelrod and Vorderwinkler 1995). These freshwater sharks, as they are popularly known (actually cyprinids) commonly display an inability to reach final oocyte maturation under conditions of captivity. There are no known reports of spawning in aquariums without some intervention. Only recently have suitable methods for their mass culture been established, based on the application of exogenous hormones (Low and Wong 1984; Shireman and Gildea 1989).

The injection of a combination of gonadotropins (carp pituitary extract, CPE; or human chorionic gonadotropin, HCG) and gonadotropin releasing hormone (GnRHa) induces final oocyte maturation, producing around 10,000 viable eggs per female (Shireman and Gildea 1989). Spawning is accomplished by hand stripping, and larval rearing is carried out effectively using newly hatched *Artemia* sp. as a starter diet. This sort of spawning-induction treatment is effective only in fish that are already in an advanced state of oogenesis, however, which is subject to the effects of the natural seasonal cycle. At least in part as a result of the limitations imposed by the seasonal reproductive cycle, reliance of the U.S. market on foreign imports of rainbow and red-tail sharks is considerable—both rank among the top 30 species of freshwater ornamental fish imported into the United States (Chapman et al. 1997).

Increasingly, the seasonal limitations of cultured fish are being overcome through manipulations of photoperiod and temperature. In the case of the Atlantic halibut, *Hippoglossus hippoglossus,* experimental shifts in the phase of the photoperiod correlated closely with patterns of circulating hormones, culminating in either an advance or a delay in the spawning season (Bjornsson et al. 1998). Similar results have been obtained for the cod (Norberg et al. manuscript submitted). In each of these studies, the day length was adjusted each day to either compress or expand the annual cyclic changes that occur in natural photoperiod, resulting in either an 8-month or 16-month cycle, compared with controls on a 12-month cycle. A simpler form of photoperiod adjustment relies on the concept that day lengths longer than a critical minimum are required for gonadal development, whereas shorter photoperiods result in gonadal regression (Watson-Whitmyre and Stetson 1988). By simply placing animals on a continuous artificial "long day" photoperiod, such as 14 hours of light and 10 hours of dark, the seasonal constraints on the reproductive system of photoperiodic organisms can be overridden, resulting in activation of the pituitary-gonadal axis. This approach has been studied extensively in rodents, birds, and other vertebrates (see Watson-Whitmyre and Stetson 1988), but the same principles are also widely applied in the commercial aquaculture of fish (Liao 1991). Some crustaceans respond similarly to environmental cues of this sort (Hedgecock 1983), although the endocrine transducers involved are different.

In the following study, the working hypothesis was that the rainbow shark can be induced to spawn at times other than its naturally occurring spawning season, by subjecting a broodstock colony to an artificial photoperiod and temperature regime.

Materials and Methods

In November 1998, a group of 216 young adult *E. frenatus* were moved into an indoor aquarium facility at the Hawaii Institute of Marine Biology laboratory in Kaneohe, Hawaii. These fish were sexually immature, and consequently they were of indeterminate sex. They were distributed into 11 glass aquariums, each 29 gallons in volume (29-gallon "tall" configuration), 18 fish per tank. Water was supplied to the colony through a recirculating system, which was flushed for one hour weekly with dechlorinated freshwater to avoid accumulation of chemical waste materials. Feeding was done ad libitum, three to five times a day, using pelleted commercial diets and a frozen paste made from homogenized beef heart, liver, and spinach. In order to keep the diet as varied as possible,

frozen *Artemia* and frozen bloodworms were fed approximately once a week. Overhead fluorescent lighting was regulated with a timer at 14 hours of light, 10 hours darkness (LD 14:10). Aquarium temperature was maintained between 20°C and 24°C by heating the enclosed environment room with a thermostatically controlled space heater. A similarly sized group of fish, from the same source, was held at the Windward Community College Fish Farm in an outdoor fiberglass tank under ambient lighting and temperature conditions. The natural and artificial day length and temperature conditions are summarized in Figure 18.1. Control fish were fed *ad libitum* on a commercial diet consisting of trout chow.

Ovarian cannulation was begun at the first indication that some fish were showing physical evidence of becoming gravid. Both control and experimental fish were sampled weekly (n = 4) from February onward. Fish were anesthetized with 2-phenoxyethanol (Shreck and Moyle 1990), and a 1-mm diameter plastic tube was inserted into the oviduct. The eggs were examined under a dissecting Nikon microscope to determine their size and the position of the germinal vesicle, as indices of ovarian readiness for spawning.

Upon observation of germinal vesicle migration in a majority of sampled ova (seen in experimental but not control females, April 26,

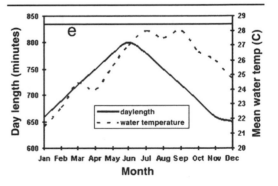

Fig. 18.1 Seasonal cycle of day length and temperature observed at Windward Community College, Oahu, Hawaii. Spawning is normally restricted to August and September, a time in which high temperatures prevail, following a photostimulatory peak in day length. Under our experimental conditions (e), day length remained constant at 840 minutes.

1999), spawning was induced by using a method similar to that used conventionally on Florida fish farms (Shireman and Gildea 1989). A trial spawn involving four males and four females was carried out on April 26, 1999. One week later, a total of approximately 160 experimental fish were induced to spawn, using approximately equal numbers of both sexes. Fish were injected with a combination of carp pituitary extract (CPE; 3 mg per kg fish) and human chorionic gonadotropin (HCG; 60 IU per kg fish) dissolved in distilled water such that the injection volume was 0.1 ml per fish. Female fish were injected with a priming dose consisting of 0.02 ml of the hormone cocktail (20% of total) at 2 a.m. in the early morning before spawning was expected to take place. A resolving dose (the remaining 0.08 ml) was given to females at 8 a.m. Male fish were injected with the entire 0.1 ml dose at 8 a.m. This treatment resulted in ovulation by about 12:45 p.m. the same day, and both eggs and milt were hand-stripped into a glass bowl. The eggs were hatched in a MacDonald jar, and the hatch rate was estimated by determining the percentages of opaque, unhatched eggs and living fry. The resultant fry were supplied to a local ornamental fish farmer for growout.

Results

It has been observed over the past 5 years that rainbow sharks *(E. frenatus)* held under natural photoperiod and temperature conditions in Hawaii become gravid in August and are available for spawning in late August through September. The results with the control fish were consistent with this pattern—ovarian samples collected and examined through the course of the experiment revealed that egg diameter began to increase several months later in control than in experimental fish. At the time spawning was induced in the experimental group (late April), germinal vesicles were still centrally located in the control samples, indicating oocyte immaturity only in the controls. In addition, the control group did not appear to be gravid and in a condition suitable for spawning in May 1999, by which time the only concern in the experimental group was that they may have been overripe. The controls showed signs of germinal vesicle migration in August 1999 and were induced to spawn at that time.

The only mortality among fish in the study was attributed to jumping by eight fish, generally during the night, shortly after they were stocked into the experimental tanks (4% mortality). PVC pipe segments were added to the aquariums for benthic shelter, and each aquarium was covered by a plastic mesh to prevent further losses. By one month after stocking, the fish in the experimental group displayed behavior suggesting that they had become fully adapted to the aquariums under the increased temperature and photoperiod paradigm. The response to the investigators approaching the aquariums changed by the end of the first month indoors, from dashing for cover to rising to the top of the water column in apparent anticipation of feeding. The physical activity lev-

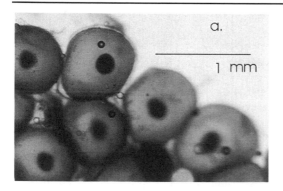 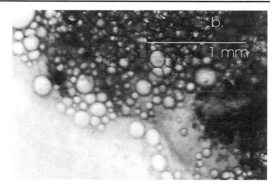

Fig. 18.2 (a) Ovarian biopsy from experimental fish collected on January 19, 1999, less than 2 months after transfer to an indoor facility under conditions of artificial day length and temperature. Oocytes are approximately 0.5 mm in diameter in this illustration. (b) Control ovaries, sampled at the same time as those shown above, uniformly showed no signs of vitellogenesis.

el, appetite, and apparent acceptance of the "paste" diet of the experimental fish increased noticeably, although none of these parameters was quantified.

By January increasing egg diameters in the experimental fish demonstrated that vitellogenesis and egg growth were well underway (Fig. 18.2a). The ovaries of the controls appeared to be quiescent at this time (Fig. 18.2b). Egg diameters increased to a uniform size of 1 mm in the experimental group during March, which coincided with the onset of a distinctly gravid appearance in females. Running milt and microscopic evidence of sperm motility were observed in experimental but not control males examined at this time. During March and through April, the percentage of ova with signs of germinal vesicle migration gradually increased in the experimental females, culminating at the time of spawning when most eggs sampled were clearly mature.

The four female fish from the experimental group used for the initial spawning trial produced an estimated total of 45,000 viable offspring and approximately 2500 infertile eggs (or 95% fertility and hatch rate). Fry fed aggressively and appeared healthy. In the second trial, approximately 80 females produced an estimated 125,000++ viable fry.

Discussion

This study was not designed to exemplify the scientific method; rather, this was a practical demonstration of the application of appropriate technology. The control group was held under normal fish farm conditions, which constituted a set of compounded variables (photoperiod, temperature, and diet); thus, it is not possible to ascribe differences in the reproductive patterns of the two groups to any one treatment. Nevertheless, spawning was accomplished successfully in the fish transferred to a long photoperiod and warm temperature approximately 4 months before the established spawning season, demonstrating that this approach certainly has potential for the production of fry out of season. The excellent survival rate of fry suggests that the experimental conditions did not compromise egg quality. This contrasts with results in the rainbow trout, in which photoperiod-manipulated fish exhibited evidence of reduced rates of fecundity and/or fertility (Duston and Bro-

mage 1988). Further experimentation and refinement of the technique may make year-round availability of domestically produced rainbow shark fry a possibility. It is anticipated that this technology will be applied in the near future for a range of ornamental marine species. The spawning of larger numbers of fish produced a crop of fry with an estimated market value of more than US$30,000 using the current wholesale price of US$0.25 per fish, although the number of surviving fry per female was relatively greater in the trial involving smaller numbers of broodstock fish. There are several possible explanations for the relatively greater production of fry per female in the smaller (first) spawning trial. This difference can be attributed at least in part to a suspicion that the eggs in the second and larger spawning trial may have been overripe. The four females used in the first trial were uniformly large and gravid (approximately 14 g body weight), and the second trial involved a more heterogeneous population; consequently, the hormone doses were more precisely calibrated to body weight in the case of the smaller group of fish. In addition, the differences in these two trials may serve to illustrate the inherent difficulties in extrapolating from small-scale experimentation to production on a commercial farming scale.

The limitations on the degree to which artificial lighting and temperature can be used effectively in this and other species including marine ornamental fish have not been ascertained. It has been shown only that it is possible to advance the onset of reproductive function using artificial conditions. The results suggest that with further experimentation, it would be reasonable to expect to be able to spawn these fish year-round on demand. The experimental design used does not allow the determination of the interactive effects of the altered lighting pattern, the increased temperature, or the nutritional consequences of varied diet and increased appetite under these conditions. Any and probably all of these variables may have affected reproductive status. Available resources did not allow a larger, more complex, or more expensive study.

It seems likely at this time that further refinement of the experimental conditions used to promote reproductive system function might result in more rapid and therefore more cost-effective out-of-season maturation. It is well established that the prior history of animals subjected to artificial conditions can have an effect on their responsiveness to seasonal cues. Our selection of immature fish for the experiment and the relatively slow maturation of ova following the onset of oogenesis in the experimental group suggest a lag that might be circumvented in the future; an earlier response to artificial summer conditions might be possible if established breeders were used. Also, despite indications of a rapid acclimation to experimental conditions, and almost immediate indications of vitellogenesis (see Fig. 18.2), ovarian maturation required several months, and the pace appeared to lag somewhat after the attainment of sufficient egg mass. Some loss of appetite was observed in the last month before spawning within the experimental group. Because maturation occurs under ambient conditions of decreasing day length, at least in Hawaii, it is possible that the prolonged exposure to an artificial LD 14:10 photoperiod may have exerted some untoward side effects on the reproductive system. Some animals become insensitive to artificial photoperiods, or photorefractory, after prolonged exposures (Watson-Whitmyre and Stetson 1988). For these reasons, a recommended next step would be to continue these trials using a long photoperiod to induce egg growth followed by exposure to a declining photoperiod during the latter phases of ovarian maturation.

The relatively less advanced hatchery methods currently available for marine ornamental fish will undoubtedly continue to develop and to gain technology from freshwater ornamental and marine food fish aquaculture. An emergence from a nearly exclusive concentration on the culture of benthic spawners with large eggs and some degree of parental care into the more difficult pelagic-spawning marine ornamentals is already being witnessed. As larval feeding requirements become better understood and technically feasible, it is likely that environmental and endocrine methods of inducing oocyte maturation will become a major focus of attention.

Acknowledgments

The authors wish to thank Bob Kern for the provision of broodstock animals and for assistance with the growout of fry. The help of Rob Carpenter and Michael Haring with the feeding of broodstock is also appreciated. Special thanks are expressed to Mr. Kimo Keohokalole for the generous contribution of his time in instruction and in welding during the setup phase of the project.

Support for this project was contributed by three agencies (1) The United States Department of Agriculture Center for Tropical and Subtropical Aquaculture (CTSA) through a grant from the U.S. Department of Agriculture Cooperative State Research, Education and Extension Service (USDA Grants # 96-38500-2743 and 97-38500-4042). (2) The University of Hawaii Sea Grant Extension Service (SGES) through the National Oceanic and Atmospheric Administration (NOAA) Project # JC9909, which is sponsored by the University of Hawaii Sea Grant College Program, School of Ocean and Earth Science and Technology (SOEST), under institutional grant number NA86RG0041 from NOAA Office of Sea Grant, Department of Commerce, UNIHI-SEAGRANT-TR-99-01. (3) The Aquaculture Development Program, Department of Agriculture, State of Hawaii, as part of the Aquaculture Extension Project with the University of Hawaii Sea Grant Extension Service, contract # 44576.

The views expressed herein are those of the authors and do not necessarily reflect the views of the funding agencies or their subagencies.

References

Axelrod, Herbert R., and Vorderwinkler, William. 1995. *Encyclopedia of Tropical Fish*. TFH Publications, Neptune, N.J. 631 pp.

Bjornsson, BjornThrandur, Olafur Halldorsson, Carl Haux, Birgitta Norberg, and Christopher L. Brown. 1998. Photoperiod control of sexual maturation of the Atlantic halibut (*Hippoglossus hippoglossus*): Plasma thyroid hormone and calcium levels. *Aquaculture* 166:117–140.

Chapman, Frank A., Sharon A. Fitz-Coy, Erik M. Thunberg, and Charles M. Adams. 1997. United States of America trade in ornamental fish. *Journal of the World Aquaculture Society* 28(1):1–10.

Duston, Jim, and Niall Bromage. 1988. The entrainment and gating of the endogenous circannual rhythm of reproduction in the female rainbow trout, Salmo gairdneri. *J. Comp. Physiol.* A 164:259–268.

Hedgecock, Dennis. 1983. Maturation and spawning of the American lobster, *Homaris americanus*. pp 261–270 In McVey, J.P. (ed.): *Handbook of Mariculture, Volume I— Crustacean Aquaculture*. CRC Press, Boca Raton, Forida.

Liao, I Chiu. 1991. Milkfish culture in Taiwan. pp.

91–115. In McVey, J.P. (ed.): *Handbook of Mariculture, Volume II—Finfish Aquaculture.* CRC Press, Boca Raton, Florida.

Low, S.J., and C.C. Wong. 1984. The growth and survival of *Labeo erythrurus* Fowler, in monoculture and in polyculture with *Pterophyllum scalare* (Lichtenstein) respectively, in cage-nets in a pond. *Singapore Journal of Primary Industries* 12(2):110–119.

Naylor, Rosamond L., Rebecca J. Goldburg, Jurgenne H. Primavera, Nils Kautsky, Malcolm C.M. Beveridge, Jason Clay, Carl Folke, Jane Lubechenco, Harold Mooney, and Max Troell. 2000. Effect of aquaculture on world fish supplies. *Nature* 405:1017–1024.

Shireman, Jerome V., and James A. Gildea. 1989. Induced spawning of rainbow sharks (*Labeo erythrurus*) and redtail black sharks (*L. bicolor*). *Progressive fish-culturist* 51:104–108.

Shreck, Carl B., and Peter B. Moyle. 1990. *Methods for fish biology.* American Fisheries Society, Bethesda, Maryland 684 pp.

Watson-Whitmyre, Marcia, and Milton H. Stetson. 1988. Reproductive refractoriness in hamsters: Environmental and endocrine etiologies. In Stetson, Milton H. (ed). Processing of Environmental Information in Vertebrates. Springer Verlag, Berlin. 219–249.

PART **IV**

Reef Fish

B. *Feeding and Nutrition*

19

Advances in the Culture of Rotifers for Use in Rearing Marine Ornamental Fish

Clyde S. Tamaru, Harry Ako, Vernon T. Sato, and Ronald P. Weidenbach

Introduction

Advances in the artificial propagation of marine food fish species have led to the belief that culturing marine ornamental fish can alleviate some of the fishing pressure on wild stocks and also create small- or large-scale industries. Because of the relatively high prices these saltwater aquarium fish command (Chapman et al. 1997), the propagation of marine ornamentals may seem like a lucrative enterprise. In reality, however, the hatchery propagation of many species is stymied because successful rearing protocols have yet to be developed in captive spawning and first feeding of the larvae. A snapshot of the number of fish species in the marine aquarium trade reveals 500–700 fish and 300 species of invertebrates, among which 98% are obtained from the wild (Moe 1999).

The common denominator when examining rearing protocols of artificially propagated marine ornamental species versus marine food fish species is the utilization of the rotifer as an initial food organism (Moe 1999; Tamaru et al. 1995). The same feeding regimen, however, has been found to be inappropriate for the larvae of some marine fish (e.g., gobies, angels, groupers, rabbitfish, surgeon fish), where the major constraint apparently is the small size of the newly hatched larvae and corresponding small mouth gape (Tavarutmaneegul and Lim 1988; Cheah et al. 1994; Tamaru et al. 1995; 2000a; Doi et al. 1997). Recently, research activities in Hawaii have delivered promising results in the larval rearing of the marine angelfish (TenBruggencate 2002), which indicate

that the identification of initial food item(s) that will result in larval survival of species that have been difficult to propagate will soon be forthcoming.

The focus of the current chapter is on the recent developments in the use of rotifers as a live food for the culture of marine fish. It is the intent of the authors to introduce readers to these recent developments that may be incorporated into hatchery operations for marine ornamental fish larvae. It is envisioned that the research activities will ultimately improve the cost-effectiveness of current production practices and lead to the expansion and diversification of marine ornamental culture activities. Information is provided about:

- Types of rotifers available
- Initial rotifer and larval stocking densities
- Manipulating the nutritional value of rotifers
- Alternative feeding strategies for culturing rotifers

Collectively, the results presented are already having a profound impact on hatchery operations and design. Commercial availability of a varied array of enrichment media for live foods, commercial availability of condensed phytoplankton (Maruyama et al. 1997), and the unprecedented improvements in overall rotifer production per unit volume (Fu et al. 1997; Yoshimura et al. 1997) lead to the conclusion that the aquaculture industry will soon re-evaluate the hatchery paradigm.

The marine rotifers of the genus *Brachionus* have been pivotal in the development of hatchery technologies for a varied number of fish (Fukusho 1989; Lubzens et al. 1989). Comparative investigations have revealed that the two main types of rotifers used in aquaculture (L- and S-types) are different species. They have distinguishing characteristics such as size, overall body shape, and shape of their anterior spines, as well as differences in their mate recognition pheromones (Seger 1995; Kotani et al. 1997). What was formerly known as the L-type rotifer (130-340 μm, lorica length) is currently classified as *B. plicatilis* and the S-type rotifer (100-210 μm lorica length) is classified as *B. rotundiformis* (Seger 1995). Of particular interest to marine fish culturists are the various strains of *B. rotundiformis* found in Thailand, Hamana, Fiji and Spain. These rotifer strains have been found to be smaller (90-150 μm, lorica length) in their overall body size (Kotani et al. 1997). There is also a general interest in applying the mass culture technologies developed for marine rotifers for use with the freshwater rotifer species (e.g., *B. calyciflorus*) as a means to expand and diversify the already large freshwater ornamental fish industry (Lim and Wong 1997; Tamaru et al. 2000b).

To demonstrate how diverse rotifers are, three rotifer types, (1) freshwater rotifer *B. calyciflorus*, (2) the S-type rotifer *B. rotundiformis* (Hawaii strain), and (3) SS-type rotifer *B. rotundiformis* (Thailand strain) are shown (Fig. 19.1). Note the differences in overall body shape, anterior spines, and body spines. Of ma-

jor interest to fish culturists is the body size of the rotifer, which largely determines whether a fish larva is capable of utilizing it as an initial food item.

For comparison, the lorica lengths of the three rotifers were measured, and the size frequency distribution of each of the rotifer types is presented (Fig. 19.2). The freshwater variety is significantly larger than the other two rotifer types with an average lorica length of 266 ± 36 μm. The differences between the two strains of *B. rotundiformis*, although statistically significant, are not as clear. Although there is considerable overlap, closer examination of the data indicate that the SS-type strain has a higher number of individuals that measure 100 to 140 μm in size. Because of the smaller size classes, there is an advantage of using the SS-type strain over the S-type strain when the larval mouth gape size is a limiting factor. One constraint with the use of the smaller-body strains is that the smaller-bodied individuals make up only a percentage of the total rotifer culture population. One means to circumvent this constraint is discussed in the following section.

First Feeding

It has long been established that the growth and survival of marine fish larvae are profoundly influenced by the ratio of larvae to their prey items (Houde 1977; Werner and Blaxter 1980). However, a puzzled look is often seen when

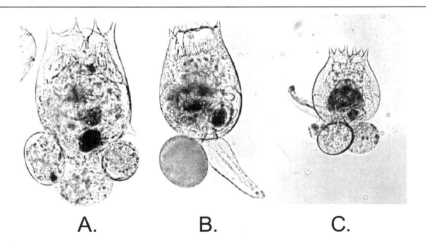

A.   B.   C.

Fig. 19.1 Photomicrographs of three rotifers. A = *B. calyciflorus*, B = *B. rotundiformis* (S-type, Hawaii strain), and C = *B. rotundiformis* (SS-type, Thailand strain).

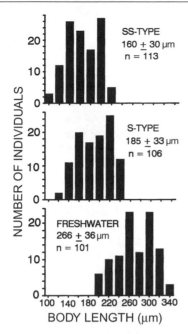

Fig. 19.2 Size frequency distribution of three types of rotifers cultured in Hawaii.

hatchery operators who are experiencing difficulties are asked, "What were the initial rotifer and larval stocking densities?" The reply most often encountered is, "You mean we have to count?" To provide an understanding of how important the relationship of the rotifer and larval densities is, a summary (from Tamaru et al. 1991) is presented utilizing striped mullet (*Mugil cephalus*) larvae. That experiment shows that for fish larvae with a large enough mouth size to take rotifers as an initial food, the hatchery operator can manipulate rotifer/larval ratios in order to influence the fish larvae to feed on the rotifers.

The experiment was designed to test the effects of two rotifer densities and three larval mullet densities on the incidence of first feeding and their growth and survival during the first eight days post-hatching. The rotifer densities tested were 1 and 10 rotifers/ml, whereas larval densities were 25, 50, and 100 larvae per liter. At predetermined intervals, larvae were sacrificed and their gut contents analyzed. The mullet larvae began to feed on the third day post-hatching, and it is apparent that the incidence of first-feeding larvae was clearly affected by the rotifer density (Fig. 19.3). The results indicate that mullet larvae initially strike randomly at

whatever prey swims into their path. This is consistent with the observations made when mullet larvae were given a variety of food organisms from which to choose (Eda et al. 1990). However, the significant increase in the percentage of larvae that were feeding by the fourth day post-hatching suggests that by this time they were actively pursuing their prey. Obviously, providing a "high" density of rotifers at initial feeding ensures a higher percentage of mullet larvae that begin to feed on the rotifers. Although the rotifer densities examined differed by an order of magnitude, there were no significant differences in survival by the eighth day post-hatching. However, when the dry weights of the larvae were determined (Fig. 19.4), treatment effects could be clearly discerned. The dry weights of rotifers from each of the treatments indicated that the most appropriate ratio of rotifer density is 10 rotifers per milliliter for densities between 25 and 50 larvae per liter. The results indicate that there is an optimal stocking density of rotifers in relation to the number of striped mullet larvae that were stocked. It is highly probable that there may be an optimal ratio for each fish species being considered for culture.

As presented earlier, the size frequency distribution of the SS-type rotifer differs from that of the S-type rotifer. Although the smaller size rotifers may make up only a percentage of the rotifer culture population, the results of the experiments with striped mullet larvae indicate

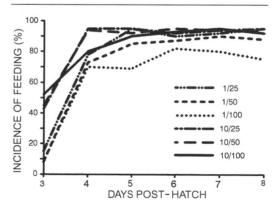

Fig. 19.3 Changes over time in the percentage of mullet larvae feeding during the first 8 days post-hatching, with the larvae stocked at various rotifer/larval ratios. Values are in units of (rotifers/ml)/(larvae/liter).

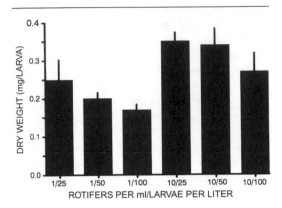

Fig. 19.4 Observed dry weights of larvae obtained from the various rotifer/larvae stocking densities.

that the predator/prey ratios may be manipulated in order to account for the difference in availability of the smaller rotifers. In light of these results, hatchery operators are encouraged to conduct a critical examination of their initial stocking densities, especially for any new species that is targeted for production.

Nutrition

Two live feeds, the rotifer *Brachionus* sp. and the nauplii of the brine shrimp *Artemia* spp., satisfy both the numerical and dimensional requirements of several larval fish species (Lubzens et al. 1989; Sorgeloos et al. 1991). Manipulation of the nutritional quality of both these food organisms to meet the specific requirements of various fish larvae has resulted in major strides in the hatchery production of fish. Enrichment of, or "boosting," the fatty acids of *Artemia* has now become a standard operating procedure in the larval rearing protocols for many fish species (Sorgeloos et al. 1991; Sorgeloos and Legger 1992).

Until recently, the nutritional profile of the rotifer was largely determined by the phytoplankton used as a culture medium (Tamaru et al. 1993a). Phytoplankton species with an elevated highly unsaturated fatty acid (HUFA) content resulted in a correspondingly more nutritious rotifer. Thus, the hatchery manager was limited to culturing a particular species of phytoplankton, and hatchery operations were dependent on continuous production of that particular phytoplankton. Advances in commercial

enrichment media now available to treat rotifers have set the stage for future transformation of hatchery design and operations for the culture of marine fish.

In a classic work (Watanabe et al. 1983), the essential fatty acid docosahexaenoate (C22:6n3) or DHA was shown to be critically important for the survival and growth of a variety of marine fish larvae. The results of our investigations of the enrichment media designed to boost the fatty acid profiles of rotifers are summarized in Table 19.1. In comparison with the algae paste control (*N. oculata*, Reed Mariculture Inc., San Diego, California), it can be seen that the commercial boosting media result in the enhancement of DHA levels in the treated rotifers. Some of the newer enrichment products such as Algamac-2000, 3010, and DO-COSA Gold are based on the heterotrophically grown fungus-like algae, *Schizochytrium* sp. They are well suited for use in high rotifer densities with minimal loss of rotifers during the enrichment process (Fig. 19.5). The benefits of boosting fatty acids in live feeds are not necessarily restricted to larval fish growth and survival. Enriched *Artemia* has been demonstrated to contribute to a marked improvement in the tolerance of marine fish larvae to physical handling (Dhert et al. 1990; Kraul et al. 1993, Ako et al. 1994). The implications of these studies are that the general health and well-being of the marine fish larvae are significantly improved by the feeding of enriched *Artemia*.

It would be a logical assumption that the benefits to fish larvae fed enriched *Artemia* would also be true for early larval forms that were fed enriched rotifers. Until recently, however, the latter was not fully demonstrated. In one study (Tamaru et al. 1998), 7-day-old Pacific threadfin, *Polydactylus sexfilis*, larvae were fed either unenriched rotifers or rotifers enriched with DOCOSA Gold and evaluated after being physically handled. The evaluation consisted of suspending the larvae out of water for 15 sec and then counting the survivors 1 hour later. Because of the age of such young larvae, an "unstressed" control was also included in the testing to rule out the effects of the netting process. The results (Fig. 19.6) show a significantly higher ability to tolerate physical handling in larvae that were fed rotifers that had been enriched. The implications are that the enriched rotifer is nutritionally superior to the un-

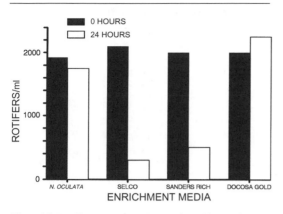

Fig. 19.5 Percent survival of rotifers during the enrichment process of various commercial media.

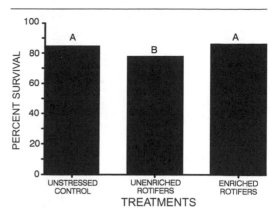

Fig. 19.6 Stress test results of 7-day-old Pacific threadfin larvae fed unenriched and enriched rotifers.

enriched rotifer, that its nutritional value can be modified independently of what it is fed, and that there are consequences to not providing a nutritionally adequate food item to first-feeding fish larvae.

To demonstrate that the nutritional value of the rotifer can be manipulated to meet the requirements of a particular fish species, a fish must be found that has an essential fatty acid requirement that differs from that of marine species that require DHA. It has been reported that the Asian sea bass, *Lates calcalifer*, differs from other marine species in that the essential fatty acid, eicosapentaenoate (C20:5n3), or EPA, is of greater importance during the early larval stages (Dhert et al. 1990; Rimmer and Reed 1994; Rimmer et al. 1994). The results of rearing trials where Asian sea bass larvae were stocked into 85-liter plastic sinks at an average density of 17 larvae per liter have been summarized (Weidenbach et al. 2002). The initial rotifer stocking density was 50 rotifers per milliliter, and two different treatments were used during the rearing trials. One treatment consisted of the larval rearing tanks being provided 5 ml of *N. oculata* paste added directly to the culture water twice daily and with a 20% *Tilapia* greenwater exchange once daily. The second treatment consisted of 5 ml of *N. oculata* paste and 0.3 g DOCOSA Gold per 2 million rotifers per feeding added directly into the larval rearing tanks twice daily with a 20% daily seawater exchange. As can be seen in the fatty acid profiles reported previously (see Table 19.1), the fatty acid profiles of rotifers treated with the al-

ga (*N. oculata*) will have a higher amount of EPA than DHA than that enriched with DOCOSA Gold. In contrast, rotifers provided DOCOSA Gold will have a higher amount of DHA than EPA. The larval-rearing trials were conducted until day 15 post-hatching, when a stress test was conducted on the larvae (Table 19.2). There were no significant differences in survival and stress resistance between the two treatment groups. However, larvae fed rotifers enriched with *N. oculata* were significantly larger in size than those larvae fed rotifers enriched with DOCOSA Gold. The overall results indicate that the nutritional value of the rotifers can be manipulated to meet the requirements of a particular fish species independently of what culture medium they are grown in. These results have far-reaching implications with regard to the future development of culture technologies for rotifers.

Culture Technologies

Methods of culturing rotifers have been well documented (Fulks and Main 1991; Tamaru et al. 1993b) and will not be reviewed here. What will be examined, however, are newer developments that impact the manner in which rotifers are cultured. The successful larval rearing of many marine fish species is dependent on the production of phytoplankton as well as zooplankton. Phytoplankton is a necessary food for the zooplankton, which in turn is used as food for the fish larvae. Phytoplankton production can be a constraint to commercial operations

Table 19.1   Fatty acid profiles of rotifers enriched with various commercial media

Fatty acid	*N. oculata* paste	Super Selco	Sander's Rich	Algamac 2000	Algamac 3010	DOCOSA Gold
C14	0.18	0.12	0.55	0.37	0.79	0.42
C16	0.83	0.70	1.45	1.05	1.76	1.85
C16:1n7	0.52	0.29	0.58	2.22	0.88	0.56
C18	0.02	0.02	0.03	0.31	0.33	0.05
C18:1n9	0.23	0.49	0.66	1.42	0.59	0.92
C18:2n6	0.50	0.86	0.99	9.64	0.09	0.43
C18:3n3	0.17	0.26	0.39	0.02	0.06	0.16
C18:4n3	0.00	0.04	0.46	0.02	0.03	0.01
C20:1n9	0.12	0.21	0.71	0.01	0.13	0.15
C20:4n6	0.25	0.27	0.22	0.12	0.12	0.20
C20:5n3	0.53	0.87	0.93	0.28	0.14	0.28
C22:1n11	0.02	0.09	0.59	nd	0.02	0.03
C22:5n3	0.08	0.30	0.13	nd	nd	0.09
C22:6n3	0.04	0.78	0.98	0.17	0.72	1.28
Total	3.48	5.31	8.66	15.63	5.44	6.42

Note: Results are in terms of mg/100 mg dry weight.

because of the associated space, labor, and energy requirements (Fulks and Main 1991). Phytoplankton culture facilities account for approximately 60% of the space utilized by a hatchery (Sato 1991). Because the nutritional value of the rotifer is no longer questionable, studies are now focused on a re-evaluation of rotifer culture methods (Hagiwara et al. 2001).

One of the first algal substitutes investigated was baker's yeast, which continues to be popular, as it is very easy to use, easy to store, and inexpensive (Fukusho 1989). Two of the inherent challenges associated with the use of baker's yeast are catastrophic culture crashes as a result of unstable water quality conditions and the limited nutritional value to the rotifers (Hagiwara et al. 2001). Another challenge is that the resulting rotifers are nutritionally deficient in many of the essential fatty acids (Tamaru et al. 1993a). Whereas the latter constraint can be addressed, the unstable cultures that result when using baker's yeast as a food for rotifers limit its widespread acceptance and use.

The use of dense algal slurries or pastes in the culture of rotifers can circumvent some of the constraints experienced with yeast, potentially completely replacing all of the facility requirements for producing phytoplankton (Maruyama et al. 1997). Two commercially available products of live or preserved phytoplankton slurries (*Chlorella* V12 and Instant Algae) are currently available from Japan and the

United States, respectively. Each product is undergoing intensive investigation for cost-effectiveness in a commercial setting, as the use of these products can potentially revolutionize hatchery operations. Use of these commercially available algal products has also led to the development of rotifer production units that result in unprecedented rotifer densities, ranging from 5,000 to 10,000 rotifers per milliliter (Fu et al. 1997; Yoshimura et al. 1997). In essence, these products have the potential not only to eliminate the need for phytoplankton production systems but to result in much more efficient rotifer production systems as well. However, the algal pastes do have associated costs and may not be available in a particular region, and hatchery operations will ultimately depend on quality control and dependability of the producers of these algal products. A technical constraint that results from the current ultra-high density rotifer cultures is the constant water quality monitoring and manipulation of dissolved oxygen and pH required to offset high ammonia levels. Thus, substantial initial capital investment in engineering and automation may be required, as the units require a high degree of maintenance and are often computerized.

Another approach to producing rotifers that is also being re-evaluated is the simplest of phytoplankton production technologies, the *Tilapia* "greenwater" system pioneered for the production of *Macrobrachium rosenbergii* post-larvae

Table 19.2   Summary of larval-rearing trials to determine effects of enrichment in Asian sea bass larvae

Trials	Survival (%)	Total length (mm)	Stress test (%)
N. oculata paste + DOCOSA Gold	29.8	7.6 ± 0.5a	100
N. oculata paste + Green Water	10.9	8.6 ± 1.1b	100
N. oculata paste + DOCOSA Gold	41.7	7.1 ± 0.8a	100
N. oculata paste + Green Water	55.7	8.9 ± 0.8b	100

Source: From Weidenbach et al. (2002).

in Hawaii (Corbin et al. 1983). *Tilapia (Oreochromis* spp.) are stocked in tanks (Fig. 19.7) at a density of 25 lb/1000 gallons of brackish water or full seawater. The fish are fed a diet of commercially available pellets ad libitum, and within a week, approximately one third of the volume of the tank can be removed and used for rotifer culture. To increase the target rotifer production, baker's yeast can be used as an algal supplement, and although rotifer production is still dependent somewhat on the use of live algae (e.g., *Tilapia* green water), 80% of the live algae can be substituted with baker's yeast without compromising production (Weidenbach et al. 2002). The inference is that either rotifer production could be increased by using existing algal production facilities, or that the volume of live algae needed to produce the same amount of rotifers could be substantially decreased. In many cases, the smaller working volumes are more appealing for space-limited marine fish hatchery operations, as this would mean a considerable reduction in the amount of floor space needed for the live feed production portion of hatchery operations.

The scenario described was field tested using the Asian sea bass as the target species, and the estimated production costs using the green water plus baker's yeast feeding scenario for producing rotifers are: US$1.45 for yeast plus US$14.00 labor equals US$15.45 per day for 125 million rotifers or approximately US$0.12 per million rotifers. The rotifers produced will also require enrichment, and as reported earlier, if *N. oculata* paste is used for the enrichment medium the amount of paste needed to enrich 125 million rotifers would be 125 ml, which equates to US$7.50 or US$0.06 per million rotifers. Production costs to produce rotifers for use in the rearing of Asian sea bass using the green water plus yeast combination are es-

timated at US$0.18 per million rotifers. If the *N. oculata* algal paste from Reed Mariculture, Inc. is used exclusively, the estimated production costs are US$0.12 per million rotifers. In either case, it is far less than traditional phytoplankton-based rotifer production systems that range from US$0.66 per million to US$4.50 per million and uses only a fraction of the spatial requirements (Fulks and Main 1991; Weidenbach et al. 2002).

In summary, advances that have been outlined in this report clearly place the hatchery manager in an exciting era. Several options and opportunities are now available for expansion and diversification of marine fish production where once there were only a few.

Acknowledgments

This chapter is funded in part by a grant from the National Oceanic and Atmospheric Administration, Project No. R/AQ-63, which is sponsored by the University of Hawaii Sea Grant College Program, SOEST, under Institutional Grant No. NA86RG0041 from NOAA Office of

Fig. 19.7   Culture tanks for the production of *Tilapia* green water at Anuenue Fisheries Research Center, Honolulu, Hawaii.

Sea Grant, Department of Commerce. The views expressed herein are those of the authors and do not necessarily reflect the views of NOAA or any of its subagencies. UNIHI-SEA-GRANT-CR-01-02. Partial support for the activities reported were through a U.S. Department of Agriculture Small Business Innovative Research award, Project #00-33610-8903 and from the State of Hawaii Department of Agriculture Aquaculture Development Program. The authors wish to express their sincere thanks and appreciation to Shelley Alexander, Lena Asano, Dr. Atsushi Hagiwara, Erie Shimizu, Christine Tamaru, Estralita Weidenbach, Jeff Wallace, and Claudia Wallace for their assistance.

References

Ako, H., C.S. Tamaru, P. Bass, and C.-S. Lee. 1994. Enhancing the resistance of physical stress in larvae of *Mugil cephalus* by feeding enriched *Artemia* nauplii. *Aquaculture* 122:81–90.

Chapman, F.A., S.A. Fitz-Coy, E.M. Thunberg, and C.M. Adams. 1997. United States of America trade in ornamental fish. *Journal of the World Aquaculture Society* 28:1–10.

Cheah, S.H., S. Senoo, S.Y. Lam, and K.J. Ang. 1994. Aquaculture of a high-value freshwater fish in [Malaysia]: The marble or sand goby (*Oxyeleotris marmoratus*, Bleeker). NAGA, ICLARM Quarterly, April 1994, pp. 22–25.

Corbin, J.S., M.M. Fujimoto, and T. Y. Iwai. 1983. Feeding practices and nutritional considerations for *Macrobrachium rosenbergii* culture in Hawaii. In J.P. McVey (ed.), CRC *Handbook of Mariculture, Volume I, Crustacean Aquaculture*, CRC Press, Boca Raton, Florida, pp. 391–412.

Dhert, P., P. Lavens, M. Duray, and P. Sorgeloos. 1990. Improved larval survival at metamorphosis of Asian sea bass (*Lates calcarifer*) using omega 3-HHFA-enriched live food. *Aquaculture* 90:63–74.

Doi, M., J.D. Toledo, M.S.N. Golez, M. Dela Santos, and A. Ohno. 1997. Preliminary investigations of feeding performance of larvae of early red-spotted grouper, *Epinephelus conoides*, reared with mixed zooplankton. *Hydrobiologica* 358:259–263.

Eda, H., R. Murashige, Y. Oozeki, A. Hagiwara, B. Eastham, P. Bass, C. S. Tamaru, and C.-S. Lee. 1990. Factors affecting intensive larval rearing of striped mullet, *Mugil cephalus*. *Aquaculture* 91:281–294.

Fu, Y., A. Hada, T. Yamashita, Y. Yoshida, and A. Hino. 1997. Developments of a continuous culture system for the stable mass production of the marine rotifer *Brachionus*. *Hydrobiologia* 358:145–151.

Fukusho, K. 1989. Biology and mass production of the rotifer, *Brachionus plicatilis* II. *Int. J. Aqu. Fish. Technol.* 1:292–299.

Fulks, W., and K.L. Main. 1991. Rotifer and microalgae culture systems. Proceedings of a U.S.-Asia Workshop. Honolulu, Hawaii January 28–31, 1991. The Oceanic Institute, Makapuu Point, Hawaii, 364 pp.

Hagiwara, A., W.G. Gallardo, M. Assavaaree, T. Kotani, and A.B. de Araujo. 2001. Live food production in Japan: Recent progress and future aspects. *Aquaculture* 200:111–127.

Houde, E.D. 1977. Food concentration and stocking density effects on survival and growth of laboratory-reared larvae of bay anchovy *Anchoa mitchilli* and lined sole *Achirus lineatus*. *Marine Biology* 43:333–341.

Kotani, T., A. Hagiwara, and T.W. Snell. 1997. Genetic variation among marine *Brachionus* strains and function of mate recognition pheromone (MRP). *Hydrobiologia* 358:105–112.

Kraul, S., H. Ako, K. Brittain, R. Cantrell, and T. Nagao. 1993. Nutritional factors affecting stress resistance in the larval mahimahi, *Coryphaena hippurus*. *Journal of the World Aquaculture Society* 24:186–193.

Lim, L.C., and C.C. Wong. 1997. Use of the rotifer *B. calyciflorus* Palla, in freshwater ornamental fish larviculture. *Hydrobiologia* 358:269–273.

Lubzens, E., A. Tandler, and G. Minkoff. 1989. Rotifers as food in aquaculture. *Hydrobiologia* 186/187:387–400.

Maruyama, I, T. Nakao, I. Shigeno, Y. Ando, and K. Hirayama. 1997. Application of unicellular algae *Chlorella vulgaris* for the mass-culture of marine rotifer *Brachionus*. *Hydrobiologia* 358:133–138.

Moe, M.A., Jr. 1999. Marine ornamentals: The Industry and the Hobby. In Marine Ornamentals '99 (J. Corbin Conference Proceedings, Waikaloa, Hawaii. November 16–19, 1999, pp. 53–63.

Rimmer, M.A., A.W. Reed, M.S. Levitt, and A.T. Lisle, 1994. Effects of nutritional enhancement of live food organisms on growth and survival of barramundi, *Lates calcarifer* (Bloch), larvae. *Aquaculture and Fisheries Management* 25:143–156.

Sato, V. 1991. The development of a phytoplankton production system as a support base for finfish larval rearing research. In W. Fulks and K.L. Main (eds.), Rotifer and Microalgae Systems, Proceedings of a U.S.-Asia Workshop, Honolulu, Hawaii, 1991. The Oceanic Institute, pp 257–273.

Seger, H. 1995. Nomenclatural consequences of some recent studies on Brachionus plicatilis (Rotifera, Brachionidae). Hydrobiologia 313/314:121–122.

Sorgeloos, P., and P. Legger. 1992. Improved larvae culture outputs of marine fish, shrimp and prawn. *Journal of World Aquaculture Society* 23:251–264.

Sorgeloos, P., P. Lavens, P. Leger, and W. Tackaert. 1991. State of the art in larviculture of fish and shellfish. In: P. Lavens, P. Sorgeloos, E. Jaspers, and F. Ollevier (eds), Larvi '91 —Fish & Crustacean Larviculture Symposium, European Aquaculture Society, Special Publication No. 15, Gent, Belgium, pp. 3–5.

Tamaru, C.S., C.-S. Lee, and H. Ako. 1991. Improving the larval rearing of striped mullet (*Mugil cephalus*) by manipulating quantity and quality of the rotifer, *Brachionus plicatilis*. In W. Fulks and K. Main (eds.), Rotifer and Microalgae Culture Systems, Proc. U.S.-Asia Workshop, The Oceanic Institute, Honolulu, Hawaii, pp. 89–103.

Tamaru, C.S., R. Murashige, C.-S. Lee, H. Ako, and V. Sato. 1993a. Rotifers fed various diets of baker's yeast and/or *Nannochloropsis oculata* and their effect on the growth and survival of striped mullet (*Mugil cephalus*) and milkfish (*Chanos chanos*) larvae. *Aquaculture* 110: 361–372.

Tamaru, C.S., W.J. FitzGerald, and V.S. Sato, 1993b. Hatchery manual for the artificial propagation of striped mullet (*Mugil cephalus*). C. Carlstrom-Trick (ed.), Guam Aquaculture Development and Training Center. 165 pp.

Tamaru, C.S., F. Cholik, J.C.-M. Kuo, and W. FitzGerald. 1995. Status of the culture for striped mullet (*Mugil cephalus*) milkfish (*Chanos chanos*) and grouper (*Epinephelus* sp.). *Reviews in Fisheries Science* 3(3):249–273.

Tamaru, C.S., H. Ako, V. Sato, and S. Alexander. 1998. Status of rotifer production as a food for marine fish culture. *International Aquafeed* Issue 4:17–20.

Tamaru, C.S., D. Abalos, and J. Collier. 2000a. Induced spawning of the Convict Tang, *Acanthurus triostegus sandvicensis*. University of Hawaii Sea Grant College Program, *Makai*. Vol. 22, No. 8.

Tamaru, C.S., H. Ako, and L. Asano. 2000b. Culturing freshwater rotifers for use in the production of freshwater ornamental fishes—Part I. *I'a O Hawai'i*, Vol. 2000, No. 2.

Tavarutmaneegul, P., and C.K. Lim. 1988. Breeding and rearing of sand goby (*Oxyeleotris marmoratus* Blk.) fry. *Aquaculture* 69:299–305.

TenBruggencate, J. 2002. Biologist score breakthrough in raising reef fish. *The Honolulu Advertiser*—Wednesday, January 23, 2002.

Watanabe, T., C. Kitajima, and S. Fujita. 1983. Nutritional values of live food organisms used in Japan for mass propagation of fish: A Review. *Aquaculture* 34:115–143.

Weidenbach, R.P., C.S. Tamaru, and C. Tamaru. 2002. Commercial culture of Asian sea bass *Lates calcarifer*: Phase I. Early larval rearing. Final Project Report. Project No. 00-33610-8903. Submitted to the U.S. Department of Agriculture Small Business Innovative Research Program.

Werner, R.G., and J.H.S. Blaxter. 1980. Growth and survival of larval herring (*Clupea harengus*) in relation to prey density. *Canadian Journal of Fisheries and Aquatic Sciences* 37:1063–1069.

Yoshimura, K., K. Usuki, T. Yoshimatsu, C. Kitajima, and A. Hagiwara. 1997. Recent developments of a high density mass culture system for the rotifer *Brachionus rotundiformis* Tschugunoff. *Hydrobiologia* 358:139–144.

PART IV

Reef Fish

C.   *Seahorses*

20

Factors Affecting Successful Culture of the Seahorse, *Hippocampus abdominalis* Leeson, 1827

Chris M.C. Woods

Introduction

Seahorses are used as an ingredient in traditional medicine (TM), particularly in Southeast Asia, where traditional Chinese medicine (TCM) and its derivatives are practiced (Vincent 1995, 1996; Lourie et al. 1999). In TCM, seahorses are credited with having a role in increasing and balancing vital energy flows within the body, as well as a curative role for such ailments as asthma, high cholesterol, goiter, kidney disorders, and various skin afflictions (Vincent 1995, 1996). In addition to the TCM trade in seahorses, there is also a significant, though smaller, global trade in seahorses as aquarium fish and curios (Vincent 1995, 1996).

Because of the increasing demand for seahorses in the medicinal and aquarium trades and observed decline in some exploited seahorse wild stocks, there is considerable interest in the potential of seahorse aquaculture (Vincent 1996; Hilomen-Garcia 1999). Seahorse aquaculture could be a means of both helping to meet a demand and possibly helping to conserve and protect wild stocks, although there are complex socioeconomic effects to consider with regard to those currently involved in the wild exploitation if aquaculture becomes a mainstream activity.

The commercial culture of seahorses, outside of Southeast Asia, is a relatively new industry the development of which has been hampered by a general lack of published information on successful culture techniques. Currently, industry output is the limiting factor

because there is a high market demand, particularly from the TCM market. Therefore, culture methodologies that can improve production need to be determined.

In 1997 the National Institute of Water and Atmospheric Research (NIWA) began an investigation into the potential for culturing the large-bellied seahorse *Hippocampus abdominalis*.

In the first captive-bred generation, breeding of adults proved to be relatively easy, given sufficient vertical tank height, and growth of juveniles was reasonable (Woods 2000a). However, juvenile mortality in the first few months was high, with an average 10.6% of juveniles per brood surviving to 1 year of age. Therefore, in the second captive-bred generation, simple techniques for improving juvenile survival were tested (Woods 2000b). The results from implementing these simple techniques have allowed survival rates in subsequent captive generations to consistently exceed 80% to 1 year of age.

Research has most recently focused on determining culture techniques to enable the commercial aquaculture of *H. abdominalis* to be both biologically and economically viable. This chapter details some of the findings from this research. Specifically, the following factors on the growth and survival of cultured *H. abdominalis* were tested and the results presented:

1. Effect of water temperature on juvenile growth and survival
2. Effect of stocking density on juvenile growth and survival
3. Use of inert foods with juveniles

Materials and Methods

Effect of Water Temperature

One hundred and twenty cultured F2 *H. abdominalis* (mean ±1 SE length = 76.69 ± 0.54 mm, and mean ±1 SE wet weight = 0.75 ± 0.01 g) were randomly allocated to one of four temperature treatments: 12, 15, 18, and 21°C, with three replicate tanks per treatment, and 10 seahorses per tank. There were no significant differences in either seahorse length or weight between treatments or replicates at the start of the experiment (ANOVA, $P > 0.05$).

Tanks were aerated opaque white 20-liter (40 × cm high × 27 cm diameter) polyethylene flat-bottomed cylinders. Flow-through seawater filtered to 20 μm was heated or chilled to the four test temperatures and delivered to the tanks at a rate of 0.5 liters per minute. Holdfasts for the seahorses to attach to were provided by synthetic aquarium plants. Light intensity at the water surface of each tank was 150 lux, provided by a central 100-watt incandescent bulb. Tanks were checked daily for any mortalities, and any excess waste/feces were siphoned out. Tanks were completely drained and cleaned (chlorine/detergent) every 2 weeks.

Seahorses were fed daily with *Artemia* (mean ± SE length = 2.938 ± 0.092 mm) enriched with Algamac-2000, and 2–6 mm amphipods (*Orchestia chilensis*) divided among two feedings (1000 and 1500 hours), except for the weekend when only one daily feed was conducted. Mean ±1 SE daily feeding rates were 639.2 ± 107.8 *Artemia* per seahorse and approximately 5–10 amphipods per seahorse.

The experiment was concluded after 120 days. Final seahorse length and wet weights were recorded. Mean specific growth rate (SGR) and individual condition factor (CF) were calculated. Mean specific growth rate was calculated as follows: SGR percent increase in body weight per day = $((\ln Wf - \ln Wi)/t) \times 100$, where Wf = final wet weight, Wi = initial wet weight, and t = number of days. Condition factor was calculated as follows: CF = (wet weight (g)/length (cm^{-3})) × 100.

Mean ±1 SE water parameters during the experiment were as follows: dissolved oxygen (7.72 ± 0.06 mg per liter), salinity (34.5 ± 0.02 ppt), and pH (8.14 ± 0.02). There were no differences between the different treatments with regard to these parameters (ANOVA, $P > 0.05$)

Effect of Stocking Density

Three stocking density treatments thought to be feasible for commercial culture of late juvenile *H. abdominalis* were tested: 1, 2, and 5 juveniles per liter, with four replicate 9-liter tanks per stocking density (i.e., 9, 18, and 45 seahorses per tank). Two hundred and eighty-eight captive-bred F2 juvenile *H. abdominalis* (mean ±1 SE length = 71.96 ± 0.29 mm, mean ±1 SE wet weight = 0.51 ± 0.01 g) were randomly allocated to one of the three stocking densities and one of each density's four replicate tanks. There were no significant differences in either seahorse length or weight between treatments or replicates at the start of the experiment (ANOVA, $P > 0.05$).

Tanks used were 9-liter transparent plastic flat-bottomed circular fish bowls surrounded by a dark blue background. Ambient seawater filtered to 20 μm entered each tank down a central inflow line to the bottom of the tank, where it then exited through a 360-degree spray nozzle (Plassay® microjet garden spray). This created a water current out from the central inflow line and across the tank bottom, with a gradual weakening of the current up the tank sides and back down the central inflow line. Water flow through the tanks was approximately 0.25 liter per minute. Individual strands of separated black shade cloth (1 mm diameter) attached to the base of the seawater inflow line provided holdfasts for the seahorses.

A photoperiod of 12L:12D was provided by two 58-watt cool white fluorescent tubes on a timer above the tanks at a water surface intensity of 480 lux. Tanks were inspected daily and excess waste/feces siphoned out. Tanks were completely drained and cleaned every two weeks. Juveniles were fed daily to satiation with *Artemia* (mean ±1 SE length = 1.16 ± 0.04 mm) enriched with DC DHA Super Selco® in three separate feeds (0900, 1200, and 1500 hours), except for the weekend when only one feeding was conducted daily.

Any mortalities were removed, recorded, and replaced with identifiable juveniles of a similar size from a single brood of juveniles whose tails had unusual kinks, in order to maintain the original stocking densities. These replacement seahorses were not included in final analyses.

On five separate days at random times the number of juveniles in each tank and treatment that were being grasped by a conspecific was recorded.

The experiment was concluded after 60 days. Final juvenile length and wet weight were recorded. Mean SGR and individual CF were calculated.

Mean ±1 SE water parameters during the experiment were as follows: temperature (14.88 ± 0.11°C), dissolved oxygen (7.92 ± 0.08 mg-per liter), salinity (34.66 ± 0.03 ppt), and pH (8.11 ± 0.01).

Use of Inert Foods

In all inert food experiments carried out, the same tank setup as used in the stocking density experiment was used. Two inert foods were tested: one frozen food and one artificial food. The frozen food tested was Cyclop-eeze® from Argent Laboratories. Cyclop-eeze® consists of flash-frozen copepods (proximate analysis: protein 60%, lipid 35%—18:3n3 10.45%, 20:5n3 11.74%, 22:6n3 11.09%—ash 3%, carbon hydrate 2%, astaxanthin 2867 ppm, canthaxanthin 15 ppm). Mean ±1 SE length of copepods was 1.05 ± 0.03 mm. The artificial food tested was "Golden Pearls" from Brine Shrimp Direct (proximate analysis: 60% protein, 8% lipids, 15% ash, 8% moisture, vitamin C—2000 ppm, vitamin E—400 ppm, Astaxanthin—500 ppm). Two size grades of Golden Pearls were used: grade I (200–300 μm) and grade II (300–500 μm). This artificial food was chosen because of its similar coloration to newly hatched Artemia nauplii and physical appearance (agglomerations of microcapsules). These characteristics were thought to be important in increasing the attractiveness of the Golden Pearls, as H. abdominalis relies primarily on its keen eyesight for prey capture and does not usually ingest food such as fish flake, that is, food that does not resemble live prey (C. Woods, personal observation).

Each experiment was run for 30 days, during which time the tanks were inspected daily for mortality and any excess food/feces siphoned to waste. Tanks were completely drained and cleaned (chlorine/detergent) each week. After 30 days the surviving seahorses were counted and their lengths and wet weights measured. In all experiments there were no significant differences in either juvenile length or wet weight between treatments or replicates at the start of each experiment (ANOVA, $P > 0.05$).

During each experiment, observations were conducted on feeding of juveniles to determine whether the foods being offered were actually being consumed and at what relative rate. Once a week one individual from each replicate tank was selected at random and observed for a period of 1 minute immediately after food was introduced into its tank. The number of feeding strikes that individual seahorse performed in the 1 minute period was then recorded.

Mean ±1 SE water parameters during the course of the experiments were as follows: water temperature 15.03 ± 0.15°C, pH 8.17 ± 0.01, salinity 33.83 ± 0.03 ppt, dissolved oxygen 7.92 ± 0.08 mg per liter.

a. One-Month-Old Juveniles Fed Cyclop-eeze® One hundred and fifty 1-month-old juveniles (mean ±1 SE length = 33.67 ± 0.09 mm, mean ±1 SE wet weight = 0.05 ± 0.001 g) were tested using Cyclop-eeze® as an alternative diet to Artemia. Juveniles were randomly allocated to one of five co-feeding treatments as follows: (1) Artemia nauplii only (control), (2) Cyclop-eeze® only, (3) both Artemia nauplii and Cyclop-eeze® initially with Cyclop-eeze® only after 5 days, (4) both Artemia nauplii and Cyclop-eeze® initially with Cyclop-eeze® only after 10 days, and (5) both Artemia nauplii and Cyclop-eeze® initially with Cyclop-eeze® only after 20 days. There were three replicates for each treatment.

Seahorses were fed at a rate of 1000 Artemia nauplii per liter and 100% of their body weight of Cyclop-eeze® three times daily (0900, 1200, and 1500 hours) except for the weekend when only one feeding was conducted. Mean ±1 SE length of Super Selco® enriched Artemia used was 0.924 ± 0.022 mm.

Surface illumination in this experiment and all other inert food trials was provided by two 58-watt cool white fluorescent tubes on a timer above the tanks. Light intensity at the water surface was 5.13 μE s^{-1} m^{-2}.

b. One-Month-Old Juveniles Fed Golden Pearls One hundred and fifty 1-month-olds (mean ±1 SE length = 33.68 ± 0.06 mm, mean ±1 SE wet weight = 0.053 ± 0.001 g) were tested using size grade I Golden Pearls as an alternative diet. Juveniles were randomly allocated to one of five co-feeding treatments as in a, but with Golden Pearls as the inert food. There were three replicates for each treatment.

Table 20.1   Mean ±1 SE total length (mm), wet weight (g), condition Factor (CF), and mean specific growth rate (SGR %/day) of juvenile *H. abdominalis* after 4 months at constant water temperatures of 12, 15, 18, and 21°C

	12°C	15°C	18°C	21°C
Initial length (mm)	79.77 ± 1.03	79.73 ± 1.16	80.37 ± 0.92	78.9 ± 1.19
Final length (mm)	107.1 ± 1.44	110.53 ± 1.34	114.93 ± 1.51	114.79 ± 1.25
Initial weight (g)	0.79 ± 0.03	0.72 ± 0.02	0.74 ± 0.02	0.74 ± 0.03
Final weight (g)	3.06 ± 1.52	3.25 ± 0.15	3.44 ± 0.15	3.5 ± 0.14
Initial CF	0.16 ± 0.01	0.15 ± 0.01	0.14 ± 0.01	0.15 ± 0.01
Final CF	0.25 ± 0.01	0.24 ± 0.01	0.23 ± 0.01	0.23 ± 0.01
SGR (%)	1.89 ± 0.09	2.1 ± 0.06	2.25 ± 0.04	2.3 ± 0.03

The feeding rate and regime were the same as in *a*. Mean ±1 SE length of Super Selco® enriched *Artemia* used was 0.912 ± 0.034 mm.

c. Two-Month-Old Juveniles Fed Cyclop-eeze®   Seventy-five 2-month-old juveniles (mean ±1 SE length = 55.43 ± 0.106 mm, mean ±1 SE wet weight = 0.239 ± 0.001 g) were tested using Cyclop-eeze® copepods as an alternative diet to *Artemia*. Juveniles were randomly allocated to one of five co-feeding treatments as used in *a*. There were three replicates for each treatment.

The feeding rate and regime were the same as in *a*. Mean ±1 SE length of Super Selco® enriched *Artemia* used was 1.31 ± 0.04 mm.

d. Two-Month-Old Juveniles Fed Golden Pearls   Seventy-five 2-month-old juveniles (mean ±1 SE length = 55.75 ± 0.13 mm, mean ±1 SE wet weight = 0.24 ± 0.001 g) were tested using size grade II Golden Pearls as an alternative diet to *Artemia*. Juveniles were randomly allocated to one of five co-feeding treatments as used in *a* but with Golden Pearls

as the inert food. There were three replicates for each treatment.

The feeding rate and regime were the same as in *a*. Mean ±1 SE length of Super Selco® enriched *Artemia* used was 1.32 ± 0.05 mm.

Results

Effect of Water Temperature

After 4 months there was a significant difference in seahorse length between the treatments ($F_{3,119}$ = 6.66, $P < 0.05$) (Table 20.1), with juveniles in the 18°C and 21°C treatments longer than those in the 12°C treatment (SNK, $P < 0.05$). There was also a significant difference in juvenile weights between the treatments ($F_{3,119}$ = 7.66, $P < 0.01$), with juveniles in the 18°C and 21°C treatments heavier than those in the 12°C treatment (SNK, $P < 0.05$). In terms of CF, there were no significant differences between the treatments (ANOVA, $P > 0.05$). Comparison of the mean SGR revealed a significant difference between the treatments (Kruskal-Wallis ANOVA, $H_{3,12}$ = 8.69, $P < 0.05$), with juveniles in the 15°C, 18°C,

Table 20.2   Mean ±1 SE total length (mm), wet weight (g), condition factor (CF), and mean specific growth rate (SGR %/day) of juvenile *H. abdominalis* after 2 months at stocking densities of 1, 2, and 5 juveniles/liter

	1/liter	2/liter	5/liter
Initial length (mm)	72.19 ± 0.93	71.94 ± 0.65	71.92 ± 0.35
Final length (mm)	86.42 ± 0.97	81.77 ± 0.78	79.21 ± 0.52
Initial weight (g)	0.49 ± 0.02	0.51 ± 0.02	0.52 ± 0.01
Final weight (g)	1.37 ± 0.09	1.14 ± 0.05	0.88 ± 0.03
Initial CF	0.14 ± 0.01	0.14 ± 0.00	0.14 ± 0.00
Final CF	0.21 ± 0.01	0.2 ± 0.01	0.17 ± 0.00
SGR (%)	1.7 ± 0.12	1.34 ± 0.06	0.87 ± 0.02

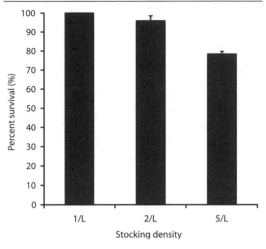

Fig. 20.1 Mean ±1 SE survival of juvenile *H. abdominalis* at stocking densities of 1, 2, and 5 juveniles/liters.

and 21°C treatments having a greater mean SGR than those in the 12°C (SNK, $P < 0.05$).

In terms of survival, there was only a single mortality in the 21°C treatment. This seahorse was accidentally killed during routine tank cleaning.

Effect of Stocking Density

After 2 months there was a significant difference in juvenile lengths between the treatments ($F_{2,245} = 18.45$, $P < 0.001$) (Table 20.2), with juveniles in the 1 per liter treatment longer than those juveniles in the 2 and 5 per liter treatments (SNK, $P < 0.05$). There was also a significant difference in juvenile weights between the treatments ($F_{2,245} = 35.86$, $P < 0.001$), with juveniles in the 1 per liter treatment heavier than those in the 2 and 5 per liter treatments, while juveniles in the 2 per liter treatment were heavier than those in the 5 per liter treatment (SNK, $P < 0.05$). In terms of CF, there was a significant difference between the treatments ($F_{2,245} = 15.68$, $P < 0.01$), with juveniles in the 1 and 2 per liter treatments having higher CFs than the juveniles in the 5 per liter treatment. Comparison of the mean SGR also revealed a significant difference between the treatments (Kruskal Wallis ANOVA, $H_{2,12} = 9.85$, $P < 0.01$), with juveniles in the 1 per liter treatment having a greater mean SGR than those in the 2 and 5 per liter treatments, with juveniles in the

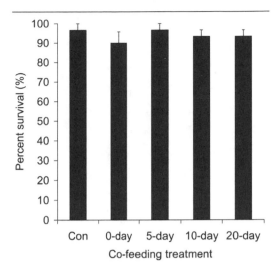

Fig. 20.2 Mean ±1 SE percent survival of juvenile *H. abdominalis* fed either *Artemia* or four different co-feeding treatments with Cyclop-eeze® for 30 days from 1 month old. Key: Con = *Artemia* only, 0-day = Cyclop-eeze® only, 5-day = Weaned onto Cyclop-eeze® at 5 days, 10-day = Weaned on to Cyclop-eeze® at 10 days, and 20-day = Weaned on to Cyclop-eeze® at 20 days.

2 per liter treatment having a greater mean SGR than the 5 per liter treatment (SNK, $P < 0.05$).

There was no significant difference in survival between the 1 and 2 per liter treatments (Fig. 20.1), with mean survival 100% and 95.83 ± 2.66%, respectively, but survival in the 5 per liter treatment at 78.33 ± 3.67% was significantly lower (Kruskal-Wallis ANOVA, $H_{2,12} = 9.12$, $P < 0.05$) (SNK, $P < 0.05$).

The five random observations on the incidence of juvenile wrestling revealed that the mean ±1 SE percentage of seahorses involved in wrestling with a conspecific were as follows: 1 per liter (2.22 ± 1.02%), 2 per liter (4.17 ± 0.68%), and 5 per liter (10.89 ± 0.56%).

Use of Inert Foods

a. One-Month-Old Juveniles Fed Cyclopeeze® After 1 month, there was no significant difference in juvenile survival (Kruskal-Wallis ANOVA, $P > 0.05$) (Fig. 20.2), with mean survival between 90% and 97% across the treatments.

There were significant differences among the treatments in terms of both juvenile length

Table 20.3 Mean ±1 SE length and wet weight of juvenile *H. abdominalis* fed *Artemia* or four different co-feeding treatments with Cyclop-eeze® for 30 days from 1 month old

Treatment length	n	Mean length (mm)	Difference (SNK)
Con	29	55.21 ± 0.35	3,2,4,5
0-day	27	49.81 ± 0.53	5,1
5-day	29	49.41 ± 0.3	5,1
10-day	28	51.11 ± 0.6	1
20-day	28	52.25 ± 0.67	3,2,1

Treatment weight	n	Mean wet weight (g)	Difference (SNK)
Con	29	0.26 ± 0.01	3,4,2
0-day	27	0.2 ± 0.03	1
5-day	29	0.17 ± 0.01	5,1
10-day	28	0.17 ± 0.01	5,1
20-day	28	0.24 ± 0.01	3,4

Note: Newman-Keuls Multiple Comparison Test (alpha = 0.05).

Key: Con = *Artemia* only, 0-day = Cyclop-eeze® only, 5-day = weaned onto Cyclop-eeze® at 5 days, 10-day = weaned onto Cyclop-eeze® at 10 days, and 20-day = weaned onto Cyclop-eeze® at 20 days.

($F_{4,141}$ = 18.24, $P < 0.01$) and wet weight ($F_{4,141}$ = 10.63, $P < 0.01$) (Table 20.3). The control juveniles fed *Artemia* were the largest and heaviest with generally increasing length and weight with the longer co-feeding periods among the other treatments.

Observation of feeding behavior revealed the incidence of feeding strikes to be higher for juveniles feeding on *Artemia* than Cyclop-eeze® (Table 20.4). The co-feeding of *Artemia* with Cyclop-eeze® appeared to slightly increase the incidence of feeding strikes on Cyclop-eeze®.

b. One-Month-Old Juveniles Fed Golden Pearls After 1 month, there was a significant difference in juvenile survival ($H_{4,15}$ = 23.05, $P < 0.01$) (Fig. 20.3), with greater survival in the control treatment fed *Artemia* than juveniles fed Golden Pearls only and juveniles weaned at 5 and 10 days (Bonferroni MCP, $P < 0.05$). Juveniles fed Golden Pearls only had lower survival than the 10- and 20-day co-feeding treatments, and juveniles weaned at 5 days had lower survival than the 20-day co-feeding treatment (Bonferroni MCP, $P < 0.05$).

There were significant differences among the treatments in terms of both juvenile length ($F_{4,109}$ = 11.39, $P < 0.01$) and wet weight ($F_{4,109}$ = 29.26, $P < 0.01$) (Table 20.5). The control juveniles fed *Artemia* were the largest and heaviest, with generally increasing length and weight with the longer weaning periods among the other treatments.

Table 20.4 Mean ±1 SE feeding strike rate (no. strikes/min) of juvenile *H. abdominalis* of different ages offered non-live foods and/or *Artemia* nauplii

Age (months)	Treatment	Food item	Food item
1		*Artemia*	Cyclop-eeze®
	Artemia only	2.61 ± 0.13	—
	Cyclop-eeze® only	—	0.47 ± 0.06
	Artemia + Cyclop-eeze®	2.28 ± 0.12	0.97 ± 0.14
1		*Artemia*	Golden Pearls
	Artemia only	2.58 ± 0.1	—
	Golden Pearls only	—	0.56 ± 0.06
	Artemia + Golden Pearls	2.2 ± 0.09	0.89 ± 0.12
2		*Artemia*	Cyclop-eeze®
	Artemia only	2.5 ± 0.12	—
	Cyclop-eeze® only	—	0.53 ± 0.06
	Artemia + Cyclop-eeze®	2.31 ± 0.07	0.92 ± 0.1
2		*Artemia*	Golden Pearls
	Artemia only	2.53 ± 0.12	—
	Golden Pearls only	—	0.49 ± 0.06
	Artemia + Golden Pearls	2.26 ± 0.07	0.94 ± 0.1

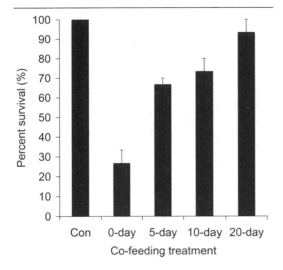

Fig. 20.3. Mean ±1 SE percent survival of juvenile *H. abdominalis* fed either *Artemia* or four different co-feeding treatments with Golden Pearls for 30 days from 1 month old. Key: Con = *Artemia* only, 0-day = Golden Pearls only, 5-day = weaned onto Golden Pearls at 5 days, 10-day = weaned onto Golden Pearls at 10 days, and 20-day = weaned onto Golden Pearls at 20 days.

Table 20.5 Mean ±1 SE length and wet weight of juvenile *H. abdominalis* fed *Artemia* or four different co-feeding treatments with Golden Pearls for 30 days from 1 month old

Treatment length	n	Mean length (mm)	Difference (SNK)
Con	30	60.03 ± 0.57	2,3,4
0-day	8	49 ± 0.23	5,1
5-day	21	51.95 ± 0.94	5,1
10-day	22	52.91 ± 0.54	5,1
20-day	28	59.86 ± 0.7	2,3,4

Treatment weight	n	Mean wet weight (g)	Difference (SNK)
Con	30	0.32 ± 0.01	2,3,4
0-day	8	0.13 ± 0.004	4,5,1
5-day	21	0.16 ± 0.01	5,1
10-day	22	0.2 ± 0.01	2,5,1
20-day	28	0.27 ± 0.01	2,3,4

Note: Newman-Keuls Multiple Comparison Test (alpha = 0.05).

Key: Con = *Artemia* only, 0-day = Golden Pearls only, 5-day = weaned onto Golden Pearls at 5 days, 10-day = weaned on to Golden Pearls at 10 days, and 20-day = weaned on to Golden Pearls at 20 days.

Observation of feeding behavior revealed the incidence of feeding strikes to be higher for juveniles feeding on *Artemia* than Golden Pearls (see Table 20.4). The co-feeding of *Artemia* with Golden Pearls appeared to slightly increase the incidence of feeding strikes on Golden Pearls.

c. Two-Month-Old Juveniles Fed Cyclop-eeze®

After 1 month, there was a significant difference in juvenile survival ($H_{4,15} = 9.09$, $P < 0.01$) (Fig. 20.4), with greater survival in the control treatment fed *Artemia* compared with juveniles fed Cyclop-eeze® only, and juveniles weaned at 5- and 10-day treatments, while juveniles fed Cyclop-eeze® only had lower survival than the 20-day co-feeding treatments (Bonferroni MCP, $P < 0.05$).

There were significant differences among the treatments in terms of both juvenile length ($F_{4,59} = 24.09$, $P < 0.01$) and wet weight ($F_{4,59} = 54.13$, $P < 0.01$) (Table 20.6). The control juveniles fed *Artemia* were the largest and heaviest, with generally increasing length and weight with the longer co-feeding periods among the other treatments.

Observation of feeding behavior revealed the incidence of feeding strikes to be higher for juveniles feeding on *Artemia* than Cyclop-eeze® (see Table 20.4). The co-feeding of *Artemia* with Cyclop-eeze® appeared to slightly increase the incidence of feeding strikes on Cyclop-eeze®.

d. Two-Month-Old Juveniles Fed Golden Pearls

After 1 month, there was no significant difference in juvenile survival (Kruskal-Wallis ANOVA, $P > 0.05$) (Fig. 20.4), with mean survival between 67% and 93% across the treatments.

There were significant differences among the treatments in terms of both juvenile length ($F_{4,58} = 23.31$, $P < 0.01$) and wet weight ($F_{4,58} = 315.17$, $P < 0.01$) (Table 20.7). The control juveniles fed *Artemia* were the largest and heaviest, with generally increasing length and weight with the longer co-feeding periods among the other treatments.

Observation of feeding behavior revealed the incidence of feeding strikes to be higher for juveniles feeding on *Artemia* than Golden Pearls (see Table 20.4). The co-feeding of *Artemia* with Golden Pearls appeared to slightly increase the incidence of feeding strikes on Golden Pearls.

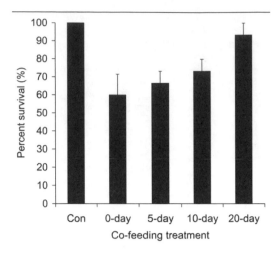

Fig. 20.4 Mean ±1 SE percent survival of juvenile *H. abdominalis* fed either *Artemia* or four different co-feeding treatments with Cyclop-eeze® for 30 days from 2 months old. Key as for Figure 20.2.

Discussion

Effect of Water Temperature

In many fish species water temperature has been observed to be a significant factor in growth and survival (Fracalossi and Lovell 1995; El-Sayed et al. 1996; Fine et al. 1996; Hale 1999). However, the exact effects of water temperature on growth and survival can vary as a result of interactions with other factors such as diet, fish age, fish acclimation, and a range of environmental parameters such as dissolved oxygen levels (Cuenco et al. 1985; Opuszynski and Shireman 1993; Fine et al. 1996).

The temperature range in the natural environment of *H. abdominalis* in New Zealand normally varies between 8°C and 24°C, from the colder southern regions in winter to the warmer northern regions in summer. Within this geographic/seasonal temperature range there is interest from prospective seahorse culturists, so it is critical to determine the effect of water temperature on the growth and survival of *H. abdominalis* in order to determine which culturists might incur water chilling and heating costs.

When tested at temperatures of 12, 15, 18, and 21°C, the growth of juvenile *H. abdominalis* was found to be reduced at 12°C in terms of length and weight increase. Although not statistically significantly different, the growth rates of juveniles at 18°C and 21°C did appear

Table 20.6 Mean ±1 SE length and wet weight of juvenile *H. abdominalis* fed *Artemia* or four different co-feeding treatments with Cyclop-eeze® for 30 days from 2 months old

Treatment length	n	Mean length (mm)	Difference (SNK)
Con	15	67.8 ± 0.44	2,3,4,5
0-day	9	61.78 ± 0.43	5,1
5-day	10	62.2 ± 0.44	5,1
10-day	11	63.73 ± 0.36	1
20-day	14	65.1 ± 0.27	2,3,1
Treatment weight	n	Mean wet weight (g)	Difference (SNK)
Con	15	0.53 ± 0.01	2,3,4,5
0-day	9	0.36 ± 0.01	5,1
5-day	10	0.37 ± 0.01	5,1
10-day	11	0.39 ± 0.01	5,1
20-day	14	0.43 ± 0.004	2,3,4,1

Note: Newman-Keuls Multiple Comparison Test (alpha = 0.05).

Key: As for Table 20.3.

to be slightly better than those at 15°C. There was no difference between 18°C and 21°C, where juvenile growth rates were virtually identical. Therefore, there is no growth advantage to heating water above 18°C (e.g., 21°C), but at the same time there is no growth disadvantage if 21°C is the ambient temperature. Depending on such factors as local electricity costs, water volumes, and heating unit efficiency, there may be an economic growth advantage to heating water from 15°C up to 18°C. However, there is no advantage to chilling water temperature down below 18°C. At NIWA, we now use 18°C as the standard rearing temperature when conducting experiments.

Hippocampus abdominalis can be maintained at temperatures up to 27°C, where some stocks in Australia naturally experience this temperature (Forteath 2000). However, in New Zealand, where the maximum temperature *H. abdominalis* normally encounters is around 24°C, captive *H. abdominalis* acclimated to this maximum temperature survive well when exposed to 24°C, but do appear to slow up in certain areas such as courtship and reproduction (J. Shirley, Paihia Aquarium, personal communication 1999). Therefore, it would be interesting to compare the effects of high temperatures on stocks acclimated to different maximum temperatures.

Table 20.7 Mean ±1 SE length and wet weight of juvenile *H. abdominalis* fed *Artemia* or four different co-feeding treatments with Golden Pearls for 30 days from 2 months old.

Treatment length	n	Mean length (mm)	Difference (SNK)
Con	14	67.79 ± 0.64	2,3,4,5
0-day	10	61.4 ± 0.37	4,5,1
5-day	11	62.91 ± 0.39	5,1
10-day	10	65.23 ± 0.48	2,3,1
20-day	13	65.23 ± 0.41	2,3,1

Treatment weight	n	Mean wet weight (g)	Difference (SNK)
Con	14	0.53 ± 0.01	2,3,4,5
0-day	10	0.36 ± 0.01	3,4,5,1
5-day	11	0.38 ± 0.01	2,4,5,1
10-day	10	0.4 ± 0.004	2,3,5,1
20-day	13	0.42 ± 0.01	2,3,4,1

Note: Newman-Keuls Multiple Comparison Test (alpha = 0.05).

Key: As for Table 20.5.

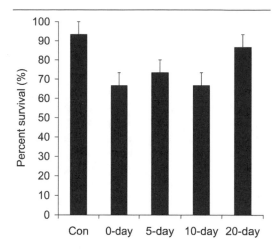

Fig. 20.5 Mean ±1 SE percent survival of juvenile *H. abdominalis* fed either *Artemia* or four different co-feeding treatments with Golden Pearls for 30 days from 2 months old. Key as for Figure

Effect of Stocking Density

In commercial aquaculture, if satisfactory growth, survival, and health of the fish species being farmed are not compromised, then the higher the stocking density, the lower the potential production cost per fish may be. However, optimal stocking densities will vary from species to species (Wallace et al. 1988), with a variety of exogenous factors influencing optimal stocking densities within a given fish species (Berg et al. 1996; Baskerville-Bridges and Kling 2000).

In this investigation, it was found that with increasing stocking density (1, 2, and 5 juveniles per liter), juvenile *H. abdominalis* growth was reduced. Survival was also lower at the highest density. It appears that the lower growth and survival in the higher stocking density treatments are not the result of food competition, or poor water quality, but rather the physical impedance of feeding and energy expenditure as the result of seahorses grasping and wrestling with each other.

Hippocampus abdominalis, like all seahorses, has a muscular prehensile tail with which it attaches itself to points of anchorage. As stocking density increases, the likelihood of anchorage points including conspecifics also increases. When this occurs, the grasped seahorse will contort violently in an effort to free itself, thus expending energy, as well as being prevented from feeding. In extreme cases, where a larger seahorse has grasped a smaller, weaker one around the snout, this can result in near-suffocation and it may take some time before the smaller seahorse can free itself.

Wrestling among seahorses was observed to be particularly pronounced during feeding when juveniles would launch themselves from their resting holdfast into the water column to feed; they would often grasp another seahorse if contact was made, while attempting to feed. At 5 juveniles per liter stocking density, grasping and wrestling was commonly observed during feeding, with reducing incidence of occurrence in the 2 per liter, and then 1 per liter treatments.

It may be possible to commercially use stocking densities greater than 1 per liter in juvenile *H. abdominalis* 70 mm+ in length without affecting growth and survival, by increasing the density of artificial anchorage substratum so that contact between juveniles is reduced. However, this will increase initial material costs and possibly labor costs through the extra cleaning effort. Unless food distribution is even throughout the stocking tanks, increased anchorage substratum may also impede feeding by making it difficult for seahorses to hunt and capture their food, so an optimal balance between stocking density and holdfast material density for the juveniles must be found.

Use of Inert Foods

Most attempts at seahorse culture have relied on cultured live foods such as brine shrimp (*Artemia*), mysid shrimp, and amphipods to feed to seahorses, in addition to collected wild foods (Correa et al. 1989; Forteath 1995; Lockyear et al. 1997; Wilson and Vincent 1998; Hilomen-Garcia 1999; Payne and Rippingale 2000). Culturing the large quantities of live food required by seahorses in commercial culture can prove difficult and costly, however, and harvesting wild live foods depends on an unpredictable resource.

Seahorses have not generally been reared on artificial foods in commercial culture because of the difficulties in getting them to accept such foods. However, there are reports of commercial seahorse culturists using artificial foods to some degree, such as diets based on shrimp and fish meal (Chen 1990; Forteath 2000). Frozen foods, such as frozen mysids, have successfully been used in many instances to feed seahorses (Garrick-Maidment 1997; Wilson and Vincent 1998; Forteath 2000), although the weaning of seahorses onto frozen foods may take time and may be affected by the appearance of the food after the freezing and thawing process. There appear to be species differences among seahorses, between those that can be readily weaned onto inert foods and those in which weaning is quite difficult, but this remains to be rigidly tested.

The ability to use inert foods in the rearing of *H. abdominalis* is vitally important to the development of commercial culture of this species. In this investigation, 1- to 2-month-old juvenile *H. abdominalis* could be weaned onto both frozen and artificial foods on an experimental scale. However, depending on the age of the juvenile seahorses and the type of inert food offered, these weaned juveniles sometimes had lower survival and growth compared with juveniles fed a control diet of enriched *Artemia*.

Inadequacies in the nutritional profile of the inert foods in relation to the dietary requirements of *H. abdominalis* could be a causative factor in the sometimes lower survival and growth of juveniles fed these diets—the optimal nutritional profile for *H. abdominalis* has yet to be determined. However, inadequacies in terms of diet presentation and how "attractive" the inert foods were to the juvenile seahorses are more likely to be the most significant causative factors in reduced juvenile survival and growth.

Although juveniles were physically capable of ingesting the size range of non-live diets offered—and did indeed ingest them as confirmed by observations of feeding—a good proportion of the inert foods remained unconsumed. This is in contrast to the juveniles offered *Artemia,* where there was virtually no wastage. Observations of feeding showed that when juveniles were presented with the inert foods they carefully scrutinized the food before striking.

On introduction of inert food into their tanks juveniles would become alert. The juveniles would then focus both eyes on the inert foods as they passed near them and move their bodies to keep the food at the optimal striking distance from their snouts while scrutinizing it and deciding whether or not to strike. Juveniles would often inspect and reject many individual Cyclop-eeze® copepods and Golden Pearls before actually striking and ingesting one. In contrast, juveniles offered *Artemia* would ingest virtually every nauplius that they visually "inspected."

The head orientation toward individual Cyclop-eeze® copepods and Golden Pearls, focusing both eyes on the inert food but not striking, and then turning away, would seem to indicate that the seahorses did not judge them acceptable to strike at or unrecognizable as food. This may be a result of physical damage to the copepods during freezing and thawing, rendering them "unattractive" or unrecognizable as prey, and in the case of the Golden Pearls simply unrecognizable as prey unless the water currents moved them in a certain manner that might make them seem alive.

Examination of the Cyclop-eeze® copepods under a stereomicroscope revealed that approximately 18% were completely intact upon thawing. Another 38% were reasonably intact (e.g., an antenna or part of the abdomen missing), while the remaining 44% were in a highly fragmented state.

The water currents within the tanks kept the copepods in suspension and moving for at least 4 hours after introduction into the tanks, so there was ample time for them all to be consumed. The water currents did not make the copepods move in a normal copepod manner, which may also have affected their "attractiveness" to the juveniles, although this assumes that the juveniles, which had never before been exposed to copepods, possessed a particular search image for copepods as prey.

Likewise the tank water currents also kept the Golden Pearls in suspension and moving for at least 2 hours after introduction into the tanks, after which time there was a gradual settling of particles around the outer edge of the tank bottoms. This settling effect could be a factor in the lower survival and growth of juvenile *H. abdominalis* exposed to this treatment, although even in 2 hours juvenile seahorses are capable of ingesting a great many prey. Lower feeding strike rates would again seem to indicate that the juveniles judged the majority of Golden Pearls as unacceptable to strike at or unrecognizable as food.

Co-feeding (combined feeding of live and inert diets) is a common strategy to wean larval fish onto inert diets and has been shown to enhance larval growth and survival beyond that achieved by feeding either type of food alone (Drouin et al. 1986; Ehrlich et al. 1989; Kolkovski et al. 1997a, b; Rosenlund et al. 1997; Daniels and Hodson 1999). One study found that the presence of *Artemia* increased dry microdiet assimilation by 30% to 50% in seabass (*Dicentrarchus labrax*) larvae (Kolkovski et al. 1997b).

In this investigation, juvenile seahorses cofed *Artemia* with inert foods generally experienced increased growth and decreased mortality the longer the co-feeding period before abrupt cessation of co-feeding. This general increase of growth and survival could be a simple reflection of longer access to a food item (*Artemia*), which was consumed in greater quantities than the inert food (as indicated by observations on feeding strike rates) and which may have been more nutritious than the non-live foods offered. However, co-feeding inert foods with *Artemia* did appear to increase feeding strike rates on the inert foods compared with feeding strike rates in juveniles exposed only to inert foods, which may also contribute to increasing growth and survival the longer the live food is present along with the inert foods. The fact that 1- to 2-month-old *H. abdominalis* could successfully feed on the inert diets without co-feeding (although usually with lower growth or survival) indicates that although co-feeding is advantageous, weaning can occur without it.

Acknowledgments

Thanks to all at NIWA who assisted at various stages with the project, particularly Phil James, Graeme Moss, Sarah Allen, Bob Hickman, Mike Tait, and Peter Redfearn. Funding was provided by the Foundation for Research, Science and Technology (Contract number CO1X0002).

References

Baskerville-Bridges, B., and L.J. Kling 2000. Larval culture of Atlantic cod (*Gadus morhua*) at high stocking densities. *Aquaculture* 181:61–69.

Berg, A.V., T. Sigholt, A. Seland, and A. Danielsberg. 1996. Effect of stocking density, oxygen level, light regime and swimming velocity on the incidence of sexual maturation in adult Atlantic salmon (*Salmo salar*). *Aquaculture* 143:43–59.

Chen, J. 1990. "Seahorse culture." In Brief introduction to mariculture of five selected species in China, edited by P. Bueno and P. Lovatelli, pp. 1–6. UNDP/FAO Regional seafarming development and demonstration project, Bangkok, Thailand.

Correa, M., K.S. Chung, and R. Manrique. 1989. Cultivo experimental del caballito de mar, *Hippocampus erectus*. *Boletin del Instituto Oceanografico de Venezuela* 28 (1–2):191–196.

Cuenco, M.L., R.R. Stickney, and W.E. Grant. 1985. Fish bioenergetics and growth in aquaculture ponds. II. Effects of interactions among size, temperature, dissolved oxygen, unionized ammonia and food on growth of individual fish. *Ecological modelling* 27(3–4):191–206.

Daniels, H.V., and R.G. Hodson. 1999. Weaning success of southern flounder juveniles: Effect of changeover period and diet type on growth and survival. *North American Journal of Aquaculture* 61:47–50.

Drouin, M.A., R.B. Kidd, and J.D. Hynes. 1986. Intensive culture of lake white fish using *Artemia* and artificial feed. *Aquaculture* 59:107–118.

Ehrlich, K.F., M-C. Cantin, and M.B. Rust. 1989. Growth and survival of larval and postlarval smallmouth bass fed a commercially prepared dry feed and/or *Artemia* nauplii. *Journal of the World Aquaculture Society* 20:1–6.

El-Sayed, A-FM, A. El-Ghobashy, and M. Al-Amoudi. 1996. Effects of pond depth and water temperature on the growth, mortality and body composition of Nile tilapia, *Oreochromis niloticus* (L.). *Aquaculture Research* 27(9):681–687.

Fine, M., D. Zilberg, Z. Cohen, G. Degani, B. Moav, and A. Gertler. A. 1996. The effect of dietary protein level, water temperature and growth hormone administration on growth and metabolism in the common carp (*Cyprinus carpio*). *Comparative Biochemistry and Physiology* 114A(1):35–42.

Forteath, N. 1995. Seahorses, *Hippocampus abdominalis*, in culture. *Austasia Aquaculture* 9(6):83–84.

Forteath, N. 2000. Farmed seahorses—a boon to the aquarium trade. *INFOFISH International* 3:48–50.

Fracalossi, D.M., and R.T. Lovell. 1995. Growth and liver polar fatty acid composition of year-1 channel catfish fed various lipid sources at two water temperatures. *Progressive Fish Culturist* 57(2), 107–113.

Garrick-Maidment, N. 1997. *Seahorses: conservation and care.* England, Kingdom Books.

Hale, R.S. 1999. Growth of White Crappies in response to temperature and dissolved oxygen conditions in a Kentucky reservoir. *North American Journal of Fisheries Management* 19(2):591–598.

Hilomen-Garcia, G. 1999. AQD's marine ornamental fish project. *SEAFDEC Asian Aquaculture* 21(2):31–38.

Kolkovski, S., A. Arieli, and A. Tandler. 1997a. Visual and chemical cues stimulate microdiet ingestion in sea bream larvae. *Aquaculture International* 5:527–536.

Kolkovski, S., A. Tandler, and M.S. Izquierdo. 1997b. Effects of live food and dietary digestive enzymes on the efficiency of microdiets for seabass (*Dicentrarchus labrax*) larvae. *Aquaculture* 148:313–322.

Lockyear, J., H. Kaiser, and T. Hecht. 1997. Studies on the captive breeding of the Knysna seahorse, Hippocampus capensis. *Aqua. Sci. Conserv.* 1, 129–136.

Lourie, S.A., A.C.J. Vincent, and H.J. Hall. 1999. *Seahorses: an identification guide to the world's species and their conservation.* Project Seahorse, London. 214 pp.

Opuszynski, K.K., and J.V. Shireman. 1993. Strategies and tactics for larval culture of commercially important carp. *Journal of Applied Aquaculture* 2(3–4):189–219.

Payne, M.F., and R.J. Rippingale. 2000. Rearing West Australian seahorse, *Hippocampus subelongatus*, juveniles on copepod nauplii and enriched Artemia. *Aquaculture* 188:353–361.

Rosenlund, G., J. Stoss, and C. Talbot. 1997. Cofeeding marine fish larvae with inert and live diets. *Aquaculture* 155:183–191.

Vincent, A.C.J. 1995. Exploitation of seahorses and pipefishes. *NAGA, The ICLARM Quarterly*, January 1995, 18–19.

Vincent, A.C.J. 1996. *The international trade in seahorses.* TRAFFIC International, 163pp.

Wallace, J.C., A.G. Kolbeinshaven, and T.G. Reinsnes. 1988. The effects of stocking density on early growth in Arctic charr, Salvelinus alpinus (L.). *Aquaculture* 73, 101–110.

Wilson, M.J., and A.C.J. Vincent. 1998. Preliminary success in closing the life cycle of exploited seahorse species, *Hippocampus* spp., in captivity. *Aquarium Sciences and Conservation* 2:179–196.

Woods, C.M.C. 2000a. Preliminary observations on breeding and rearing the seahorse *Hippocampus abdominalis* (Teleostei: Syngnathidae) in captivity. *New Zealand Journal of Marine and Freshwater Research* 34:475–485.

Woods, C.M.C. 2000b. Improving initial survival in cultured seahorses, *Hippocampus abdominalis* Leeson, 1827(Teleostei: Syngnathidae). *Aquaculture* 190:377–388.

21

Rearing the Coral Seahorse, *Hippocampus barbouri,* on Live and Inert Prey

Michael F. Payne

Abstract

Seahorses are popular specimens in the marine aquarium trade. Cultured seahorses are now appearing in retail outlets and are proving easier to maintain in home aquariums than wild-caught animals. Despite the successful culture of seahorses, more convenient culture techniques for rearing newborn juveniles may be more cost effective. This study compared the growth and survival of newborn *Hippocampus barbouri* fed diets of frozen calanoid copepods and live enriched *Artemia* for 21 days. Diets comprising copepods that had been enriched with T-Iso for 8 days prior to freezing and copepods that had no prefreezing enrichment were each tested against a diet of enriched *Artemia* in two separate trials. The growth and survival of seahorses fed unenriched frozen copepods was poor compared with those fed enriched live *Artemia*. Seahorse survival on enriched frozen copepods and enriched live *Artemia* was the same (94% at day 21), although growth was slower on the copepod diet. Whereas copepods did have significantly more DHA and less EPA compared with *Artemia*, enrichment of copepods had no effect on their fatty acid profile. The growth and survival of seahorses fed either diet were generally lower in the first trial compared with the second. This was attributed to the poor quality of juvenile seahorses used in the first trial. This study shows that newborn *H. barbouri* juveniles can be reared with a high degree of success by using inert copepods as the sole food.

Introduction

It is estimated that approximately 1 million seahorses are captured annually from 65 countries for the aquarium trade. Of these less than 10% survive longer than 6 months, with the majority dying within 6 weeks of capture. Lower estimates of the number of seahorses that are captured specifically for the aquarium trade are likely to be underestimations, as seahorses that die in holding facilities are often sold to the traditional Chinese medicine or curio trade (N. Garrick-Maidment, The Seahorse Trust, personal communication 2002).

Poor rates of survival in wild-caught seahorses may be explained partly by the general stress associated with handling and transport but also by their unusual food requirements. In the wild, seahorses feed almost exclusively on live crustaceans, a habit that makes it difficult to train them onto frozen foods. In addition, seahorses do not have a true stomach and will not thrive unless they receive frequent feedings with small quantities of food. Few collectors or wholesalers are in a position to either provide live food to seahorses in their care or train them onto frozen foods. Thus, the few seahorses that survive long enough to reach a retail outlet are often too weak to feed, and most eventually die.

Despite these poor rates of survival, wild-caught seahorses are still in demand. Experienced hobbyists tend to avoid wild-caught seahorses, whereas enthusiasts who are new to the hobby are attracted to their unusual appearance. Unfortunately, many retail outlets in Australia continue to sell wild-caught seahorses to hobbyists who have little chance of keeping them alive.

The alternative to wild capture is the culture of seahorses. Captive-bred seahorses have many advantages compared with their wild-caught counterparts. Cultured specimens have all been trained to eat frozen foods and are adapted to the tank environment. In addition, these seahorses have generally been well fed right up to the time that they are sold to hobbyists. With the appearance of captive-bred seahorses on the market, information on their care in captivity has also become more widely available, thus further improving the likelihood that inexperienced hobbyists will be able to care for their captive-bred seahorses at home. The widely held belief that seahorses are "just too hard" to maintain in aquariums is now diminishing.

Captive-bred seahorses first appeared on the Australian retail market around 1999. The first species available was the pot-bellied seahorse (*Hippocampus abdominalis*), which occurs in temperate waters of southern Australia and New Zealand. Although a large and quite attractive species, it is unable to survive the high temperatures that occur during the summer in marine aquariums throughout most of Australia. Because methods of maintaining cool temperatures in home aquariums are generally expensive, this species has found a limited market in Australia. This has led to a number of tropical and subtropical seahorse species being cultivated and sold in the aquarium trade. At present (2002), cultured species commonly available in Australia are *H. abdominalis*, *H. barbouri*, *H. kuda*, *H. procerus,* and *H. whitei*. *H. barbouri* and *H. kuda* are species that have been imported into Australia, while the others are found in Australian waters (Lourie et al. 1999; Kuiter 2001).

Of all the seahorse species, the Coral seahorse, *H. barbouri*, is one of the more commonly available and easiest to keep in aquariums (Kuiter 2000). The experience of the author is that adults readily accept dead prey and have voracious appetites. In addition, they are unusual among seahorses in that they regularly use hard corals and other solid substrates as attachment sites and are reported to be less prone to cnidarian stings than other species (Kuiter 2000). This may make them suitable for display in reef tanks that do not contain strong currents.

Coral seahorses give birth to relatively large young that attach themselves to substrate soon after birth. Juveniles will readily feed on nauplius stages of *Artemia* spp. (hereafter referred to by the common name of *Artemia*) and when these nauplii are suitably enriched, survival of juveniles is sufficiently high to enable commercial production. However, hatching and enrichment of *Artemia* nauplii must be continuous while newborn seahorses are being produced, and this is very time consuming. An inert diet such as frozen copepods can be batch-cultured and then frozen in large quantities at times that are convenient to hatchery staff, thereby reducing work load during peak periods of production. Production costs for occasional large batch cultures of copepods are also lower than for frequent small batch cultures of *Artemia*.

This study compared the growth and survival of newborn coral seahorses on diets of enriched *Artemia* nauplii and copepods collected from intensive culture and stored frozen. Diets comprising copepods that had been enriched with T-Iso for 8 days prior to freezing and copepods that had no prefreezing enrichment were each tested against a diet of enriched *Artemia* in two separate trials. Fatty acid contents of the trial diets were recorded in an attempt to correlate seahorse growth and survival with dietary fatty acid content.

Material and Methods

Broodstock *H. barbouri*, comprising a combination of wild-caught and first generation seahorses, were maintained in a 2,000-liter recirculating seawater system. Juvenile rearing trials were also conducted in this system, which comprised eight 190-liter glass aquariums, 500-liter sump, trickle filter, venturi foam fractionator, and UV filter. Aquariums were illuminated by single overhead fluorescent lights on a 14:10 light:dark cycle. Approximately 1.5% of the system's water volume was exchanged daily. Water quality parameters before, and during, the experiment were as follows: ammonia (TAN) <0.05 mg/liter, nitrite <0.1 mg/liter, pH 7.9–8.2, salinity 33–36 g/liter, and temperature 25.1–26.8°C. Prior to mating in captivity and giving birth to juveniles that were used in the present study, broodstock were fed a mixed diet comprising frozen mysid shrimp (Hikari®), frozen adult *Artemia* (Hikari®) and chopped shrimp mixed with multivitamins.

Two separate trials were conducted using newborn *H. barbouri* juveniles reared in 2-liter experimental vessels (Fig. 21.1). These vessels were floated in a glass aquarium (incorporated in the recirculating system) with a polystyrene collar and were designed to maintain inert food items in suspension with gentle aeration (1–2 bubbles per second). Water exchange occurred via passive flow through two 150-μm mesh windows (6 cm × 6 cm) on opposite sides of each vessel. Unwound rope substrate was attached to a plastic crossbar wedged at the top of the conical bottom. One entire batch of juveniles released from a single adult male was collected and divided evenly between two rearing vessels. One of two experimental diets was then allocated to these vessels. An unfed control group was also cultured. For each trial three separate batches of juveniles from different male seahorses were treated in this manner to provide three replicate treatments of each diet. Between 32 and 50 newborn seahorses were stocked into each 2-liter experimental vessel.

In both trials the control diet comprised instar II *Artemia* nauplii (length 570–580 μm) enriched with Algamac 2000® and the microalga *Tetraselmis suecica*. In the first trial the treat-ment diet consisted of frozen mixed-age calanoid copepods, *Gladioferens imparipes* (length 620–980 μm) cultured in a 5,000-liter vessel using the microalga *Isochrysis galbana* (T-Iso). Copepods were collected from this vessel by slow draining through a 150-μm bag filter, and the concentrated copepods were then rinsed with clean seawater, spread out on flat trays, and placed in a domestic freezer. Throughout this process copepods were handled as gently as practical to minimize physical damage. In the second trial mixed-age *G. imparipes* copepods (length 660–910 μm) were cultured and collected as above. However, rather than immediate freezing, the concentrated live copepods were placed at a density of 1 per milliliter into a 200-liter vessel filled with clean seawater. These copepods were then fed T-Iso at the rate of 100,000 cells per milliliter per day for 8 days, after which they were collected as described above and frozen.

In both trials juvenile seahorses were provided live and dead prey on a regular basis throughout the day such that densities remained above 1 per milliliter. Prior to the first feed being administered each day, all uneaten food and waste was removed from the vessels. The length (tip of snout to tip of tail) of four seahorses from each rearing vessel was recorded weekly for 3 weeks, and survival was recorded

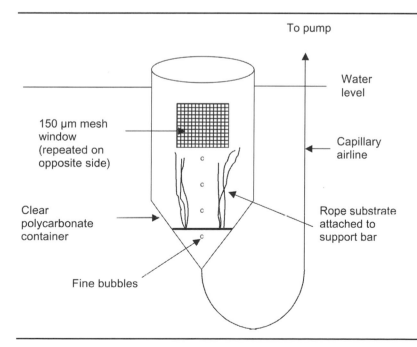

Fig. 21.1 Two-liter vessel used to rear newborn seahorses. Vessels were suspended in a glass aquarium using polystyrene collars. Water exchange occurred via passive flow through two 150-μm mesh windows on opposite sides of the vessel.

To pump

Water level

150 μm mesh window (repeated on opposite side)

Capillary airline

Clear polycarbonate container

Rope substrate attached to support bar

Fine bubbles

daily. As a comparison a single and separate batch of juveniles was left unfed. Fatty acid content of duplicate samples of live enriched *Artemia* and both unenriched and enriched frozen copepods was determined using published methods (Payne and Rippingale 2001a).

Results

Figures 21.2 and 21.3 show that growth and survival were significantly greater in those seahorses fed enriched *Artemia* compared with those fed frozen copepods that had not been enriched with T-Iso. Length and survival on day 21 was 20.2 ± 1.3 mm and 75 ±16% in *Artemia*-fed seahorses compared with 16.6 ± 0.6 mm and 11 ± 7% in copepod-fed animals. No mortality was recorded in copepod-fed seahorses

before day 8, whereas all unfed juveniles died after 7 days.

Figures 21.4 and 21.5 show that while growth was faster in *Artemia*-fed seahorses compared with those fed copepods that had been enriched with T-Iso, survival was not significantly different between the two groups. Length on day 21 was 26.3 ± 1.3 mm in *Artemia*-fed seahorses compared with 21.1 ± 0.9 mm in copepod-fed animals. Survival in both groups on day 21 was 94 ± 3%.

Growth and survival in both *Artemia*- and copepod-fed animals were greater in the second trial compared with the first. Newborn juveniles were significantly longer in the second trial (15.9 ± 0.7 mm) compared with the first (14.4 ± 0.8 mm).

Table 21.1 shows that both unenriched and

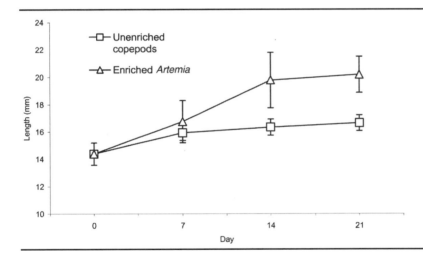

Fig. 21.2 Length of newborn seahorses fed unenriched frozen copepods and live *Artemia* enriched with Algamac 2000® and the microalga *Tetraselmis suecica* for 21 days. Each data point represents the mean of three replicates (four seahorses per replicate) ±1 sd.

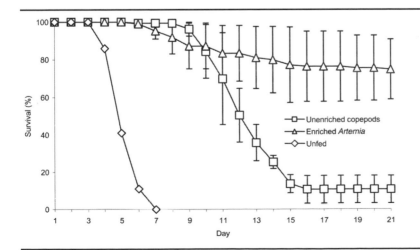

Fig. 21.3 Survival of newborn seahorses fed unenriched frozen copepods and live *Artemia* enriched with Algamac 2000® and the microalga *Tetraselmis suecica* for 21 days.

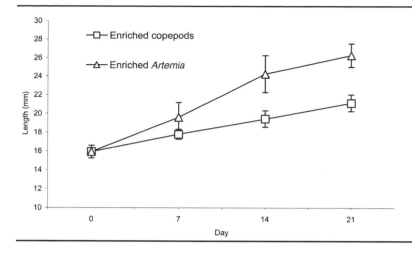

Fig. 21.4 Length of newborn seahorses fed copepods enriched with the microalga *I. galbana* for 8 days prior to freezing and live *Artemia* enriched with Algamac 2000® and the microalga *Tetraselmis suecica* for 21 days. Each data point represents the mean of three replicates (four seahorses per replicate) ±1 sd.

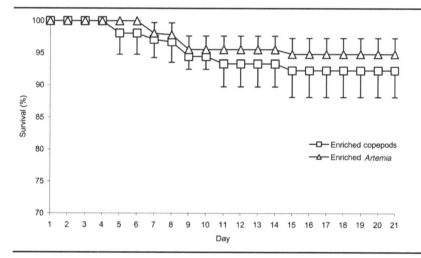

Fig. 21.5 Survival of newborn seahorses fed copepods enriched with the microalga *I. galbana* for 8 days prior to freezing and live *Artemia* enriched with Algamac 2000® and the microalga *Tetraselmis suecica* for 21 days.

enriched copepods contained significantly more C22:6n-3 (DHA) than enriched *Artemia*. In contrast, *Artemia* contained significantly more C20:5n-3 (EPA) than copepods. The DHA:EPA ratio was 9.8, 6.8, and 0.1 in unenriched copepods, enriched copepods, and enriched *Artemia*, respectively. Enrichment had no significant effect on total fatty acid content of copepods. Total fatty acid content was significantly greater in *Artemia* compared with copepods.

Discussion

The high rate of survival in newborn *H. barbouri* seahorses fed enriched frozen copepods clearly indicates that this technique has some potential for use in commercial seahorse culture. Live food production is a substantial ex-

pense in most marine aquaculture operations, and the use of inert prey such as frozen copepods has the potential to reduce this expense. Whereas the culture of large numbers of copepods does require significant amounts of time and infrastructure, the cost of managing the occasional large scale copepod culture (from which copepods can be collected and stored for use over a prolonged period) is certainly less than the effort required to produce enriched live prey such as *Artemia* on a continual basis. Further, hatcheries will find it more convenient to produce large numbers of copepods during periods when seahorse juveniles are not being produced than to constantly produce *Artemia* during busy periods.

Despite increased interest in copepod culture over the past few years, it is still not common in the aquaculture industry. Thus, little informa-

Table 21.1 Fatty acid content (mg/g dry weight) of diets fed to newborn *H. barbouri*

Fatty acid	Unenriched copepods (Trial 1)	Enriched copepods (Trial 2)	Enriched *Artemia* (Trial 1 & 2)
C14:0	2.7 ± 0.3	3.1 ± 0.2	1.3 ± 0.3
C15:0	0.1 ± 0.0	0.2 ± 0.0	0.3 ± 0.2
C16:0	7.8 ± 0.9	6.9 ± 0.6	14.9 ± 2.7
C16:1n-7	0.4 ± 0.1	0.6 ± 0.1	3.8 ± 0.9
C17:0	0.6 ± 0.2	0.6 ± 0.1	1.0 ± 0.2
C18:0	2.5 ± 0.4	1.7 ± 0.2	9.8 ± 1.4
C18:1n-9	3.1 ± 0.2	4.0 ± 0.5	31.1 ± 5.2
C18:2n-6	0.7 ± 0.0	2.6 ± 0.2	4.0 ± 0.6
C18:3n-3	0.8 ± 0.1	1.3 ± 0.2	17.0 ± 2.5
C20:1n-9	0.1 ± 0.1	0.1 ± 0.0	0.0
C20:2	0.0	0.2 ± 0.0	0.0
C20:3	0.0	0.1 ± 0.0	0.0
C20:3n-6	0.5 ± 0.1	0.7 ± 0.1	3.3 ± 0.3
C20:5n-3	1.8 ± 0.3	2.6 ± 0.2	8.6 ± 0.9
C22:4n-6	1.3 ± 0.5	0.7 ± 0.1	1.5 ± 0.6
C22:6n-3	17.7 ± 3.8	17.7 ± 1.3	1.2 ± 0.2
Total FA	40.2 ± 6.3	43.1 ± 3.4	98 ±16
DHA:EPA	9.8	6.8	0.1

Note: Diets were as follows: frozen copepods without post-collection enrichment (unenriched copepods); frozen copepods with 8 days of post-collection enrichment with T-Iso (enriched copepods), live instar II *Artemia* nauplii enriched with Algamac 2000®, and *T. suecica* (enriched *Artemia*). Data expressed as mean of 2 replicates ±1 sd.

tion is available on the subject compared with *Artemia* production, which has been the standard (yet not always the most successful) requirement for growing juvenile marine fish. At present the options for hatcheries wishing to make use of copepods are to collect from the wild, purchase commercial products, or culture their own. Collection of copepods from the wild is neither reliable nor safe, as copepod populations tend to fluctuate and there is the risk of introducing parasites and pathogens. A commercial product known as Cyclop-eeze®, available in deep-frozen and freeze-dried forms, is currently available in the United States (but not in Australia) and shows some promise as a diet for juvenile seahorses. Culturing copepods is becoming increasingly viable, with a number of publications providing detailed instruction (e.g. Ohno and Okamura 1988; Sun and Fleeger 1995; Støttrup and Norsker 1997; Schipp et al. 1999; Payne and Rippingale 2001b).

Cyclop-eeze® has been used to rear older

seahorse juveniles with some success. This product has been used as a supplement to live *Artemia* to achieve greater than 70% survival in 1-month-old *H. abdominalis, H. kuda,* and *H. erectus* (J. Gomezjurado, National Aquarium in Baltimore, personal communication 2002). C. Woods achieved greater than 93% survival in 1-month-old *H. abdominalis* juveniles weaned onto this diet (C. Woods, National Institute of Water and Atmospheric Research, personal communication 2002). However, Woods found that the growth rates of *H. abdominalis* fed Cyclop-eeze® were slower than those fed *Artemia,* as occurred with frozen copepods in the present study. Neither Gomezjurado nor Woods fed Cyclop-eeze® to newborn *H. abdominalis,* as these copepods were considered too large (~1 mm) for the newborns. Woods also observed that the feeding strike rate by 1-month-old *H. abdominalis* juveniles was lower on Cyclop-eeze® than on live *Artemia.* He attributed this to the high degree of fragmentation of the copepods in the commercial product (18% intact, 38% mostly intact, 44% highly fragmented) and the inability of newborn seahorses to recognize these fragments as food. The fact that frozen *G. imparipes* used in the present study were of mixed size (620–980 μm) and remained intact when thawed was probably essential to the success of this diet for rearing newborn *H. barbouri.*

One of the main difficulties encountered in the present study, and by C. Woods when using Cyclop-eeze®, was discovering a method that would effectively maintain dead copepods in suspension rather than letting them settle. This must be achieved without too much turbulence that would prevent newborn seahorses from targeting their food. In the present study the use of fine capillary tubing to deliver air to the conical base of the rearing vessel appeared to be an effective method, although the correct rate of air delivery was difficult to achieve as even a small adjustment in air flow greatly affected water turbulence in the small rearing vessels. Woods (unpublished data) used water flow to maintain copepods in suspension, and this may well be more effective than air, as water flows are likely to need less fine tuning to achieve the desired effect. This certainly requires more work.

Unenriched and enriched frozen copepods contained the same amount of total fatty acids and had similar profiles, with the exception of the DHA:EPA ratio. However, it is very unlikely that the DHA:EPA ratio alone could account for the lower rate of survival of copepod-fed seahorses in the first trial compared with the second trial. In the first trial newborn seahorses were significantly smaller compared with those in the second trial, suggesting that they were of inferior quality. That broodstock nutrition is quite closely correlated to larval quality has been well documented in many commercial finfish species (Bromage 1995), and there is abundant anecdotal evidence to indicate that the same applies for seahorses. Thus, the results of the second trial should be reproducible if newborn seahorses of sufficient quality are produced.

Fatty acid contents of both unenriched and enriched frozen *G. imparipes* were considerably lower than expected. A total fatty acid content of 438 mg per gram dry weight in well-fed *G. imparipes* adults has been recorded (Payne et al. 2001), which is 10 times the amount recorded in the present study. Copepods that were not enriched with T-Iso prior to freezing were collected from a 5,000-liter culture that contained 2.2 copepods per milliliter and were being fed T-Iso at the approximate rate of 40,000 cells per milliliter per day. This is considerably lower than the rate of approximately 100,000 cells per milliliter per day recommended for intensive cultures of *G. imparipes* stocked at 1 per milliliter (Payne and Rippingale 2001b). Thus, unenriched copepods in the present study were food-limited for some time before collection and freezing. Enriching copepods with T-Iso for a period before freezing was an attempt to boost fatty acid content in these food-limited copepods. However, this did not occur in the present study, despite appropriate copepod stocking densities and feed rates. It appears that these copepods did not ingest T-Iso during the enrichment process, probably because the water quality in the original culture was poor, or their feeding appendages had been damaged during collection. Ensuring that intensive copepod cultures remain healthy, as well as taking greater care during collection prior to enrichment, will certainly increase the fatty acid content of copepods and may result in faster growth rates in juvenile seahorses.

The present study does not allow any speculation regarding the fatty acid requirements of newborn *H. barbouri*. This requires feeding trials in which the same food organism is given

different fatty acid profiles before being fed to seahorses. This was the same approach (Chang and Southgate 2001) that has been used to achieve different commercial formulations to enriched *Artemia* nauplii. These workers clearly showed that growth and survival of newborn seahorses was greatest when fed diets high in DHA and EPA, although the relative importance of each of these fatty acids was not clear.

In the only published study on rearing *H. barbouri*, (Wilson and Vincent 1998), a length was recorded of approximately 40 mm in 40-day-old seahorses that were reared for the first 29 days on *Artemia* and copepods. However, comparison with the present study cannot be made, as these authors did not measure seahorses for the first 39 days of rearing. These authors also recorded 100% survival to day 106 for a small group of animals selected out of a larger batch of newborns and reared in scrupulously clean conditions that would be impractical to maintain on a commercial scale. Further recent studies on rearing newborn seahorses of other species show a clear trend toward increased rates of growth and survival as knowledge of seahorse culture improves (e.g. Payne and Rippingale 2000; Woods 2000; Chang and Southgate 2001).

This study has shown that newborn *H. barbouri* can be reared using inert food. There is little doubt the same will be possible for other seahorse species, particularly those that give birth to larger juveniles and are therefore relatively easy to rear. Further research by the author has indicated that newborn *H. barbouri* reared on frozen copepods can be weaned earlier and more successfully onto the larger frozen foods that are used for grow-out. Thus, it is now possible to rear *H. barbouri* to adulthood with high rates of survival without the use of any live food. This work will continue with investigations on the use of manufactured foods to rear seahorses.

Acknowledgments

Most of all I wish to thank Wendy Payne for taking over the seahorse workload so that I could write this chapter. Thanks also to the Aquaculture Development Unit at Challenger TAFE in Fremantle, Western Australia for the provision of copepods and T-Iso used in this trial.

References

Bromage, N. 1995. Broodstock management and seed quality—general considerations. In *Broodstock Management and Egg and Larval Quality*. N.R. Bromage and R.J. Roberts (editors). Cambridge, Blackwell Science Ltd: 1–24.

Chang, M., and P. Southgate. 2001. Effects of varying dietary fatty acid composition on growth and survival of seahorse, *Hippocampus* sp., juveniles. *Aquarium Sciences and Conservation* 3:205–214.

Kuiter, R.H. 2000. *Seahorses, pipefish and their relatives*. Chorleywood, U.K., TMC Publishing.

Kuiter, R.H. 2001. Revision of the Australian seahorses of the genus *Hippocampus* (Syngnathiformes: Syngathidae) with descriptions of nine new species. *Records of the Australian Museum* 53:293–340.

Lourie, S.A., A.C.J. Vincent, and Heather J. Hall. 1999. *Seahorses: An identification guide to the world's species and their conservation*. London, Project Seahorse.

Ohno, A., and Y. Okamura 1988. Propagation of the calanoid copepod, *Acartia tsuensis*, in outdoor tanks. *Aquaculture* 70:39–51.

Payne, M.F., and R.J. Rippingale. 2000. Rearing West Australian seahorse, *Hippocampus subelongatus*, juveniles on copepod nauplii and enriched *Artemia*. *Aquaculture*. 188:353–361.

Payne, M.F., and R.J. Rippingale. 2001a. Effects of salinity, cold storage and enrichment on the calanoid copepod *Gladioferens imparipes*. *Aquaculture* 201:251–262.

Payne, M.F., and R.J. Rippingale. 2001b. Intensive cultivation of the calanoid copepod *Gladioferens imparipes*. *Aquaculture* 201:329–342.

Payne, M.F., R.J. Rippingale, and J.J. Cleary. 2001. Cultured copepods as food for West Australian dhufish (*Glaucosoma hebraicum*) and pink snapper (*Pagrus auratus*) larvae. *Aquaculture* 194:137–150.

Schipp, G.R., J.M.P. Bosmans, and A.J. Marshall. 1999. A method for hatchery culture of tropical calanoid copepods, *Acartia* spp. *Aquaculture*. 174:81–88.

Støttrup, J.G., and N.H. Norsker. 1997. Production and use of copepods in marine fish larviculture. *Aquaculture*. 155:231–248.

Sun, B., and J.W. Fleeger. 1995. Sustained mass culture of *Amphiascoides atopus* a marine harpacticoid copepod in a recirculating system. *Aquaculture* 136:313–321.

Wilson, M.J., and A.C.J. Vincent. 1998. Preliminary success in closing the life cycle of exploited seahorse species *Hippocampus* spp., in captivity. *Aquarium Science and Conservation* 2:179–196.

Woods, C. 2000. Preliminary observations on breeding and rearing the seahorse *Hippocampus abdominalis* (Teleostei: Syngnathidae) in captivity. *New Zealand Journal of Marine and Freshwater Research* 34:475–485.

22

The Copepod/*Artemia* Tradeoff in the Captive Culture of *Hippocampus erectus*, a Vulnerable Species in Lower New York State

Todd Gardner

Abstract

Trade in traditional Chinese medicine, curios, and aquarium fish has contributed to declines in seahorse populations worldwide. Aquaculture may prove to be an important conservation measure for seahorses; however, the development of cost-effective production techniques has been slowed by several bottlenecks. One of these is early diet. The investigation described herein addresses the use of wild copepods versus *Artemia* nauplii as a first food for *Hippocampus erectus*, a vulnerable seahorse species in New York State. In preliminary experiments, it was determined that higher survival could be achieved by rearing *H. erectus* on wild copepods than on *Artemia* nauplii. In this experiment juvenile *H. erectus* were offered wild copepods for 0–5 days before being switched to a diet of enriched *Artemia franciscanis*. In the first two of five planned replicates, a significant increase in survival was observed beginning with 2 days (replicate 1) and 4 days (replicate 2) of copepod feeding. Total survival for the first two replicates was 72% and 11%, respectively. No difference in growth was observed between feeding treatments.

Introduction

Natural History

The family Syngnathidae comprises 52 genera and about 215 species. It includes the seahors-es, pipefish, sea dragons, and the pipehorse, an apparent morphological intermediate between seahorses and pipefish (Nelson 1994). There are approximately 35 recognized species of seahorse, all belonging to the genus *Hippocampus* (Vincent 1996).

Seahorses are a group of highly specialized fish characterized by a tubular mouth, an articulating neck, a prehensile tail, and rings of bony armor covering the entire body (Bigelow and Schroeder 1953). Locomotion is primarily by means of undulations or oscillations of the dorsal fin, whereas the pectoral fins serve in steering and lift generation (Fritzsche 1983). The anal fin has been greatly reduced but may facilitate expulsion of feces, eggs, and fry (personal observation 2000). The pelvic and caudal fins have been lost entirely. One of the most extraordinary features of the family is their unique reproductive biology. Males have a specialized patch or pouch (marsupium) wherein eggs are deposited, fertilized, incubated, and possibly nourished for the entire gestation period (Fritzsche 1983). The exact physiological role of the marsupium remains under debate, and its nutritive function has never been proven (Azzarello 1991); nevertheless, it appears to be involved in osmoregulation of the fluid surrounding the developing embryos (Azzarello 1991; Carcupino et al. 1997).

Seahorses occur worldwide in tropical and temperate coastal marine waters. Their primary habitats are seagrass beds, mangroves, and coral reefs (Vincent 1996). Despite the extensive distribution of the genus, no single species is very wide-ranging and populations within species tend to be unevenly distributed. Seahorses may be locally abundant in some suitable habitats, but completely absent from others (Gardner unpublished data; Vincent 1996). The narrow species ranges and patchy distributions are probably due to seahorses' limited locomotory ability, low fecundity, and lack of a pelagic larval stage (Vincent 1996).

Exploitation

The unique morphology of seahorses has earned them considerable recognition in folklore and mythology for thousands of years (Whitley and Allen 1958, as cited in Vincent 1996) and has contributed to their popularity in the aquarium industry and curio trade. The greatest demand for seahorses appears to come from traditional Chinese medicine (TCM) markets. TCM is recognized by the World Health Organization as a viable health-care option and is practiced in China, Taiwan, Singapore, Japan, Indonesia, and ethnic Chinese communities worldwide. Seahorses are used to treat a variety of health problems including asthma, sexual dysfunction, and kidney disorders (Vincent 1996).

Beginning in the 1980s, a period of economic prosperity in many Asian countries, coupled with a growing interest in alternative health treatment options in Western nations, fueled unprecedented sales in TCM products. The growth of this industry has contributed significantly to the near extinction of such animals as the rhinoceros, tiger, and leopard. At least 430 species of threatened or endangered organisms are used in TCM, necessitating a largely illegal trade. The black-market nature of the industry makes it extremely difficult to monitor and regulate (Gaski and Johnson 1994)

The first and only detailed account of the world trade in seahorses was published in 1996 (Vincent 1996). Vincent's exhaustive research included investigations of customs reports as well as interviews with fishermen, pharmacists, and traders of seahorses at all levels throughout the global community. Vincent found that seahorse consumption in China had increased 10-fold between 1986 and 1996 and cited China's surging economy and the decline of other fisheries as probable causes. At least 32 nations

were found to be involved in seahorse trading. Consumption within Asian countries (excluding Japan, Korea, Malaysia, and Singapore—all believed to be important consumers) was calculated at about 45 tonnes of dried seahorses (approximately 16 million individuals) annually. This estimate includes only imports and does not consider trade routes through unofficial channels. The consensus among fishermen interviewed by Vincent throughout the Pacific is that seahorse populations appear to be declining. They also agree that they cannot catch enough seahorses to meet the demand.

Customs reports from various countries seemed to indicate that the United States is an important importer and exporter of live and dead seahorses, but no meaningful numbers for total U.S. landings could be obtained. Of the 15 states where seahorses are known to occur, only Florida has attempted to record landings. Florida's Department of Environmental Protection reported a substantial increase in annual seahorse landings between 1990 (6,504 individuals) and 1994 (112,367 individuals).

Several alarming facts have come to light as a result of Vincent's 1996 report: (1) the world trade in seahorses has still not been well quantified, but it almost certainly exceeds 20 million individuals annually and is growing; (2) the sustainability of this fishery is virtually unknown, yet preliminary data suggest that seahorses are being overfished; and (3) population decline due to habitat loss and incidental bycatch in trawl fisheries has not been quantified, but it may easily exceed decline due to directed-effort harvesting.

The World Conservation Union (IUCN) currently lists three species of *Hippocampus* (*H. bargibanti, H. breviceps,* and *H. minotaur*) as endangered. Seventeen species, including *H. erectus*, are listed as vulnerable (IUCN 2000).

Hippocampus erectus

The lined seahorse, *Hippocampus erectus,* is one of three recognized seahorse species occurring in the western Atlantic. It inhabits shallow coastal waters from Nova Scotia to Argentina and reaches a length of at least 15 cm (Robins et al. 1986). Locally the lined seahorse can be found in eelgrass beds and around artificial structures in embayments along the south shore of Long Island and New York Harbor (Gardner unpublished data). Although there is no commercial fishery for *H. erectus* on Long Island, local populations may be vulnerable for the following reasons:

- *H. erectus* has a typically low fecundity and patchy distribution.
- New York has a high concentration of tropical-fish dealers and a large Asian constituency, two of the most important consumers of seahorses.
- The large size and relatively smooth texture characteristic of *H. erectus* are the two most desirable characteristics of a seahorse in TCM (Vincent 1996).
- Crucial eelgrass habitat already has been lost to excessive coastal development, pollution, and blooms of the harmful chrysophyte, *Aureococcus anophagefferens*.
- Seahorse collection is neither monitored nor regulated in New York (Jill A. Olin, U.S. Fish and Wildlife Service, personal communication 2001).

Aquaculture

The objective of this investigation is to take a step toward the development of an effective, reliable protocol for the captive culture of *H. erectus*. This would enable aquaculture to be used as an economically viable alternative to wild harvesting and, if necessary, as a tool in future stock-enhancement initiatives. It should be added that this is a report on an ongoing study, to which additional replicates and results will be added at a later date.

Although seahorses, including *H. erectus*, have been spawned and reared in captivity for many years, they are notoriously difficult to culture, often suffering high juvenile mortality (Scarratt 1995; Forteath 1996; Vincent 1996). Despite these difficulties, a few investigators have experienced high juvenile survival in laboratory-rearing attempts (Correa et al. 1989), yet the reasons for these successes are poorly understood. One study reports an 85% survival in rearing *H. abdominalis* but does not provide the technical details of the protocol (Forteath 1997). Another estimated 70% survival 100 days after hatching but did not count the fry on day 0 (Scarratt 1995). Additionally, the rearing system was probably too complex to be easily scaled up.

Newborn seahorses will feed only on live zooplankton, and to date, most rearing attempts have relied solely on the nauplii of *Artemia* sp. (Lunn and Hall 1998). The availability and convenience of *Artemia* cysts have made them extremely popular in aquaculture (Hoff and Snell 1999). Yet *Artemia* nauplii have repeatedly proven to be nutritionally deficient as a food source for larval and juvenile marine fish compared with copepods, which have produced much better results in terms of growth, survival, and overall health (Kraul et al. 1993; Shields et al. 1999; Gardner 2000; Payne and Rippingale 2000). This is probably because of their relatively high levels of highly unsaturated fatty acids (HUFAs), particularly docosahexaenoic acid (DHA) and eicosapentaenoic acid (EPA), both of which are crucial to fish development (Sargent et al. 1997).

Although there are now a number of enriching products on the market intended to increase the HUFA content of *Artemia* and other live foods, the use of copepods continues to produce better fish yields (Kraul et al. 1993; Shields et al. 1999; Gardner 2000; Payne and Rippingale 2000; Stottrup 2000). This suggests that either the ideal levels or proportions of the HUFAs are not being attained or that there are other factors affecting growth and survival of fish reared on copepods versus those reared on *Artemia*. The latter hypothesis is supported by an experiment (Payne et al. 1998) in which two groups of the pipefish *Stigmatopora argus* were reared on copepods containing either high or low HUFA levels. No difference in growth or survival was observed.

The use of copepods in aquaculture can be quite expensive. Copepods do not reach nearly as high a density in culture conditions as other live foods. Therefore, they require larger volumes of water and larger culture vessels. The use of wild-caught copepods can eliminate this problem, but wild populations are subject to a high degree of fluctuation, and their use can introduce parasites and fouling organisms into the rearing tank. Furthermore, the collection process can be labor-intensive, and if a boat is used to pull the plankton net, the cost of running the boat must also be taken into account. In the 1970s, using copepod-dominated wild plankton (CDWP), Moe and Young became the first and only people ever to rear the highly valued French angelfish, *Pomacanthus paru*, in captivity. However, they never achieved commercial success, in part because of the high cost of collecting the copepods (Forrest Young, Dynasty Marine Associates, personal communication 1997).

Three preliminary seahorse-rearing trials were conducted at the Hofstra University Aquaculture Laboratory over the past year. The purpose of these trials was to test some filtration designs and feeding regimens. All trials were conducted in standard 38-liter (10-gallon) tanks. In trials 1 and 2, small air-driven foam filters were used. In trial 3, one end of the tank was sectioned off with 500-μm nylon netting, placed at an angle of 20 degrees to the wall. The resulting compartment was filled with small, plastic beads, kept in motion with aeration to form a fluidized bed filter. In this tank, a 20-cm (8-inch) air diffuser was placed on the end with the filter compartment to prevent seahorses from being drawn into the netting. This method of aeration created a vertical circulation pattern in the tank that kept the newly hatched seahorses from becoming caught on the surface of the water as was experienced in trials 1 and 2. In each trial 1 liter of *Isochrysis galbana* in standard seawater medium was added to the tank each day for the first 10 days. In the first trial nauplii of *Artemia salina*, enriched with Super HUFA (Salt Creek, Inc.) were offered as the exclusive food for an entire brood of *H. erectus*. After 2 weeks 100% mortality was observed in this tank. In the second trial, copepod-dominated wild plankton (CDWP) was substituted as a live food for the first 2 weeks, after which time the diet was abruptly switched to enriched *Artemia salina*. At the end of the 60-day trial, 190 seahorses were counted. The percent survival could not be calculated because the initial brood size was not counted on day 0. In the third trial CDWP was offered for only 3 days, followed by nauplii of *Artemia franciscanis*, enriched with Ratio HUFA (Salt Creek, inc.). After 60 days, 214 seahorses were counted and moved into a 150-liter grow-out tank.

Considering the well-established importance of copepods as an early food for seahorses, and the costs associated with obtaining them, it would be useful to know the minimum number of days of copepod feeding necessary to achieve a reasonable rate of survival. A greater number of days on copepods seems to result in higher juvenile survival but may also increase the overall cost of production. This investigation will address the copepod/*Artemia* tradeoff in the culture of *Hippocampus erectus*.

Methods

The wild-caught broodstock animals are kept in 208-liter (55-gallon) and 416-liter (110-gallon) aquariums with undergravel and carbon filtration. Individual pairs are removed and placed in a 227-liter (60-gallon) spawning tank with similar filtration.

A rearing system consisting of fourteen 38-liter (10-gallon) aquariums with central filtration was constructed for the experiment. Each tank is equipped with a divider of 500-μm nylon netting and a 20-cm (8-inch) air diffuser as described above, except that no plastic beads are used. All water exchange takes place behind the divider. The central filtration consists of a foam prefilter, a 100-cm^3 biological filter with plastic bio-ball media, a 60-watt UV sterilizer, and a protein skimmer. The biological filter, UV sterilizer, and rearing tanks are fed in sequence, via a 113 liter/min (30 gallon/min) external pump. A 38 liter/min (10 gallon/min) submersible pump within the 150-liter sump operates the protein skimmer. The system is capable of housing two broods simultaneously. Water quality parameters are maintained as follows: pH: 8.0–8.3, NH_4: 0.0 ppm, NO_2: 0.0 ppm, NO_3: 0–20 ppm, salinity: 25 ppt, temperature 23–28°C.

Although the experimental design calls for a total of five replicates, only two broods have been reared in the system to date. Only broods containing at least 600 individuals are used. One hundred individuals from a brood are placed in each of six aquariums. In each replicate, one tank is fed only *Artemia* from day 1 to day 28 (treatment 1). Each of the other five tanks is fed CDWP for 1–5 days (treatments 2–6, respectively) before being switched to a diet of *Artemia*. Plankton is collected by hanging a 60-μm plankton net in a tidal current under one of several local bridges. Cysts of *Artemia franciscanis* are decapsulated, hatched, and enriched with Ratio HUFA. Food organisms are maintained at 0.5/ml in all rearing tanks. *Isochrysis galbana* is cultured in standard seawater medium according to a published protocol (Hoff and Snell 1999). For the first 5 days of each replicate *Isochrysis* is maintained at a density of approximately 2×10^5 cells/ml.

Beginning on day 6, all rearing tanks are connected to central filtration, and algae is no longer added. The reasons for using microalgae are to maintain high water quality until the juvenile seahorses are old enough to resist the current generated by filtration and to serve as a nutritious food source for the zooplankton. *Isochrysis galbana* was chosen for its high HUFA content and the relative ease with which it can be cultured.

Each replicate proceeds for 28 days. A random sample of five juvenile seahorses is removed and measured weekly. In order to minimize stress while taking measurements, live seahorses and a small ruler are placed in a shallow Petri dish with 20 ml of seawater and photographed. Using Scion Image software, each photo is calibrated to the ruler. Seahorses are measured by drawing a digital line from the coronet to the tip of the tail. After 28 days all juveniles are removed and counted. Average survival and growth is determined for each feeding regime. Costs and benefits associated with providing wild copepods versus enriched *Artemia* nauplii are determined and discussed.

Results

Survival for the two replicates is shown in Figures 22.1 and 22.2, respectively. Chi-square contingency tests demonstrated a significant increase in survival beginning with treatment 3 (2 days on CDWP) for replicate 1, and treatment 5 (4 days on CDWP) for replicate 2. In the first replicate 432 seahorses (72%) survived to day 28, versus only 67 (11%) in replicate 2.

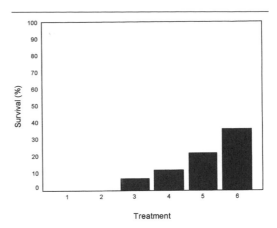

Fig. 22.2 Percent survival among seahorses, replicate 2.

Mean length is illustrated in Figures 22.3 and 22.4. The average growth rate was calculated to be 0.07cm/day. No difference in growth rate was observed between the two replicates.

Discussion

The data collected, to date, clearly demonstrate that the use of copepods as a first food for *H. erectus*, even for only a few days, can have a significant impact on survival but not necessarily on rate of growth. Significantly higher survival as well as growth rate in *H. subelongatus*, reared on copepods versus those reared on *Artemia*, has been observed (Payne and Rippingale 2000). However, in their investigation copepods were offered for the duration of the experiment. Long-term effects of early diet on the health of seahorses have not been studied but would certainly make for an interesting follow-up to these investigations.

The reason for the significant difference in survival between replicates 1 and 2 was not determined. It is possible that this disparity is a reflection of the relative health and, consequently, gamete quality of the respective parents. It is also possible that other health or fecundity factors may have come into play. Continued study and additional replicates of the trials reported herein should contribute to our growing understanding of the dietary requirements and appropriate culture technology for these seahorses.

A number of pests were introduced to rearing tanks as a result of the reliance on wild plankton. These included barnacles, hydroids,

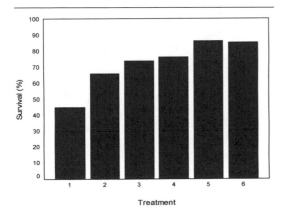

Fig. 22.1 Percentage survival among seahorses, replicate 1.

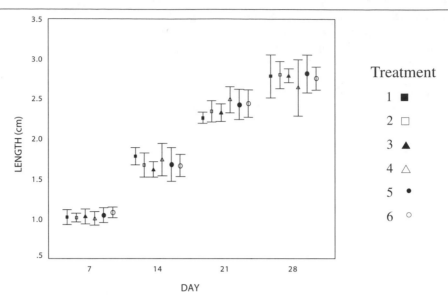

Fig. 22.3   Growth, expressed as average length of seahorses, measured weekly for replicate 1.

amphipods, gastrotrichs, and polychaetes. The polychaetes were least abundant and appeared to be harmless. Gastrotrichs were observed crawling on the skin of seahorses. Although they were probably not parasitic, as most gastrotrichs are free-living, the affected seahorses were often observed scratching, suggesting physical irritation. When barnacles, hydroids, and amphipods reached high numbers, they of-fered strong competition for food. Hydroids were the most problematic of the contaminants. In addition to competing for food, they are also capable of stinging young seahorses. Hydroid stings are suspected of playing a role in the initiation of cutaneous *Vibrio* infections on seahorse tails in earlier trials. Additionally, hydroids were the most difficult contaminant to remove. Whereas most of the other organisms

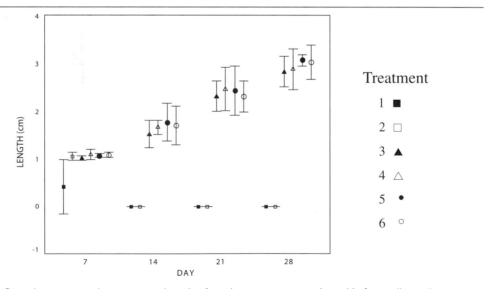

Fig. 22.4   Growth, expressed as average length of seahorses, measured weekly for replicate 2.

were eliminated by simple mechanical means, such as siphoning or scraping, these methods appeared to stimulate reproduction by fragmentation when applied to hydroids. No parasites or other disease organisms were associated with the use of wild plankton in this experiment; however, this is a risk that must be considered.

Approximately 3 hours/day was required to collect and process copepods; however, considerably more effort may be required in a tropical climate where plankton is typically less abundant or in areas lacking the strong tidal currents characteristic of coastal embayments. Natural fluctuations in copepod populations can also influence catch per unit effort (i.e., cost).

Because of its relative availability, low cost, and high protein content, *Artemia* is probably the most widely used live food in aquaculture. When used as an exclusive food source for juvenile *H. erectus*, however, lower survival can be expected than when used as a second food, following several days on copepods.

Acknowledgments

Thanks to Dr. Eugene Kaplan, Dr. Robert Johnson, Donald Axinn, Jill Olin, Donna McLaughlin, Dr. John Morrissey, Dr. Peter Daniel, Dr. Christopher Sanford, Forrest Young, Tom Frakes, Aquarium Systems, Penn Plax, Inc.

References

Azzarello, M.Y. 1991. Some questions concerning the Syngnathidae brood pouch. *Bulletin of Marine Science* 49(3):741–747.

Bigelow, H.B., and Schroeder, W.C. 1953. Fishes of the Gulf of Maine. *Fishery Bulletin of the Fish and Wildlife Service*. U.S. Government Vol. 53.

Carcupino, M., Baldacci, A., Mazzini, M., and Franzoi, P. 1997. Morphological organization of the male brood pouch epithelium of *Syngnathus abaster* before, during and after egg incubation. *Tissue and Cell* 29(1):21–30.

Correa, M., Chung, K.S., and Manrique, R. 1989. Cultivo experimental del caballito de mar, *Hippocampus erectus*. *Boletin del Instituto Oceanografico de Venezuela, Universidad Oriente* 28(1 y 2), 191–196.

Forteath, N. 1996. Seahorses, *Hippocampus abdominalis*, in culture. *Austasia Aquaculture* 9(6):83–84.

Forteath, N. 1997. The large bellied seahorse, *Hippocampus abdominalis*: A candidate for aquaculture. *Austasia Aquaculture* 11(3):52–54.

Fritzsche, R.A. 1983. Gasterosteiformes: Development and relationships. Ontogeny and Systematics

of Fishes—Special publication number 1 of the American Society of Ichthyologists and Herpetologists. pp 398–405.

Gardner, T. 2000. Spawning and rearing the yellow dottyback, *Pseudochromis fuscus*. *Freshwater and Marine Aquarium* 23(4):126–132.

Gaski, A.L., and Johnson, K.A. 1994. Prescription for extinction: Endangered species and patented oriental medicines in trade. TRAFFIC USA. 300 pp.

Hoff, F.H., and Snell, T.W. 1999. *Plankton Culture Manual, fifth edition*. Florida Aqua Farms, Inc., Dade City, Florida. 160pp.

IUCN. 2000. List of threatened and endangered species [online] Available: http://www.org.uk/data/database/rl_anml_combo.html.

Kraul, S., Brittain, K., Cantrell, R., Nagao, T., Ako, H., Ogasawara, A., and Kitagawa, H. 1993. Nutritional factors affecting stress resistance in the larval mahimahi, *Coryphaena hippurus*. *Journal of the World Aquaculture Society* 24:186–193.

Lunn, K.E., and Hall, H.J. 1998. Breeding and management of seahorses in aquaria: in Briefing documents for the First International Aquarium Workshop on Seahorse Husbandry, Management and Conservation, Project Seahorse. Chicago. 98 pp.

Nelson, J.S. 1994. *Fishes of the World, third edition*. John Wiley & Sons, Inc., New York.

Payne, M.F., and Rippingale, R.J. 2000. Rearing the West Australian seahorse, *Hippocampus subelongatus* on copepod nauplii and enriched Artemia. *Aquaculture* 188(3–4):353–361.

Payne, M.F., Rippingale, R.J., and Longmore, R.B. 1998. Growth and survival of juvenile pipefish (*Stigmatopora argus*) fed live copepods with high and low HUFA content. *Aquaculture* 167:237–245.

Robins, R.C., Ray, G.C., and Douglass, J. 1986. *A Field Guide to Atlantic Coast Fishes*. Houghton Mifflin Company, New York.

Sargent, J.R., McEvoy, L.A., and Bell, J.G. 1997. Requirements, presentation and sources of polyunsaturated fatty acids in marine fish larval feeds. *Aquaculture* 155:117–128.

Scarratt, A.M. 1995. Techniques for raising lined seahorses (*Hippocampus erectus*). *Aquarium Frontiers* 3(1):24–29.

Shields, R.J., Bell, J.G., Luizi, F.S., Gara, B., Bromage, N.R., and Sargent, J.R. 1999. Natural copepods are superior to enriched Artemia nauplii for halibut larvae (*Hippoglossus hippoglossus*) in terms of survival, pigmentation and retinal morphology: Relation to dietary essential fatty acids. *Journal of Nutrition* 129(6):1186–1194.

Stottrup, J.G. 2000. The elusive copepods: Their production and suitability in marine aquaculture. *Aquaculture Research* 31(8–9):703–711.

Vincent, A.C.J. 1996. The international trade in seahorses. TRAFFIC International. 163 pp.

Whitley, G., and Allen, J. 1958. *The Seahorse and Its Relatives*. Griffin Press, Melbourne.

Stakeholder Perspectives

A. *Museums and Public Aquariums*

23

The Role of Public Aquariums in the Conservation and Sustainability of the Marine Ornamentals Trade

Heather Hall and Douglas Warmolts

Background

Since the opening of the first public aquarium display in the Fish House at the London Zoo in 1826, there has been enormous growth of public aquariums, particularly in the past 20 years. New developments such as the National Aquarium in Baltimore, the Florida Aquarium, and the Oceanario de Lisboa, have been the cornerstone of the revitalization of run-down areas of cities. Older aquariums such as the Oceanographic Institute in Monaco and Berlin Zoo Aquarium have maintained their leading roles in the industry through the renovation and introduction of new exhibits. Advancing technology from filtration systems to acrylic viewing windows has drastically increased the scale and scope for aquarium displays and the species that can be kept.

Worldwide public aquariums and zoos enjoy immense popularity, with a collective annual attendance estimated in excess of 600 million visitors, representing approximately 10% of the world's population. These organizations are experienced in using this popularity to educate and influence their visitors, local communities, and other target groups about wildlife conservation concepts and issues. Increasingly, the direct and indirect participation of public aquariums in field conservation programs has become an integral component of institutional missions.

Public aquariums recognize that they are one highly visible component of the aquarium trade. As such, they have a responsibility to encourage and support the sustainable use of marine ornamental species and associated conser-

vation of habitats. Current initiatives range from individual institutional projects toward sustainable trade, to global programs such as the Marine Aquarium Council's certification scheme.

This chapter presents the current role of public aquariums in the conservation and sustainability of the marine ornamentals trade through a wide variety of examples from individual institutions, as well as international collaborative efforts.

Coordination of Programs in Zoos and Aquariums

Over the past 30 years the international zoo community has developed coordinated captive management programs for their living collections. Although individual zoos and aquariums are independent entities, their conservation impact is greatest when they are willing and able to act collectively. Globally the World Zoo Organization (WZO) helps coordinate and promote links between the regional associations of zoos and aquariums. Regionally based associations, such as the American Association of Zoos and Aquariums (AZA), the European Association of Zoos and Aquariums (EAZA), the European Union of Aquarium Curators (EUAC), South East Asian Zoo Association (SEAZA), Australian Regional Association of Zoological Parks and Aquariums (ARAZPA), and the Pan African Association of Zoological Gardens, Aquariums, and Botanical Gardens (PAAZAB), set standards and organize their members in the areas of conservation, education, and science. These regional associations are also responsible for establishing membership standards and policies and the collaborative management of institutional collections.

Cooperative management requires planning. Because zoos and aquariums hold such a wide array of species, many regional zoo and aquarium associations have developed Taxon Advisory Groups (TAGs) to examine the management and conservation needs of entire taxa or groups of related species. TAGs are made up of representatives from member zoos and aquariums that have an interest and/or expertise in particular taxa. Since the late 1980s TAGs in Australia, Europe, and North America have worked to identify priority species that warrant specific population management and conservation efforts, such as studbooks or species survival plans (SSPs). The overall objectives of this process are to create self-sustaining captive populations through cooperative management and to maximize the collective conservation impact of regional zoological associations and their members (Hutchins et al. 1995). TAGs also assist in selecting species for conservation programs; develop regional collection plans; establish priorities for exhibition, management, research, and conservation; examine animal management techniques on the basis of scientific studies; and assist program coordinators in developing animal care guidelines.

One of the primary responsibilities of a TAG is the development of a regional collection plan (RCP). An RCP contains recommendations for the number and type of taxa to be maintained in a particular region (e.g., Canada, Australia). The RCP also recommends specific regional objectives for individual species, ranging from the phasing out of a captive population to the development of formal cooperative management programs (Hutchins et al. 1995; Hutchins et al. 1996; Hutchins et al. 1998). In some cases, the RCP also includes the functional role of each taxon in the collection (e.g., captive breeding for reintroduction, research, or education). To produce an effective RCP, a TAG must assess the total space available for a given taxonomic group and recommend how it should be allocated. TAGs must take into account both the limited amount of space available and the need to maintain animals in populations large enough to satisfy husbandry and population management concerns (Hutchins et al. 1995). A TAG must also review the conservation status of all species within its purview and consider the potential of each species to contribute to conservation (Hutchins and Conway 1995; Hutchins et al. 1995; Hutchins et al. 1998). A TAG can also define and summarize its priority conservation goals in an action plan, published in conjunction with the RCP (Wiese et al. 1994; Hutchins and Conway 1995). An action plan summarizes a series of realistic goals a TAG aims to accomplish for the species of interest. Action plans outline specific research, education, field conservation, and other appropriate projects to be undertaken or supported by the regional association and participating member institutions. By defining specific goals and devising a means to accomplish them, a TAG's action plan can also guide fund-raising and networking efforts and can begin to generate support for conservation action (Smith et al. in press)

Databases and Information Sharing

The European Fish and Aquatic Invertebrate Taxon Advisory Group (FAITAG) initiated a taxonomic database for species held in public aquariums following a workshop in 1999. This had identified that the first step toward a cooperative breeding program was to understand the diversity and numbers of species held in aquariums, as well as the limitations of the collection. Led by the Oceanario de Lisboa, a Microsoft Excel database was developed. This program was chosen as one that is internationally known and available, accepts a wide range of data file types, is easy to use, and can potentially be readily converted in the future. The information compiled includes taxonomic details, provenance, and information on any focused breeding, husbandry, or research effort for that species at that particular institution.

Information is available on an enormous number of species and individuals in public aquariums in Europe (Table 23.1). The database is constantly expanding, with additional contributions being submitted from organizations. The next phase of this program is to incorporate data on invertebrates and aquatic plants, to integrate with other databases such as the American

Elasmobranch Society Census, as well as conservation and research programs.

As part of the contribution to global programs, public aquariums are playing a role in the Census of Marine Life (CoML). This 10-year international research program aims to assess and explain the diversity, distribution, and abundance of marine organisms throughout the world's oceans. The emphasis of the program is field studies, which are to be conducted in poorly known habitats as well as those assumed to be well known. In both coastal and deep waters projects aim to identify new organisms and collect new information on marine life. The CoML aims to enable scientists to compare what once lived in the oceans to what lives there now and to project what will live there in the future through field studies and other projects, some of which are based in aquariums. These range from analyzing historical documents to modeling future ecosystems.

Marine Ornamental Species Programs

Conservation programs specific to the marine environment have been developed by aquariums, which target flagship species, as well as threatened marine habitats, such as coral reefs. Marine ornamental species currently managed under both North American and European TAG programs include seahorses, elasmobranchs, cardinalfish, corals, and jellyfish. These programs occur at international, regional, and institutional levels.

Seahorses

Seahorses remain extremely popular with zoos and aquariums, and new special exhibit areas focusing on seahorses are opening every year.

Table 23.1 Information held in the taxonomic database of fish in European aquariums in 2001

Number of aquariums	18
Number of families	209
Number of species	1,523
Number of specimens	48,445
Breeding	162 spp.
Husbandry	19 spp.
Research	16 spp.

Of the 32 known seahorse species, approximately 18 are kept in zoos and aquariums (Bull 2001). In the World Conservation Union (IUCN) Red List of Threatened Animals (2000), 31 seahorses are listed as vulnerable, and one of these, *Hippocampus capensis,* has recently been upgraded to endangered (Hilton-Taylor 2000).

Project Seahorse (www.projectseahorse.org) is an international marine conservation initiative that has been instrumental in organizing the efforts of public aquariums worldwide in syngnathid conservation. In 1998 the John G. Shedd Aquarium (Chicago, United States) hosted a "Seahorse Husbandry, Management, and Conservation" workshop. Thirty-five invited participants attended, representing 29 institutions from 9 countries, including directors, nutritionists, veterinarians, researchers, and husbandry staff. The workshop addressed the specific role of public aquariums in seahorse conservation and identified areas that required action to improve population management and effective conservation actions. These included communication, taxonomy, record keeping, tagging, acquisitions/deaccession, and an evaluation of current techniques. A Project Seahorse Aquarium Research Co-coordinator position was developed, based at the Shedd Aquarium, to help coordinate the resources and efforts of public aquariums internationally and implement the workshop actions. Much of this has been implemented through an email list server (Syngnathidae Discussion Group), which currently has more than 130 members, representing over 40 aquariums in 11 countries, and other related professionals.

The AZA Marine Fishes TAG (MFTAG) and the European FAITAG have initiated steps to formalize some of these priority actions and have included syngnathids in their first pilot RCPs for marine fish. MFTAG has identified the following seahorses as priority species for actions in North American aquariums: *Hippocampus abdominalis, H. comes, H. erectus, H. ingens, H. kuda, H. reidi*, and *H. zosterae.* FAITAG has identified an additional four priority species on which to focus their efforts: *H. barbouri, H. capensis, H. guttulatus*, and *H. hippocampus.* The two species of sea dragons, *Phyllopteryx taeniolatus* and *Phycodurus eques,* are included in both regions. Institutions from both regions work together on all species to combine rather than duplicate efforts. Each TAG plans to address the research, management, and conservation priorities with their respective species through the outputs of various shared working groups. The first of these outputs has been the development of a husbandry manual for all 11 priority seahorse species (Bull 2001).

Population management of seahorses in zoos and aquariums is critical to conservation strategies. Well-managed populations are a priority, as is research, record keeping, and education programs. The goals for population management are twofold:

- To provide purebred species lines to zoo and aquarium partners for conservation research and education purposes, to assist collection planning, and to reduce taxonomic confusion.
- To maintain minimum genetic diversity within each captive population, improving health and fitness of the captive strains.

The appointed North American and European chairs of the seahorse specialist groups oversee these captive breeding programs, with individual coordinators working on the management and conservation of each priority species.

Elasmobranchs

Sharks and rays are an important attraction for public aquariums, where they provide an interesting and valuable educational tool. Elasmobranchs are also maintained in aquariums for the purposes of scientific investigation, and much of what we know about these enigmatic animals has been learned through observing them in aquariums.

Elasmobranchs exhibit a K-selected life history strategy characterized by low fecundity, slow growth rates, and late sexual maturity. Unfortunately, this life history strategy makes sharks and rays susceptible to overexploitation. Reproduction of elasmobranchs in aquariums is restricted to only a few species, and predisposing environmental parameters are frequently unknown. In addition, unless appropriate husbandry practices are adopted, elasmobranch survivorship in aquariums can be lower than in their natural habitat. Hence, sharks and rays have been captured in increasingly large num-

bers to supply the growing demands of the aquarium community.

As a basic conservation measure, the public aquarium community needs to increase its level of peer review and facilitate the dissemination of relevant husbandry information to interested parties. To date there has been no published handbook providing guidelines for the keeping of elasmobranchs in aquariums. In order to maximize the survival and welfare of these animals and reduce the need for the capture of replacement stock, such a document would play a valuable role. With this in mind, a project was developed to produce an elasmobranch husbandry manual. The objective of the manual is to provide a single reference source book that may be used to answer frequently asked questions, assist in the development of new exhibits, train employees responsible for the care of elasmobranches, and act as a general guide for the captive maintenance of this important taxonomic group.

An international elasmobranch husbandry symposium, co-organized by the Oceanario de Lisboa and the Columbus Zoo & Aquarium, was convened in October 2001. The symposium was attended by more than 180 people from public aquariums, universities, nongovernmental organizations, state and federal agencies, collectors, and private individuals representing 16 countries. The principal purpose of the symposium was to act as a focus for the communication and open discussion of each chapter of the manual. Authors presented their findings and discussed implications with symposium delegates. Manuscripts presented at the symposium by invited speakers were subjected to a peer review and will be published as a hardbound volume constituting the core of the elasmobranch husbandry manual. The manual's sections and individual chapters include:

- Elasmobranchs and Public Aquariums: Conservation, research, public education, community relations, census of elasmobranchs in public aquariums, ethics, legislation, permitting, and commercial collectors.
- Exhibit Design: Species description, selection, and compatibility; quarantine and isolation facilities: design and construction; exhibit design and construction.
- Collection, Acclimation, and Quarantine: Water parameters and life support system

(LSS) design; capture techniques; transportation techniques and equipment; prophylaxis and quarantine; specimen introduction and acclimation; identification of individual; captive behavior and enrichment.

- Husbandry: Diving with elasmobranchs; exhibit maintenance; diet, nutrition, vitamins, and feeding techniques; age and growth; record keeping; physiological/behavioral changes in controlled environments.
- Reproduction: Reproductive anatomy and modes of reproduction in sharks; reproduction and development; captive breeding and sexual conflict; genetics and captive elasmobranchs.
- Veterinary Care: Physical examinations/inspections; diagnostic imaging techniques; anesthesia; clinical blood parameters; Metazoan parasitology; bacteriology; protozoal diseases of elasmobranchs; viruses; fungi; medications and dosages; histology; tumors; goiters; spinal deformities.

This first International Elasmobranch Husbandry Symposium included a discussion session dedicated to the development of a plan of action. The aim was to produce a general reference document identifying shared priorities and serving as a collective and global strategic plan for aquariums maintaining elasmobranchs. TAGs and individual institutions could then use this document as a guide when prioritizing their objectives, acquisition strategies, and conservation programs to be funded. Issues covered by the plan of action include: (1) legislation, permitting, and collection strategies; (2) captive breeding programs; (3) public education and outreach; (4) conservation and advocacy; (5) research; and (6) industry communication.

Breeding Marine Ornamentals

Marine Ornamental Fish

One of the initiatives developed by the European FAITAG and led by the Rotterdam Zoo has been the development of a database of breeding records for marine fish. This was established from a comprehensive questionnaire circulated to European aquariums but has been extended to include both scientific and gray literature. Currently the database contains 1,406 records. Review papers of specific species are now being conducted, particularly as student projects in collaboration with local universities. Of the 541 species in the database, more than half have been bred in captivity. There are some discrepancies to resolve, however, with data reported at the Elasmobranch Husbandry Symposium held in Florida in 2001. The database lists 27 species of elasmobranchs as being bred in captivity, whereas the symposium reports list 91 species. The database will be held and updated on the website of the European Union of Aquarium Curators (www.euac.org).

A list of species illustrates the range of marine ornamental fish that have been bred in public aquariums and the significant role a number of aquariums have played in establishing the first breeding records (Table 23.2). The motivation for breeding marine ornamental fish in aquariums has evolved over time. Historically aquariums were often the only places that had access to exotic marine species from collecting trips, and there was interest in advancing knowledge of these fish and breeding them. Although this interest remains there is also an increased awareness of the need for aquariums to be sustainable and to review what can and should be bred for exhibit and conservation purposes. This, combined with the increase in collaborative breeding programs, has advanced the success of breeding many species because of the open exchange of information and coordinated advancement of techniques.

Monterey Bay Aquarium, for example, focuses breeding efforts on local marine species, such as tubesnouts, sailfin sculpins, and grunt sculpins. Techniques have been developed so that survival is now relatively high (typically at least 50%), and with a big demand from other public aquariums, excess juveniles can be supplied to outside facilities in exchange for other species. In order to advance basic scientific knowledge, work has also been done to rear the smooth ronquil (*Ronquilus jordani*). Although this species may not have much exhibit value, the larval description has never been done. Skills have been advanced in another program to rear bonito as one of only two scombrids that have ever been cultured (the other being yellowfin tuna in Panama). Current survivorship to the end of the larval stage is still low (about 1%), but culture techniques learned with bonito may also be applied to both yellowfin and bluefin tuna.

Table 23.2 List of marine ornamental fish species propagated in public aquariums

Species	Common name	Organization	Notes
Apogon nigripinnis	Bullseye	Berlin[1]	World first
Premnas biaculeatus	Spinecheek anemonefish	Berlin	World first
		Waikiki	
Amphiprion akallopisos	Skunk clownfish	Wilhelma, Stuttgart	World first
		Berlin	
A. clarkii	Yellowtail clownfish	Berlin	
		Wilhelma, Stuttgart	
A. ephippium	Saddle anemonefish	Wilhelma, Stuttgart	World first
		Berlin	
A. frenatus	Tomato clownfish	Wilhelma, Stuttgart	
		Berlin	
A. melanopus	Cinnamon clownfish	Waikiki	
A. nigripes	Maldive anemonefish	Wilhelma, Stuttgart	
A. ocellaris	Common clownfish	Berlin	
		Waikiki Aquarium	
		Wilhelma, Stuttgart	
A. percula	Percula clownfish	Berlin	
		Monterey Bay	
		Waikiki	
		Wilhelma, Stuttgart	
A. perideraion	Pink anemonefish	Berlin	
		Wilhelma, Stuttgart	
A. polymnus	Saddleback clownfish	Wilhelma, Stuttgart	
A. sebae	Sebae anemonefish	Wilhelma, Stuttgart	
Pterapogon kauderni	Banggai cardinalfish	Artis Zoo	
		Berlin	
		Columbus Zoo	
		Monaco	
		Monterey Bay	
		New Jersey State Aqm	
		OceanPark	
		Waikiki	
		Wilhelma, Stuttgart	
Hippocampus abdominalis	Big-bellied seahorse	Monterey Bay	
		Wilhelma, Stuttgart	
H. capensis	Cape seahorse	Monterey Bay	
		London Zoo	
H. erectus	Lined seahorse	Berlin	
		Columbus Zoo	
		Monterey Bay	
		Waikiki	
		Wilhelma, Stuttgart	
H. fisheri	Hawaiian seahorse	Waikiki	
H. fuscus	Sea pony	Monaco	
H. guttulatus	Long-snouted seahorses	Monaco	
		Wilhelma, Stuttgart	
H. hippocampus	Short-snouted seahorse	Berlin	
		Monaco	
		Wilhelma, Stuttgart	
H. kuda	Yellow seahorse	Berlin	
		Monaco	
		London Zoo	
		Wilhelma, Stuttgart	
H. reidi	Slender seahorse	Wilhelma, Stuttgart	
S. acus	Greater pipefish	Wilhelma, Stuttgart	
Sygnathus leptorhynchus	Bay pipefish	Cabrillo Marine Aqm	
S. typhle	Broad-nosed pipefish	Wilhelma, Stuttgart	

312

Table 23.2. *(continued)*

Species	Common name	Organization	Notes
Doryrhamphus dactyliophorus	Ringed pipefish	Berlin	World first
Syngnathoides biaculeatus	Alligator pipefish	Berlin	World first
Acanthochromis polyacanthus	Spiny chromis	Berlin	
Opistognathus aurifrons	Yellowheaded jawfish	Berlin	World second
Zoarces viviparous	Viviparous blenny	Berlin	
Cryptocentrus sp.	Shrimp-goby	Berlin	
Synchiropus splendidus	Green mandarinfish	Waikiki	
Coryphaena hippurus	Mahimahi (dolphinfish)	Waikiki	
Genicanthus personatus	Masked angelfish	Waikiki	World first
Hypsypops rubicundus	Garibaldi	Cabrillo Marine Aqm	
Aulorhynchus flavidus	Tubesnout	Monterey Bay	
Nautichthys oculofasciatus	Sailfin sculpin	Monterey Bay	
Rhamphocottus richardsoni	Grunt sculpin	Monterey Bay	
Gibbonsia montereyensis	Crevice kelpfish	Monterey Bay	
Rathbunella hypoplecta	Smooth ronquil	Monterey Bay	
Sarda chiliensis	Bonito	Monterey Bay	
Gymnomuraena zebra	Zebra moray	Wilhelma, Stuttgart	
Carcharhinus melanopterus	Blacktip reef shark	Ocean Park	
Triaenodon obesus	Whitetip reef shark	Ocean Park	
Atelomycterus marmoratus	Coral catshark	Ocean Park	
Haploblepharus pictus	Dark shy shark	Ocean Park	
Poroderma pantherium	Leopard catshark	Ocean Park	
Hemiscyllium ocellatum	Epaulette shark	Columbus Zoo	
Chiloscyllium plagiosum	Whitespotted bambooshark	Columbus Zoo Ocean Park	
Chiloscyllium punctatum	Brownbanded bambooshark	Ocean Park	
Cephlaloscyllium pseudoumbratile	Pygmy Swellshark	Ocean Park	
Cephaloscyllium ventriosum	Swell shark	Cabrillo Marine Aqm	
Heterodontus francisci	Horn shark	Artis Zoo Cabrillo Marine Aqm	
Poroderma africanum	Striped catshark	Ocean Park	
Chiloscyllium griseum	Grey bamboo shark	Ocean Park	
Stegostoma fasciatum	Zebra shark	Ocean Park	World first
Taeniura melanospila	Black spotted stingray	Ocean Park	

Note: This list is not comprehensive, but demonstrates the wide range of species that have been successfully reared.

[1]Coral fish species bred successfully at Berlin Aquarium since 1978.

The New Jersey State Aquarium is working on one of the only three known species of sea ravens, *Hemitripterus americanus*. This species inhabits continental shelf waters from North Canada to the Chesapeake Bay, and although described in 1789, little is known about the basic aspects of its natural history. The studies have expanded basic breeding techniques for this species to a research program that encompasses its reproductive biology, feeding ecology, population dynamics, community structure, parasitism, cytogenetics, and phylogenetic relationships. These studies will help to evaluate the present status of its populations that, according to survey records, seem in decline and to understand its function in the continental shelf community, which is poorly known.

In conjunction with a new special syngnathid exhibit, the National Aquarium in Baltimore has developed a seahorse husbandry and breeding program. To date 14 of the 22 species of syngnathids have produced young. Besides supplying captive-born animals for the exhibit, this program is generating new information on the captive care and rearing requirements of these delicate species. This program links with the international breeding program for seahorses and sea dragons.

In May 2000 at the Aquarium of the Pacific (Long Beach, California), the male weedy sea dragon (*Phyllopteryx taeniolatus*) was the first recorded in North America to successfully receive a clutch of eggs from a female. The eggs from this male did not hatch successfully, but a second male held approximately 50 eggs from a transfer in May 2000 that hatched a month later. Young were raised in a pseudokreisel tank, and 16 survived. A second pregnancy and hatching took place is June 2001, and 52 were raised from this clutch (Trautwein 2001). Other institutions are actively working on the husbandry and breeding of the weedy sea dragon as well as the leafy sea dragon, *Phycodurus eques,* including Dallas World Aquarium (Texas, United States), Wilhelma Aquarium, Stuttgart (Germany), and Oceanario de Lisboa (Portugal).

The success in continuing to breed new species relates to advances in feeding technology, rearing tanks, and particularly collaborations within the aquarium community, as well as with research and aquaculture groups. An enormous amount of research remains to be done, however, particularly with conservation concerns raised over the 15–30 million marine ornamental fish of around 1,000 species estimated to be traded each year (Wood 2001).

The Waikiki Aquarium is working on the reproduction and larval rearing of marine ornamental fish that spawn naturally in the exhibit tanks and holding tanks. Recent research has focused on using commonly cultured planktonic animals with various enrichment products on the larvae of the angelfish, *Genicanthus* spp., golden goby, *Priolepis aureoviridis,* and various *Centropyge* spp. *P. aureoviridis* have been raised to day 11 to date. Work has been initiated with the Oceanic Institute that will involve collecting wild plankton of various sizes and types as first foods for the above larval fish. Examples of species that have been spawned but not yet successfully reared in public aquariums are listed (Table 23.3).

Marine Ornamental Invertebrates

The propagation of marine ornamental invertebrates has advanced in recent years, with improved lighting and filtration systems, as well as understanding of the conditions required for successful propagation. The main emphasis of invertebrate breeding and propagation in aquariums has been for jellyfish and corals.

Jellyfish   Jellyfish exhibits have achieved unexpected popularity of displays in public aquariums, particularly considering they are such simple animals. In spite of their lack of eyes, brain, or backbone, the delicate shape and graceful movements of these animals have touched the sense of beauty and wonder in millions of visitors.

Monterey Bay Aquarium has acquired a reputation for its spectacular exhibits of jellyfish, but several species had been exhibited in Japan at the former Ueno Aquarium for a number of years and as far back as 1964 at the Enoshima Aquarium (Powell 2000). Berlin Zoo Aquarium has also played a major role in advancing techniques for breeding jellyfish in aquariums (Kaiser 1994; Lange and Kaiser 1997). Culturing jellyfish has been developed in order to ensure the successful display of these animals, as well as to improve the sustainability of aquarium exhibits. A key element of culturing cnidarian jellyfish is the establishment of polyp colonies. With a healthy population of polyps numerous newly released medusae can be collected for many years. Successful maintenance of polyps requires an appropriate diet and keeping them free of fouling growths. The extent of species and organizations actively involved in the culture of jellyfish is listed (Table 23.4).

The Jellyfish Culture Facility at Monterey Bay Aquarium is mainly devoted to the culture and holding of the display species of Hydromedusae and Scyphomedusae, although other species are cultured, including various phytoplankton, rotifers, copepods, and *Artemia* nauplii. The New England Aquarium's jellyfish culturing facility was established in 1993 to support an on-site special exhibit that opened in 1995. Since 1996, that exhibit has traveled to six institutions, while specimens and husbandry advice have been provided to more than 35 museums and aquariums worldwide. The culturing room provides animals for the traveling exhibit, as well as training and husbandry advice for the remote caretakers of the exhibit. The culturing room is also a place of research and development for improving the current exhibit and for future exhibits.

New Jersey State Aquarium is involved in more detailed research on the life cycle of the moonjelly, *Aurelia aurita*. These studies, together with the discovery of a new asexual reproductive mechanism (gemmation), will help

Table 23.3 Examples of species successfully spawned but not yet reared in public aquariums

Species	Common name	Organization
Abudefduf abdominalis	Black-spot sergeant	Waikiki
Abudefduf sordidus	Hawaiian sergeant	Waikiki
Chromis vanderbilti	Blackfin chromis	Waikiki
Chromis viridis	Green chromis	Waikiki
Centropyge bicolor	Bicolor angelfish	Waikiki
Centropyge bispinosus	Two-spined angelfish	Waikiki
Centropyge flavissimus	Lemonpeel angelfish	Waikiki
Centropyge loricula	Flame angelfish	Waikiki
Centropyge potteri	Potter's angelfish	Waikiki
Dascyllus albisella	Hawaiian dascyllus	Waikiki
Dascyllus aruanus	Humbug damsel	Waikiki
Gobiodon okinawae	Yellow clowngoby	Waikiki
Histrio histrio	Sargassum frogfish	Waikiki
Istiblennius zebra	Zebra rockskipper	Waikiki
Priolepis aureoviridis	Golden goby	Waikiki
Pseudanthias bartlettorum	Bartlett's fairy basslet	Waikiki
Pseudanthias squamipinnis	Lyretail fairy basslet	Waikiki
Pseudochromis porphyreus	Purple dottyback	Waikiki
Ptereleotris zebra	Zebra dart goby	Waikiki
Sphaeramia nematoptera	Pyjama cardinal	Waikiki
Zebrasoma flavescens	Yellow tang	Waikiki
Leuresthes tenuis	Grunion	Cabrillo
Heterostichus rostratus	Giant kelpfish	Cabrillo
Zaniolepis frenata	Shortspine combfish	Cabrillo
Lythrypnus dalli	Blue banded gobies	Cabrillo
		Monterey Bay
Hexagrammos decagrammus	Kelp greenling	Monterey Bay
Oxylebius pictus	Painted greenling	Monterey Bay
Aeoliscus strigatus	Razorfish	Berlin
Synanceja verrucosa	Stonefish	Berlin
Taenionotus triacanthus	Leaf scorpionfish	Berlin
Monocentrus japonicus	Pineconefish	Berlin

to understand its evolutionary success and to clarify the mechanisms of sex development, metagenesis, larval dispersal, and adaptability of this species.

Corals

ASEXUAL REPRODUCTION The asexual propagation of corals using various budding and culturing techniques in public aquariums is a method for providing colonies and supplying live displays in the public exhibits. The main motivation for these programs has been to optimize the survival and growth of corals in public aquariums as well as to decrease the pressure on the collection of wild colonies. The range of species being propagated is illustrated (Table 23.5), using examples from Waikiki Aquarium and Columbus Zoo and Aquarium. Through the coordinated efforts of the AZA Aquatic Inverte-

brate TAG and European FAITAG, there is an expanding network for the exchange of husbandry information and fragments.

In many cases propagation is carried out in facilities behind the scenes, but several facilities have included propagation units as part of the public display (e.g., Columbus Zoo and Aquarium). At Monterey Bay Aquarium, two displays of completely captive-cultured corals and clams within an area of the aquarium that has been purposely built for children demonstrates the potential for reducing collection stress on natural populations. More than 20 species are now being cultivated, all by asexual fragmentation.

The Oceanographic Institute in Monaco has operated an extensive facility since 1990, where 70 species of hard corals and 10 species of soft corals are maintained and bred. A major new exhibit in 2000 was stocked entirely with

Table 23.4 Jellyfish species successfully propagated in public aquariums

Species	Common name	Organization
Aurelia aurita	Moon	Berlin
		Cabrillo
		Genoa
		Monterey Bay
		Ocean Park
		Waikiki
A. labiata		Berlin
		Cabrillo
		Monterey Bay
Cassiopeia andromeda	Upside-down	Berlin
		Monterey Bay
		Waikiki
Pelagia colorata	Purple striped	Monterey Bay
Chrysaora fuscescens	Pacific sea nettle	Cabrillo
		Monterey Bay
C. quinquechirra		Berlin
		Monterey Bay
		Oceanario de Lisboa
C. melanaster		Berlin
		Monterey Bay
C. achlyos	Black sea nettle	Cabrillo
		Monterey Bay
Phacellophora camtschatica	Fried egg	Berlin
		Cabrillo
		Monterey Bay
Cyanea capillata	Lion's mane	Cabrillo
		Monterey Bay
Phyllorhyza punctata		Berlin
		Oceanario de Lisboa
Mastigis papua	Palau Lake	Berlin
		Monterey Bay
Aequorea victoria	Crystal	Monterey Bay
Eutonina indicans	Umbrella	Monterey Bay
Catostylus mosaicus	Blue	Monterey Bay
Tripidalia sp.	Box	Monterey Bay
Cotylorhiza tuberculata		Berlin
Nestrostoma setouchiana		Berlin
Cephea cephea		Berlin
Rhizostoma pulmo		Berlin
Sanderia malayensis		Berlin
		Monterey Bay
Caribdea marsupialis		Berlin
Cladonema sp.		Berlin
Euchilota sp.		Berlin
Tima formosa		Berlin
		Monterey Bay

corals cultured in the coral facility within the aquarium.

SEXUAL REPRODUCTION   In addition to the increasing skills and success with the asexual reproduction of a wide variety of Caribbean and Indo-Pacific coral species, newer projects are under way on sexual reproduction. Waikiki Aquarium has been working for a number of years to learn more about the mode and timing of reproduction in scleractinian corals. Biologists from the Audubon Aquarium of the Americas (AAOA) have collected coral gametes at the Flower Garden Banks National Marine

Table 23.5 Examples from Waikiki Aquarium and Columbus Zoo and Aquarium of corals propagated asexually in public aquariums

Coral species	Soft/stony	Organization
Discosoma sp.	Soft	Waikiki
Nephthea sp.	Soft	Waikiki
Protopalythoa spp.	Soft	Waikiki
Rhodactis sp.	Soft	Waikiki
		Columbus
Rhodactis sp.	Soft	Columbus
Rumphella sp.	Soft	Waikiki
Sinularia sp.	Soft	Waikiki
Tubipora sp.	Soft	Waikiki
Xenia sp.	Soft	Waikiki
		Columbus
Zoanthus spp.	Soft	Waikiki
		Columbus
Acropora spp.	Stony	Waikiki
Acropora austera		Columbus
Acropora cervicornis		
Acropora elsyi		
Acropora formosa		
Acropora hyacinthus		
Acropora kirstyae		
Acropora micropthalma		
Acropora yongei		
Acropora millepora		
Anacropora sp.	Stony	Waikiki
Caulastrea sp.	Stony	Waikiki
Euphyllia sp.	Stony	Waikiki
Hydnophora sp.	Stony	Waikiki
Hydnophora excesa		Columbus
Merulina sp.	Stony	Waikiki
Merulina ampliata		Columbus
Montipora spp.	Stony	Waikiki
Montipora digitata		Columbus
Pavona sp.	Stony	Waikiki
Pocillopora damicornis	Stony	Waikiki
		Columbus
Porites spp.	Stony	Waikiki
Seriatopora hystrix	Stony	Columbus
Tubastraea sp.	Stony	Waikiki

Sanctuary (FGBNMS), Gulf of Mexico, in 2000 and 2001 to begin preliminary study on coral planulae growth rates and viability in captivity. The primary species of interest is the stony brain coral *Diploria strigosa*. In both years gamete packets were collected during the annual spawning event and successfully fertilized in specimen jars. All collecting techniques were noninvasive for the parent colonies. The free-swimming planulae were transported back to AAOA and allowed to settle on ceramic tiles. The "newborn" corals began forming hard skeletons and acquired zooxanthellae from their culture system. The AAOA plans to send gametes to several other public aquariums that have coral propagation systems including New England Aquarium, the Florida Aquarium, Moody Gardens Aquarium, Columbus Zoo and Aquarium, and Underwater Adventure-Mall of America to help in the study. The AAOA has developed a partnership with the FGBMNS to develop an exhibit within the aquarium that showcases the unique biology of the marine sanctuary, making note of the proximity of the oil industry to the area.

The University of Essen (Germany) and Rotterdam Zoo (Netherlands) have also recently begun a research project to study ex situ reproduction of reef-building corals. The project aims to highlight key processes for future breeding and restoration efforts and to reduce the in situ collection to supply public aquariums and the aquarium trade. One focus of the research is the investigation of induction of planulation and the settlement of planulae in closed-system aquariums. The experimental work in the marine laboratory of Rotterdam Zoo is being carried out in conjunction with field studies in Curaçao (Netherlands Antilles). The development of methods that enable the exchange of coral larvae among field stations, research institutions, and aquariums will contribute to establishing coordinated breeding programs for endangered species. Studies have already been conducted that show that it is possible to transport larvae effectively and economically over long distances (Dirk Petersen, Rotterdam Zoo, Netherlands/University of Essen, Germany, personal communication 2001). The project is carried out in cooperation with Sea Aquarium Curaçao and the reef restoration project "PortoMari" Curaçao, the aquariums of Zoo Cologne, Zoo Hagenbeck Hamburg, Aquazoo Duesseldorf (Germany), with the London Zoo (United Kingdom) and Burger's Zoo, Arnhem (Netherlands).

Other Invertebrate Species Whereas the predominant focus of effort on marine ornamental invertebrates has been on the reproduction of corals, a number of other species have also been propagated. Coordinated by Rotterdam Zoo, an international database is being compiled of invertebrate species that have been successfully bred in aquariums. Examples of some of these species are shown in Table 23.6.

Efforts are under way with other species that have been successfully spawned but not yet reared. At Waikiki Aquarium, these include the bulb-tipped anemone, *Entacmaea quadricolor,* and several species of cuttlefish; flamboyant (*Metasepia tullbergi*), flashback (*Sepia latimanus*), pharoah (*Sepia pharaonis*), Mediterranean (*Sepia officinalis*). Monterey Bay Aquarium is extending the species of jellyfish it is propagating through attempts with the flower hat (*Olindias formosa*) and blue jellyfish (*Catostylus mosaicus*). Aquarists at Cabrillo Marine Aquarium (California, United States) have attempted the propagation of market squid (*Loligo opalescens*), two-spotted octopus (*Octopus bimaculoides*), and red rock shrimp (*Lysmata californica*) with varying degrees of success and are about to try abalone (*Haliotis* spp., including the endangered *H.sorenseni*).

Research Programs

Public aquariums are actively engaged in a number of research programs that are usually driven by husbandry or management needs and increasingly for conservation programs. Many aquariums have research institutions that are part of the same governing institution (e.g., London Zoo and the Institute of Zoology are part of the Zoological Society of London), or have close affiliations with local universities (e.g., Columbus Zoo and Aquarium with Ohio State University), or have dedicated research staff (e.g., New Jersey State Aquarium). As such, aquariums offer invaluable, yet often underutilized, opportunities for undergraduate and postgraduate researchers for short- and long-term research opportunities. The following examples are intended to represent the range and

diversity of research projects being conducted with public aquariums around the world.

New Jersey State Aquarium is involved in a project on the systematics, reproductive biology, embryology, and ecology of a large group of gobies (tribe Gobiosomini), most of which are found in the Atlantic Ocean and in the Caribbean Sea. The main goals are to clarify their phylogenetic relationships, to understand their ecological role in coral reef ecosystems, and to promote their conservation.

With an eye to alleviating stress on wild seahorse populations while still meeting the demands of the market, the Vancouver Aquarium Marine Science Centre has been working on a seahorse aquaculture project. The goal of the project is to selectively breed seahorses adapted to eat a formulated, pelleted diet containing all necessary nutrients. The seahorse species chosen for the project are *Hippocampus abdominalis* and *H. erectus*. So far, the clearest signs of domestication have occurred with *H. erectus*. Selected juveniles not only eat the formulated pellets but actually prefer them. The next phase of the project will be to create a pellet that will remain intact and sink more slowly during feeding, then gradually work toward developing economic potential.

The 1998 workshop on seahorse husbandry and management (Lunn et al. 1999) identified a number of research goals, and Project Seahorse/John G. Shedd Aquarium subscquently appointed an Aquarium Research Co-coordinator to facilitate these objectives. Small-scale projects have looked at the effects of feeding frequency on seahorse growth (London Aquarium, United Kingdom) and the development of tagging methods for use in both captive and wild situations (London Zoo, United Kingdom).

Table 23.6   Examples of invertebrate species propagated in public aquariums

Species	Common name	Organization
Entacmaea quadricolor (asexual)	Bulb-tipped anemone	Waikiki
Hymenocera picta	Harlequin shrimp	Waikiki
Lysmata amboinensis	Cleaner shrimp	Waikiki
Lysmata californica		Monterey Bay
Mastigias papua	Lagoon jelly	Waikiki
Nautilus belauensis	Nautilus (Palauan)	Waikiki
Nautilus pompilius	Nautilus (Fijian)	Waikiki
Sepia officianalis	Cuttlefish	Monterey Bay
		Oceanario de Lisboa

The Vancouver Aquarium Marine Science Centre's latest research into the breeding of cold water ornamentals has centered on the dietary requirements of larvae and juveniles. Using an experimental diet for black cod, a pellet feed with a high marine lipid content, lower nutritionally induced mortality in white-spotted greenlings (*Hexagrammos stelleri*), Pacific spiny lumpsuckers (*Eumicrotremus orbis*), and a number of seahorses species has been achieved. In addition the Vancouver Aquarium has also been conducting research on line-breeding, out-crossing, and inbreeding depression on several cold-water species prized as ornamentals, including grunt sculpin (*Rhamphocottus richardsoni*) and sailfin sculpin (*Nautichthys oculofasciatus*), and tubesnouts (*Pallasina barbata*). The experiments provide an empirical test of the role of larval drift in protecting against inbreeding depression.

Studies are being conducted at the Wildlife Conservation Society (New York Aquarium, United States) to investigate the symbioses between scleractinian corals and photosynthetic dinoflagellate algae, particularly in relation to the loss of these algae during bleaching events (Baker 2001). A number of North American aquariums have contributed samples to a study of these *Symbiodinium* algae in corals. This research is part of a broader study on the types of *Symbiodinium* found in corals from disturbed reefs around the world (Baker 2001). Molecular techniques are used to isolate *Symbiodinium* DNA from coral tissues that are then analyzed (Rowan and Powers 1991; Rowan 1998) and so far, one type (type C) of *Symbiodinium* seems to be dominant in long-term cultured coral, but more samples are needed before any conclusion can be drawn.

Since 1998 the National Aquarium in Baltimore (Maryland, United States) has been involved in research with Johns Hopkins University to establish sustainable cell lines of the hard corals *Acoropora micropthalma* and *Pocillopora damicornis*. Studies have focused on inducing cell division as a prelude to establishing cell lines and further optimizing growth conditions. This research program has recently reported, for the first time, the production of aragonite crystals in cell culture. The presence of these crystals, the building blocks of the calcium carbonate skeleton, indicates that the cell cultures retain many properties of intact corals (Kopecky and Ostrander 1999). The successful development of coral cell lines will lead to laboratory studies to determine the effects of various diseases, environmental conditions, and pollutants on corals in the natural environment.

In Situ Conservation

Increasingly, aquariums around the world recognize their importance in conserving species in situ as well as focusing on breeding and research programs ex situ. The following examples illustrate the involvement of aquariums in local to international efforts and from single institution to multi-institutional collaborations.

Glovers Reef Survey, Belize

In 1994, the Wildlife Conservation Society established the Middle Cay Marine Research Station in Belize at the southern end of Glover's Reef, one of the few atolls in the Caribbean. Opening the station was an important step in creating the necessary infrastructure for conducting research needed to manage the newly created Marine Reserve. In order to monitor Glover's Reef ecology and to provide essential information for effective management decisions, a baseline faunal and flora survey was conducted of the key reef inhabitants. The goals of the survey were to determine abundances and distributions of algae, invertebrates (especially corals), and fish living in the patch reef area of Glover's Reef (more than 700 patch reefs; 312 square kilometers) and to store and analyze the data in a geographic information system (GIS). This was the beginning of a long-term study of this important reef ecosystem. Forty-four scientists from 22 North American public aquariums spent 722 man-days collecting information required for the database. One hundred eighty patch reefs were sampled, each consisting of five fish surveys and four benthic transects. In addition, 112 bounce dives were conducted to collect sand and to quickly access substrate between patch reefs. Several water samples were also collected to study nutrient distributions across the reef. Differences in algal growth and diversity patterns between reefs appear to be related to current and wave energy and location on the atoll. Nutrients also may be a factor. Patch reefs in the center of the atoll have significantly higher abundances of macro-algae and fewer healthy corals. Fewer fish also are associated with these algae dominated reefs. Future surveys will be conducted to study the effectiveness of the reserve and to access changes that result from natural phenomena such as hurricanes and extreme temperature fluctuations.

The Florida Aquarium is developing a Caribbean coral propagation facility. The coral fragments are being acquired from the Florida Keys that have been damaged from ship groundings, construction, or storms, in order to rehabilitate fragments. The objective is to grow them to a size that could be used in other reef restoration activities and to provide a facility to conduct research on captive Caribbean corals. Other benefits include the future supply of coral fragments to AZA institutions and coral researchers. Oceanario de Lisboa (Portugal) is participating in a similar "coralture" project, which focuses on coral culturing in conjunction with a field restoration and recovery program for damaged reefs in Israel.

Project Seahorse

Project Seahorse is a team of biologists, social workers, and other professionals committed to conserving and managing seahorses, their relatives, and habitats while respecting human needs. The project was established and is run by McGill University (Canada) and the Zoological Society of London (United Kingdom). The in situ component involves running community-based conservation in Philippines fishing villages, including research, establishment of marine sanctuaries, fisheries modifications and management, enforcement of legislation, education programs, and the development of alternative livelihoods for seahorse fishers. The ex situ component includes the support for aquarium-based breeding and research programs for seahorses, including educational initiatives. Because seahorses are a group of species that are threatened by the aquarium trade, there is also a responsibility to improve the sustainability of that trade and the husbandry and breeding successes within aquariums.

Banggai Cardinalfish

New Jersey State Aquarium has initiated a program to aid the conservation of the Banggai cardinalfish, *Pterapogon kauderni*. This species, endemic to the Banggai island region in Indonesia, became popular in the fish hobby and aquarium trade after its rediscovery by Western science in 1994. In 1996 a holistic research project began including studies on its reproductive biology, behavior, and morphology. Recently, fieldwork was completed that involved studies on its geographic distribution, trophic ecology, population and community structure, and conservation.

Field researchers from the Zoological Society of London conducted a complementary study on the trade in Banggai cardinalfish. This research revealed a significant and growing trade for this species, which considering the biology and limited distribution of this species, were a cause of concern.

Local in Situ Efforts

Many public aquariums are coastal and are increasingly becoming engaged in the conservation of their local aquatic fauna and associated habitats. The Oceanographic Institute, Monaco, has participated in the work and actions of the Undersea Reserve of Monaco, an experimental red coral farm that ran from 1990 to 1998, currently being pursued in Tunisia. Staff has also helped produce an inventory of populations of Mediterranean black grouper (*Epinephelus marginatus*) with the GEM (Groupe d'Etude sur le Mérou), based on dive surveys.

The National Aquarium in Baltimore's Conservation Department is leading efforts to restore seagrass beds in the Chesapeake Bay that provide critical habitat for the lined seahorse, *Hippocampus erectus*. Through exhibits and educational programs, the aquarium has been teaching people about the Chesapeake Bay for years. Members of the Aquarium Conservation Team (ACT!), a volunteer corps dedicated to the restoration of estuarine habitats in the Chesapeake Bay, are now learning first hand about tidal wetland ecosystems. ACT! members demonstrated their skills at several Chesapeake Bay restoration events during the year. They also participated in outreach programs designed to educate the general public, school groups, and concerned citizens about the importance of watersheds and wetlands and the need to conserve and protect these valuable resources.

The Bermuda Aquarium has initiated a survey program to identify and map local seahorse populations. Led by staff within the aquarium, it has attracted significant local interest and awareness as well as support for seahorse conservation.

Policy and Best Practice

The AZA and the AZA MFTAG have been closely involved with the Marine Aquarium Council (MAC) since its inception, and many individual aquariums have been directly involved in MAC's efforts to develop certification and labeling to ensure the marine ornamentals trade is sustainable and environmentally sound. Linkages between MAC and EUAC and several individual aquariums in Europe have also been initiated. MAC is an independent, multistakeholder effort that brings together representatives of the aquarium industry, hobbyists, conservation organizations, government agencies, public aquariums, international organizations, and others with a shared interest in the sustainability of marine aquarium organisms and the marine ornamentals trade.

Public aquariums play several key roles in partnership with MAC to ensure that certification becomes established and the marine ornamentals trade is sustainable:

- Raising public awareness about issues in the marine ornamentals trade and the need to buy only marine aquarium organisms that are certified and labeled as sustainable;
- Establishing purchasing practices/policy to buy only certified organisms whenever possible;
- Ensuring collecting, handling, and holding practices and policies undertaken by aquariums are linked to certification standards.

An informed public, consumer (hobbyist), and aquarium industry are fundamental to transforming the marine ornamentals trade through market forces. Aquariums, zoos, and museums have a unique and unparalleled potential to communicate marine issues to the public. We have millions of visitors worldwide, a diversity of scientific and other resources, and

are a highly trusted and respected source of information. Aquariums are perhaps the single most important conduit to the ocean-conscious public and to the many aquarium hobbyists and industry operators who visit them. Aquariums provide a unique opportunity to inform and educate the public about the marine aquarium trade. However, the educational and awareness-raising potential of aquariums has not yet been developed in relation to the trade, certification, and MAC. Therefore, a major awareness-raising campaign with and through the public aquariums in partnership with MAC has now begun.

Aquariums are also key sources of information to the media on marine and aquarium-related issues, and they are often contacted when stories are prepared on the marine aquarium trade. The media focus provided by aquariums on the aquatic environment (e.g., through the launch of new exhibits) creates communications opportunities to proactively put out positive messages on issues, such as sustainable practices, the marine aquarium trade, and certification.

Most public aquariums purchase marine organisms from the same sources that supply retailers and hobbyists. If aquariums preferentially purchased certified marine organisms as a matter of policy, they would contribute considerably to the "critical mass" of consumer demand essential for certification to become established. The commitment of aquariums to buy only certified marine organisms whenever possible will be a significant communication and conservation opportunity, for example, through the display of the MAC label and promotion of the aquarium's policy to purchase only organisms that are certified.

Aquarium professionals, such as those from Antwerp Zoo, are their country representatives for CITES (Convention for the International Trade in Endangered Species). In this role they have actively contributed to discussions on the control of trade in seahorses and corals.

The Monterey Bay Aquarium has developed a list of seafood products that it will and will not purchase for its restaurant based on current information from fisheries scientists and managers. The Monterey Bay Aquarium has made this information available to the public through its publication *Seafood Watch—A Guide for Consumers* to help others make their own informed choices about fish and seafood. The John G. Shedd Aquarium has adopted *The Audubon Guide to Seafood* in a similar effort and promotes these practices as part of its conservation outreach to the public. London Zoo has a policy of serving no IUCN Red-Listed species in its catering outlets.

Education Programs

Worldwide aquariums attract in the region of 600 million visitors per year. People come for an enjoyable day out, and often the primary motivation for the visit is the entertainment value of the institutions. However, the visit provides the opportunity for education and awareness programs addressing specific issues relating to the aquatic environment. Specific exhibits can encourage learning about species, their biology, their conservation, and the environment in which they live. Cabrillo Marine Aquarium, for example, has an exhibit that is devoted to the subject of aquaculture, including captive breeding, species survival issues, and growing fish for human food. Students help and conduct research in full public view, and it is described as a place to "grow young sea organisms and young scientists."

Education programs increasingly target a broader audience than the visitors to the institution. Ocean Park in Hong Kong, for example, has an education program targeting teachers, including games that help in the understanding of biodiversity, and uses origami to make marine creatures. Since 1997 the Oceanographic Institute in Monaco has been conducting environmental education in the schools of the region of Nice in the form of a traveling interactive exhibition called *Expomer, à la découverte de la Méditerranée* (Expomer, discovering the Mediterranean Sea). In 1999 the Columbus Zoo and Aquarium began offering distance learning programs to schoolrooms throughout the United States via live teleconferencing from its exhibits. Schools were provided program options in advance that were tailored to their science curriculum needs. Students viewed the exhibit in real time while interacting with an on-line educator. During the 2000–2001 school year, 337 sessions reached approximately 9,850 students in 13 states in the United States. This emerging technology has become quite popular,

and the demand for expanded programming is growing rapidly.

Materials developed for marine education within public aquariums have also been applied to in situ programs. Marine education kits developed at Columbus Zoo and Aquarium are being used by Project Seahorse staff in the Philippines and Vietnam as the basis for the interactive presentations to children in fishing communities, providing novel ways for presenting marine conservation topics in developing countries where materials are limited.

Acknowledgments

Europe: Zoological Society of London (United Kingdom) <<www.zsl.org>>, Chester Zoo (United Kingdom) <<www.demon.co.uk/chester zoo/>>, Oceanario de Lisboa (Portugal) <<www.oceanario.pt/Inicio.htm>>, Artis Zoo (Netherlands) <<www.artis.nl/>>, Rotterdam Zoo (Netherlands) <<www.rotterdamzoo.nl/>>, Oceanographic Institute (Monaco) <<www. oceano.org/>>, Berlin Zoo Aquarium (Germany) <<www.aquariumberlin.de/>>, Genoa Aquarium (Italy) <<www.acquario.ge.it/>>.

USA: Aquarium of the Pacific <<www. aquariumofpacific.org>>, Audubon Aquarium of the Americas <<www.auduboninstitute. org/aoa.htm>>, Cabrillo Marine Aquarium <<www.cabrilloaq.org/>>, Columbus Zoo and Aquarium <<www.colszoo.org/>>, Florida Aquarium <<www.flaquarium.net/>>, John G. Shedd Aquarium <<www.sheddnet.org/>>, Monterey Bay Aquarium <<www.mbayaq. org>>, National Aquarium in Baltimore <<www.aqua.org/>>, New England Aquarium <<www.neaq.org/>>, New Jersey State Aquarium <<www.njaquarium.org>>, Waikiki Aquarium <<www.mic.hawaii.edu/aquarium/>>, Wildlife Conservation Society <<http://wcs. org/>>.

Rest of the World: OceanPark (Hong Kong) <<www.oceanpark.com.hk>>, Vancouver Aquarium (Canada) <<www.vanaqua.org/>>.

References

Baker, Andrew C. 2001. Reef corals bleach to survive change. *Nature* 411:765–766.

Bull, Colin. 2001. Seahorse husbandry in public aquaria 2001 Manual. Available from John G. Shedd Aquarium, Chicago, United States.

Hilton-Taylor, Craig (compiler). 2000. 2000 IUCN red list of threatened species. IUCN Gland, Switzerland and Cambridge, United Kingdom. 310 pp.

Hutchins Michael, and William G. Conway. 1995. Beyond Noah's ark: The evolving role of modern zoological parks and aquariums in field conservation. *International Zoo Yearbook* 34:117–130.

Hutchins Michael, Kevin Willis, and Robert Wiese. 1995. Strategic collection planning: Theory and practice. Zoo *Biology* 14:5–25.

Hutchins Michael, Kevin Willis, and Robert Wiese. 1996. Why we need captive breeding. 1996 American Zoo and Aquarium Association Regional Conference Proceedings. Wheeling, West Virginia, pp 77–86.

Hutchins Michael, Miles Roberts, Karen Cox, and Michael Crotty. 1998. Marsupials and monotremes: A case study in regional collection planning. *Zoo Biology* 17:433–451.

Kaiser, Rainer. 1994. The maintenance of pelagic jellyfish in the aquarium. Congress EUAC, Leipzig 1994. Mem. Inst. oceanogr. Paul Ricard, 47–51.

Kopecky, L.J., and G.K. Ostrander. 1999. Isolation and primary culture of viable multicellular endothelial isolates from hard corals. *In Vitro Cellular & Developmental Biology*. 35:616–624.

Lange, Jurgen, and Rainer Kaiser. 1997. Further experiences with the maintenance of pelagic jellyfish in the Zoo-Aquarium Berlin. Congress EUAC.

Munich, Salzburg & Innsbruck 1996. Mem. Inst. Oceanogr. P. Ricard 1997, 39–42.

Lunn, K.E., J.R. Boehm, H.J. Hall, and A.C.J. Vincent (eds.). 1999. Proceedings of the First International Aquarium Workshop on Seahorse Husbandry, Management and Conservation. John G. Shedd Aquarium: Chicago, United States.

Powell, David. 2000. Jellies…a sense of wonder. AZA Communiqué. November: 8–10.

Rowan, Rob, and Dennis A. Powers. 1991. A molecular genetic classification of zooxanthellae and the evolution of animal-algal symbioses. *Science* 251:1348–1351.

Rowan, Rob. 1998. Diversity and ecology of zooxanthellae on coral reefs. *Journal of Phycology* 34:407–417.

Smith, Brandie, Ruth Allard, Michael Hutchins, and Doug Warmolts. In press. Regional Collection Planning for Speciose Taxonomic Groups. *Zoo Biology*.

Trautwein, Sandy. 2001. Sea dragons make history at Aquarium of the Pacific. AZA Communiqué, October 12–13.

Wiese, Robert J., Kevin Willis, and Michael Hutchins. 1994. Is genetic and demographic management conservation? *Zoo Biology* 13(4):297–299.

Wood, Elizabeth M. 2001. Collection of coral reef fish for aquaria: Global trade, conservation issues and management strategies. Marine Conservation Society, United Kingdom, 80 pp.3

Stakeholder Perspectives

B. *Collectors*

24

Trends Determined by Cyanide Testing on Marine Aquarium Fish in the Philippines

Peter J. Rubec, Vaughan R. Pratt, Bryan McCullough, Benita Manipula, Joy Alban, Theo Espero, and Emma R. Suplido

Abstract

Cyanide has been demonstrated to kill corals, and there is scientific evidence that it contributes to the high delayed mortality of marine fish in the aquarium trade. The International Marinelife Alliance (IMA) has conducted cyanide detection testing (CDT) on marine aquarium and food fish in the Philippines since 1993. The testing was conducted under contract with the Philippine Bureau of Fisheries and Aquatic Resources. A network consisting of six CDT laboratories and four regional Marine Inspection and Sampling offices were established. The CDT network conducted random sampling of fish from collectors, middlemen, and exporters. Samples were also collected by law enforcement personnel and voluntarily submitted by fish exporters to determine whether or not the fish were cyanide-free. More than 48,000 aquarium and food fish specimens were tested from 1993 to 2001. The CDT laboratories in conjunction with law enforcement efforts have served to deter cyanide fishing in the Philippines. The proportion of marine aquarium fish tested with cyanide present dropped from 43% in 1996 to 8% in 1999, then rose to 29% in 2000. The data are presented by family to indicate the proportion of specimens with cyanide present or absent.

Introduction

The Cyanide Problem

The use of cyanide for the capture of fish has spread throughout Southeast Asia in association with the live fish trades, for marine aquarium fish exported worldwide, and for food fish (mostly groupers) exported to Hong Kong and mainland China (Rubec and Pratt 1984; Rubec 1986, 1987a, 1988; Johannes and Riepen 1995; Barber and Pratt 1997, 1998). Fishermen squirt concentrated cyanide solution on coral heads to capture fish that seek cover in the reef. Cyanide is extremely toxic to fish (Duodoroff 1980; Leduc 1984), yet if rapidly moved to clean water about 50% survive the initial exposure, despite being exposed to cyanide concentrations most likely ranging from 1,500 to 25,000 mg/liter (Dempster and Donaldson 1974; Johannes and Riepen 1995; Rubec et al. 2001a).

A major toxic action of cyanide is the inhibition of electron transport in the cytochrome transport chain, effectively arresting aerobic respiration. The cyanide ion (CN^-) has an affinity for metal-centered biomolecules and will bind to various enzymes and molecules (e.g., methemoglobin) to form relatively stable, but reversible, complexes. Effects on oxygen metabolism result in the disruption of physiological functions in fish and damage to the liver, spleen, heart, and brain (Dempster and Donaldson 1974; Dixon and Leduc 1981; Leduc 1984; Hall and Bellwood 1995; Hanawa et al. 1998). Even if fish survive the initial pulse-exposure to high levels of cyanide, they are subject to a high delayed mortality (>80%) during export to other countries (Rubec 1986, 1987b, 1987c, 1987d, 1988; Rubec and Sundararajan 1991; Rubec et al. 2001a, 2001b). This delayed mortality is greatly reduced with net-caught fish.

The application of 5,200 mg/liter CN^- [present as hydrogen cyanide (HCN) in seawater] for 10, 20, or 30 minutes resulted in the mortality of hard corals (genera *Pocillopora* and *Porites*) within 7 days (Jones and Steven 1997). Lower concentrations (520 mg/liter) resulted in the loss of zooxanthellae and impaired photosynthesis (Jones 1997; Jones and Steven 1997; Jones and Hoegh-Guldberg 1999; Jones et al. 1999). Hard and soft corals (genera *Scolymia, Goniopora, Euphyllia, Acropora, Heliofungia, Plerogyra, Favia, Sarcophyton*) exposed for 3 minutes to HCN concentrations ranging from 50 to 600 mg/liter exhibited loss of color associated with the expulsion of zooxanthellae (bleaching), disruption of protein synthesis, and altered rates of mitosis (Cervino et al., In press). Mantle tissue died and peeled off the skeleton with most genera tested, and the majority of test specimens died within 2 months. These experiments confirm earlier reports from collectors/fishers that cyanide fishing destroys coral reefs (Rubec 1986; Johannes and Riepen 1995).

The U.S. Coral Reef Task Force (USCRTF) was created by an executive order from President Clinton in 1998 to develop policies and strategies that support coral reef conservation and develop means to regulate the aquarium trades in coral reef fish and invertebrates (USCRTF 2000). The USCRTF held six public hearings to consider regulation of the aquarium trade in order to protect coral reefs (USCRTF 1999). A ban on U.S. imports was proposed to regulate the trade in aquarium fish, corals, and other reef invertebrates originating from countries lacking sustainable management plans (USCRTF 2000). Because collection of fish with cyanide is one of the most destructive

methods of fish collection, one recommendation made by the task force was that cyanide testing be implemented in the United States to help regulate imports of marine aquarium fish caught using cyanide.

Cyanide-Testing Laboratories in the Philippines

Rising concern about the detrimental effects of cyanide fishing induced the Philippine Bureau of Fisheries and Aquatic Resources (BFAR) to issue contracts to the International Marinelife Alliance (IMA) to manage and operate six cyanide detection test (CDT) laboratories. The first CDT laboratory was established at the BFAR headquarters in Quezon City in 1992. Additional CDT laboratories were created in Puerto Princesa (Island of Palawan) in 1993, near the Manila airport (Island of Luzon) in 1994, Zamboanga City (Island of Mindanao) in 1995, Palo (Island of Leyte) in 1996, as well as Davao City (Island of Mindanao) and Cebu City (Island of Cebu) in 1998. The CDT laboratory located at the BFAR headquarters in Quezon City (Metro Manila) closed after the creation of the laboratory situated at the Manila airport in 1994. The IMA operated the remaining six laboratories from 1993 to 2001 under contract with BFAR.

There were 48,689 fish and invertebrate specimens tested for the presence of cyanide (CN⁻) in the Philippines from 1993 to 2001 (Table 24.1). Total numbers by year and for each CDT laboratory are presented. At present, there are no CDT laboratories in other countries, such as Indonesia and Vietnam, where cyanide use is widespread.

Table 24.1 Summary of the number of fish and invertebrates tested for the presence of cyanide by the BFAR/IMA CDT laboratories in the Philippines from 1993 to 2001

CTD laboratory	1993	1994	1995	1996	1997	1998	1999	2000	2001	Total number
Central-Manila	0	610	556	1131	1616	3132	4443	3749	2002	17139
Puerto Princesa	68	418	495	1704	1600	2112	1335	1481	1737	10950
Palo, Leyte	0	0	0	291	532	1564	1878	1409	905	6579
Zamboanga City	0	0	261	794	573	1374	1805	1291	657	6755
Cebu City	0	0	0	0	0	974	1521	632	322	3499
Davao City	0	0	0	0	0	382	1397	1072	966	3817
Total number by year	68	1028	1312	3920	4321	9538	12379	9634	6589	48739

Methods

Sampling

Originally, most marine aquarium fish collected in the Philippines were exported from Manila. In the early 1990s, regional and international airports were established at other locations, such as Cebu City (Island of Cebu), as well as Davao City and Zamboanga City (Island of Mindanao). Hence, it was necessary to monitor and test aquarium fish near airports throughout the Philippines.

IMA operated four monitoring, inspection, and sampling (MIS) stations located in Coron (northern Palawan), Subic (Zambales area, Island of Luzon), Virac (Island of Catanduanes) and Batangas (Island of Luzon). The MIS monitoring network provided the government with the data needed to effectively monitor and evaluate the live reef fish trades in the Philippines (Alban et al. 2001).

MIS officers deputized as fish wardens by BFAR collected samples on a daily basis and forwarded them to the CDT laboratories for testing (Alban et al. 2001). Randomly selected fish samples were collected from fishers/collectors at sea or at village holding localities, from fish shipment points at piers or airports, and from the fish holding facilities of middlemen and fish exporters. Fish samples were also collected from fishers and the facilities of exporters associated with the live food fish trade. Fish specimens were sampled and identified to species using published references (Lieske and Myers 1996; Myers 1999).

The sampling program served two purposes. First, sampling for enforcement purposes was done in collaboration with law enforcement personnel (e.g., Philippines National Police-Maritime group, the Coast Guard, or municipal government officials). The MIS staff accompanied law enforcement officials in the field. Fish were confiscated from fishing boats at sea or at fishing ports, along with destructive fishing gear (e.g., cyanide tablets, cyanide squirt bottles, illegal trawl nets, and explosives). The results derived from cyanide testing of seized specimens were used as evidence in cases filed against persons charged with illegal destructive fishing.

The second purpose was to monitor the extent of cyanide fishing, to pinpoint the areas where cyanide fishing exists, and to help evaluate causes of fish kills in specific bodies of water. Random sampling for monitoring purposes was concentrated on the export facilities of live fish exporters and middlemen and on areas of conveyance.

In addition to fish sampled at random from export facilities, some marine aquarium fish exporters voluntarily submitted samples to the CDT laboratories to determine which collectors could provide cyanide-free fish. Several companies (Asian Marine Resources International, Aquarium Habitat) have marketed marine aquarium fish as being cyanide-free on the basis of CDT certificates obtained from the BFAR/IMA CDT laboratories. Aquarium Habitat also participated in a pilot program, conducted by the International Marinelife Alliance (IMA), Marine Aquarium Council (MAC)-Aquarium Fisheries Directorate (AFD), to evaluate environmentally friendly (net-caught) collection and transport procedures associated with the MAC certification program (Resor 1997, IMA-MAC 2001, MAC 2001). Hence, the CDT testing has been used to support efforts by exporters to supply net-caught cyanide-free fish (Rubec et al. 2001a).

Smaller aquarium fish (≤30 g) were kept alive in plastic bags (Alban et al. 2001). Larger fish specimens were brought back dead, packed in ice and then frozen at the MIS office. Living and/or dead frozen specimens were shipped by the most rapid means (often by air) to the nearest CDT laboratory.

Sample Preparation

Upon arrival at the CDT laboratory, the specimens were either frozen for later analyses or they were processed immediately to identify the species, measured and weighed, and then dissected to determine signs of exposure to cyanide and/or explosives prior to cyanide analyses (Manipula et al. 2001a). Gonads and otoliths (inner ear bones) were removed from groupers for population dynamics studies associated with the live food fish trade.

Fish specimens were identified from photographs and drawings in books (Rau and Rau 1980; Burgess et al. 1990; Heemstra and Randall 1993). For less common species, scientific publications with taxonomic keys were utilized.

For specimens ≤30 g total weight, the whole fish was weighed using a triple beam balance or a digital top loading balance. For larger fish, the internal organs (heart, liver, kidney, anterior intestine) and gills were removed, weighed in aggregate, and then either frozen or immediately processed for CDT analyses. Careful records were kept, and the samples were labeled for future use.

Before cyanide testing, approximately 10 g of fish tissue was first weighed and then blended with sodium hydroxide (NaOH) solution before the sample was added to a distillation flask situated in a fume hood.

Cyanide Testing

The IMA adopted and implemented the standard operating method for the determination of total cyanide ion published by the American Society of Testing and Materials (ASTM 1997), and the American Public Health Association (APHA 1998). The test procedure has undergone extensive evaluation by U.S. government agencies. Likewise, the IMA conducted quality assurance/quality control procedures that demonstrated that the CDT procedure was reliable (Manipula et al. 2001b).

The method uses reflux-distillation apparatus into which chemicals are added to deal with interfering substances. Sulfamic acid is added to reduce interference from nitrates and/or nitrites. Magnesium chloride is added as a catalyst. Concentrated sulfuric acid is added and the solution heated (using a heating mantle under the flask) for about an hour to digest the fish tissue. HCN gas passes through the reflux condenser and CN^- is captured in an absorption tube containing concentrated NaOH solution (Manipula et al. 2001b). Lead carbonate added to the absorption tube is used to eliminate sulfide interference. The CN^- concentration is then measured using an ion selective electrode (ISE) linked to a pH/ISE meter (Manipula et al. 2001c). An overview of the sample preparation and cyanide testing procedures is depicted in Figure 24.1.

Results

Whereas 48,689 aquarium fish and food fish specimens were tested for cyanide in the Philippines from 1993 to 2001 (see Table 24.1), not all of the CDT data were available when the second Marine Ornamentals Conference was held in Lake Buena Vista, Florida, November

Procedure for Cyanide Detection Test

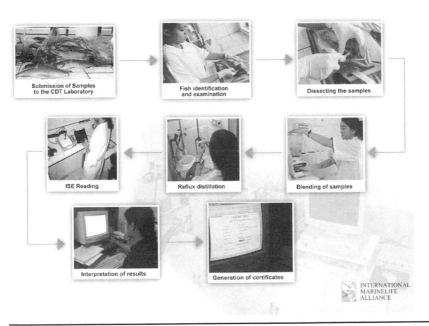

Fig. 24.1 Steps associated with processing samples and testing fish for the presence of cyanide in the BFAR/IMA CDT laboratory. (See also color plate 37.)

26 to December 1, 2001. Total numbers by year and for each CDT laboratory are presented. The Central Laboratory in Manila tested the most specimens overall (35.2%), followed by the laboratory in Puerto Princesa (22.39%).

Cyanide test results for 20,555 food and aquarium fish specimens tested from 1996 to 1999 were available in time for the conference (Table 24.2). There were 7,703 specimens of aquarium fish and 12,852 food fish specimens tested during this time period. With aquarium fishes, 44 families of fishes representing 178 genera and 625 species were tested by the CDT laboratories. With food fish, 34 families of fishes, 90 genera, and 348 species were tested. The totals assume that aquarium fish and food fish genera do not overlap in the aquarium fish and food trades. This is somewhat incorrect, because there are some genera with species found in both trades.

Overall cyanide was found to be present in 25% of the aquarium specimens tested from 1996 to 1999. Cyanide was found to be present in 44% of the food fish specimens tested (see Table 24.2). The overall percentage of specimens (aquarium and food fish combined) found with cyanide present was 37% (see Table 24.2).

On the basis of the 1996–1999 data (Table 24.3 A and B), cyanide was found to be present in 44% of the surgeonfish (Acanthuridae), 43% of the squirrelfish (Holocentridae), 34% of the triggerfish (Balistidae), 30% of the cardinalfish (Apogonidae), 29% of the angelfish (Pomacanthidae), 29% of the scorpionfish (Scorpionidae), 28% of the sweetlips (Haemulidae), 22% of the wrasses (Labridae), 21% of the butterflyfish (Chaetodontidae), 19% of the damselfish (Pomacentridae), 13% of the blennies (Blenniidae), and 12% of the parrotfish (Scaridae) (Fig. 24.2).

An interannual trend (Fig. 24.3) for the presence of cyanide was determined for marine aquarium fish from the CDT database for the years 1996 to 2000 (Table 24.4). The proportion of fishes with cyanide present across all species declined from 43% in 1996, 41% in 1997, 18% in 1998, to 8% in 1999, then increased to 29% in 2000. A similar trend was found with food fish from 1996 to 2000 (Table 24.5).

To examine the relative proportion of the number of fish tested by each CDT laboratory, the numbers of fish tested (see Table 24.1) were converted to percentages by year and percentages for each laboratory by year and across years (Table 24.6). The Manila laboratory tested the highest proportion for all years. The proportion of the total numbers tested by year remained relatively constant from 1996 to 2001, ranging from 28% to 38%. The proportion of fish tested by the laboratory in Puerto Princesa declined from 43% in 1996 to 10.8% in 1999, and then increased to 15.5% in 2000 and 26% in 2001. An increasing trend in the proportion of fish tested by the Davao laboratory also occurred. No apparent changing trend was noted with the proportion of the number of fish tested by the other laboratories.

The number of samples submitted by law enforcement officials for testing by the CDT laboratories was 37 in 1994, 19 in 1995, 95 in 1996, 121 in 1997, 270 in 1998, 346 in 1999, 181 in 2001, and 110 up to the end of September 2001.

Discussion

As previously discussed, not all of the CDT data (see Table 24.1) were available in time to be summarized and presented at the Second Marine Ornamentals Conference. Additional data for specimens tested during the years 2000 and 2001 were checked and entered into the database after the conference. A more detailed analysis of the CDT database is currently being conducted. The IMA plans to publish a comprehensive report when the work is completed.

Examination of the 1996 to 1999 data records summarized by individual species (not presented) indicates that almost all of the fish species exported from the Philippines have

Table 24.2 Summary of cyanide testing of aquarium fish and food fish conducted by BFAR/IMA CDT laboratories in the Philippines from 1996 to 1999

	Food fish	Aquarium fish	Both groups
No. of families	34	44	78
No. of genera	90	178	268
No. of Species	348	625	973
No. CN absent	7138	5754	12892
No. CN present	5714	1949	7663
Total No. tested	12852	7703	20555
Percent CN absent	56%	75%	63%
Percent CN present	44%	25%	37%

Table 24.3A Summary of cyanide testing on marine aquarium fish by family including the number of genera, and number of species tested, along with the number and percentage of fish specimens determined with cyanide present or absent from 1996 to 1999

Family	Group	No. of genera	No. of species	No. CN^- absent	No. CN^- present	Total number	Percent CN^- absent	Percent CN^- present
Acanthuridae	Surgeonfish	5	38	393	309	702	56	44
Anabantidae	Gourami	1	1	0	4	4	0	100
Anomalopidae	Lanterneye	1	1	2	0	2	100	0
Antennaridae	Frogfish	1	3	5	1	6	83	17
Apogonidae	Cardinalfish	6	15	48	21	69	70	30
Ariidae	Seacatfish	1	1	0	1	1	0	100
Balistidae	Triggerfish	10	24	166	87	253	66	34
Blenniidae	Blenny	7	11	46	7	53	87	13
Callionymidae	Dragonet	2	6	84	25	109	77	23
Centiscidae	Shrimpfish	1	1	1	0	1	100	0
Chaetodontidae	Butterflyfish	9	59	1011	264	1275	79	21
Cirrhitidae	Hawkfish	5	8	42	20	62	68	32
Dasyatidae	Ray	1	1	1	0	1	100	0
Ephippidae	Spadefish	1	5	32	12	44	73	27
Gobiidae	Goby	13	30	60	19	79	76	24
Haemulidae	Sweetlips	1	11	149	57	206	72	28
Hemirhamphidae	Halfbeak	1	1	0	2	2	0	100
Holocentridae	Squirrelfish	1	5	8	6	14	57	43
Labridae	Wrasse	30	135	849	245	1094	78	22
Lutjanidae	Snapper	3	3	9	3	12	75	25
Macrodesmidae	Firefish	1	2	28	12	40	70	30
Subtotal-A		21	101	359	2934	1095	4029	

Table 24.3B Summary of cyanide testing on marine aquarium fish by family including the number of genera, and number of species tested, along with the number and percentage of fish specimens determined with cyanide present or absent from 1996 to 1999

Family	Group	No. of genera	No. of species	No. CN^- absent	No. CN^- present	Total number	Percent CN^- absent	Percent CN^- present
Malacanthidae	Tilefish	2	5	5	4	9	56	44
Microdesmidae	Dartfish	1	5	19	5	24	79	21
Monacanthidae	Filefish	5	7	28	7	35	80	20
Mullidae	Goatfish	2	22	221	88	309	72	28
Muraenidae	Moray/Ribbon Eel	2	2	5	1	6	83	17
Ophichthidae	Snake Eel	1	1	1	0	1	100	0
Ostraciidae	Trunkfish	2	5	8	2	10	80	20
Pingulpedidae	Sandperch	1	10	46	19	65	71	29
Plesiopidae	Commet	1	1	7	3	10	70	30
Plotosidae	Eel Catfish	1	1	3	1	4	75	25
Pomacanthidae	Angelfish	7	34	594	242	836	71	29
Pomacentridae	Damselfish	19	111	1462	345	1807	81	19
Priacanthidae	Bigeye	2	2	2	1	3	67	33
Pseudocromidae	Dottyback	2	7	36	7	43	84	16
Scaridae	Parrotfish	2	3	15	2	17	88	12
Scorpaenidae	Scorpionfish	6	9	66	27	93	71	29
Scyliorinidae	Catshark	1	1	1	0	1	100	0
Serranidae	Anthias/basslets	7	19	227	74	301	75	25
Syngnathidae	Seahorse/pipefish	2	3	6	0	6	100	0
Tetradontidae	Puffer	3	10	20	10	30	67	33
Triacanthidae	Triplespine	1	1	1	0	1	100	0
Trypterygiidae	Triplefin	2	2	4	0	4	100	0
Zanclidae	Moorish Idol	1	1	43	16	59	73	27
Subtotal-A		21	101	359	2934	1095	4029	
Subtotal-B		23	75	266	2820	854	3674	
Total		44	176	625	5754	1949	7703	75	25

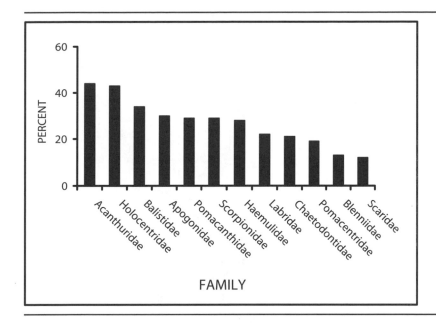

Fig. 24.2 Percent cyanide present by fish family determined by BFAR/IMA CDT laboratories in the Philippines.

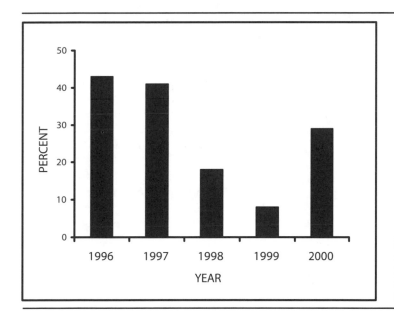

Fig. 24.3 Trend in percent cyanide present in marine ornamental fish by year determined by BFAR/IMA CDT laboratories from 1996 to 2000.

Table 24.4 Proportion of marine aquarium fish tested nationwide found to contain cyanide from 1996 to 2000 in the Philippines

Year	No. CN^- present	No. CN^- absent	Percent CN^- present	Percent CN^- absent
1996	454	601	43	57
1997	329	570	41	59
1998	728	3167	18	82
1999	285	2128	8	92
2000	1080	2681	29	71

Table 24.5 Proportion of food fish tested nationwide found to contain cyanide from 1996 to 2000 in the Philippines

Year	No. CN^- present	No. CN^- absent	Percent CN^- present	Percent CN^- absent
1996	1750	640	73	27
1997	1907	893	68	32
1998	2365	3745	39	61
1999	217	2534	8	92
2000	1389	3249	30	70

Table 24.6 Percent of the total numbers by year of marine food and aquarium fishes tested for cyanide in the Philippines by BFAR/IMA CDT Laboratories from 1993 to 2001

CDT laboratory	1993	1994	1995	1996	1997	1998	1999	2000	2001	Percent all years
Central-Manila	0	59.3	42.4	28.8	37.4	32.9	35.9	38.3	30.4	35.2
Puerto Princesa	100	40.7	37.7	43.5	37	22.1	10.8	15.5	26.4	22.5
Palao	0	0	0	7.4	12.3	16.4	15.2	14.8	13.7	13.5
Zamboanga City	0	0	19.9	20.3	13.3	14.4	14.6	13.5	10	13.9
Cebu City	0	0	0	0	0	10.2	12.3	6.6	4.9	7.1
Davao City	0	0	0	0	0	4	11.2	11.3	14.6	7.8
Total percent by year	100	100	100	100	100	100	100	100	100	

been captured with cyanide at one time or another. Both species prevalent in shallow water and species occurring in deeper water were determined to contain cyanide in their tissues. In the Philippines, the main form of cyanide placed in squirt bottles for fish collecting is sodium cyanide (Rubec et al. 2001a). In Indonesia, potassium cyanide also is used for cyanide fishing. It is apparent from the data summarized (see Tables 24.2 through 24.5) that cyanide fishing is still prevalent in the Philippines.

Because of the large number of genera and species of marine fish in the aquarium trade, it was difficult to collect a large number of any one species using random sampling. Hence, the data were aggregated by family to make meaningful comparisons of the percentages of aquarium fish specimens with cyanide present or absent (see Tables 24.3 A and B).

It is not clear from the data (see Tables 24.3 A and B) whether or not higher-priced deepwater species like angelfish (Pomacentridae) are being targeted with cyanide (S. Robinson, Cortez Marine Inc., personal communication 2001). Not enough angelfish specimens were sampled and tested to provide an adequate assessment of this assertion.

Certain species of groupers in the genus *Epinephalus* are being targeted by live food fishers using cyanide. For example, 83% of *Epinephalus sexifasciatus*, 82% of *E. caeruleopunctatus*, 61% of *E. ongus*, 57% of *E. malabaricus*, 63% of *E. microspilos*, 71% of *E. coioides*, and 64% of *E. areolatus* were found to have cyanide present.

Hypotheses

Several hypotheses have been suggested to explain the decrease in the prevalence of cyanide in marine aquarium and food fish from 1996 to 1999 and the marked increase in 2000 (see Tables 24.4 and 24.5):

(a) Cyanide testing by the CDT laboratories in conjunction with strong law enforcement may have been effective in curbing cyanide fishing up to 1999. A decline in law enforcement in 2000 may account for a rise in the use of cyanide.

(b) Changes in the proportion of fish tested regionally by different CDT laboratories may have influenced interannual countrywide trends.

(c) Declining economic conditions may have forced more people to resort to cyanide fishing in recent years.

(d) Legislation under review in 1997 by the Philippine Congress to completely ban the export of live food fish and aquarium fish may have prompted exporters and middlemen to stop supplying cyanide to the collectors. Cyanide fishing may have increased in 2000, after it became apparent that the Congress would not enact the ban.

(e) Laws passed by the Philippine Congress to control the sale, distribution, and use of cyanide may have been effective in limiting cyanide fishing up to 1999 but not thereafter.

(a) Testing and Law Enforcement   The increase in the number of samples submitted by law enforcement officials to the CDT laboratories from 1994 (37) to 1999 (346) indicates that law enforcement officials were actively enforcing laws against destructive fishing during those years. The chemists employed by IMA were frequently called to testify in court concerning the CDT testing to support enforcement of laws against cyanide fishing. About 100 persons were convicted and incarcerated for cyanide fishing. The drop in the number of enforcement samples submitted in 2000 (181) and 2001 (110) may indicate less effort to enforce laws against destructive fishing during the past two years. Hence, cyanide testing in combination with vigorous law enforcement efforts appears to have acted as a deterrent helping to reduce cyanide fishing from 1996 to 1999 (see Tables 24.4 and 24.5 and Fig. 24.3).

(b) Regional Testing Changes It has been suggested that more sampling was conducted in areas where cyanide fishing was prevalent in 2000 than in preceding years (e.g., Coron area of Palawan, Zambales area of northwestern Luzon). It should be noted that the Puerto Princesa CDT facility on the Island of Palawan tested more food fish than aquarium fish, while the Central CDT laboratory was focused more on testing aquarium fish sampled from Manila-based exporters and other areas like Zambales.

The proportion of food fish and aquarium fish combined sampled from Manila-based exporters was relatively constant from 1996 to 2001 (see Table 24.6). This tends to negate the hypothesis that more sampling regionally accounts for the higher proportion of fish with cyanide present in 2000 compared with 1999. The number of fish tested by the regional CDT laboratories (see Table 24.1) does not show a large increase in the numbers of fish sampled that would explain higher proportions of cyanide present for both aquarium fish and food fish in 2000 (see Tables 24.4 and 24.5).

More samples (see Table 24.1) and a higher proportion of samples were obtained by the Puerto Princesa CDT laboratory in relation to other CDT laboratories (see Table 24.6) in 2000 compared with 1999. The numbers of fish involved (see Table 24.1) were relatively small in relation to the numbers used to calculate trends nationwide (see Tables 24.4 and 24.5). Likewise, the number of marine aquarium fish obtained from Zambales and tested by the Central

laboratory in Manila was small (N = 801 in 1999, N = 384 in 2000) in comparison with the total numbers tested nationwide used to determine the trend for marine aquarium fish (see Fig. 24.3 and Table 24.4). Hence, it is unlikely that changes in the proportion of fish sampled regionally would affect the national trends for the presence of cyanide in either marine aquarium fish or food fish (see Tables 24.4 and 24.5).

(c) Effect of Economic Conditions Economic conditions in the Philippines have become more unstable in recent years, since the Asian Economic crisis (August 1998). The exchange rate of the Philippine peso in relation to the U.S. dollar declined from 29 pesos in 1997 to 41 pesos in 1998, 39 pesos in 1999, 44 pesos in 2000, and about 51 pesos in 2001 and 2002 (Source: Central Bank of the Philippines). This may have induced more poor Filipinos to resort to using destructive fishing methods. The number of aquarium fish collectors is believed to have increased from 4,000 to about 8,000 in recent years (F. Cruz, International Mainelife Alliance, personnel communication 2000). Likewise, the number of food fishers using destructive methods is believed to have increased along with rising human population numbers and declining economic conditions.

(d) Potential Legislation During 1997, the Philippine Congress reviewed legislation that would completely ban the export of live marine organisms, including marine aquarium fish and food fish. At about the same time, the Philippine Department of Environment and Natural Resources (DENR) attempted to completely ban exports of live marine organisms by means of a presidential decree from President Fidel Ramos. The Philippine Tropical Fish Exporters Association (PTFEA) became actively involved in opposing these measures to impose bans on the live fish trades. It enlisted the support of the IMA by promising to support sustainable harvesting methods such as the use of barrier nets and actively denounced cyanide fishing. The IMA drafted a position paper and supported the PTFEA at public hearings, since a ban would be detrimental to the aquarium fish collectors and small-scale food fishermen (IMA 1997).

(e) Effect of Laws Diminishes Republic Act 6969 was passed by the Philippine Congress in 1998 to control the availability of cyanide (Cardenas and Liao 1999). The Act specifies the requirements pertaining to the importation, manufacture, and use of cyanide and cyanide compounds in the Philippines, as well as their proper storage and waste disposal. RA 6969 also provided for the issuance of a Chemical Control Order, which states, "At no instance should cyanide and cyanide-related compounds be used or made available to the fishery sector."

Because a large part of the approximately 6,000 metric tonnes of sodium cyanide imported annually to the Philippines is used by the mining industry, the Environmental Management Bureau (EMB) and the Mines and Geosciences Bureau (MGB) within DENR, and BFAR within the Department of Agriculture, agreed to complement their efforts in the issuance of Inspection Certificates (IIC) for cyanide (Cardenas and Liao 1999). The EMB also signed an agreement with the Bureau of Customs to regulate the entry of chemicals including cyanide and cyanide compounds.

The EMB issued a Departmental Administrative Order (DAO 29) which stipulates that all manufacturers, importers, and distributors of cyanide and cyanide-related compounds need to register and secure a license from EMB before the purchase and usage of such chemicals. Users must also submit to EMB a Cyanide Management Plan and keep records of how the cyanide was used.

The regulations described by Cardenas and Liao (1999) may have made cyanide more difficult to obtain by fishers during 1999. Despite the laws and regulations, fishers increased their usage of cyanide in 2000 (see Tables 24.4 and 24.5).

Summary Of the hypotheses discussed, (a), (c), (d), and (e) appear to have influenced the trends in cyanide fishing (see Tables 24.4 and 24.5). On the basis of the discussion above, the trends presented appear to accurately monitor the usage of cyanide by fishers for aquarium fish and food fish. The cyanide testing by the BFAR/IMA CDT laboratories combined with law enforcement efforts significantly reduced cyanide fishing up to 1999. A reduction in law enforcement efforts against those involved in cyanide fishing may account for the increase in cyanide usage in 2000.

Need for Legal Measures

BFAR allowed its contract with IMA to expire at the end of September 2001. BFAR intends to continue to run the CDT laboratories using Philippine government employees. The IMA provided a training course for BFAR staff to prepare them to manage the CDT laboratories and distributed manuals documenting the sampling (Alban et al. 2001), sample preparation (Manipula et al. 2001a) and cyanide testing procedures (Manipula et al. 2001b, 2001c).

It is hoped that government agencies and the court system will continue to enforce existing Philippine fisheries laws (e.g., Fisheries Legislation enacted in 1998, Republic Act 8550) against the use of fish poisons. Philippine law enforcement concerning cyanide fishing has been directed against poor aquarium-fish collectors and to some extent against food-fish fishers. Legal steps are required against those who provide cyanide to fishermen and those who export cyanide-caught fish from the Philippines.

There also is the problem of dealing with cyanide fishing in other Southeast Asian countries and preventing the spread of cyanide use to other countries situated in Southeast Asia and in the South Pacific. Cyanide testing is needed by other countries to regulate the aquarium and food fish trades.

Need for New Testing Procedure

Part of the problem with testing for total cyanide ion (ASTM 1997; APHA 1998) is that the procedure is time consuming and labor intensive. Another problem is that cyanide is metabolized, making it more difficult to detect the presence of CN^- over time. The IMA was able to detect CN^- in tissues of marine fish 2–3 weeks after they were collected. Generally speaking, the CN^- concentrations detected were quite low (<0.2 mg/kg) by the time aquarium fish were sampled from the facilities of Manila-based fish exporters. Hence, it does not appear likely that this method could be applied to test for the presence of CN^- in marine aquarium fishes being imported from other countries into the United States.

Most of the cyanide present in fish coming into the United States is likely to be thiocyanate ion (SCN^-) rather than CN^-. The major pathway for metabolic cyanide detoxification is the conversion from CN⁻ to SCN⁻ by the enzyme rhodanese (Leduc 1984), with the reaction being essentially chemically irreversible (equilibrium constant >10¹⁰) (Westley 1981). Estimates of the ratio of CN⁻:SCN⁻ in biological fluids are very low, ranging from 1:664 in fish plasma (Heming et al. 1985) to 1:10⁴ in human plasma (Pettigrew and Fell 1973). Although SCN⁻ is much less toxic than CN⁻ (Solomonson 1981) and is readily excreted in renal filtrate, reabsorption by the kidney tubules is quite efficient, resulting in a relatively slower, more irregular excretory rate than would be expected of this type of anion (Westley 1981).

The link between waterborne CN⁻ exposure and plasma SCN⁻ has been established in a freshwater fish, the rainbow trout (*Oncorhynchus mykiss*) (Raymond et al. 1986; Speyer and Raymond 1988; Heming and Blumhagen 1989; Lanno and Dixon 1993, 1996a, 1996b). In situations of chronic cyanide exposure, detectable levels of SCN⁻ were present in the plasma of trout after one day of exposure to 0.01, 0.02, or 0.03 mg HCN/liter and increased over the 20-day exposure period (Raymond et al. 1986). Lanno and Dixon (1996a) also observed the accumulation of SCN⁻ in the plasma of trout exposed to 0.006 or 0.03 mg CN⁻/liter over a 16 week period. Bioconcentration factors (SCN⁻/CN⁻) were 170 and 88, respectively, for the two exposure concentrations.

The pharmacokinetics of plasma SCN⁻ in rainbow trout exposed to 40 mg SCN⁻/liter have been examined (Brown et al. 1995). Using depuration rate constants (k_2) ranging from 0.29 to 0.34 per day, a depuration half-life of about 4 days was estimated. Significant levels of SCN⁻ were detectable at 8 days and declined to below detection limits by 16 days. These experiments illustrate that exposure to waterborne cyanide resulted in the accumulation of SCN⁻ in the plasma of rainbow trout, that depuration was moderate, and that SCN⁻ levels in the plasma could be determined by high-pressure liquid chromatography (HPLC). The studies discussed in the last two paragraphs suggest that the determination of plasma SCN⁻ levels may be a useful biomarker of cyanide exposure.

The conversion rate of CN⁻ to SCN⁻ facilitated by the enzyme rhodanese is not simply a function of enzyme kinetics (Leduc 1984). The conversion rate appears to be limited by the availability of sulfur present in the fish. Likewise, the metabolism and excretion of SCN⁻ from the fish may be related more to osmoregulation than to temperature-mediated enzyme kinetics. Freshwater fish have a higher blood ion concentration (hyperosmotic) in relation to the surrounding water; while marine fish have a lower blood ion concentration (hypoosmotic) in relation to seawater. Hence, freshwater fish have a high rate of urinary excretion and marine fish have a low rate of urinary excretion, which helps the fish to maintain osmotic equilibrium with surrounding aquatic environments (Smith 1982).

Because the physiology of freshwater and marine fish is drastically different with regard to the maintenance of osmotic balance (Smith 1982), it stands to reason that the regulation of plasma cyanide anions may also differ. It is believed that marine fish retain SCN⁻ for a longer time period than freshwater fish because of the lower rate of urinary excretion. Hence, the interpretation of results from cyanide exposure studies conducted with freshwater fish to the toxicity and kinetics of cyanide in marine fish should be conducted with caution. Scientific studies are needed concerning the physiology and pharmacokinetics of CN⁻ and SCN⁻ with marine fish.

The development and validation of a reliable test for measuring metabolic SCN⁻ in marine fish would be of tremendous value in detecting cyanide exposure, possibly at times much longer than CN⁻ would be detectable. Research curently being undertaken by Dr. Roman Lanno of Ohio State University will determine whether a reliable HPLC test for detecting SCN⁻ in the blood of marine fish is feasible.

Management Implications

Because CN⁻ concentrations decline over time in live fish, it does not appear feasible to measure CN⁻ in marine aquarium fish being imported to the United States. It may be feasible to measure SCN⁻ for a longer time period. The HPLC method (Brown et al. 1995) may provide a practical method of rapidly measuring SCN⁻ in the blood of marine fish being imported into the United States. This would give the U.S. government a means of prosecuting companies that knowingly import fish collected with cyanide. The worst offenders are vertically integrated with business partners who have export companies in the Philippines and/or Indonesia.

Because exporting countries have been slow to implement controls in their jurisdictions, steps by the U.S. government are necessary to ensure compliance with U.S. laws, such as the Lacey Act, and the United Nations Convention on International Trade in Endangered Species of Flora and Fauna (CITES) regulations. The HPLC method for the detection of SCN⁻ may give both exporting and importing countries a more effective tool to enforce laws that help to protect coral reef habitats from cyanide fishing.

References

Alban, Joy, Benita E. Manipula, and Peter J. Rubec. 2001. Standard operating procedures for sampling marine fish and invertebrates used by the Philippines Cyanide Detection Test (CDT) Laboratory Network. Philippine Department of Agriculture—Bureau of Fisheries and Aquatic Resources/International Marinelife Alliance, 24 pp.

APHA. 1998. Method 4500-CN·E. *Standard methods for the examination of water and wastewater, 20th Edition*. American Public Health Association, Washington, D.C.

ASTM. 1997. Standard test methods for cyanides in water. In R.A. Storer et al. (eds.) *1997 Annual Book Of Standards*, Vol. 11.02, Water (ll), Designation D2036-91. American Society of Testing and Materials (ASTM), Philadelphia, Penn., pp. 76–92.

Barber, Charles V., and Vaughan R. Pratt. 1997. Sullied seas: Strategies for combating cyanide fishing in Southeast Asia and beyond. Report prepared by World Resources Institute and International Marinelife Alliance, 73 pp.

Barber, Charles V., and Vaughan R. Pratt. 1998. Poison for profits: Cyanide fishing in the Indo-Pacific. *Environment* 40(8):5–9, 28–34.

Brown, D.G., Roman P. Lanno, Mark R. van Den Heuval, and D. George Dixon. 1995. HPLC determination of plasma thiocyanate concentrations in fish blood: Application to laboratory pharmakinetic and field monitoring studies. *Ecotoxicology and Environmental Safety* 30:302–308.

Burgess, Warren E., Herbert R. Axelrod, and Raymond E. Hunzinger. 1990. Atlas of marine aquarium fishes, second edition. Neptune, N.J.: TFH Publications, 768 pp.

Cardenas, Marlito L., and Teresita Liao. 1999. Control of the use of cyanide in fishing: A Philippine management policy. In Proceedings of the first Asia-Pacific seminar/workshop on the live reef fish trade, held 11–14 August 1998, Manila, Philippines, sponsored by United States Agency for International Development Philippines, International Marinelife Alliance Proceedings No. 1, Manila. pp 46–50.

Cervino, James M., Raymond L. Hayes, Marinella Honovitch, Thomas J. Goreau, Sam Jones, and Peter J. Rubec. In press. Changes in zooxanthellae density, morphology, and mitotic index in hermatypic corals and anemones exposed to cyanide. *Marine Pollution Bulletin*.

Dempster, Robert P., and Melvin S. Donaldson. 1974. Cyanide-tranquilizer or poison? *Aquarium Digest International Tetra* 2(4): 21–22. Issue No.8.

Dixon, D. George, and Gérard Leduc. 1981. Chronic cyanide poisoning of rainbow trout and its effects on growth respiration and liver histopathology. *Archives of Environmental Contamination and Toxicology* 10:117–131.

Duodoroff, Peter. 1980. A critical review of recent literature on the toxicity of cyanides on fish. Washington D.C.: American Petroleum Institute, 71 pp.

Hall, K.C., and David R. Bellwood. 1995. Histological effects of cyanide, stress, and starvation on the intestinal mucosa of *Pomacentrus coelestis*, a marine aquarium fish species. *Journal of Fish Biology* 47: 438–454.

Hanawa, Maki, Leanna Harris, Mark Graham, Anthony P. Farrell, and Leah I. Bendall-Young. 1998. Effects of cyanide exposure on *Dascyllus aruanus*, a tropical marine fish species: Lethality, anaesthesia and physiological effects. *Aquarium Sciences and Conservation* 2:21–34.

Heemstra, Phillip C., and John E. Randall. 1993. FAO Species Catalogue, Groupers of the World (Family Serranidae, Subfamily Epinephilinae: an annotated and illustrated catalogue of the grouper, rockcod, hind, coral grouper, and lyretail species). Food and Agriculture Organization, Rome, FAO Synopsis No. 125, FIR/S125, Vol. 16, 382 pp.

Heming, Thomas A., and Karen A. Blumhagen. 1989. Factors influencing thiocyanate toxicity in rainbow trout *Salmo gairdneri*. *Bulletin of Environmental Contamination and Toxicology* 43:363–369.

Heming, Thomas A., Robert V. Thurston, Elizabeth L. Meyn, and Richard K. Zajdel. 1985. Acute toxicity of thiocyanate. *Transactions American Fisheries Society* 114: 895–905.

IMA. 1997. Position paper re the proposed ban on the collection of, and commerce in, live reef fish. International Marinelife Alliance, Manila, Philippines, 11 pp.

IMA-MAC. 2001. Feasibility study implementation for core ecosystems management (EM), collection and fishing (CF) standards, and handling and transport (HT) standards. Report prepared by International Marinelife Alliance (IMA), Aquarium Fisheries Directorate, and Marine Aquarium Council (MAC)—funded by the Marine Aquarium Council, June 2001, 20 pp. plus Attachments 1–9, Marine Aquarium Council, Honolulu, Hawaii, United States.

Johannes, Robert E., and Michael Riepen. 1995. Environmental, economic, and social implications of the live fish trade in Asia and the Western Pacific. Honolulu: Nature Conservancy Report, Honolulu, Hawaii, United States, 87 pp.

Jones, Ross J. 1997. Effects of cyanide on corals. South Pacific Commission, Noumea Cedex, New Caledonia, *Live Reef Fish Information Bulletin* 3:3–8.

Jones, Ross J. and Andrew L. Steven. 1997. Effects of cyanide on corals in relation to cyanide fishing on reefs. *Marine and Freshwater Research* 48:517–522.

Jones, Ross J., and Ove Hoegh-Guldberg. 1999. Effects of cyanide on coral photosynthesis: Implications for identifying the cause of coral bleaching and for assessing the environmental effects of cyanide fishing. *Marine Ecology Progress Series* 177:83–91.

Jones, Ross J., Tim Kildea, and Ove Hoegh-Guldberg. 1999. PAM Chlorophyll Fluorometry: A new in situ technique for stress assessment in scleratine corals, used to examine the effects of cyanide from cyanide fishing. *Marine Pollution Bulletin* 38(10):864–874.

Lanno, Roman P., and D. George Dixon. 1993. Chronic toxicity of waterborne thiocyanate to the fathead minnow (*Pimephales promelas*). *Environmental Toxicology and Chemistry* 13(9):1423–1432.

Lanno, Roman P., and D. George Dixon. 1996a. The comparative chronic toxicity of thiocyanate and cyanide to rainbow trout (*Oncorhynchus mykiss*). *Aquatic Toxicology* 36:177–187.

Lanno, Roman P., and D. George Dixon. 1996b. Chronic toxicity of waterborne thiocyanate to rainbow trout (*Oncorhynchus mykiss*). *Canadian Journal Fisheries and Aquatic Sciences* 53:2137–2146.

Leduc, Gérard. 1984. Cyanides in water: Toxicological significance. In L.J. Weber (ed.), *Aquatic Toxicology*, Vol. 2, New York: Raven Press, pp. 153–224.

Lieske, Ewald, and Robert Myers. 1996. *Coral reef fishes Caribbean, Indian Ocean, and Pacific Ocean including the Red Sea.* Princeton University Press, Princeton N.J., 400 pp.

MAC 2001. International performance standards for the marine aquarium trade, core standards and best practice guidance documents. Marine Aquarium Council (MAC), Honolulu, Hawaii (United States), Issue 1, July 2001.

Manipula, Benita E., Emma R. Suplido, and Nhilda M. Astillero. 2001a. Standard operating procedures for preparation of marine fish and invertebrate samples for cyanide testing, and assessment of blast fishing. Philippine Department of Agriculture—Bureau of Fisheries and Aquatic Resources/International Marinelife Alliance, Manila, Philippines, 19 pp.

Manipula, Benita E., Emma R. Suplido, and Nhilda M. Astillero. 2001b. Standard operating procedures for cyanide testing used by the Philippines Cyanide Detection Test (CDT) Laboratory Network. Philippine Department of Agriculture—Bureau of Fisheries and Aquatic Resources/International Marinelife Alliance, Manila, Philippines, 33 pp.

Manipula, Benita E., Peter J. Rubec, and Martin Frant. 2001c. Standard operating procedures for use of thermo orion Ion-Selective Electrode (ISE) and the ISE/PH meter to assess cyanide concentrations. Philippine Department of Agriculture—Bureau of Fisheries and Aquatic Resources/International Marinelife Alliance, Manila, Philippines, 12 pp.

Myers, Robert F. 1999. Micronesian reef fishes: A comprehensive guide to the coral reef fishes of Micronesia, 3rd revised and expanded edition. Coral Graphics, Guam Main Facility, Barrigada, Territory of Guam, 330 pp.

Pettigrew, A.R., and G.S. Fell. 1973. Microdiffusion method for estimation of cyanide in whole blood and its application to the study of conversion of cyanide to thiocyanate. *Clinical Chemistry* 19:466–471.

Rau, Norbert, and Anke Rau. 1980. *Commercial marine fishes of the Central Philippines (bony fishes).* Dentsche Geselschaft für Technische Zusammenarbeit GmbH (German Agency for Technical Cooperation), 6236 Eshbon, Republic of Germany, ISBN 3-88085-089-1, 316 pp.

Raymond, Pierre, Gérard Leduc, and Jack A. Kornblatt. 1986. Investigation sur la toxicodynamique du cyanure et sur la biotransformation chez la truit arc-en-ciel (*Salmo gaidneri*). *Canadian Journal of Fisheries and Aquatic Sciences* 43:2017–2024.

Resor, Jamie. 1997. The Marine Aquarium Fish Council: Certification and market incentives for ecologically sustainable practices. In Marea E. Hatziolos, Anthony J. Hooten, and Martin Fodor (eds.), Coral Reefs: Challenges and Opportunities for Sustainable Management, Washington, D.C., The World Bank, pp. 82–84.

Rubec, Peter J. 1986. The effects of sodium cyanide on coral reefs and marine fish in the Philippines. In Jay L. Maclean, L.B. Dizon, and L.V. Hosillos, (eds.), Proceedings The First Asian Fisheries Forum, Manila, Philippines: Asian Fisheries Society, pp. 297–302.

Rubec, Peter J. 1987a. The effects of sodium cyanide on coral reefs and marine fish in the Philippines. *Marine Fish Monthly* 2(2):7–8, 17, 20, 27, 34–35, 39, 44, 46–47, and 2(3):8–10, 14, 24, 44, 47.

Rubec, Peter J. 1987b. Cyanide and the first Asian Fisheries Forum—IMA Philippines Visit: Part 1. *Marine Fish Monthly* 2(6):36–41.

Rubec, Peter J. 1987c. Fish capture methods and Philippine coral reefs—IMA Philippines Visit: Part II. *Marine Fish Monthly* 2(7):26–31.

Rubec, Peter J. 1987d. Export of Philippine marine fish—IMA Visit: Part III. *Marine Fish Monthly* 2(8):12–13, 16–18, 20.

Rubec, Peter J. 1988. The need for conservation and management of Philippine coral reefs. *Environmental Biology of Fishes* 23(1–2):141–154.

Rubec, Peter J., and Vaughan R. Pratt (1984) Scientific data concerning the effects of cyanide on marine fish. *Freshwater and Marine Aquarium* 7(5):4–6, 78–80, 82–86, 90–91.

Rubec, Peter J., and Rengarajan Sundararajan. 1991. Chronic toxic effects of cyanide on tropical marine fish. In P. Chapman et al. (eds.), Proceedings of the Seventeenth Annual Toxicity Workshop: November 5–7, 1990, Vancouver, B.C. *Canadian Technical Report of Fisheries and Aquatic Sciences* 1774(1):243–251.

Rubec, Peter J., Ferdinand Cruz, Vaughan Pratt, Richard Oellers, Brian McCullough, and Frank Lallo. 2001a. Cyanide-free net-caught fish for the marine aquarium trade. *Aquarium Sciences and Conservation* 3:37–51.

Rubec, Peter J., Vaughan R. Pratt, and Ferdinand Cruz. 2001b. Territorial use rights in fisheries to manage areas for farming coral reef fish and invertebrates for the aquarium trade. *Aquarium Sciences and Conservation* 3:119–134.

Smith, Lynwood S. 1982. Osmoregulation. In *Introduction to Fish Physiology*, TFH Publications, Inc., Neptune, N.J. Chapter 2, pp 19–86.

Solomonson, Larry P. 1981. Cyanide as a metabolic inhibitor. In: Birgit Vennesland, Eric E. Conn, Christopher J. Knowles, John Westley, and Frode Wissing (eds.), *Cyanide in Biology*. Academic Press, London, pp. 11–28.

Speyer, Menno R., and Pierre Raymond. 1988. The acute toxicity of thiocyanate and cyanate to rainbow trout as modified by temperature and pH. *Environmental Toxicology and Chemistry* 7:565–571.

USCRTF. 1999. Draft national action plan to conserve coral reefs. U.S. Virgin Islands, November 2, 1999: Report of U.S. Coral Reef Task Force Working Groups, 46 pp.

USCRTF. 2000. International trade in coral and coral reef species: The role of the United States. Report to the Trade Subgroup of the International Working Group to the U.S. Coral Reef Task Force, March 2, 2000, Washington, D.C., http://www.coralreef.gov/wgr.html.

Westley, John. 1981. Cyanide and sulfane sulfur. In: Birgit Vennesland, Eric E. Conn, Christopher J. Knowles, John Westley, and Frode Wissing (eds.), *Cyanide in Biology*. Academic Press, London, pp. 61–76.

Stakeholder Perspectives

C. *Hobbyists*

25

Cultured Marine Ornamentals—
Retail Consumer Perspectives

Scott E. Clement

What is an Ornamental?

When we think of marine ornamentals, we typically think of all the brightly colored reef fish on display at the local pet store. But a reasonable argument can be made for a much broader working definition. Any organism that is collected from the wild or raised in captivity, put in an environment where it can be viewed and studied, and is protected from being eaten by other organisms or humans themselves becomes an ornamental. But one person's ornamental may be a food source for another individual, especially in tropical climates.

For example, a recent night-collecting trip on the Gulf of Mexico produced a range of post-larval fish, many of which do not meet the conventional definition of ornamental fish. A pinfish, two spadefish, and a flounder were netted, transported back to southern Illinois, and feed-trained on 700 µm pellets. Other species collected in the past include American eels, barracuda, and puffers. None of these fish are recognized as standards in the ornamentals trade, although some have become available in limited quantities recently. Several are considered edible at larger sizes. Nevertheless, the effectiveness of these species in display aquariums exemplifies the fact that ornamental fish are not necessarily brightly colored, as long as the species under consideration have some capacity to attract interest or to serve an educational purpose. Indeed, the freshwater ornamentals trade includes a range of species endeared for qualities other than color, often called "oddballs," "primitives," and so on. The spadefish are now on display in a local university office. Also collected were some hermit crabs and mangrove propagules, six of which have now germinated in the zoology department invertebrate aquarium.

In a sense then, anything from the ocean can potentially become a cultured ornamental. This fact is evident when one visits a public aquarium, where entire marine ecosystems are replicated and put on display with a great deal of attention to accuracy and realism. In the past decade this trend has reached the marine aquarium hobby. The bare white coral skeleton tank of the past has been replaced with models of the rocky shoreline, sand and mud flats, estuaries, coral reefs, and mangrove forests (Fig. 25.1).

Even in fish-only aquariums, live rock, realistic-looking plastic plants, and artificial corals can all provide a close facsimile to a patch reef. A complete reef aquarium can be set up with entirely cultured organisms.

Two Types of Cultured Ornamentals: Tank-Raised and Captive-Reared

Wild-caught ornamentals are just as the name suggests, animals captured directly from the wild, sent to a wholesaler, possibly to a transshipper, then on to the retailer, all hopefully in a matter of a couple weeks' time at most. Tank-raised animals are spawned in captivity either from wild-caught or cultured parentals, and then raised from eggs. Captive-reared organisms are collected as larvae or post-larvae from the ocean with plankton collectors or light traps. Larvae are cultured past metamorphosis near the point of collection, while post-larvae are feed-trained. The animals are then shipped

Fig. 25.2   Tank-raised clownfish and gobies along with captive reared tangs on display at the trade show of Marine Ornamentals 2001. These fish were raised in Puerto Rico at C-Quest's hatchery. (Photo by J. Lichtenbert; see also color plates.)

Fig. 25.1   Captive-reared Hawaiian red mangrove tree (*Rhizophora mangle*) grown in the Fisheries office at SIU from a propagule the author collected in January 1999. Six other propagules were traded for a 12.7 cm (5 inch) cultured giant clam (*Tridacna derasa*) with a local marine aquarium store. (Photo by the author; see also color plates.)

to another location to be grown out. Although they are taken from the wild, they are collected at a time when they have an extremely low survival rate, so there is minimal impact on the wild population. In contrast to many wild-caught ornamentals, the broodstock adults of the population are not removed (Fig. 25.2).

Tank-raised and captive-reared animals have many advantages over wild-caught animals. They grow up in tanks and aquariums, so they are artificially selected for an enclosed environment. They tolerate changes in water chemistry much better than their wild-caught cousins. Implicit in this selection process is the crafting of a domesticated or semi-domesticated genome, in which disease resistance, ability to grow and reproduce under artificial conditions, and other advantages are incorporated. These qualities also make these animals better able to thrive in the aquarium setting and during transport and transfer along the chain of custody. Aquaculture makes rare species such as the Banggai cardi-

nalfish more readily available to the hobbyist than they have been before and endangered species like giant clams available without breaking international law. Cultured fish of some aggressive species such as the dottybacks are less temperamental than wild-caught specimens (Fig. 25.3).

New deepwater species like the peppermint angel (*Centropyge boylei*) could become affordable and available to the average hobbyist through culture. Sensitive and difficult to keep fish such as the corallivorous butterflies can be feed trained as post-larvae and are now available for purchase on the Internet, while hardy tank-raised seahorses are also available (Fig. 25.4).

Fig. 25.3   Tank-raised orchid dottybacks (*Pseudochromis fridmani*) cultured at ORA and awaiting shipment. (Photo by the author; see also color plates.)

Fig. 25.4 Tank-raised seahorses on display at the trade show of Marine Ornamentals 2001. They were born and bred in Kona, Hawaii at Ocean Rider's hatchery. (Photo by J. Lichtenbert; see also color plates.)

Fig. 25.5 Tank-raised "reverse tomato ocellaris" (*Amphiprion ocellaris*), showing some possibilities for strain production. These designer fish were treated with thyroid hormones to examine their role in early larval development. (Photo by the author; see also color plates.)

Because cultured marine fish are already selected for captivity, they tend to be more comfortable in aquariums and are easier to spawn than wild-caught fish. Recently, new strains of cultured species have begun to appear on the market, and with selective breeding more will become available, just as has occurred with the freshwater ornamentals (Fig. 25.5).

The biggest disadvantage to cultured ornamentals is the higher price paid by the hobbyist. This may range from pennies on the dollar above the wild-caught specimens, to a prohibitively higher cost. The price increase is generated by the labor and feed costs incurred during larval rearing and growout. For some species this translates to a situation where they are not profitable for culture, especially those that have larval stages that may last as long as a year. There have been some problems with mis-barring and off-coloration in cultured clownfish, and a few have been reported not to establish the classic symbiotic relationship with anemones for which these species are so widely known. Some markets may not have access to cultured animals, or their availability is sporadic compared with wild-caught individuals of the same species. And at this time relatively few cultured species are on the market compared with the wild-caught organisms. It is possible that the status of cultivation efforts involving many of the species currently considered unprofitable could change as a consequence of changes in the availability of wild specimens.

Either a diminution of wild supplies or the imposition of increasingly prohibitive restrictions on collection could have this result, and both possibilities seem consistent with current trends.

Cultured Live Rock and Live Sand

Cultured live rock and live sand are now available on the market, but only in a limited number of pet shops. Live rock is being cultured in Fiji, Florida, and a few other locales, but wild-harvested Fiji rock still dominates the hobby and is the only live rock found for sale in southern Illinois. Cultured live rock is extremely scarce in the U.S. heartland at the time of this writing; even in major cities such as Chicago, this product is generally unavailable. Most hobbyists must settle for collected specimens, typically originating in Tonga, Samoa, or the Marshall or Solomon Islands. Distribution of live rock from these sources generally goes no farther than urban pet shops in the Midwest, and cultured live rock is considered a special-order item. Some direct marketing of these assemblages from Internet and hobby magazine sources allows determined hobbyists to obtain this sort of material. Cultured Fijian rock is quite close to the wild-collected rock both in quality and price, while the Florida rock may be a little more expensive. It can have more diversity than its wild Caribbean counterpart though, sometimes including baby corals. Some shops have cultured sand of various types, but that can be grown in the store with little difficulty. The diversity of scavengers and sand-stirrers within the cultured sand may be low or nonexistent, however. The purchase of detritivores and microfauna should alleviate this problem.

Designer live rock has come on the market the past few years. This may be in the form of special shapes such as bridges, arches, or caves, or assemblages of cultured invertebrates, such as multicolored button polyps and mushroom anemones. Cultured marine algae such as *Caulerpa*, *Gracilaria*, or *Ulva* can be purchased attached to pieces of rock. Several recipes also exist for the hobbyist to make live rock forms from concrete and coral sand. This rock can then be cultured in the reef aquarium for a year and be colonized by the coralline algae and invertebrates in the system. Cuttings of hard and soft corals, as well as the previously mentioned anthozoans, can be added to speed the in-tank culture process.

Cultured Corals

Cultured corals help to relieve some collecting pressure from wild coral reefs. There will most likely always be a wild harvest from somewhere in the world, especially if sustainable collecting methods are used, but cultured animals do have at least one advantage over their wild cousins. Generally the cultured corals tend to be hardier and more disease resistant than those collected from the reef. This is important when it comes to changing water quality conditions. The coral will go through at least four different lighting and water quality regimes in its travel from farm to wholesaler to retailer to hobbyist, unless the coral is purchased directly from the farm or from the Internet. A particular specimen must be able to withstand these changes, or it will succumb to a number of infections recently documented. One of the most sensitive corals being cultured and slowly finding its way to retail shops is the blue staghorn coral (*Acropora* spp.). These are difficult for even public aquariums to keep and propagate (Charles Delbeek, Waikiki Aquarium, personal communication 1999), but the cultured specimens are much more disease resistant, especially smaller colonies (Fig. 25.6).

Cultured corals at present are started from

Fig. 25.6 Aqua Jewels cultured corals by Walt Smith International on display at the trade show of Marine Ornamentals 2001. These corals are captive-reared in the Fiji Islands. (Photo by J. Lichtenbert; see also color plates.)

small fragments that are attached with epoxy putty or cyanoacrylate glue to a synthetic calcium base or piece of live rock (Bowden-Kerby 1999; Gaines and Main 1999; Liu et al. 1999). Hobbyists have been trading small fragments and cuttings of their hard and soft corals this way for almost as long as live corals have been offered in the trade (Fig. 25.7).

In the future it is conceivable that captive rearing of wild-caught coral eggs or larvae will take place. The timing of mass coral spawns on reefs around the world has been documented (Babcock et al. 1994; Gittings et al. 1994; Itano and Buckley 1988; Oliver et al. 1988), and one merely has to be present at the moment of release or the morning after with a plankton net. In Hawaii, several coral spawns occurred each summer in spawning events that were typically synchronized with a full moon (Clement, unpublished). These events typically produced visible and olfactory evidence that could readily be detected the following morning in nearby inshore locations. The cultivation of corals from their zygotes, as opposed to the use of fragments, may be possible by the collection of material from naturally occurring spawns as described. It is also possible that a hatchery technology could become established in which the induction of spawning is accomplished chemically as is done by potassium chloride injection of sea urchins, or by replication of lunar cycles.

Wild coral dominates the reef aquarium market and will probably continue to do so in the

Fig. 25.7 Captive-reared soft corals. (Photo by the author; see also color plates.)

case of many of the large polyp stony corals. A trend may be seen toward mostly cultured small polyp stony corals in future trade, as a result of specific colors and their hardiness relative to their wild counterparts. As is the case with live rock, wholesalers in St. Louis and Chicago have not yet begun to offer cultured corals, and considerable effort is needed on the part of retail customers seeking such. A 2- or 3-hour drive is not unusual for Midwestern hobbyists desiring to stock their aquariums with cultured corals. This inconvenience, coupled with uncertainty about the availability of preferred specimens, has led an increasing share of the Midwestern hobby market to patronize direct marketing sources. Although this adds the expense of overnight shipping charges, the consumer can typically obtain the species or variety desired, while avoiding the undesirable side effects of prolonged captivity under suboptimal conditions, such as rapid tissue necrosis (RTN) or bleaching (loss of zooxanthellae).

It is possible that some of the market demand for live corals may eventually be met by the limited distribution of Australian corals from the Great Barrier Reef. Although this particular trade is currently prohibited, some discussion has been focused on the establishment of sustainable harvest parameters, which could eventually lead to the availability of these corals to consumers. We may also see designer corals, in which pleasing color or morphological combinations are made available by grafting.

New color strains of coral such as brick red and bright orange staghorns could be created by manipulating the incorporation of selected strains of zooxanthellae. Corals respond to temperature increases with the ejection of zooxanthellae ("bleaching"), but new data suggest that this reaction may be one step in a physiological process that results in the acquisition of new symbionts. It has been demonstrated that corals can acquire better temperature-adapted symbionts from the water column after bleaching events (Kinzie et al. 2001). Similar phenomena have been observed casually (Clement, unpublished), as exemplified by a bleaching event that occurred in the summer of 1995 at the Hawaii Institute of Marine Biology. A 30-gallon holding tank with a crowded assortment of live rock, sun corals, five species of pygmy angelfish, two small maroon clownfish, and some button polyps (*Protopalythoa* spp.) was inadvertently exposed to supraoptimal water temperatures. Bleaching and an overgrowth of red slime algae (cyanobacteria) developed as temperatures reached the upper 80s (degrees Fahrenheit) by mid-summer. As temperatures dropped, the polyps began to open more, and an unusual change in pigmentation was observed. The incorporation of reddish pigment resembling the color of the cynobacteria was observed, and this persisted to some extent over the ensuing months. Although anecdotal, this evidence suggests that these and other soft-bodied anthozoans could be deliberately manipulated in such a way as to alter coloration, perhaps by using cultured zooxanthellae from sources such as red mushroom anemone or red brain coral. Whole new color strains of many species could then be produced, offering a range of appearances not available in naturally occurring corals.

Cultured Giant Clams

Giant clams have been cultured for almost two decades (Heslinga and Fitt 1987) but have been marketed in the aquarium trade for only about the past 10 years. Certain species had been extirpated in various localities throughout the Pacific, but they are now being reintroduced to their respective ranges. Because all *Tridacna* species are still listed on Appendix II of the CITES agreement, it is safe to say that without aquaculture these animals would not be obtainable by hobbyists. Now they might even be delisted in the future (Fig. 25.8).

Giant clams continue to be in the realm of the advanced hobbyist, so the pet stores in

Fig. 25.8 Cultured blue and turquoise *Tridacna maxima* awaiting transshipment at ORA. (Photo by the author; see also color plates.)

southern Illinois rarely carry them. Clams have been available in this area for only a few years. Once again a trip to the larger cities is required, but only three shops in St. Louis have any on hand. *Tridacna derasa* is by far the cheapest, hardiest, and most abundant clam available, followed by *Tridacna maxima* and finally *Tridacna crocea*. None of the other five species in the family are available for purchase anywhere, except possibly the point of culture. With the reduced cost and greater use of digital cameras, it is now possible to shop on the Internet for clams and view the exact color and species desired before placing an order for overnight delivery. Each clam has a unique mantle patterning so the consumer has a great deal of choice.

Other Cultured Invertebrates

Many other small cultured invertebrates are now available, mostly only through the Internet, but also in the hobby magazines. Many of the microfauna are utilized for scavenging leftover fish food and cleaning up detritus. These include amphipods, copepods, chitons, spaghetti and bristle worms, brittle starfish, various snails, and mysid shrimp. Queen and fighting conchs from Florida are now marketed for algae control, as well as tropical abalone. Cultured nudibranchs that eat glass anemones (*Aiptasia* spp.) can be purchased, and peppermint shrimp (*Lysmata wurdemanni*) that reportedly do the same job are also available. Many of these organisms would not be available without aqua-

culture, unless the hobbyist collected his or her own specimens in the Gulf of Mexico or on one of the coasts.

MAC Certification

The Core Standards of the Marine Aquarium Council were introduced at Marine Ornamentals 2001 and are discussed elsewhere in this volume. The Core Standards currently apply to captive-reared fish and inverts, especially at capture, during transport, and at wholesale and retail facilities. The MAC Full Standards will be released later and will include Mariculture and Aquaculture Management guidelines for tank-raised organisms. The Core Standards and the Full Standards will go a long way to ensuring the survival of both captive-reared and tank-raised organisms in the whole chain of custody from farm to hobbyist. This will mean a slight price increase passed on to the consumer, but in the long run it will save the hobbyist more money, time, and headache.

A telephone survey was conducted of the five local pet stores in southern Illinois that sell saltwater fish and/or invertebrates. Five out of the five managers had never heard of the Marine Aquarium Council, nor were they familiar with the certification process. One wholesale manager in St. Louis was familiar with MAC and agreed that a markup as much as 20% was acceptable to guarantee net-caught and healthy animals handled in accordance with MAC standards. However, this company is not currently in the practice of offering any cultured marine ornamentals at either the wholesale or retail level. A manager at the wholesale company in Chicago that supplies southern Illinois also agreed that the markup was acceptable. This wholesaler, in contrast to the St. Louis company, does offer cultured marine fish and corals, and a few reach the retail trade in southern Illinois. The retail fraction of the marine aquarium trade will probably be the last of the chain of custody to be educated about MAC, and this will have to come about by pressure from hobbyists and other wholesalers as certified organisms become available. Consumer education and increasing awareness of MAC certification throughout the marine ornamentals distribution network will probably lead to significant changes in both the supply and demand sides of the market. Eventually, it is the desire of hobbyists for better specimens that will ultimately dictate the growth of the market for MAC-certified marine ornamentals, and it is the ability to respond to consumer demand that will determine the success or failure of the various competing wholesalers.

Conclusions

From the hobbyist perspective, cultured marine ornamental organisms have more advantages than disadvantages. Their higher initial costs are perceived as justifiable, based at least in part on the implicit improvements in the areas of health and hardiness. Some organisms are not available at all without aquaculture, unless the hobbyist collects them personally, and it is an atypical hobbyist who can travel to remote reefs to collect his or her own specimens. At present, the most significant obstacles seem to arise in the areas of production and marketing. Too few cultured organisms are being produced to keep up with demand, and those available are not reaching all the markets adequately. For these reasons, consumer markets for this sort of value-added product have been slow to establish. This was a problem early in the freshwater ornamentals trade, when wild fish supplied most of the market; eventually mass culture became the dominant source of animals in the freshwater ornamentals trade. Similarly, it is likely that a greater percentage of tank-raised and captive-reared marine ornamentals will be seen in local retail markets. Probably the wild-caught component of this market will be reduced but not eliminated, as the concept of sustainable reef management is increasingly put into practice.

References

Babcock, R.C., B.L. Wills, and C.J. Simpson. 1994. Mass spawning of corals on a high latitude coral reef Coral-Reefs 13(3):161–169.

Bowden-Kerby, Austin. 1999. Working towards sustainable and environmentally sound "green" methods of coral culture. Marine Ornamentals '99 Program and Abstracts, p 18.

Gaines, Kevin E., and Kevan L. Main. 1999. Conservation of Caribbean sea fans through closed-system culture. Marine Ornamentals '99 Program and Abstracts, p 38.

Gittings, S.R., W.A. Inglehart, G.S. Rinn, and Q.R. Dokken. 1994. Coral mass spawning on the Flower Garden Banks, NW Gulf of Mexico. In Prospero, J.M., and C.C. Harwell, eds. Symp. on Florida Keys Regional Ecosystem, Miami, Florida (United States), Nov 1992. 54(3):1076.

Heslinga, Gerald A., and William K. Fitt. 1987. The domestication of reef-dwelling clams. *BioScience* 37 May 1987:332–339.

Itano, D., and T. Buckley. 1988. Observations of the Mass Spawning of Corals and Palolo (*Eunice viridis*) in American Samoa. Report to the Department of Marine and Wildlife Resources, American Samoa. 14 pp.

Kinzie, R.A., III, M. Takayama, S.R. Santos, and M.A. Coffroth. 2001. The adaptive bleaching hypothesis: Experimental tests of critical assumptions. *Biol. Bull. Mar. Biol. Lab Woods Hole* 200(1):51–58.

Liu, Li-Lian, Ta-Yu Lin, Lee-Shing Fang, and Jan-Jung Li. 1999. Effect of substrata on budding of the corals of *Pachyclavularia violacea* and *Galaxea* sp. in aquaria. Marine Ornamentals '99 Program and Abstracts, p 57.

Oliver, J.K., R.C. Babcock, P.L. Harrison, and B.L. Willis. 1988. Geographic extent of mass coral spawning: Clues to ultimate causal factors. In Choat, J.H., D. Barnes, M.A. Borowitzka, J.C. Coll, P.J. Davies, et al. eds. Proceedings of the Sixth International Coral Reef Symposium, Townsville, Australia, August 8–12, 1988. Vol. 2: 803–810.

Stakeholder Perspectives

D. *Government*

26

Balancing Collection and Conservation of Marine Ornamental Species in the Florida Keys National Marine Sanctuary

Billy D. Causey

Background

The coral reefs of the Florida Keys have been the focus of fishing and collecting activities since before the invention of SCUBA in the 1940s. These activities have increased in intensity over the past few decades, and today many Keys residents simply talk about what it used to be like in the "old days." Stories about beds of queen conch, rafts of sea turtles, huge schools of grouper, snapper, abundant tropical reef fish, and so many lobster all you had to do was wade out from shore to catch them are common.

The collection of marine ornamental species has a long history in the Florida Keys. Commercial collecting began in the mid to late 1950s, as public aquariums were stocking their display tanks for public viewing. Collecting activities included the taking of truckloads of living coral and other marine life from the coral reefs of the Upper Keys. It was, in part, a result of this and other pressures on the coral reefs that prompted a group of conservationists and environmentalists to meet at Everglades National Park in 1957 to consider what actions should be taken to mitigate for these human impacts. This effort eventually led to the designation of the John Pennekamp Coral Reef State Park in December 1960.

My wife and I have had academic, business, and personal experiences with all aspects of ornamental marine life, including degrees in marine biology, as owners and operators of wholesale and retail marine life businesses, and

myself as manager of one of the largest marine sanctuaries in the United States. Two years as the owner and operator of an exclusively saltwater retail aquarium store and 10 years as the owner and operator of a wholesale marine life collecting and research business in the Florida Keys taught me that the environmental and conservation community held some negative views about the use of marine life in collection and sale. As long as marine ornamentals have been collected in the Florida Keys, there has been a philosophical debate between conservationists and collectors over the value of the removal of marine life for use in marine aquariums. Although it was probably not the feeling of the majority of Keys residents, many questioned the value of the business, with its perceived impacts to the marine environment. Interestingly, few people have questioned the value of catching sport and food fish in Keys waters. In response to these growing concerns, a group of marine life collectors began meeting in 1974 and soon organized into the Florida Keys Marinelife Association. The intent was to be proactive in initiating management actions that would help maintain a sustainable marine life fishery in the Florida Keys. The group also served in a watchdog role to some extent, but the primary purpose was to ensure sustainability of the industry.

Although the career of a marine life collector has its rewards, it is a difficult way to make a living. The hours are long and physical demands high. A massive cold-water fish die-off in 1977 and a warm-water fish die-off in June and July of 1980 demonstrated that many stressors were affecting the health of the marine environment in the Florida Keys. Besides these natural impacts, wire fish traps were introduced into the Keys in the mid-1970s, and within months their impacts on nontarget species such as butterflyfish, surgeonfish, and angelfish and others were obvious. The State of Florida took quick action to ban fish traps; however, other fisheries management actions were not so quickly implemented. Additionally, rapid development in the Keys was destroying marine habitats, nutrients from wastewater and storm water runoff were affecting nearshore water quality, and global warming was impacting the coral reef environment.

My 12 years of business experience has been followed by 18 years' experience as a sanctuary manager. Eight and a half years was spent as the manager of the Looe Key National Marine Sanctuary (LKNMS), a 5.3 square nautical mile marine protected area that surrounded the popular spur and groove reef formation of Looe Key Reef, located off Big Pine Key, Florida. The benefits and knowledge gained while working alongside other commercial fishers and marine life collectors in the marine ornamentals trade were immeasurable in managing the LKNMS, where there was a prohibition on the collecting of all marine life by divers, except for spiny lobsters. Working in this field for more than a decade, combined with earlier business experience, gave me a unique perspective as the LKNMS manager and later as the manager of the Florida Keys National Marine Sanctuary.

Balancing Collection with Conservation

In November 1990, the 2,900 square nautical mile Florida Keys National Marine Sanctuary (FKNMS) was designated (Fig. 26.1). The Sanctuary boundary extended around all of the Florida Keys islands, beginning at the mean high tide mark. With the designation of the larger FKNMS, marine life collectors were concerned that their activities would be prohibited throughout the Sanctuary as they had been in Pennekamp Park and the Key Largo and Looe Key National Marine Sanctuaries. Members of the Florida Keys Marinelife Association were already in the process of working with state fisheries managers to develop size limits, bag limits, and gear restrictions for the collection of the diverse marine life that sustained their industry. The philosophical debate over the conservation value of removing marine ornamentals from their natural environment and shipping them hundreds or thousands of miles to landlocked destinations continued, causing concern among those whose livelihoods were dependent on the trade.

Public and private marine aquariums are important in promoting marine conservation values and ethics. It is understandable that people learn to care about marine organisms through intimate contact with them. My own personal experiences while engaged in the marine life collection business showed me that many aquarium hobbyists in any one of a dozen landlocked cities in middle America know the scientific name of every species on a 400-plus species price list. However, many have never seen the ocean. This is an incredible testimony to the educational value of living marine organisms. Today, it is those same people who often have a small piece of the coral reef environment in their living rooms, that educators, managers, and the industry are attempting to reach with the message of marine conservation and a new ocean ethic. More than 40% of North America's watershed drains into the Gulf of Mexico, eventually making its way past the coral reefs of the Florida Keys. People who live hundreds of miles from the ocean contribute to the origin of problems affecting the health of the oceans and coral reefs. Therefore, it is critical that broad scale water quality problems be addressed where the problem begins.

Despite the arguable educational and conservation influence of live marine species, concerns still exist about many issues and practices in the business of collecting marine ornamentals. The use of chemicals, especially cyanide and bleach (which are not used in the Florida Keys), the physical destruction of coral reef habitat during collection operations, and the manner in which fish and invertebrates are held during collection and transit between locations are significant problems in the industry. Good management practices must be used in every step of the marine ornamentals trade, from capture, through multiple handlers, to their final destination.

Florida Keys National Marine Sanctuary

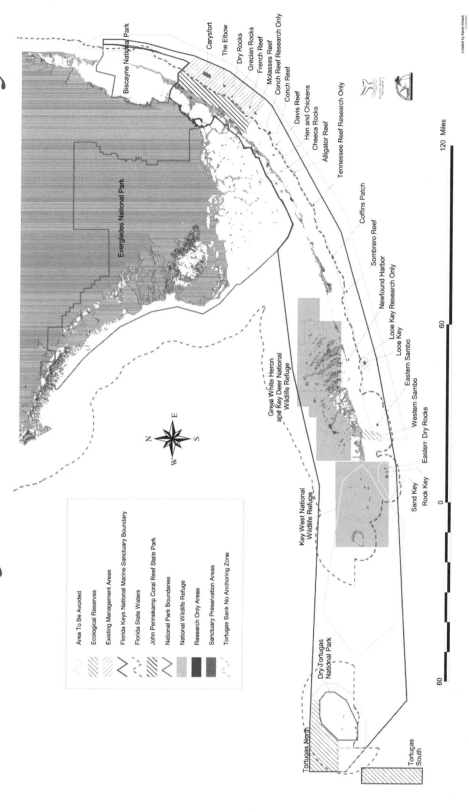

Fig. 26.1 The Florida Keys National Marine Sanctuary and marine zones used to manage the sanctuary for various uses. (See also color plates.)

Most important are the sustainability and conservation of marine life species themselves. Marine ornamental collectors in the Florida Keys capture and ship hundreds of species of fish, invertebrates, and plants. These organisms have complex life cycles and specific ecological and physiological requirements to reproduce and simply survive in their natural habitats. Although major concerns do not exist about the total numbers of plants collected and shipped from the Florida Keys, the sustainability of some of the highly desired species of fish and invertebrates is questionable. One reason is that their complex life cycles are not yet studied and known.

The State of Florida responded to one of these concerns by prohibiting the collection of the long-spined sea urchin (*Diadema antillarum*) through passage of the Marine Life Rule that went into effect January 1, 1991. This action was taken after the massive die-off of the urchin throughout its range in the wider Caribbean in 1983. Whereas marine life collecting was not responsible for the disappearance of 90% to 95% of the individuals of this species, the die-off did make apparent the lack of understanding regarding the physiology, biology, and ecology of some of these complex organisms. Today the FKNMS is working with marine life collector Ken Nedimyer and marine expert and author Martin Moe to experiment with restoration techniques that might enhance the re-establishment of wild populations of *Diadema* urchins. Their knowledge of the biology of these urchins and vast field expertise is being utilized in the Sanctuary.

Unlike other commercial fisheries, little is known about the reproductive survivorship of many of the targeted marine life species. How many adult, spawning individuals in the population does it take to guarantee that some spawn or juveniles will survive to reproductive age? How many juvenile fish of a certain species does it take to assure survivorship of a single reproductively viable adult? Where do these fish spawn, and how far are the larvae broadcasted? What are the environmental factors that affect the survivorship of annual spawn? These questions can be asked about nearly every species of fish collected.

At a minimum, the number of marine life collectors in the Florida Keys and the landings they have reported to the state are known,

thanks to the efforts of the Florida Keys Marinelife Association and the State of Florida. These data show various trends, but with so many questions about the complex life cycles of hundreds of species, it remains difficult to scientifically respond with traditional fisheries management solutions. If the landings for a species of angelfish are showing a decline, how are the environmental impacts on these species that have precise physiological requirements separated from the impacts of overfishing? It becomes quite complicated to tease out species decline due to water quality degradation compared with species decline due to the number of fish being removed from the marine environment.

There are far more questions than answers regarding the complex biology and physiology of the diverse marine life of the coral reef environment. Until investigations aimed at these and other questions are undertaken, management solutions must err on the side of conservation if managers are to be successful in sustaining these populations for the use and enjoyment of future generations. More importantly, these species must be sustained so that they may fulfill their critical ecological role in maintaining a healthy coral reef community.

Conservation Tools

Traditional fisheries management measures such as gear types, seasons, bag limits, size limits, and other management measures have demonstrated limited success in the coastal waters of the United States. Many fisheries, including snapper and grouper in the Florida Keys, are showing steep declines in landings. Many fisheries are serially overfished and continually face more stringent management actions despite our considerable knowledge of life histories, biology, ecology, and landings statistics. As discussed above, no comparable data exist for angelfish, butterflyfish, surgeonfish, jawfish, anemones, or mollusks. Therefore, how can it be guaranteed or even speculated that marine ornamental collecting is occurring in a sustainable manner?

In order to address concerns about the health of all 6,000 species of fish, invertebrates, and plants that inhabit the Florida Keys National Marine Sanctuary, a comprehensive management plan for the Sanctuary was developed. The management plan contains some of the most innovative tools available for protecting America's coral reef and its surrounding marine communities for the use and enjoyment of future generations. It also balances common sense and practical solutions. The management plan represents the most inclusive approach ever attempted at protecting a marine community as diverse, and a socioeconomic setting as complex, as that found in the Florida Keys.

A major component of the management plan has been the implementation of the first broad-scale marine zoning plan in the United States. Among other objectives, marine zoning is designed to protect the biodiversity of Sanctuary resources, including the habitats and food sources for all of the 6,000 marine species found in the Sanctuary. Much like zoning on land, marine zoning is the setting aside of areas for specific activities, which balances commercial and recreational interests with resource protection and conservation. This reduces user conflicts and focuses management activities on critical and threatened marine habitats. Marine zoning is being used in the Sanctuary to protect biological diversity and reduce human pressures on important natural resources. Five separate zone types have been established in the Sanctuary to achieve these goals: Sanctuary Preservation Areas, Ecological Reserves, Special Use Areas, Wildlife Management Areas, and Existing Management Areas. A description of each of these zone types follows (See Fig. 26.1).

Sanctuary Preservation Areas

Sanctuary Preservation Areas (SPAs) have been established to protect shallow, heavily used reefs where conflicts occur between user groups and where concentrated activity by visitors leads to resource degradation. These zones encompass discrete, biologically important areas and are designed to both reduce user conflicts in high-use areas and sustain critical marine species and habitats. Regulations for SPAs are designed to meet the objectives of these zones by limiting consumptive activities while continuing to allow activities that do not threaten resource protection. SPAs, therefore, restrict consumptive activities with two exceptions. The first exception is that catch and release fishing by trolling is allowed in four SPAs, and the second is that the taking of ballyhoo (baitfish) using nets is currently allowed by permit in all SPAs. Noncon-

sumptive activities such as snorkeling and SCUBA diving are allowed in all SPAs. There are currently 18 SPAs in the Sanctuary.

Ecological Reserves

Ecological Reserves (ERs) are areas of high habitat and species diversity that are representative of the Florida Keys marine ecosystem. These reserves have been established in the Sanctuary to protect biodiversity by setting aside areas with minimal human disturbance. ERs encompass large, contiguous diverse habitats, and by doing so they protect and enhance natural spawning, nursery, and permanent residence areas for the replenishment and genetic protection of fish and other marine life. Allowing certain areas to evolve in or return to a natural state preserves the full range of diversity of resources and habitats found throughout the Sanctuary. ERs protect the food and home of commercially and recreationally important species, as well as the hundreds of marine organisms not protected by fishery management regulations. Regulations for ERs are designed to meet the objectives of these zones by limiting consumptive activities while continuing to allow activities that do not threaten resource protection. ERs, therefore, restrict all consumptive activities and allow nonconsumptive activities in some zones where such activities have been determined to be compatible with resource protection. There are currently two ERs in the Sanctuary, the Western Sambo Ecological Reserve and the Tortugas Ecological Reserve, totaling approximately 160 square nautical miles (548 square kilometers).

Special Use Areas

Special Use Areas (SUAs) are zones that set aside areas for scientific research and educational purposes or to provide for the recovery or restoration of injured or degraded Sanctuary resources. SUAs may also be established to facilitate access or use of Sanctuary resources or to prevent user conflicts. Because SURs are designated to facilitate special Sanctuary management programs such as habitat recovery, restoration, and research, or to minimize impacts on sensitive habitats, access to these areas is restricted to entry by permit only. There are currently four permanent SURs in the Sanctuary. All have been designated for scientific research and monitoring.

Wildlife Management Areas

Wildlife Management Areas (WMAs) have been established in the Sanctuary to minimize disturbance to especially sensitive or endangered wildlife populations and their habitats. These zones typically include bird nesting, resting, or feeding areas, turtle nesting beaches, and other sensitive habitats. Regulations governing access are designed to protect endangered or threatened species or their habitats, while providing opportunities for public use. Access restrictions include no-access buffer zones, no-motor zones, idle speed only/no wake zones, and closed zones. Some restrictions specify time periods when use is prohibited. There are currently 27 WMAs in the Sanctuary. Twenty of these WMAs are under joint management of the Sanctuary and the U.S. Fish and Wildlife Service as part of its plan for managing backcountry portions of the Key West, Key Deer, Great White Heron, and Crocodile Lake National Wildlife Refuges.

Existing Management Areas

Existing Management Areas (EMAs) simply delineates areas that are managed by other agencies where restrictions already exist, such as state parks, aquatic preserves, sanctuaries, and other restricted areas. Their purpose is to recognize established management areas, complement these existing programs, and ensure cooperation and coordination with other agencies in the Florida Keys. Because EMAs are already managed by other agencies, regulations already exist under those authorities. Sanctuary regulations supplement these authorities for the comprehensive protection of resources. There are currently 21 EMAs within the Sanctuary.

Other Conservation Solutions

Another conservation solution related to the marine life trade in the Sanctuary followed the prohibition against the collection of "live rock" in state and federal waters of Florida. The state took the lead and prohibited the collection of live rock in state waters in 1989, and the two federal fisheries management councils followed

with closures in the early 1990s. These prohibitions were due to the early recognition that the collection of live rock (pieces of living coral reef substrate) was not only unsustainable but was extremely detrimental to the marine habitats of the region. Early after the prohibition went into effect, the Sanctuary began working with marine life collectors in the Keys to establish live rock aquaculture sites. These are areas where quarried rock could be placed on the sea floor to receive natural growth and settlement of marine life before being sold. This alternative to the collection of wild live rock has met with mixed success, largely as a result of storms and hurricanes that have destroyed the live rock sites and live rock being cultured. For the most part, marine life collectors who have invested time and energy in maintaining stock at their sites have had good success and now have a valuable product for sale on the market.

The skills of professionals in the marine life industry and the strong educational value of local public aquariums are assets the Sanctuary is relying upon to implement another important conservation program in the Florida Keys, called Reef Medics. Reef Medics was initiated to respond to the more than 600 boat groundings that occur on living marine habitats of the Sanctuary annually. The majority of these groundings are in sea grass habitats, but a significant number occur on living coral reefs. Reef Medics is working with volunteers, marine life collectors, and public aquariums in the immediate response to small boat groundings. Trained volunteers will stabilize corals that can be restored on site, marine life experts may assist in collecting and shipping smaller fragments to a public aquarium or holding facility, and the aquariums or educational institutions will work to stabilize the fragments. The ultimate goal is for coral pieces to be returned to the area of reef that was impacted by the boat grounding. The multiphase coral restoration operation of Reef Medics will achieve many objectives, including scientific and educational goals while increasing the public's awareness of the scope of the boat-grounding problem in the Florida Keys National Marine Sanctuary.

Conclusion

Conserving marine species and habitats through the extraordinary pressures and multiple stresses that our coastal communities face today is a challenge for any resource manager. Pollution, habitat degradation, massive algal blooms, serial overfishing, and global warming are killing coral reef communities around the world. Effective management of human impacts, including marine ornamental collecting, is therefore necessary for long-term sustainability.

In the Florida Keys National Marine Sanctuary, success has been achieved by using a variety of tools, including education and outreach, water quality management, research and monitoring, and enforcement to protect resources. As with other marine protected areas around the world, the Sanctuary relies on marine zoning as an important strategy for protecting and conserving critical marine habitats and species. Fully protected areas, such as the Ecological Reserves in the Sanctuary, take the guesswork out of answering complicated resource management questions. This also relieves the pressure to understand the complex and intricate biology of the marine organisms that live and reproduce in these areas in order to ensure their survivorship. By setting marine areas aside where human intervention is reduced as much as possible, the full diversity of marine life has a chance to survive and flourish.

Balancing the collection of marine ornamentals with marine conservation can be achieved. However, long-term success and the sustainability of marine life species depend on the success of science and management in responding to some of the enormous issues affecting the global tropical environment. Impacts to coral reefs and their inhabitants are occurring at the local, regional, and global levels. Now is the time to focus combined strengths and abilities on dealing with these serious threats at every level. The marine ornamental industry can be a powerful influence by promoting the tool of fully protected reserves around the world in order to address some of the challenges that face the industry today. The increased collaboration between marine life collectors and Sanctuary managers in the Florida Keys has been critical in striking the balance between collection and conservation. This kind of collaboration and communication between marine protected area managers and marine ornamental collectors around the world may be the key to the long-term sustainability of the inhabitants of the marine environment.

Stakeholder Perspectives

E. *NGOs/Environmental Management*

27

Wild-Caught Marine Species and the Ornamental Aquatic Industry

John Dawes

Abstract

Although the number of fish and invertebrate species bred or cultured to cater to the international marine aquarium hobby is expanding rapidly, the actual percentage of non-wild-caught species is still quite low. Some estimates place the percentage of captive-bred fish species at around 10%, but it is generally accepted within the industry that the actual figure is considerably lower. This figure will undoubtedly increase over the coming decade and probably not in a linear fashion, because major breakthroughs in husbandry and, particularly, larval nutrition are likely to result in periods of significant acceleration.

Despite this, the majority of species within the marine aquarium industry and hobby will continue to be wild caught for the foreseeable future. This has caused concern in a number of quarters, with accusations of overexploitation, irreversible reef damage, and unethical methods of collecting being leveled at the ornamental marine aquarium industry over the years.

This chapter will review and discuss the validity, or otherwise, of some of these beliefs and assumptions and their merits. It will also attempt to place them in perspective with other aspects of marine harvesting, among them the construction and food fish industries, whose practices are sometimes attributed to the marine ornamentals industry.

It will also highlight the philosophies of the modern-day ornamental aquatic industry with regard to collecting and handling methods, the conservation of reef systems, participation in— and support for—sustainable and ethical harvesting programs, the need for education and
correct information (as opposed to current levels of misinformation), and the importance of maintaining a well-monitored (certified) sector of the industry specializing in wild-caught marines.

Overview

In order to assess the significance of the wild-caught sector of the industry and hobby and place it in some form of perspective, it may prove of value to review, relatively briefly, some general statistics relating to the ornamental aquatic industry.

Some estimates place the value of the entire global aquarium industry at US$15 billion (Table 27.1). This includes the value of both the freshwater and marine sectors and the value of all supporting sectors of the industry, including equipments sales, food sales, and so on. Whereas this set of data was issued a few years ago, the latest statistics (not fully available for every sector) would not show an overall increase. In fact, trade in live fish and invertebrates has generally declined.

From the marine perspective, two values are of particular relevance: (a) the total value of the marine sector is approximately 10% of its freshwater counterpart; (b) the values given for captive-bred marines represent those that have traditionally been quoted many times (see Table 27.1). If a survey were to be carried out today, however, the percentage would more likely range between 2% and 5%, with many well-informed (unpublished) sources tending to agree that the lower end of the estimate is closer to correct. This does not mean, of course, that fewer species of marines are being bred today

Table 27.1 Ornamental aquatic industry: some global figures

Approximate export value of ornamental fish	US$206,603,000
Approximate import value of ornamental fish	US$321,251,000
Approximate import value of marine species	US$24,000,000–34,000,000
Approximate import value of freshwater species	US$287,251,000– 297,251,000
Entire industry worth about	US$15,000,000,000
Annual growth rate since 1985	c.14%
Exports from developing countries	c.63%
Exports from developed countries	c.37%
Captive-bred freshwater species/varieties	90–95%
Captive-bred marine species	5–10%

Sources: FAO 1998; Cheong 1996.

than a few years ago. On the contrary, the number is increasing steadily. The traditional figure is therefore more likely to represent an inaccuracy in the actual survey or surveys on which the percentage was based.

The global market in live ornamental fish has declined since 1996, which was the peak year (Table 27.2; Table 27.3). Trends in the data presented show how the global market in ornamental fish has been performing over the past few years in both gross figures for value and weight (see Table 27.2) and in terms of percentage increases and/or decreases (see Table 27.3). Note that, in terms of value, the 1996 estimate differs slightly from that presented earlier in Table 27.1. This demonstrates just how difficult it sometimes is to obtain absolutely accurate worldwide figures. In fact, even accurate national figures can prove difficult to obtain. For example, a survey carried out in 2001 by the American Pet Product Manufacturer's Association (APPMA) indicates that in the United States, the number of freshwater fish kept by hobbyists is 159 million, with marine species accounting for just 6 million (APPMA 2001). Thus, marine species represent only 3.6% of the total 165 million fish kept by U.S. aquarists. Most estimates in the past have placed this figure closer to 10%.

The percentage analysis is particularly interesting (Table 27.3). Total market value in-

Table 27.2 Global market in live ornamental fish

Year	Value (000US$)	Weight (metric tons)
1994	169,053	13,879
1995	202,523	24,802
1996	207,100	16,203
1997	193,601	15,014
1998	173,870	14,757

Source: FAO 1998; Olivier 2000.

creased 2.3% from 1995 to 1996, but total weight of fish marketed declined 34.7%. From 1996 to 1997, value decreased 6.5% while weight decreased 7.3%. This indicates a decline in the total number of fish marketed. Table 27.3 is particularly interesting in that while market and, hence, hobby demand declined during the latter half of the 1990s, the magnitude of that decline, at least in terms of weight of fish—an indication of actual numbers—decreased progressively during this time.

It may be true to say that the marine hobby may have suffered to an even greater extent than its freshwater counterparts, probably owing to the generally higher levels of costs involved. Today, though, we are repeatedly being told via the news media, that some telecommunications companies are cutting jobs owing to saturated markets or falling demand for items like mobile phones. Equally, there is a growing feeling (albeit unquantified and unpublished) within the ornamental aquatic industry that as a result, families once more appear to be investing in hobbies, with aquarium keeping being among these.

The largest single supplier of ornamental fish to the hobby is Singapore (Table 27.4). Of particular interest is that after a decline of three years (1996–1998), the fourth year of decline was quite small (1.2%), while an increase oc-

Table 27.3 Global market in live ornamental fish

Year	Value (US$)	Weight (metric tons)
1994–95	19.8% increase	78.7% increase
1995–96	2.3% increase	34.7% decrease
1996–97	6.5% decrease	7.3% decrease
1997–98	10.2% decrease	1.7% decrease

Source: FAO 1998; Olivier 2000.
Notes: (a) Between 1995 and 1998, there was a 40.5% decrease in weight of fish traded. (b) Between 1995 and 1998, there was a 14.1% decrease in value of fish traded.

Table 27.4   Sales of Singapore ornamental fish

Year	Millions $		Change (%)
	Singapore	U.S.	
1991	70	41.2	—
1992	72	42.4	2.9% increase
1993	73.8	43.4	2.4% increase
1994	80.3	47.2	8.8% increase
1995	84.2	49.5	4.9% increase
1996	83	48.8	1.4% decrease
1997	79	46.5	4.7% decrease
1998	73	42.9	7.7% decrease
1999	72	42.4	1.2% decrease
2000	76	44.7	5.4% increase

Source: Lim Lian Chuan, Lawrence Chia, and Jang Jing Loo, Agri-food and Veterinary Authority (AVA), Singapore personal communication 2000, 2001.

curred in 2000 (5.4%). It also appears that the first three quarters of 2001 were generally optimistic. Then came 11 September 2001 and, with it, the acceptance that this optimism may not be fully realized in the immediate future.

In summary, the marine aquarium hobby, along with its freshwater counterpart, has experienced a challenging few years in terms of popularity but has appeared—more recently—to be showing some encouraging signs of recovery. Whether this will be maintained over the coming year, or years, remains to be seen.

Captive-Bred and Wild-Caught Marines

Regardless of any changes that the future may bring, two aspects of the sector can be fairly confidently predicted: (a) the number of species of both marine fish and invertebrates bred in captivity is certain to increase; and (b) the percentage of wild-caught marine fish and wild-harvested invertebrates will continue to dominate the sector, at least, for the foreseeable future.

The past decade has witnessed major breakthroughs in the captive breeding sector of the industry. Large-scale commercial breeding of clownfish (*Amphiprion* spp.), neon goby (*Gobiosoma oceanops*), and other popular species is well known and has been well publicized. Over the past few years, however, other, more challenging species have also been bred in captivity. These include some pomacanthids, like the

Arabian Angelfish (*Pomacanthus asfur*) and the Yellow-band Angelfish (*P. maculosus*) in the Far East (Anon. 1999).

Confirmation of the actual numbers currently being produced, along with clarification of some of the techniques being employed, is still not available. However, the fact remains that captive breeding of these species is not only feasible but has actually been achieved. Others will, undoubtedly, follow.

Confirmation does exist about some other species, such as the large-bellied or pot-bellied seahorse (*Hippocampus abdominalis*) of southeast Australian waters, Tasmania and New Zealand. Sufficient numbers are currently being bred in Tasmania to meet much of the demand for this attractive species (Nigel Forteath, Emeritus Professor and Aquaculture Consultant, personal communication 2001). Reportedly, as many as 97% of all the fry survive to selling size—a much higher percentage than would be expected under prevailing predation and other conditions in the wild. Further, these captive-bred specimens do not require live foods (after the initial stages). As long as the food is rich in some omega oils, a wide range of commercial formulations will be found acceptable. In addition, these fish are reported as being "robust and disease-free," capable of courting and mating at 4 months and able to live for 9 years.

These results may seem too optimistic, since *Hippocampus abdominalis* is a temperate species. It is therefore likely not to be ideally suited to long-term exposure to temperatures above approximately 26°C (79°F) or so, although it is reported to be able to tolerate a temperature of 28°C (82°F) with careful acclimatization. Alternatively, anyone contemplating keeping this fish in an aquarium could consider the use of an aquarium chiller to maintain a water temperature in the low- to mid-20°C (68–75°F).

This is not the only seahorse species that has been bred in captivity. Successes have been reported with many others—including the Pacific Seahorse (*H. ingens*) (Galindo 1998). Pipefish, too, have been bred in commercial hatcheries, so the future of this sector of the industry and hobby appears to hold exciting possibilities.

Captive breeding will undoubtedly continue to expand and is likely to do so in surges, rather than in a linear fashion, as major advances are made, particularly with regard to larval nutrition. As this occurs, how will the industry view such progress with its inevitable consequences on the wild-caught sector?

The short answer is that the ornamental aquatic industry anticipates that the breeding of marine species in captivity will represent a major positive contribution and one that will grow in importance over the coming years. It does not, however, anticipate that it will, or should, replace the wild-caught sector, at least, not in the foreseeable future. The industry sees captive breeding as being complementary to, but not a replacement for, collection from the wild. It also fears that if the expansion in captive breeding were to escalate, with the majority of this activity being based in countries or regions located away from existing collecting zones (i.e., ex situ), the effects on both local human communities and the reefs on which they depend for their survival could be serious indeed. The following questions are typical of those asked by industry: (a) Will it eventually be possible to breed all popular reef fish species so that the marine aquarium fish industry becomes fully self-supporting? (b) Is it required, or even desirable, that all marine aquarium fish should be bred so that no claim at all will be placed on natural reefs (Brons 2000)?

The answers to these questions are not straightforward. The inescapable fact is that, on the basis of current technology, it is neither possible, nor economically viable, to breed more than around 30 popular species in captivity, whether this is carried out ex situ or in situ.

Ex situ production has the undoubted advantage because it does not involve long-distance transportation. However, production costs are relatively high in temperate and (even) subtropical regions, where labor, heating, and hardware costs can be considerable. In situ production in tropical regions will generally involve lower production costs, but the advantages that this may provide are partly or wholly nullified by higher transportation costs when compared with ex situ production. In fact, tropical marine fish bred near their place of origin can often turn out more expensive than those bred in, say, Europe. Therefore, even where environmental and some other considerations may be favorable near to the source, other factors may well dictate against, at least, some in situ captive-breeding projects.

The only other option is collection from the wild where production costs are, obviously, generally much lower than for either of the captive breeding alternatives mentioned above. Transportation costs to final destinations are, equally obviously, the same as for fish bred in situ, but, even so, wild-caught fish can usually compete with ex situ–produced fish in terms of final price to the consumer.

The decision to breed or not to breed is therefore a multifaceted one. All the factors affecting this decision must be finely balanced and assessed before a final choice is made. Quoting Brons (2000), "One possible future scenario could be a marine aquarium fish industry which breeds relatively cheap mass-produced species within, or close to, its markets, with more expensive species being produced in tropical open-system hatcheries." Species that are less in demand, or occur sufficiently abundantly, could be wild caught, using sustainable harvesting techniques, thereby stimulating efforts to preserve the reef ecosystem.

Some Wild-Caught Perspectives

All three strategies of ex situ production, in situ captive breeding, and collection from the wild, are currently in operation within the industry. Wild collection is the dominant of these (by far), a situation that has resulted in numerous accusations over the years. Accusations of over-exploitation and of physical and/or chemical destruction of reef habitats have been (and are) the most frequently encountered. Undoubtedly, some malpractice has occurred in the past, and some still does. It would be both improper and inaccurate to proclaim otherwise. This is, however, almost by definition, inevitable in any area of human activity; it most certainly is not a characteristic of just the ornamental aquatic industry. Having said this—and while emphasizing that such malpractice must not and cannot be condoned—it would be equally improper and inaccurate to proclaim or imply that malpractice is the norm rather than the exception. Those who malpractice (especially without analysis of the facts, or in ignorance of them or with intentional disregard for them) do themselves, the industry, and the hobby a gross injustice and disservice.

Table 27.5 Marine food/ornamental fish: comparative weights global annual catches/harvest (tons)

(a) Food fish harvest	100 million
(b) Food fish by catch	17 million
(c) Ornamental fish harvest	70–100

Source: Figures were obtained from calculations carried out by OATA (Ornamental Aquatic Trade Association), OFI-UK (Unpublished).

Note: Weight of ornamentals excludes transportation water.

Comparisons: (a) The marine ornamental hobby and industry harvest represents 0.00007–0.0001% of the food fish harvest, that is, a maximum of one ten thousandth of one percent (or one millionth) of the food fish harvest. (b) The marine ornamental hobby and industry harvest represents 0.0004–0.0006% of the food fish by catch or "waste," which is discarded, that is, a maximum of six ten thousandth of one percent (or a six hundred thousandth part) of the food fish by catch.

Fish

A brief look at some statistics relating to the wild-caught/harvested sectors of the marine industry in comparison with some other extractive practices will help place them in some form of perspective.

The wild-caught sector of the ornamental aquatic industry is a classic high-value/low-impact one when compared with its food fish counterpart. Misinterpreting or misquoting these figures would seem difficult. Yet this is precisely what happened after the author submitted printed copies of all the relevant statistics to a Reuters reporter following the Keynote Address delivered at the 2nd World Conference in Ornamental Fish Aquaculture at the 2001 Aquarama in Singapore.

Somehow, the 70–100 tons of marine fish collected annually for the marine aquarium trade became transformed into 70–100,000

tons, that is, figures that represent quantities one thousand times higher than the real ones! The misinterpretation was totally unintentional, but these figures, nevertheless, appeared around the world, including on the worldwide web— along with a comment attributed to the author that wild-caught fish represent a high-value/low-impact sector of the industry. To say that such misinformation is unhelpful would be putting it mildly. With commendable credit, the reporter in question issued an immediate correction, but the damage was already done, and uncorrected versions of the report continued to appear for several weeks subsequent to the original publication.

Corals

In terms of the weight of hard corals harvested for the ornamental aquatic industry, the situation is not dissimilar to that outlined above for fish. A few items of relevant data (a small fraction of total world usage) illustrate this point (Table 27.7). Also highlighted in these figures is the relatively high value per unit weight of corals collected for aquariums compared with those extracted for other purposes.

When statistics like these are universally available, it is difficult to understand how some legislators can end up concluding that the only way to save reefs is to ban all harvesting of hard corals for home aquariums. A good example of such a decision—which appears to be based on a number of criteria other than biological or conservation ones—is a decision likely (according to unconfirmed reports) to be taken in Australia banning all collection of corals from the entire Great Barrier Reef (GBR). Estimated production (i.e., natural growth) of hard corals in the GBR is some 50 million tons. Around 15 tons of these

Table 27.6 Marine food/ornamental fish: comparative values

About 3% of the weight of an ornamental fish consignment consists of fish
The remainder is water, that is, approximately 30 kg/ton = fish; 970 kg/ton = water
The average marine ornamental fish weighs approximately 8 g, that is, 125 fish per kilogram; 3,750 fish per 30 kg; 125,000 fish per ton (wet weight)
Average retail price of food fish = approximately $14,500–$16,500 per ton
Average retail price of ornamental fish = approximately $1,800,000 per ton

Sources: Ornamental Aquatic Trade Association (OFI-UK Unpublished); Ornamental Fish International database 1999.

Table 27.7.   Some coral comparisons

	Tonnages
1997 Global "aquarium" harvest	: 687 Tons (Green and Shirley 1999)
Jakarta area (coral mined for construction purposes)	: 15–37,500 tons per year (Polunin 1983)...OR 5,000 tons per year (Bentley 1998)
Maldives (coral mined for construction purposes)	: 25,000 tons per year (Brown and Dunne 1988)
Indonesia (damage caused by dynamiting)	: 15,000 tons per year (From: Cesar 1996; Spalding and Grenfell 1998)
West Lombok, Indonesia (coral mined by 60 families for lime extraction)	: 1,600 tons per year (Cesar 1996)

	Value
1997 Global "aquarium" harvest	: US$7,000 per ton
Lime produced from coral	: US$60 per ton

Sources: Green and Shirley 1999; Ornamental Fish International database 1999.

are being harvested annually (although licenses allow for up to 50 tons, that is, one millionth of the annual growth), from a total area amounting to just 0.003% of the GBR. Further, the decision to ban all coral harvesting appears to run contrary to statements from several scientists that the Australian coral fishery is ecologically sustainable and poses no threat to the GBR.

Misinformation

The "culture of misinformation," whether of the unintended kind or of a more deliberate nature, unfortunately still thrives today. It makes good headlines and is sometimes presented under the apparently contradictory, but in practice, widespread premise that "bad" news is "good" news. In other words, "bad" news is good for sales, for example, of newspapers.

There are two major damaging consequences for the industry when misinformation is published, whether it is done intentionally or not. First, there is the immediate, sometimes "shocking," impact of the news itself. In many instances, however, such impact is relatively short-lived. More ominous, though, are the long-term effects when such incorrect statements are repeated as facts in subsequent reports, articles, books, papers, newscasts, and so on. When this happens, the myths are promulgated to the extent that they become accepted as truths, which they most definitely are not.

In addition to misquotes regarding the actual size of the ornamental aquatic industry, the two other most frequently encountered areas of misinformation concern the use of dynamite

and cyanide. The industry is accused of using both fish-collecting techniques indiscriminately and to disastrous effect. Reportedly, reefs are being subjected to wholesale destruction in the name of the aquarium industry and hobby. It is sometimes even inferred that, by buying marine fish and invertebrates, aquarists are actually aiding and abetting reef destruction.

Dynamiting is undoubtedly used but not to collect fish or invertebrates for aquariums; it is mainly employed to collect fish for human consumption and other purposes. When there was a spate of such accusations some years ago, Ornamental Fish International (OFI) mounted a search for documented evidence of dynamiting being employed within the marine ornamentals wild collection sector of the industry. No documented evidence was found. This was presented to Don McAllister, the President of Ocean Voice International (OVI) (recently deceased) and he responded:

> I hear this old chestnut from aquarists from time to time. Usually it is aquarists who have just heard it, not thought about it, or confuse capture of food fishes with explosives with collecting marine aquarium fishes. No one to my knowledge collects aquarium fishes with explosives. They should not be repeating misinformation in regard to aquarium fishes. Yes, explosives are (mis)used around the world for getting food fishes. (Don McAllister, Ocean Voice International, personal communication 1999)

Dynamiting and dredging are also used to extract corals for purposes other than home aquariums, the main one being construction, whether of airport runways, houses, shrines, or a multitude of other purposes.

Turning to cyanide use, it is perfectly accurate to say that it is still being used to collect fish for aquariums. However, this fact needs to be placed in some form of perspective. It must not be forgotten that cyanide use for aquarium fish collecting is much reduced compared with what it was some years ago and that the vast majority of fish harvested with sodium cyanide are destined, not for the marine aquarium hobby, but for human consumption. It is also relevant to point out that many thousands of collectors—primarily in the Philippines—have now been trained in the use of nets instead of cyanide. This long-running project has been carried out by both Ocean Voice International (OVI)[1] and the International Marinelife Alliance (IMA)[2] with the latter now being the only one involved in this activity, as OVI concentrates on other important reef-protection programs.

Despite these efforts—which receive enthusiastic and continued support from the ornamental aquatic industry—cyanide use still occurs, although at a significantly lower and progressively decreasing level than in the past. Some may argue that the main cause of mortalities in cyanide-caught fish is not the cyanide itself but mishandling of the fish subsequent to capture. Even if this were the case (which is open to debate), the undeniable fact is that sodium cyanide does not distinguish between a target fish and the surrounding reef. It is indiscriminate in its damage, and leading trade associations the world over therefore condemn its use.

Industry Certification, Data Gathering, and Modern-Day Philosophies

Tangible proof of the condemnation of the use of cyanide may be seen in the active role played by a number of ornamental aquatic industry organizations, including (OFI), within the board of the Marine Aquarium Council (MAC). It is not, however, the purpose of this chapter to discuss the magnificent work that MAC is doing. Suffice it to say that today's marine ornamental aquatic industry regards MAC's work and the establishment of its standards as vital compo-

nents in the future development of the ornamental marine aquarium industry and hobby.

The same goes for the work being undertaken to establish the Global Marine Aquarium Database (GMAD). Again, this is being covered elsewhere in this book, so GMAD will not be discussed in any detail. It must be stressed, however, that the support that the industry is giving MAC and GMAD is indicative of its modern-day philosophy regarding the wild-caught sector. Some of the top criteria in this philosophy are: (a) a belief that the wild-caught sector is vitally important, not just for the industry itself, but for the hobby services, as well as for the protection of reefs themselves; (b) an equally firm belief that accurate data are essential for a well-monitored and well-managed industry; (c) a total commitment toward the use of collecting techniques that are both ethical and eco-friendly; (d) an equally total commitment to a certified wild-caught industry that is wholly sustainable.

Some of these philosophies are today reflected, not just in statements made by the industry itself but by some conservation organizations as well. For example, Bio-Amazonia Conservation International—responsible for Project Piaba based on the Rio Negro—has two slogans: "Buy a Fish, Save a Tree" and, perhaps more descriptively (as stated in its 1999 mission statement): "Buy an Ornamental Fish: Help Save the Rainforest" (Dawes 1999). Having actively supported Project Piaba for many years, OFI is hugely encouraged by such statements. More recently, OFI learned that one of the keynote presentations at the 2nd International Marine Ornamentals 2001 conference carried the optimistic title: "Buy a Fish, Buy a Coral, Save a Reef."

There is no denying that the prevailing mood today with regard to the wild-caught sector of the aquarium industry is a much more constructive, positive, and mutually supportive one than that which existed in the past. OFI pledges its wholehearted commitment to this ongoing "evolution" and looks forward to playing a full part in its fulfillment.

Notes

1. OVI—Ocean Voice International—is an Ottawa-based nonprofit, nongovernmental organization that "encourages harmony between people, marine

life and environment" and that is "environmental, humanitarian and global in its concerns." It works through "education, research, and economic and technical co-operation." (Website: www.ovi.ca)

2. IMA—International Marinelife Alliance—has its main headquarters in the Philippines. It is a nonprofit, nongovernmental marine conservation organization founded in 1985 "to help conserve marine biodiversity, protect marine environments and promote the sustainable use of marine resources for the benefit of local people." (Website: www. imamarinelife.org)

Acknowledgments

Svein Fosså, Aquatic Consultant and Norwegian Advisor to OFI; Elwyn Segrest, Segrest Farms, Gibsonton, Florida; Paul Holthus, Executive Director, Marine Aquarium Council, Hawaii; Don McAllister (late), President, Ocean Voice International, Ottawa, Canada; Alf Jacob Nilsen, Author, Consultant and Photographer, Norway; Nigel Forteath, Aquaculture Consultant, Tasmania; Audun Lem, Fish Utilization and Marketing Services, Food and Agriculture Organization, Rome; Stefania Vannucini, Fishery Information, Data and Statistics Unit, Food and Agriculture Organization, Rome; Lim Lian Chuan, Agri-Food and Veterinary Authority, Singapore; Lawrence Chia, Agri-Food and Veterinary Authority, Singapore; Jang Jing Loo, Agri-Food and Veterinary Authority, Singapore.

References

American Pet Products Manufacturer's Association. 2001. The 2001/2002 APPMA national pet owners survey.

Anon. 1999. Taikong Corporation. Taikong's angelic success. OFI Journal Issue 29, 2.

Bentley, N. 1998. An overview of the exploitation, trade and management of corals in Indonesia. *TRAFFIC Bulletin* 17(2), 67-68.

Brons, Robert. 2000. Coral reef fish: culture or capture? *OFI Journal* Issue 30, 8-10.

Brown, B.E. and R.P. Dunne. 1988. The impact of coral mining on coral reefs in the Maldives. *Env. Cons.* 15, 159-165

Cesar, H. 1996. Economic analysis of Indonesian coral reefs. The World Bank, Washington D.C.

Cheong, Leslie. 1996. Overview of the current international trade in ornamental fish, with special reference to Singapore. Rev. sci. tech. Off. int. Epiz., 15(2), 445-481.

Dawes, John. 1999. Crystal ball gazing. *OFI Journal*, Issue 29, 6-10.

Food and Agriculture Organization of the United Nations (FAO). 1998. *FAO Yearbook: Fishery Statistics*. Vol. 87.

Galindo, José Luis Ortíz (1998). Reef fish aquaculture (paper presented - but not subsequently published - at: First Workshop on Management Strategies for the Marine Ornate Species of the Gulf of California. La Paz, Baja California, 18-19 November 1998).

Green, Edmund and Francis Shirley. 1999. The global trade in coral. WCMC Biodiversity Series, No. 9.

OFI-UK 1999. (Now OATA - Ornamental Aquatic Trade Association), Annual Report, (1995/96), 8.

Olivier, Katia. 2000. The ornamental fish market. FAO/GLOBEFISH Research Programme. Vol. 67, 88p.

Polunin, N.V. 1983. The marine resources of Indonesia. *Ocean and Mar. Biol. Rev.*, 21, 455-531.

Spalding, M. and A. Grenfell. 1998. New estimates of global and regional coral reef areas. *Coral Reefs*, 16, 225-230.

28

Transforming the Marine Ornamentals Industry: A Business Approach

Andreas Merkl, Darcy L. Wheeles, John Claussen, and Heather F. Thompson

Introduction

The reefs of Indonesia, the Philippines, and the Western Pacific constitute one of the world's richest genetic storehouses. The region contains more than 600 of the 800 reef-building species and an overall range of more than 100,000 square miles of reef (Burke et al. 2002). This is "ground zero" for marine biodiversity, the habitat from which virtually all tropical fish and coral speciation developed. Unfortunately, the aquarium and live fish trade is contributing significantly to the destruction of these reefs, with rampant overfishing and highly damaging fishing methods, employing sodium cyanide and explosives.

What's more alarming, the reefs of the Indo-Pacific region are being wiped out for minimal economic gain. For example, Indonesian fishers collecting aquarium fish collectively net less than US$5 million per year; total aquarium fish exporter revenues are estimated at less than US$50 million per year.[1] And yet, because the destructive poison fishing practices are so widespread and damaging, these small dollar amounts threaten the existence of entire reef ecosystems. In addition, inefficiencies in the current distribution model account for tremendous costs and fish mortality at every step of the supply chain, from reef to the hobbyist's tank, furthering the destruction.

This unfortunate situation exists because the current industry business model relies on a condition of negative economic incentives and inefficient distribution practices. Under this model, the exporters and importers realize sizeable profits, while the reef, the local community, and even the consumer (who ends up with poor quality fish) suffer.

After researching the industry for most of 2001, the Conservation & Community Investment Forum (CCIF) developed a new economic model that will encourage all industry participants to promote the long-term health of coral reefs in the Indo-Pacific region. This new model for doing business is described in two documents. The first is a detailed, investable business plan to implement a for-profit business that imports only sustainably caught, certified fish and other reef products. The second document is a "companion piece" that describes the results of the fieldwork and articulates the steps industry, nongovernmental organizations (NGOs), government, and so on can take to make the industry more sustainable. This chapter will discuss the findings and recommendations of these two documents.

Field Research Summary

The problem of unsustainable reef fish harvesting has been known for some time, and a number of multilaterals and NGOs, such as the U.S. Agency for International Development (US-AID) and Coral Reef Rehabilitation and Management Program (COREMAP), have invested heavily in reform efforts. Programs include training fishers in sustainable practices, resource mapping and monitoring, and captive breeding. Whereas these programs are worthwhile, the underlying fundamental economics must also be addressed in order to solve the problem; local fishers must be able to make a living through reef preservation rather than reef destruction. In addition, the industry must adopt better management practices such as net-collecting, appropriate handling and shipping, and husbandry techniques. The Marine Aquarium Council (MAC) has recently introduced the first set of standards for industry certification; these standards provide a framework for making the industry more sustainable.

With the need to better understand the underlying fundamental economics of the industry in mind, CCIF[2] was commissioned by the David and Lucile Packard Foundation and the International Finance Corporation (IFC) to determine if an innovative, private-sector-led reform of the aquarium fish trade in the Indo-Pacific region could be implemented, and if so, how to structure the reform.

The goals of the initiative were to: (1) outline how to best implement sustainable methods of collecting, handling, transporting, and holding of aquarium fish for trade, thereby reducing the high mortality rates typical under the current business model; and (2) provide incentives for fishers to replace cyanide fishing practices with far more benign methods. The basic premise was to dramatically increase the economic value of coral reefs to local communities by creating an international distribution channel for sustainably harvested high-quality marine aquarium fish.

In April and May 2001, the CCIF team conducted 6 weeks of fieldwork in Indonesia, the Philippines, Australia, and the South Pacific. This fieldwork included interviews with government, industry, and nonprofit stakeholders, as well as extensive site visits. This fieldwork was complimented by months of thorough research in the United States on the status of the industry as well as U.S. stakeholder interviews (including NGOs, industry, and government). The findings are described in Table 28.1.

These findings lead to the primary final conclusion: Only a coordinated approach that reforms the economics of the aquarium industry, creates actively enforced marine managed areas, and provides massive technological assistance to fisheries, NGOs and communities alike is likely to succeed and protect large areas of coral reefs in the Indo-Pacific.

Economics of Aquarium Fishing

Coral reefs can generate significant returns. On the basis of CCIF field observations, depending on the intensity of fishing, aquarium exports from source countries could generate between US$1,300 and US$8,000 in net profits per square kilometer of reef per year.

However, this economic value is concentrated away from the local fishers and middlemen. To understand the economics involved, it is helpful to "follow the dollar" as an aquarium fish moves through the industry chain from collector to the U.S. wholesaler's tank.

Table 28.1   Findings from CCIF research, field work, and analysis

1	The current aquarium fishing economics pay off principally the most sophisticated exporters and importers, while local collectors and middlemen earn so little that only massive collection rates, aided by the use of cyanide and other poisons, ensure their economic survival.
2	No single organization owns or controls the entire chain of aquarium fish from collector to the hobbyist's tank, making it impossible for ecologically conscious collectors or hobbyists to track their fish across the entire chain.
3	The current economic structure of the aquarium industry's value chain invites reform. The economic gains realized from integrating operations (allowing for reduced mortality and improved species mix) will more than pay for the costs associated with upgrading to sustainable practices. However, to be successful on an industrywide scale, these improvements will require capital and technical assistance.
4	Whereas the economics of the aquarium industry lend it to profitable reform, this reform must be part of a larger marine managed conservation scheme to successfully protect the reefs in Indonesia and the Philippines, the countries that have the highest marine biodiversity.
5	Although it is certainly possible to prepare and enforce a management plan that ensures the reef's long-term survival (barring further massive global bleaching events), such an approach is unlikely to take hold in Indonesia and the Philippines before the majority of reefs are irreversibly damaged.
6	Existing attempts by NGOs to provide local alternatives to reef destruction, while extremely helpful in places such as Komodo Island National Park, are unlikely to solve the problem as a whole.

Collectors

Collector wages in developing countries range from US$2 per day for a free-diving fisher in Bali, to US$5 per day for a hookah collector on an exporter's payroll.[3] Although this compares favorably to an unskilled labor rate of about US$1 per day, unsafe diving practices prevent most fishers from pursuing this line of work past the age of 35. Additionally, most independent fishers have seen their income decrease as the price for lower-value species (damsels, zebras, etc.) has been eroded by oversupply, the availability of more profitable species has decreased, and bribes required to local authorities and export "screening personnel" are increasing. In many cases, the collectors are terminally indebted to middlemen and/or exporters who provide loans for boats and engines. This is not a good life. By contrast, a skilled Australian collector (after a year of training) can make US$3,000 per month, with full benefits, and the use of state-of-the-art equipment such as dive computers.[4]

Middlemen

Middlemen get a bad reputation as profiteers and prime drivers of the cyanide trade. In fact, most of them barely get by. They typically have very small operations, often consisting of little less than a shack on the beach, some holding cages off the beach, one or two boats, and (sometimes) a truck. The gross margins realized by middlemen, none of whom have an export license on their own, are extremely low, ranging from an estimated 14% to 11%[5] (see Table 28.2), and many are in unsolvable debt situations with exporters in Bali, Jakarta, or Manila. It is doubtful that any of them at this time can realize sufficient profits to professionalize their operations. The resulting primitive methods of handling and storage contribute greatly to fish mortality. In terms of propagation of cyanide practices, many middlemen are aware that a shift to net-caught techniques is necessary. However, without the capital to buy nets, hire import experts, train collectors, and absorb the costs of the inevitable initial reduction in volume as the collectors learn new techniques, this simply cannot be done.

Exporters

Exporters, by contrast, can make considerable profits. This is primarily driven by their ability to market the fish to U.S. and European markets at considerable markups. Typical markups for representative species of fish at every level of the industry value chain are shown in Table

Table 28.2 Middleman economics (in US$)

	Bali		Sulawesi	
	Middleman A	Middleman B	Middleman C	Middleman D
Revenue	15,470	12,514	43,951	79,133
COGS	10,099	6,443	20,348	30,541
Sales costs				
Packaging	1,488	1,373	1,351	5,490
Direct sales	287	539	457	2,413
Transportation	—	455	904	18,867
Gross margin	1,820	1,792	19,085	13,919
Percent	12%	14%	43%	18%
SGA				
Boat costs	4,250	—	12,750	4,000
Facility costs	360	2,363	240	4,646
Labor	122	870	2,761	2,522
Depreciation	1,000	—	3000	1565
Net margin	−2,136	472	2,141	9,089
Percent	−14%	4%	5%	11%

Note: COGS = cost of goods sold; SGA = sales and general administrative (costs).

Table 28.3   Typical markups, Indonesia (in US$)

Species	Collector	Middleman	Exporter	Importer
Three spot damsel	$0.01	$0.04	$0.27	$1.35
Blue devil damsel	$0.01	$0.04	$0.26	$1.25
Anemonefish	$0.04	$0.14	$0.78	$12.50
Clown triggerfish	$3.00	$6.00	$14.90	$49.95
Emperor angelfish	$3.00	$6.00	$15.25	$64.95
Koran angelfish	$0.40	$1.00	$6.32	$23.95

28.3.[6] Whereas middlemen can charge relatively small markups on a small base, the value accumulates at the exporter level.

The exporter's profit margins depend highly on the quality of supply because demand for high-end fish is virtually assured. The exporter has two basic options: rely on the network of middlemen for supply (nonintegrated model) or build a proprietary fleet of collection boats.

Net margins for nonintegrated exporters are lower. Relying on middlemen is less profitable because of the inevitable oversupply of low-end fish, large fluxes in capacity utilization, poor fish quality, and no hope of verifying chain of custody for certification purposes. In Indonesia, the competition for nonintegrated exporters has become quite fierce, as the number of exporters alone in Bali, Indonesia has grown from 5 to more than 20 in the past 5 years alone. For large, well-run *partially* integrated operations, net margins can range from zero to more than 30%.[7]

Exporters who have captive fleets avoid these problems. Although few exporters have been able to provide the necessary capital and management skills to successfully run a fleet of 20-plus collection boats that are required to fill a full-scale export facility, many run a small number of boats and/or local collection stations to fill a base volume of supply.

For the small number of large operations who have the capital and know-how to run their own fleets, aquarium fishing is a highly profitable business. Field observations showed that returns for these integrated exporters could be up to 36% (Table 28.4).

Importers

There are many types of "importers": consolidators, jobbers, transshippers, wholesalers, and so on. The majority of fish will go through a wholesaler's tank. For the wholesale importer,

the quality and quantity of supply is the most critical profit driver. By relying on a proprietary (or at least fully controlled) collection operation, an importer can control fish quality and species assortment, eliminate exporter margins, and develop brand recognition. However, this is quite difficult to do, and only two operations in the United States today have built a successful fully integrated operation. The major barrier to integration is capital. In an industry where even the largest importers have sales below US$8 million, the capital required to build a (semi) proprietary collection infrastructure is difficult to come by; a fully scaled in-country collection operation will cost about US$1,000,000, with another US$500,000 required for working capital. In the absence of an integrated operation,

Table 28.4   Integrated exporter economics in Indonesia (in US$)

Revenue	1,531
Collector costs	120
Skiffs	33
Collection stations	282
Transportation to central facility	11
COGS	446
Gross profit	1,085
Salaries	227
Travel	18
Facilities	98
Packaging	59
Transportation	12
Insurance	15
Permits	153
Monitoring	52
Training of collectors	3
Other	30
Operating expenses	667
Operating income	418
Net margin	27%
EBITDA	557
ROIC	36%

Note: COGS = cost of goods sold; EBITDA = earnings before interest, taxes, depreciation, and amortization; ROIC = return on invested capital.

importer net margins are between 2% and 10%.[8]

Resulting Economic Drivers

Unfortunately, the economics of the aquarium trade are tailor-made for destructive fishing methods for the reasons discussed in the following sections:

Mortality Matters Little

In the aquarium trade, the exporter's profits are relatively insensitive to the cost/mortality of the fish; the cost of goods sold for the lower-value fish (about 80% of total sales) accounts for less than 5% of revenues.[9] The major profit drivers for exporters are turnover, capacity utilization, and species mix. Therefore, from a strictly economic standpoint, it does not make sense to switch to nondestructive methods, at least in the short-term. Nondestructive practices save little in terms of mortality costs and cost plenty in terms of direct training and infrastructure expenses, as well as indirect costs associated with temporary reductions in volume. Although cyanide prices have recently increased sharply, they have not risen to the degree that would be necessary to directly offset the costs of switching to nondestructive fishing methods.

Outsiders Profit Most

Nonlocal collectors do a significant portion of cyanide fishing. Live food fish syndicates, often financed by Hong Kong importers, fan out all over Indonesia. Indonesian Bali- and Java-owned aquarium collection boats are similarly peripatetic. In the Philippines, traveling collector boats probably account for more than 50% of the total harvest.[10] This effectively divorces the local community from any economic interest in the well-being of "its" reefs. In Indonesia, the free access issue looms large; communities have no legal basis for keeping outsiders out. This makes the development, implementation, and enforcement of strategies to ban destructive fishing practices, even if pursued at the national level, exceedingly difficult (Pet-Soede et al. 2000). In Fiji, a quite different model applies: Local communities own reefs, and exporters must contract with local collectors for their fish. This way, the economic value of the reef is rec-

ognized locally, and it is no surprise that the Fijian reefs are in far better shape than their Indonesian counterparts.

Destructive Reef Fishing Is Well Organized

As discussed above, capital accumulates in relatively few hands in the aquarium trade; the harvesting, handling, and long-distance transportation/distribution of aquarium fish and corals are reasonably complex and capital intensive. Exporters are therefore well organized, highly profitable, and able to dictate terms and practices. Where blatantly illegal practices come into play, such as the non-CITES (the Convention on International Trade in Endangered Species of Wild Fauna and Flora) export of live coral, operators and authorities often work together in organizing highly profitable, protected cartels.[11]

Distribution Channels Are Undifferentiated and Ambivalent

The current chain of custody for the average reef fish involving any number of independent fishers, middlemen, exporters, transshippers, consolidators, and so on, is virtually impossible to control. A single fish, or batch of fish, from one harvest operation simply cannot be traced through the supply chain. As a result, fish caught in a nondestructive fashion are almost always comingled with "cyanide" fish. U.S. consumers have no way to purchase fish that are guaranteed cyanide free, even if they are willing to pay more for a more vigorous fish that has lower mortality. Unable to reap returns on their investment in nondestructive harvest training and equipment, even anticyanide collectors are often forced to "backslide" and return to destructive practices. In addition, a spot-check of U.S. retailers has shown that consumers often do not get the full truth about the origin of the fish; many retailers claim that their Indonesian and Philippine fish are cyanide free, which is plain and simply wrong.[12]

Opportunities for Reform

Fortunately, the structure of the aquarium fish value chain invites profitable reform. The basic strategy is to offset the considerable cost of conversion to sustainable operations (capital upgrades, collector training, reef monitoring, sharply higher earnings for local fishers, ramp-up costs, etc.) with the even more considerable savings provided by a fully integrated distribution channel. These savings include mortality reduction and improved species mix.

Mortality Reduction

The current distribution model allows no quality control over harvest, handling, and husbandry procedures. Consequently, the mortality rates can be astronomical; field estimates ranged from 40% (for high-end fish) to over 80% (for low-end fish such as banggai cardinals in Northern Sulawesi).[13] By controlling all aspects of fish harvest and transportation, this mortality rate can be significantly reduced, if not eliminated. Although mortality costs for lower-end fish are relatively low, they are an important consideration, principally for the rarer, higher-end fish (such as *Pomacanthus imperator*, the emperor angelfish).

Improved Species Mix

Independent collectors often "bundle" the most profitable, sought-after "high-end" fish with a significant number of "low-end" fish of lower profitability. Exporters depending on these collectors have no choice but to buy, inventory, and ship these low-end, low-margin fish.[14] The market therefore becomes oversaturated with low-end fish, which has the following negative implications: (1) more fish are removed from the reef than is necessary, and (2) the price of low-end fish is depressed throughout the supply chain. These problems can be avoided by improving the species mix to include more middle and high-end fish and fewer low-end fish.

Integrated versus Traditional Approach

When developing a new economic model, the CCIF team compared three types of exporters: average (traditional), partially integrated, and fully integrated. The average exporter buys all fish from middlemen, and the species mix exported represents the average Philippine tropical fish export for 2000. The "partially integrat-

ed" exporter has made some investments in controlling supply but remains somewhat dependent on middlemen, and the species mix is more heavily weighted toward high-end fish. The "fully integrated" exporter has invested in proprietary collection stations, boats, and salaried collectors, and thus has complete control over fish supply.

The integrated exporter also realizes a number of indirect advantages that, while potentially significant, were not included in the analysis. These include optimizing inventory management, improved marketing, and increasing fleet efficiency.

To illustrate the economics involved in the switch toward integrated operations, CCIF first contrasted the profitability of a "traditional" Manila-based exporter with that of an "integrated" exporter. Second, the analysis showed how full control of this integrated exporter affects the profitability of a typical U.S. importer.

Exporter Economics

A traditional Manila exporter does not control the collection of fish. Instead, fish are bought from independent contractors (middlemen or independent collectors), with all the attendant complications in terms of mortality and nonoptimal species mix. Whereas, in some cases, traditional exporters finance the boats of the contractors in order to gain some measure of control over supply, they do not own their own collection stations. A hypothetical "integrated" exporter, by contrast, owns all boats as well as the collection stations, and collectors earn salaries. This integrated exporter can count on a reduction in mortality from current levels to less than 10% resulting from superior fish handling and nondestructive harvest methods (i.e., no cyanide) and can adjust species mix to fit market conditions.[15]

The economic advantages of the improved species mix and lessened mortality are considerable. They have the potential to not only fully offset the costs of converting operations to a fully sustainable, nondestructive set of practices (i.e., establishment of local collection stations, training of fishers, reef baseline assessment and monitoring, and ramp-up costs), but also to dramatically improve exporter profitability. The income statements for these three types of ex-

Table 28.5   Benefits of integration

Annual income statement (US$ in 000's)			
	Average exporter	Partially integrated exporter	Integrated exporter
Sales			
Fish—extra low end	80.2	81.2	78.9
Fish—low end	45.0	65.4	91.6
Fish—medium end	118.1	164.8	240.7
Fish—high end	27.6	94.9	267.4
Fish—very high end	61.5	165.9	415.3
Others & discounts	-3.2	-6.1	0.0
Total sales	329	566	1094
COGS			
Collection costs			
Collector costs	80	80	120
Mother boats	0.0	0.0	0.0
Skiffs	14.8	14.8	33.4
Collection stations	0.0	141.1	282.1
Middlemen	37.8	15.1	0.0
Transportation to central facility	14.3	14.4	11.6
Other costs	0.0	0.0	0.0
Total COGS	146.8	265.3	447.2
Gross profit	182.3	300.8	646.7
Operating expenses			
Central station			
Salaries & benefits	135.3	135.3	144.1
Travel	12.0	12.0	18.0
Facility	20.2	24.5	90.0
Packaging	12.6	21.7	42.0
Transportation to International airport	5.3	6.2	10.2
Insurance	3.3	5.7	10.9
Permits	32.9	56.6	109.4
Monitoring	12	12.0	52.0
Training—collectors	0.0	0.0	3.1
Other	24.0	24.0	30.0
Total operating expenses	257.6	298.0	509.7
Operating income	-75.3	2.8	137.0
Other income (expense)	0.0	0.0	0.0
EBIT	-75.3	2.8	137.0
EBITDA-	45.3	29.1	275.9
EBITDA margin	-13.8%	5%	25%

Note: EBITDA = earnings before interest, taxes, depreciation, and amortization. It is the best representation of the pure operational economics of a business.

porters are contrasted (Table 28.5). The costs were built "bottom up" from field data.

On the basis of income statements derived from field interviews, it is clear that, despite the considerable additional amortization, salary, monitoring, and training expenses, the integrated model is far more profitable. An average exporter's earnings before interest, taxes, depreciation, and amortization (EBITDA) was estimated at 13.8%, a partially integrated exporter at 5%, and a fully integrated exporter at 25%.

Importer Economics

At the importer level, the benefits of the integrated approach are equally pronounced, since the issues of inventory management, product quality, transportation cost control, and so on are amplified with volume. The economics of an importer who can rely on a "captive" exporter in the Philippines for the great majority of his product will see estimated margins of about 42% (this analysis is based on field observations and assumes that a U.S. importer fi-

Table 28.6   Integrated vs. traditional importer

Annual income statement (US$, in 000's)		Traditional importer	Integrated importer
Sales			
	Fish	7,383.5	17,168.3
	Complimentary sales	0.3	0.3
	Dry goods	279.2	279.2
	Freight sales	156.0	156.0
	Publication	80.3	80.3
	Others & discounts	80.3	80.3
Total sales		7,980.0	17,764.4
COGS			
	Fish	4,257.0	9,898.5
	Discount	—	(490.4)
	Other sales	387.8	387.8
	Inventory loss	124.0	20.0
	Publishing	69.5	69.5
	Other costs	413.7	413.7
Total COGS		5,252.0	10,299.1
Gross profit		2,728.0	7,465.3
		34%	42%
Operating expenses			
	G&A	2,057.9	1,646.3
	Depreciation	22.3	22.3
	Marketing	25.9	25.9
	Other	30.0	30.0
Operating expenses		2,136.1	1,724.5
Operating income		591.9	5,740.8
Other income (expense)		-	-
EBIT		591.9	5,740.8
EBITDA		769.5	7,463.3
EBITDA margin		10%	42%

Note: EBIT = earnings before interest, taxes; EBITDA = earnings before interest, taxes, depreciation, and amortization; G&A = general and administration.

nances the conversion of a Filipino exporter in exchange for a locked-in discount of 15%). In contrast, a "traditional" importer will see margins of only 10%.[16] The majority of the savings result from the cheaper supply, reduced ordering and inventory complexity, and of course, improved species mix (see Table 28.6).

This financial model does not assume a price premium for sustainably collected fish, which could add to the profitability of the model. The economic value is derived exclusively from improved species mix, lower mortality, and streamlined distribution and transportation economics.

These financial figures demonstrate the central strategy of CCIF's new business model: to use the considerable value added created by integrating import and export operations to finance the conversion of both to fully sustain-

able practices. Doing so will, of course, require that reef fish be harvested at a level that creates sufficient income for this operation.

Sustainable Harvest?

The art and science of sustainable reef product harvesting have not been fully established; this can be done only experimentally by running a number of alternative harvest protocols through a tightly controlled and monitored experimental reef fishing operation.

To date, no operator has conducted this type of systematic research, and therefore, no current aquarium fishing operation can claim to be 100% sustainable. However, a properly structured aquarium fishing operation can certainly make a legitimate claim that it is vastly more sustainable than the operation(s) that it replaces, particularly if it has been based on carefully prepared and scientifically based fishery management plans. Such operations can be systematically monitored by a third party under a protocol approved and/or certified by a major institution such as Reef Check, Marine Aquarium Council (MAC), the International Marinelife Association, the World Wildlife Fund, or other similar organizations.

As sustainable catch has not yet been defined, the CCIF team started with the following two most important ecological issues when considering sustainable harvest requirements. First, harvest levels must generate sufficient revenues to finance the operating and capital requirements of a sustainable export operation (the capital upgrades, collection stations, training, and monitoring). Secondly, the sustainability of harvests at these levels must be determined.

CCIF used conservative assumptions to make a "best guess" at sustainable harvest for business planning purposes. In addition, the following requirements were built into the business model: development of site-specific fisheries management plans/harvest protocols, development and implementation of reef monitoring protocols, maintaining a liaison with the scientific community, full chain of custody verification, and consulting with conservation groups.

A number of mitigation measures can be used to further reduce impact. For example, in the Philippines, the great concentration of reefs may make it possible for a single collection area to rotate their harvesting among three separate reef areas to allow for 2-year fully protected regeneration periods. Fishery management plans can call for reef zoning that permanently sets aside source reefs to prohibit any fishing in spawning aggregation areas. Key species with unknown reproductive habits and cycles can be exempted from the catch. The CCIF business plan requires all investments to include full baseline assessments of reef health, ongoing monitoring, and the continued scientific development of truly sustainable reef harvest principles.

CCIF's business plan also limits investments to areas where exclusive access to the reef fishing rights have been obtained. This is relatively easy to achieve in some places, such as Papua New Guinea (PNG) and difficult in others, such as Indonesia. CCIF is therefore working with the

foundation and NGO communities to help obtain private parks management rights to a number of "paper parks" in Indonesia: officially designated marine management areas under varying levels of protection, which are now neither effectively enforced nor funded by the Indonesian government. Once site control has been established, an integrated sustainable development plan can be developed, of which aquarium fishing will be a part. In the Philippines, the situation is somewhat easier. It is possible to obtain and enforce maritime leases as is done frequently by major pearl farming operations.

Because a clear definition of sustainable harvest does not yet exist, protocols will have to be tested in the field and improved over time.

Aquarium Fishing Reform

A fully reformed aquarium fishing industry would greatly contribute to the health of the Indo-Pacific reefs. However, it would by no means guarantee the full protection of the reefs from other destructive forces, such as dynamite and live food fishing. Aquarium fish harvesting at sustainable levels simply does not generate sufficient profits to pay for an army charged with the enforcement of all maritime protection and fisheries laws, and at this rate of destruction, such an army will eventually be required.

It is still extremely important, however, that the reform be undertaken. For some reefs, it will mean the difference between survival and destruction. For other reefs, it will be one of the factors among others (conservation concessions, enlightened local government, ecotourism, etc.) that will make the difference. Most importantly, it will demonstrate that it is possible to generate local income without destroying the resource.[17]

Having said that, it is also exceedingly important that integrated, site-specific conservation plans be realized for at least some of the richest reefs in Indonesia and the Philippines. Aquarium fishing will be an important contributor to such endeavors, but a strong overall management approach will be required.

CCIF is laying out a foundation to respond effectively and comprehensively on both of these fronts through two efforts. The first is the Reef Product Alliance (RPA) business plan that calls for the formation of a for-profit limited liability investment corporation managed by

Table 28.7   Objectives for long-term protection of reef resources

1	Establish a social process for developing effective fisheries and conservation management systems.
2	Establish local knowledge of available resources and foster community "ownership" of these resources.
3	Determine threats and set use regulations to manage these threats.
4	Create recognition by community, outsiders, and local and central governments of area status.
5	Encourage and use regulation and management authorities.
6	Identify and develop sustainable enterprise activities to offset the continuous costs of conservation and, in some cases, concession payments.

professional venture capitalists and tropical fisheries experts. The second is the development of specific site-based management strategies to ensure long-term protection of reef resources.

RPA's objective is to finance the conversion of leading companies in the international aquarium fish and marine ornamentals trade to fully sustainable fish collection, handling, holding, transporting, and marketing practices. RPA investments in the Philippines and Indonesia will compliment CCIF's site-specific strategy to ensure long-term protection of reef resources through community-focused marine area management plan approaches. Although these long-term protection schemes will necessarily differ between Indonesia and the Philippines, they will have the same objectives (Table 28.7).

Conclusion

Current regulations of coral reefs in the Indo-Pacific vary widely, and protection in the most biodiverse areas is not effective. The economics of destructive reef fishing are such that they will destroy these reefs well after the point of no (ecological) return. Existing attempts by NGOs to provide local alternatives to reef destruction are unlikely to solve the problem as a whole; comprehensive management is key to preserving these precious marine resources. Although it is certainly possible to use traditional methods to prepare and enforce a management plan that ensures the reef's long-term survival, such an approach is unlikely to take hold in Indonesia and the Philippines before it is too late. Only a coordinated approach that reforms the economics of the aquarium industry, creates actively enforced marine managed areas, and provides massive technological assistance to fisheries, NGOs, and communities alike is likely to succeed. CCIF plans to coordinate and facilitate these types of coordinated efforts in the coming years.

Notes

1. Conservation Community & Investment Forum (CCIF) fieldwork (April/May 2001) and analysis.

2. CCIF, a project of the Canopy Trust, is a 501(c)(3) nonprofit organization. Support for this project was provided by the Packard Foundation and the IFC. CCIF researches and identifies specific opportunities where targeted private investment can effectively stop destructive or natural resource extraction practices, address major social problems, and do so profitably. CCIF focuses on small, targeted equity investments. Activities include developing new business models, coordinating nonprofit and for-profit capital investors, conducting research and analysis (due diligence) on organizations that will lead local efforts, and facilitating initiation of international projects.

3. CCIF fieldwork (April/May 2001) and analysis.

4. Ibid.

5. Ibid.

6. Ibid.

7. Ibid.

8. Ibid.

9. Ibid.

10. Ibid.

11. CCIF fieldwork (April/May 2001) and analysis.

12. CCIF interviews with U.S. retailers, summer 2001.

13. CCIF field observation and interviews in Tumbuc, Sulawesi, and Bali Barat, CCIF, June 2001.

14. CCIF fieldwork (April/May 2001) and analysis.

15. Ibid.

16. Ibid.

17. The government of Papua New Guinea is requiring commercial flights in and out of Manus island to make available a fixed percentage of its freight areas for aquarium fish, because it is currently the only industry that employs local fishers in a way that does not harm the reefs.

References

Burke, Lauretta, Liz Selig, an Mark Spaulding. 2002. Reefs at Risk in Southeast Asia. <http://www.wri.org/reefsatrisk>.

Pet-Soede, C, H.S.J. Cesar, and J.S. Pet. 2000. Economic issues related to blast fishing on Indonesian coral reefs. *Indonesian Journal of Coastal and Marine Resources.* Vol. 3, No. 2. pp 33–40.

General Index

A

Acanthuridae, nutritional management of, 101–105
Aeromonas spp., 99
Algae, species successfully propagated, 15, 23
Amyloodinium, in fish, 98
Anemones, species successfully propagated, 15, 23
Anesthesia, and disease diagnosis, 95
Angelfish, Queen (*Holacanthus ciliaris*), 79–80, 125–137
Annelids, species successfully propagated, 15, 24
Aqualife Research, 56
Aquarium hobby. *See* Hobbyists
Aquarium of the Pacific, sea dragon breeding efforts, 314
Aquariums, public
   fish species successfully propagated by, 312–313
   public education and MAC, 321–322
   role in conservation, 8, 307–322
   species spawned but not yet reared, 315
Artemia nauplii, 290–295, 297–303

B

Bacterial diseases, in fish, 99–100
Banggai cardinalfish, research on, 320
Best Practice Guidance documents, and certification, 120
Betel lime, from coral, 158
Biopsy techniques, in fish, 95–96
Blood, collection of, in fish, 96
Breeders, of ornamentals. *See* Culturing, of ornamentals
Brooklynella, parasitic in fish, 98–99

C

Census of Marine Life (CoML), 309
Certification, of product. *See also* Marine Aquarium Council (MAC)
   benefits of, 89, 121–122
   and community-based management of coral reefs, 141–162
   and coral reef conservation, 118–122
   development of, MAC system, 119
   and labeling, 118, 121
   MO conference priority recommendations for, 6
   prices, break-even, for certified Queen Angelfish, 125–137
   programs for, 126–127
   recommendation of authors regarding, 121–122
   response to, retail, 348–349
   standard preference analysis, of certification program, 128–134
   standards, 120, 127–128
   survey of industry on value of certification, 128–137
Chesapeake Bay, research effort and public education, 320–321
Chloramine, and water quality, for fish, 94–95
CITES (Convention on International Trade in Endangered Species of Flora and Fauna)
   appendices of, 76
   and data collection, 33–35
   import licenses of, 52
Clam, giant, culturing of, 347–348
Cleaning behaviors, of shrimp, 223
Climate, global change in, impact on coral reefs, 113
Cnidarians
   imported to U.S., 41
   successfully propagated by researchers, 23
Collection, of ornamentals. *See also* Conservation; Fishing, methods
   balancing, with conservation, 353–358
   corals
      impact of, 31–32, 65, 141–142, 169–171
      live, taxa and methods, 168–169
      propagated on live rock, 217
      recommended practices, 161–162, 179
   cyanide, use of in, 327–338
   education, public, value of, 354
   fish
      collectors' role in distribution network, 50
      versus culturing of fish, statistics, 54–56
      cyanide and, 327–328
      recommended collection and management practices, 161
      value of landings, by species type, 80
      wild-caught, 366
   in Florida Keys National Marine Sanctuary, 353–359
   Global Positioning System used for, 88

Species Index